HANDBOOK OF OPERATIONS RESEARCH IN NATURAL RESOURCES

Recent titles in the **INTERNATIONAL SERIES IN OPERATIONS RESEARCH & MANAGEMENT SCIENCE**
Frederick S. Hillier, Series Editor, *Stanford University*

Gass & Assad/ *AN ANNOTATED TIMELINE OF OPERATIONS RESEARCH: An Informal History*
Greenberg/ *TUTORIALS ON EMERGING METHODOLOGIES AND APPLICATIONS IN OPERATIONS RESEARCH*
Weber/ *UNCERTAINTY IN THE ELECTRIC POWER INDUSTRY: Methods and Models for Decision Support*
Figueira, Greco & Ehrgott/ *MULTIPLE CRITERIA DECISION ANALYSIS: State of the Art Surveys*
Reveliotis/ *REAL-TIME MANAGEMENT OF RESOURCE ALLOCATIONS SYSTEMS: A Discrete Event Systems Approach*
Kall & Mayer/ *STOCHASTIC LINEAR PROGRAMMING: Models, Theory, and Computation*
Sethi, Yan & Zhang/ *INVENTORY AND SUPPLY CHAIN MANAGEMENT WITH FORECAST UPDATES*
Cox/ *QUANTITATIVE HEALTH RISK ANALYSIS METHODS: Modeling the Human Health Impacts of Antibiotics Used in Food Animals*
Ching & Ng/ *MARKOV CHAINS: Models, Algorithms and Applications*
Li & Sun/ *NONLINEAR INTEGER PROGRAMMING*
Kaliszewski/ *SOFT COMPUTING FOR COMPLEX MULTIPLE CRITERIA DECISION MAKING*
Bouyssou et al/ *EVALUATION AND DECISION MODELS WITH MULTIPLE CRITERIA: Stepping stones for the analyst*
Blecker & Friedrich/ *MASS CUSTOMIZATION: Challenges and Solutions*
Appa, Pitsoulis & Williams/ *HANDBOOK ON MODELLING FOR DISCRETE OPTIMIZATION*
Herrmann/ *HANDBOOK OF PRODUCTION SCHEDULING*
Axsäter/ *INVENTORY CONTROL, 2nd Ed.*
Hall/ *PATIENT FLOW: Reducing Delay in Healthcare Delivery*
Józefowska & Węglarz/ *PERSPECTIVES IN MODERN PROJECT SCHEDULING*
Tian & Zhang/ *VACATION QUEUEING MODELS: Theory and Applications*
Yan, Yin & Zhang/ *STOCHASTIC PROCESSES, OPTIMIZATION, AND CONTROL THEORY APPLICATIONS IN FINANCIAL ENGINEERING, QUEUEING NETWORKS, AND MANUFACTURING SYSTEMS*
Saaty & Vargas/ *DECISION MAKING WITH THE ANALYTIC NETWORK PROCESS: Economic, Political, Social & Technological Applications w. Benefits, Opportunities, Costs & Risks*
Yu/ *TECHNOLOGY PORTFOLIO PLANNING AND MANAGEMENT: Practical Concepts and Tools*
Kandiller/ *PRINCIPLES OF MATHEMATICS IN OPERATIONS RESEARCH*
Lee & Lee/ *BUILDING SUPPLY CHAIN EXCELLENCE IN EMERGING ECONOMIES*

**** A list of the early publications in the series is at the end of the book ****

HANDBOOK OF OPERATIONS RESEARCH IN NATURAL RESOURCES

Edited by

Andres Weintraub
University of Chile, Santiago, Chile

Carlos Romero
Technical University of Madrid, Spain

Trond Bjørndal
CEMARE, University of Portsmouth, UK

Rafael Epstein
University of Chile, Santiago, Chile

with the collaboration of

Jaime Miranda
Diego Portales University, Santiago, Chile

Springer

Andres Weintraub
University of Chile
Santiago, Chile

Carlos Romero
Technical University of Madrid
Madrid, Spain

Trond Bjørndal
CEMARE, University of Portsmouth
Portsmouth, United Kingdom

Rafael Epstein
University of Chile
Santiago, Chile

Series Editor:
Fred Hillier
Stanford University
Stanford, CA, USA

With the collaboration of:
Jaime Miranda
Diego Portales University,
Santiago, Chile

Library of Congress Control Number: 2007924350

ISBN 978-0-387-71814-9 ISBN 978-0-387-71815-6 (eBook)

Printed on acid-free paper.

9 8 7 6 5 4 3 2 1

springer.com

Contents

Contributing Authors

Christopher Alford
University of Queensland
c.alford@smi.uq.edu.au

Ragnar Arnason
University of Iceland
ragnara@hi.is

Manuel Arriaza
Andalusian Institute of Agricultural, Fisheries and Food Research
manuel.arriaza.ext@juntadeandalucia.es

Frank Asche
University of Stavanger
frank.asche@tn.his.no

Enrique Ballestero
Technical University of Valencia
eballe@esp.upv.es

Julio Berbel
University of Córdoba
berbel@uco.es

Pete Bettinger
University of Georgia
pbettinger@forestry.uga.edu

Trond Bjørndal
University of Portsmouth
t.bjorndal@imperial.ac.uk

Houcine Boughanmi
Sultan Qaboos University
boughanh@squ.edu.om

Marcus Brazil
University of Melbourne
brazil@unimelb.edu.au

Joseph Buongiorno
University of Wisconsin
jbuongio@wisc.edu

Jeroen Buysse
Ghent University
J.Buysse@UGent.be

Louis Caccetta
Curtin University of Technology
caccetta@cs.curtin.edu.au

Alejandro Caparrós
Spanish Council for Scientific Research (CSIC), Spain
acaparros@ieg.csic.es

Rodrigo Caro
University of Chile
rcaro@dii.uchile.cl

Carmen Castrodeza
University of Valladolid
ccch@eco.uva.es

Emilio Cerdá
University Complutense of Madrid
ecerdate@ccee.ucm.es

Richard L. Church
University of California at Santa Barbara
church@geog.ucsb.edu

Roberto Cominetti
University of Chile
rcominet@dim.uchile.cl

Jon M. Conrad
Cornell University
jmc16@cornell.edu

Luis Diaz-Balteiro
Technical University of Madrid
luis.diaz.balteiro@upm.es

Rafael Epstein
University of Chile
repstein@dii.uchile.cl

Bruno Fernagut
Centre for Agricultural Economics
bruno.fernagut@ilvo.vlaanderen.be

José A. Gómez-Limón
University of Valladolid
limon@iaf.uva.es

Daniel V. Gordon
University of Calgary
dgordon@ucalgary.ca

Eldon A Gunn
Dalhousie University
eldon.gunn@dal.ca

Robert Haight
North Central Research Station
rhaight@fs.fed.us

Olivier Harmignie
Université Catholique de Louvain
harmignie@ecru.ucl.ac.be

Kiyotada Hayashi
National Agricultural Research Center
hayashi@affrc.go.jp

Bruno Henry de Frahan
Université Catholique de Louvain
henrydefrahan@ecru.ucl.ac.be

John Hof
Rocky Mountain Research Station
johnhof@comcast.net

Ruud Huirne
Wageningen University
Ruud.Huirne@wur.nl

Lynn Huntsinger
University of California (Berkeley)
buckaroo@nature.berkeley.edu

Veijo Kaitala
University of Helsinki
veijo.kaitala@hut.fi

Jenny Karlsson
Linköping University
jekar@mai.liu.se

Mark Kuchta
School of Mines
mkuchta@mines.edu

Daniel E. Lane
University of Ottawa
dlane@uottawa.ca

Ludwig Lauwers
Centre for Agricultural Economics
ludwig.lauwers@ilvo.vlaanderen.be

David H. Lee
University of Melbourne
dhlee@bigpond.net.au

Marko Lindroos
University of Helsinki
marko.lindroos@helsinki.fi

Peter Lohmander
Department of Forest Economics
peter.lohmander@sekon.slu.se

Hamish Marshall
New Zealand Forest Research Institute Ltd.
Hamish.Marshall@ensisjv.com

David L. Martell
University of Toronto
martell@smokey.forestry.utoronto.ca

Michael Martinez
Colorado School of Mines
michael.martinez@usafa.af.mil

Miranda Meuwissen
Wageningen University
miranda.meuwissen@wur.nl

Gordon Munro
University of Portsmouth
munro@econ.ubc.ca

Glen Murphy
Oregon State University
Glen.Murphy@orst.edu

Alan T. Murray
Ohio State University
amurray@geography.ohio-state.edu

Alexandra M. Newman
Colorado School of Mines
anewman@mines.edu

Linda Nøstbakken
Cornell University
linda.nostbakken@nhh.no

Sean Pascoe
University of Portsmouth
sean.pascoe@csiro.au

Teresa Peña
University of Valladolid
maitepe@eco.uva.es

Philippe Polomé
Université Catholique de Louvain
polome@ecru.ucl.ac.be

Ronald Raunikar
University of Wisconsin
raunikar@students.wisc.edu

Carlos Romero
Technical University of Madrid,
carlos.romero@upm.es

Mikael Rönnqvist
The Forestry Research Institute
of Sweden
mikael.ronnqvist@nhh.no

Jaime San Martín
University of Chile
jsanmart@dim.uchile.cl

Pablo Santibañez
University of Chile, Santiago
psantiba@dii.uchile.cl

John Sessions
Oregon State University
john.sessions@oregonstate.edu

Richard B. Standiford
University of California (Berkeley)
standifo@nature.berkeley.edu

Diana Tingley
University of Portsmouth
diana.tingley@port.ac.uk

Marcel Van Asseldonk
Wageningen University
marcel.vanasseldonk@wur.nl

Guido Van Huylenbroeck
Ghent University
Guido.VanHuylenbroeck@UGent.be

Jef Van Meensel
Centre for Agricultural Economics
jef.vanmeensel@ilvo.vlaanderen.be

Andres Weintraub
University of Chile
aweintra@dii.uchile.cl

Slim Zekri
Sultan Qaboos University
slim@squ.edu.om

Preface

Operations Research/Management Science (OR/MS) approaches have helped people for the last 40 years or so, to understand the complex functioning of the systems based upon natural resources, as well as to manage this type of systems in an efficient way. The areas usually viewed within the natural resources field are: agriculture, fisheries, forestry, mining and water resources.

Even though, the above areas are usually viewed as separate fields of study, there are clear links and relations between them. In fact, all of them share the common problem of allocating scarcity along time in an optimal manner. The scale of time or length of the planning horizon is very different. Thus, we have almost a continuous renewal in the case of the fisheries, periodic cycles in the case of agriculture and forestry (ranging from some few months in the case of a horticultural crop to more than a century for some forest species), and enormous periods of time much beyond the human perception in the case of mining resources. But in all the cases, the key matter is to obtain an efficient use of the resource along its planning horizon.

Another element of connection among the different natural resources is due to the interaction between the use of the resource, and the environmental impact caused by its extraction or harvest. This type of interaction implies additional complexities in the underlying decision-making process, making the use of OR/MS tools especially relevant.

The above views are corroborated by the massive use of quantitative approaches in the management of natural resources. It can be said that this broad field was one of the first where the OR/MS discipline was successfully applied.

The papers presented correspond to invitations made to the specialists we considered the most distinguished in each area, and we are extremely satisfied with the positive response we obtained from them. In defining the subject matters, we tried to cover comprehensively the most relevant topics in each area, from the application point of view, as well as consideration of the operations research techniques involved. In particular, we wished to highlight the successes of the OR approach to deal with problems, which involves a conceptual view of problems, modelling of complex realities, and development of algorithms for problems increasingly difficult to solve. Issues of large scale, uncertainty, multiple objectives appear increasingly in these decision processes. Also, we view the integration in multidisciplinary approaches, where specialists in the specific areas need to interact with operations research specialist, and the need to incorporate information technologies for implementations is also present.

The set of papers compiled in this volume attempts to provide readers with significant OR/MS contributions in each one of the applied areas previously defined. In this way, we hope to encourage the use of quantitative techniques in order to manage the use of the different natural resources efficiently from an economic as well as an environmental point of view.

The papers are divided by area of application: agriculture, fisheries, forestry and mining.

Acknowledgments

The preparation of the volume was a very long process that exceeds considerably the initial target. Hence, we thank all the authors for their co-operation and patience. All the papers were assessed following a blind reviewing process. Our gratitude to all the anonymous referees.

The following funding is acknowledged. The work of Carlos Romero was supported by the Spanish "Ministerio de Educación y Ciencia" under research grant SEJ2005-04392. The work of Rafael Epstein and Andres Weintraub was supported by Nucleo Milenio "Complex Engineering Systems".

PART I

AGRICULTURE

In the area of agriculture we have eight chapters with different concerns such as conceptual problems related with risk analysis, the interaction between agriculture and the environment, water resources planning, agroforestry systems management, simulation of effects on agriculture of changes in the common agriculture policy, and so on. OR/MS techniques used are basically the following: linear programming, multi-objective fractional programming, goal programming, multi-attribute utility theory and control dynamic optimization.

The chapter "Importance of whole-farm risk management in agriculture", by Huirne, Meuwissen and Van Asseldonk, deals with the problems associated with the definition and measurement of risk at the whole-farm level. The conceptual framework is tested through a questionnaire survey among livestock and arable farmers in the Netherlands.

The chapter "Dealing with multiple objectives in agriculture", by Hayashi, presents state-of-the-art of multiple criteria decision-making approaches applied to the selection problems in agricultural systems. The chapter pays special attention to matters related with attributes definition and problem structuring, in order to build suitable models for agri-environmental decision-making.

The chapter "Modelling multifunctional agroforestry systems with environmental values: Dehesa in Spain and woodland ranches in California", by Campos, Caparrós, Cerdá, Huntsinger and Standiford, deals with modelling agroforestry systems ("dehesas") with the help of optimal control techniques. Two studies, one in California and the other one in Spain, are accomplished under a comparison basis.

The chapter "Environmental criteria in pig diet formulation with multi-objective fractional programming", by Peña, Castrodeza and Lara, incorporates environmental criteria in pig diet formulation. The proposed model is satisfactorily solved by resorting to an interactive multigoal programming model that allows the incorporation of goals of fractional nature.

The chapter "Modelling the interactions between agriculture and the environment", by Zekri and Boughanmi, reviews the integration of different OR/MS approaches for modelling the interaction between agriculture and the environment. In this way, the authors propose a decision support system based upon multi-criteria techniques and geographical information systems within a participatory decision-making perspective.

The chapter "MCDM farm system analysis for public management of irrigated agriculture", by Gómez-Limón, Berbel and Arriaza, proposes a multi-criteria approach to assist policy decision-making on water management for irrigated agriculture. The methodology is a hybrid between multi-attribute utility theory and goal programming. The methodology is applied to several Spanish case studies within the recent European Water Framework Directive.

The chapter "Water public agencies agreeing to a covenant for water transfers: How to arbitrate price-quantity clauses", by Ballestero, deals with inter-basin water covenants guided by the principle of arbitration and implemented through public agencies. The methodology is illustrated with the help of a realistic example in a maritime region near the Mediterranean Sea.

Finally, the chapter "Positive mathematical programming for agriculture and environmental policy analysis: Review and practice", by de Frahan, Buysse, Polomé, Fernagut, Harmignie, Lauwers, van Huylenbroeck and van Meensel, introduces a farm-level sector model, called SEPALES, based upon the approach known as a positive mathematical programming. After this, SEPALES is used to simulate several economic and environmental effects on Belgium agriculture, due to some possible changes in the European Common Agricultural Policy.

Chapter 1

IMPORTANCE OF WHOLE-FARM RISK MANAGEMENT IN AGRICULTURE

Ruud Huirne, Miranda Meuwissen, and Marcel Van Asseldonk
Institute for Risk Management in Agriculture, Wageningen University, The Netherlands

Abstract Risk management is an increasingly important topic. At the farm level, it received little attention in Europe. Research indicates that whole-farm risk-management approaches, that is approaches in which multiple risks and farm activities are considered simultaneously, seem more efficient than 'single risk and commodity strategies'. This chapter first gives an overview of risk management and then it discusses the results of a questionnaire survey among livestock and arable farmers in the Netherlands. The survey deals with farmers' perceptions of risk and risk-management strategies. Risk-management strategies include both 'single risk' strategies as well as strategies for simultaneously covering multiple risks. The latter are restricted to the type of strategies currently available in the Netherlands. Next, opportunities for broadening the scope of risk-management strategies covering multiple risks are discussed. The paper concludes by identifying areas for further research in the field of whole-farm risk management.

Keywords: Risk management, agriculture, whole-farm approach, multiple risks, questionnaire survey

1 INTRODUCTION

The agricultural firm is constantly developing. The farm is and remains an essential player in the agricultural supply chain and in the rural area. The differences between the agricultural sector and the rest of the industry are getting smaller and smaller. Increasing farm sizes result in a more industrialized way of undertaking such operations. Important 'new' characteristics of such bigger, industrialized farms include: importance of manufacturing processes (vs. commodities); a systems approach to production and distribution; separation and realignment of the stages in the food chain for the purpose of efficiency and low cost-price; negotiated coordination among those stages and with the environment (rural area); concern about system power and control; and new kinds of risk combined with a more important role for information. This implies that risk considerations are becoming more important and should be addressed in a more formal way.

Income from farming is usually considered rather volatile because of a whole series of stochastic factors, that is risk. Over the years, a range of risk-management strategies has been used to reduce, or to assist farmers to absorb, some of these risks (see later). Risk-management strategies, especially risk-sharing strategies, generally deal with only one type of risk at a time. For instance, futures market contracts deal with price risks, hail and storm insurance schemes cover weather-related production risks, and livestock insurance schemes cover the death of animals. Even disaster relief programs in such events as droughts and floods consider only one type of risk (which, in itself, is relevant if the whole — or a notable part of the — crop or herd is destroyed).

This chapter first discusses risk management in general (definition, sources of risk, risk-management strategies) and then the results of a questionnaire survey among livestock and arable farmers in the Netherlands. Because Dutch farms are not really representative compared to farms in many other countries, the results of the survey should be seen as an example. The survey deals with farmers' perceptions of risk and risk-management strategies. Risk-management strategies include both 'single risk' strategies as well as strategies for simultaneously covering multiple risks. The latter are restricted to the type of strategies currently available in the Netherlands. Next, opportunities for broadening the scope of risk management strategies covering multiple risks are discussed. The chapter concludes by identifying areas for further research in the field of whole-farm risk management.

2 RISK MANAGEMENT

The concepts of 'risk' and 'uncertainty' have already been referred to several times. It is time to elaborate upon them. The meanings of 'risk' and 'uncertainty' come close (Hardaker *et al.*, 2004). Uncertainty is the result of incomplete knowledge. Risk can be defined as uncertain consequences or results at the time of making decisions. Risk particularly concerns exposure to unwanted, negative consequences. Risk management concerns the way in which managers deal with risk and uncertainty (Meuwissen *et al.*, 1999, 2001; Huirne *et al.*, 2000; Van Asseldonk *et al.*, 2001; Huirne, 2002).

2.1 Types of Risk

The current government policy has increasingly been aimed at creating an open market system. This results in, amongst other things, the fact that agriculture in the Netherlands is increasingly confronted with price-making in international markets, such as the world market, which generally means lower and definitely more fluctuating prices (Huirne *et al.*, 1997; Meuwissen *et al.*, 1999). Further modernization of the sector has resulted in increasing economic consequences. Dealing with such risks, that is risk management, is gaining more and more importance, not only for individual farmers, but for all firms in the agricultural supply chain.

Many activities of an agricultural firm take place outdoors and are weather-dependent. The agricultural sector also deals with live material. Because of this the sector is an outstanding example being exposure to risks (Anderson *et al.*, 1977; Barry *et al.*, 2000; Van Asseldonk *et al.*, 2001; Hardaker *et al.*, 2004). Production risks are caused by the unpredictable character of the weather and hence uncertainty as to the physical yield of animals and crops. Diseases and infestations can have a great influence on farm results, as the classical swine fever outbreaks in 1997/1998 and the foot-and-mouth disease outbreaks in 2001 clearly showed.

Moreover, the prices of production means most often purchased (such as concentrates, fertilizer, pesticides and machines) and of products sold (such as milk, tomatoes and cut flowers) are not known, at least not at the time decisions on these have to be taken. As already mentioned, farmers are increasingly exposed to price-making forces in unpredictable markets. Thus, market and price risks are important factors.

Governments form another source of risk to farmers. Changes in laws and regulations with respect to running the farm can have far-reaching consequences for farm results. Examples are the continuing changes in the regulations regarding environment, pesticides, animal diseases and animal welfare. On the other hand, governments have also set off particular risks (up to now).

Farmers working on their farms are themselves a risk to the profitability and continuity of the farm. The farm's survival may be threatened by death of the owner, or by divorce of a couple together running the farm. Long-term illness of the owner or employees can also cause considerable losses or can increase the costs considerably. Such risks are called human or personal risks.

There are also financial risks involved (Belli *et al.*, 2001). These are related to the financing of the farm. Using borrowed capital (such as mortgages and the like) means that first the interest needs to be paid before increasing one's equity capital. For farms with relatively much debt capital (for example, as a result of large investments), little will be left as a reward to one's equity capital at times of high interest rates. Only farms that are entirely equity-financed are not subject to such financial risks, but yet can sustain capital loss. Other risks connected to the use of credit and loans are uncertain interest rates and inability to obtain a loan or mortgage.

2.2 Reducing and Sharing Risk

Risks are thus unavoidable and influence almost any decision the farmer takes. That is to say risks are present, but can be counteracted. The farmer should anticipate such risks by his management. But how? In what way can risks be reduced? There are two categories of measures to reduce risks: taking measures within the farm and sharing risks with others (Huirne *et al.*, 1997; Belli *et al.*, 2001; Huirne, 2002; Hardaker *et al.*, 2004).

During many uncertain events (extra) information can be obtained easily. For example, asking for the weather forecast, analyzing feed or soil samples and consulting experts. Also particular risks can possibly be avoided or prevented. It is known that certain activities carry more risks than others. Reducing farm contacts can, for example, reduce the risk of disease introduction considerably. Another good strategy to minimize risks is not to invest all of one's money on a single farm activity. By selecting a combination of activities, risks can be considerably reduced. The same holds for having various suppliers and buyers. Flexibility can be mentioned as a last measure at the farm level. Flexibility refers to how well a farm can anticipate changing conditions. For example, by investing in multipurpose machines and buildings.

The second set of measures refers to sharing risks with others (Huirne *et al.*, 1997; Hardaker *et al.*, 2004). One possibility here is buying insurance. At present, there are several types of insurance available, with which, by payment of a premium, risks can be reduced or even eliminated. The farmer can also conclude contracts for example with suppliers and buyers in which price agreements are laid down. Agreements can be made on the duty to deliver and to buy as well as on the quality of the products or raw materials. Lastly, by using the futures market, price risks can largely be eliminated. The futures market is not yet well known in the Netherlands, but in the USA it is popular for a number of agricultural products.

Most farmers try to reduce risks when they face decisions that may have a considerable influence on their income or wealth (Anderson *et al.*, 1977; Belli *et al.*, 2001; Hardaker *et al.*, 2004). Examples of such decisions are sizeable investments in milk quotas or in a second farm enterprise. The attitude of reducing exposure to risks is called risk aversion. A risk-averse person is willing to sacrifice part of his income to reduce risks. This consideration serves as a means to make a choice among the above measures. However, reducing risks will generally involve a cost.

2.3 Risk Perception

Managers, policy makers and researchers alike often have a binary way of dealing with risk and uncertainty. One either assumes certainty and an exactly predictable future, or uncertainty and an entirely unpredictable future. In the latter case further analyses are often omitted and decisions are made either intuitively or not made at all. Under- as well as overestimating the risks is potentially dangerous. Further analysis reveals that there are at least four levels of risk and uncertainty (Courtney *et al.*, 1997):

1. A clear-enough future; a single forecast precise enough for the purpose of decision making
2. Alternate futures; a few discrete outcomes that define the future
3. A range of futures; a whole range of possible outcomes
4. True ambiguity; no basis to forecast the future

Levels 1 and 4 do not occur very often in practice; they are extreme situations. Therefore, it is all the more distressing that many managers and advisors regularly operate at these levels of risk. Particularly working at level 1 where calculations are carried out and advice is given under the assumption of complete information and certainty, is alarming.

3 FARMERS' PERCEPTIONS OF RISK MANAGEMENT

3.1 Materials

The questionnaire survey included questions on: (i) the farm, (ii) farmers' risk attitude, (iii) farmers' perception of risk-management strategies, (iv) their perceptions of risks and the extent to which risks are managed on their own farm, (v) farmers' ability to define 'risk management', and (vi) farmers' interest into assistance for setting up a whole-farm risk-management plan for their own farm. Most questions were closed questions, mainly in the form of Likert-type scales ranging from 1 to 5 (Churchill, 1995). In total, the questionnaire included 177 variables. The (pretested) questionnaire was sent in July 2001 to 390 clients of the Rabobank (major agricultural bank in the Netherlands). These included cattle, pig, poultry and arable farmers. After screening on completeness, the questionnaires of 101 farmers were available for statistical analyses, that is, the effective response rate was 26%.

3.2 Results

The majority of respondents has more than one type of farming: 44 farmers have dairy cattle on their farm, 58 have pigs, 9 respondents have poultry and 84 of the respondents are (also) crop farmers. In order to get insight into farmers' risk attitudes, 5 statements were rated. Table 1 shows the results.

Table 1. Farmers' attitude towards risks, $n = 101$ (1: don't agree; 5: fully agree).

	1 (%)	2 (%)	3 (%)	4 (%)	5 (%)	Average	Std
I am willing to take more risks than other farmers	7	16	44	22	11	3.14	1.04
I need to take risks to be successful	9	15	26	40	10	3.27	1.12
I am reluctant to introducing new ideas	14	27	29	25	5	2.79	1.12
New technologies first need to be proved at other farms	16	23	27	26	8	2.88	1.20
I am more concerned about losses than forgoing some profits	20	17	40	18	5	2.71	1.14

From the scores in Table 1 it can be concluded that based on these questions respondents have a risk-seeking attitude. It is noteworthy that this holds for all statements.

Table 2 shows farmers' perceptions of risk-management strategies. We subdivided the strategies into strategies that cover single risks and strategies that simultaneously cover multiple risks. In making this subdivision we assumed that new technologies are primarily implemented to deal with production risks, that leasing machinery has mainly to do with financial risks and that leasing milk quota mostly deals with production risks. In the category 'multiple risk strategies', we assumed that vertical and horizontal cooperation deal with both price and production risks. In relation to spatial diversification we supposed that this has not only to do with diversifying production risks but most likely also with diversifying institutional risks (e.g. in case of environmental requirements) and/or price risks.

Table 2 shows that, in general, farmers perceive the single risk-management strategies as more relevant than the strategies covering multiple risks: of the ten strategies ranked highest (see last column 'overall rank') only four strategies are within the multiple risk category. These strategies include increasing the solvency rate, comprising financial reservations, on-farm diversification and vertical cooperation. Popular risk-management strategies in 'single-risk strategies' are strict hygiene rules, business insurance, personal insurance and the application of new technologies.

Table 2. Perception of risk-management strategies, n = 101 (1: not relevant at all; 5: very relevant).

	Average	**Std**	**Overall rank**
Single-risk strategies			(1)
Strict hygiene rules	4.08	0.96	1
Business insurance	3.80	0.98	4
Personal insurance	3.71	1.09	5
Application of new technologies	3.64	0.93	6
Manure delivery contracts	3.54	1.35	7
Leasing/renting machinery	3.44	1.24	8/9
Price contracts for farm input	2.90	1.10	12
Price contracts for farm output	2.88	1.10	13
Leasing/renting milk quota	2.43	1.09	15
Futures and options market	2.35	0.92	16
Multiple risk strategies			(2)
Increase solvency rate	4.02	0.96	2
Comprise financial reservations	3.81	0.99	3
On-farm diversification	3.44	1.21	8/9
Vertical cooperation	3.40	1.20	10
Horizontal cooperation	3.27	1.20	11
Off-farm investments	2.75	1.21	14
Off-farm employment	2.27	1.31	17
Spatial diversification	2.15	1.00	18

Asking respondents for their 'top 3' risk-management strategies resulted in the following answers (the percentage of respondents indicating a particular strategy is given in parentheses):

1. Increase solvency rate (16%), on-farm diversification (16%), comprise financial reservations (14%);
2. Increase solvency rate (11%), comprise financial reservations (10%), strict hygiene rules (10%); and
3. Vertical cooperation (14%), on-farm diversification (12%), application of new technologies (12%).

From these answers it can be seen that from the 'top 4 strategies' from Table 2 (i.e. 1: strict hygiene rules; 2: increase solvency rate; 3: comprise financial reservations; 4: business insurance) multiple risk strategies (option 2 and 3) are favourite in among top 3.

Table 3 illustrates farmers' perceptions of risks and the extent to which they believe that the risks are dealt with on their own farm. There are seven risk categories. Besides the ones distinguished by Hardaker *et al.* (2004) we added the categories 'liability risks' and 'risks related to immovable objects'.

Table 3. Perception of risk and the extent to which risks are managed on own farm, $n = 101$.

	Relevance of risk (1: not relevant; 5: very relevant)			**Risk is managed on my farm (%)**				
	Average	**Std**	**Overall rank**	**No**	**Not yet**	**Yes partly**	**Yes**	**n.a.**[1]
PRODUCTION RISK			(1)					
Variability in technical results	4.22	1.07	2	6	2	37	55	–
Epidemics (livestock and crop)	3.98	1.22	5	2	4	39	55	–
Bad product quality	3.95	1.05	6	2	1	34	63	–
Diseases (non-epidemic)	3.76	1.21	8	15	2	34	47	2
Suffocation and decay	3.41	1.41	12	11	4	16	68	1
PRICE OR MARKET RISKS			(2)					
Price variability	4.00	1.20	4	47	13	29	10	1
Dependency on Dutch suppliers or buyers	3.50	1.21	11	42	13	33	12	–

[1] Not applicable

Dependency on foreign suppliers or buyers	2.99	1.33	15	35	5	39	18	3
INSTITUTIONAL			(3)					
Regulations and sanctions	4.32	0.91	1	2	4	45	49	–
Elimination of government support	2.61	1.26	20	51	30	12	7	–
PERSONAL RISKS			(4)					
Death	4.19	1.18	3	12	3	17	67	–
Illness and disability	3.88	1.14	7	5	4	26	64	1
Personnel	2.32	1.42	23	24	8	24	40	4
RISKS RELATED TO 'IMMOVABLE OBJECTS'			(5)					
Fire and ignition	3.73	1.13	9	9	10	45	36	–
Burglary	3.07	1.13	4	7	7	41	45	–
LIABILITY RISKS			(6)					
Products and services sold	3.59	1.24	10	11	7	40	42	–
Buildings	2.96	1.28	16	5	1	33	61	–
Contracts (supply and delivery)	2.95	1.37	17	23	13	36	28	–
Environment	2.79	1.11	18	6	4	44	46	–
Traffic	2.55	1.27	22	23	5	30	41	1
Personnel	2.11	1.43	24	23	10	38	24	5
FINANCIAL RISKS			(7)					
Changes in interest rates	3.16	1.23	13	14	8	51	27	–
Decrease in farm's collateral value	2.64	1.29	19	35	23	19	23	–
Decrease in retirement provisions because of declining farm values	2.61	1.13	21	28	14	16	42	–

Table 3 shows that farmers perceive production and price risks as very important. Liability risks and financial risks are ranked 6th and 7th respectively. With respect to the management of the risks, farmers are convinced that they (largely) handled the production risks, institutional risks (as far as it concerns governmental regulations), personal risks, risks related to immovable objects, liability risks and financial risks. Note that for some type of risks the numbers in the column 'yes I managed the risk partly' are higher than for other risks. This is for instance the case for liability risks. Not (yet) adequately dealt with are price risks, risks related to the elimination of government support (e.g. in case of droughts and livestock epidemics) and the decrease in farms' collateral value.

The two remaining parts of the questionnaire, that is farmers' ability to define risk management and farmers' interest into assistance for setting up a whole-farm risk management-plan for their own farm, led to the following results:

1. About 70% of the respondents was able to adequately define risk management.
2. About 62% of the respondents showed interest in assistance for developing a risk management plan for their own farm.

There was a significant positive relationship between farmers being able to define risk-management and those interested in a risk management plan ($P \leq 0.05$).

4 MULTIPLE RISK STRATEGIES

The 'multiple risk strategies' included in the chapter so far are classical examples of on-farm diversification, off-farm employment, increasing the solvency rate, etc. Vertical and horizontal cooperation are more recent examples (Boehlje and Lins, 1998). This section discusses three further opportunities for simultaneously covering multiple risks: certification, revenue insurance and stabilization accounts. Certification can be categorized as an 'on-farm strategy'; revenue insurance and stabilization accounts are 'risk-sharing strategies'.

Certification is already widely available in the Netherlands. Examples include KKM (Chain Quality Milk) for dairy farms, PVE/IKB (Integrated Chain Control) for pig farms, Safe Quality Food for primary producers (SQF-1000) and Good Agricultural Practices as defined in Eurep-GAP.

Certification reduces production risks (through, among others, improved internal efficiency and less failure costs), liability risks (since certification effectuates due diligence) and price risks – if markets for certified products are more stable than other markets (Unnevehr *et al.*, 1999; Velthuis *et al.*, 2003; Meuwissen *et al.*, 2003b).

Revenue insurance is not (yet) available in the Netherlands. It simultaneously covers price and yield risks of a particular commodity. If the correlation between both parameters is negative (i.e. lower yields result in higher prices, and vice versa) revenue insurance should be less expensive than insurance for yields only. The concept has existed in the USA for many years (see for instance Goodwin and Ker, 1998). Schemes are highly subsidized by the US government (Skees, 1999). However, since these type of insurance schemes seem legitimate in the WTO-framework (i.e. they fit into the "green box" representing allowed forms of support), the European Commission is now considering similar tools (Meuwissen *et al.*, 2003a).

Stabilization accounts not only cover multiple risks but (if relevant for a particular farm) also multiple commodities. The principle of stabilization accounts is that farmers can put money into the account in high-income years (when marginal taxes are high) while withdrawing it in low-income years (when marginal taxes are low). Examples of stabilization accounts (currently not available in the Netherlands) include the Canadian Net Income Stabilization Accounts (NISA) and the Australian Farm Management Deposits. NISA is a whole-farm program in which farmers put money into a bank account, government matches the farmer's deposits ("dollar for dollar"), and each farmer can withdraw from the account in adverse times. Also NISA is legitimate under WTO regulations. The Canadian government is currently reconsidering their program in order to also include on-farm food safety issues and environmental programs. The Australian scheme equals the Canadian one but without the matching contributions from the government (European Commission, 2001).

5 FUTURE OUTLOOK

This chapter was set up around 'whole-farm risk management'. Results from the questionnaire indicate that farmers perceive that they have managed their farm risks quit well (with some exceptions, mainly in the field of price risks and risks related to the elimination of government support). Farmers generally prefer 'single-risk and commodity strategies'.

Following a whole-farm risk-management approach, that is an approach in which multiple risks and farm activities are considered simultaneously, may be more efficient, but probably also more complicated. Of the respondents 62% indicated that they were interested in assistance in setting up a whole-farm risk-management plan. This percentage may even have been higher if respondents had not known that the survey was initiated by the Rabobank (which has some direct interest in such risk-management plans).

The multiple risk strategies discussed (i.e. certification, revenue insurance and stabilization accounts) have some features of a whole-farm risk-management approach. For instance, when designing revenue insurance schemes it is relevant to have insight into the correlation between prices and yields. When setting up (subsidized) stabilization accounts, insight is needed into the correlation of revenues among various farm activities. Certification programs require the identification of critical control points of a farm, for example with respect to food safety.

From the above, we define the following areas for further research in the field of whole-farm risk management:

1. An analysis of (the dynamics in) correlations between prices and yield of various agricultural activities.
2. An analysis of the critical control points of a farm from the perspective of the overall farm viability.

After these steps have been taken, whole-farm risk-management plans can be designed – and the ideal partners for advising about them can be identified.

6 REFERENCES

Anderson, J.R., Dillon, J.L., Hardaker, J.B., 1977, *Agricultural Decision Analysis*. Iowa State University Press, Ames, Iowa.

Barry, P.J., Ellinger, P.N., Hopkin, J.A., Baker, C.B., 2000, *Financial Management in Agriculture*. Interstate, Danville, Illinois.

Belli, P., Anderson, J.R., Barnum, H.N., Dixon, J.A., Tan, J.-P., 2001, *Economic Analysis of Investment Operations*. WBI Development Studies, The World Bank, Washington, D.C.

Boehlje, M.D., Lins, D.A., 1998, Risks and risk management in an industrialized agriculture. *Agric. Finance Rev.* **58**: 1–16.

Churchill, G.A., 1995, *Marketing Research Methodological Foundations*. The Dryden Press, New York.

Courtney, H., Kirkland, J., Viguerie, P., 1997, Strategy under uncertainty. *Harvard Bus. Rev.* **75**: 67–79.

European Commission, 2001, *Risk Management Tools for EU Agriculture, with a Special Focus on Insurance*. Working Document, Agriculture Directorate-General.

Goodwin, B.K., Ker, A.P., 1998, Revenue insurance—a new dimension in risk management. *Choices* **13**(4): 24–27.

Hardaker, J.B., Huirne, R.B.M., Anderson, J.R., Lien, G., 2004, *Coping with Risk in Agriculture*, second edition. CAB International, Wallingford.

Huirne, R.B.M., 2002, Strategy and risk in farming. *Neth. J. Agr. Sci.* **50**: 249–259.

Huirne, R.B.M., Hardaker, J.B., Dijkhuizen, A.A. (eds), 1997, *Risk Management Strategies in Agriculture: State of the Art and Future Perspectives*. Mansholt Studies, No. 7, Wageningen Agricultural University, Wageningen.

Huirne, R.B.M., Meuwissen, M.P.M., Hardaker, J.B., Anderson, J.R., 2000, Risk and risk management in agriculture: an overview and empirical results. *Int. J. Risk Assess. Manage.* **1**: 125–136.

Meuwissen, M.P.M., Huirne, R.B.M., Hardaker, J.B., 1999, *Income Insurance in European Agriculture*. Scientific Report EU-Project, European Economy, No. 2, DGII, Brussels, pp. 95.

Meuwissen, M.P.M., Huirne, R.B.M., Hardaker, J.B., 2001, Risks and risk management strategies; an analysis of Dutch livestock farmers. *Livest. Prod. Sci.* **69**: 43–53.

Meuwissen, M.P.M., Huirne, R.B.M., Skees, J.R., 2003a, Income insurance in European agriculture. *EuroChoices* (Spring), 12–17.

Meuwissen, M.P.M., Velthuis, A.G.J., Hogeveen, H., Huirne, R.B.M., 2003b, Traceability and certification in meat supply chains. *J. Agribusiness* **21**: 167–181.

Skees, J.R., 1999, Agricultural risk management or income enhancement? Regulation. *Cato Rev. Bus. Gov.* **22**: 35–43.

Unnevehr, L.J., Miller, G.Y., Gómez, M.I., 1999, Ensuring food safety and quality in farm-level production: emerging lessons from the pork industry. *Am. J. Agric. Econ.* **81**: 1096–1101.

Van Asseldonk, M.A.P.M., Meuwissen, M.P.M., Huirne, R.B.M., 2001, Stochastic simulation of catastrophic hail and windstorm indemnities in the Dutch greenhouse sector. *Risk Anal.* **21**: 761–769.

Velthuis, A.G.J., Unnevehr, L.J., Hogeveen, H., Huirne, R.B.M. (eds), 2003, *New Approaches to Food-Safety Economics*. Kluwer Academic Publishers, Dordrecht, The Netherlands.

Chapter 2

DEALING WITH MULTIPLE OBJECTIVES IN AGRICULTURE

Kiyotada Hayashi
National Agricultural Research Center, 3-1-1 Kannondai, Tsukuba, Ibaraki 305–8666, Japan

Abstract This chapter reviews the multicriteria analysis applied to the selection problems in agricultural systems. It also discusses life cycle impact assessment (LCIA) from the decision analytic framework. Special attention is paid to the attributes (evaluation criteria) used for evaluating agricultural systems by considering their impacts on the environment. Problem structuring in decision analysis, which is related to the definition of impact categories in LCIA, is also discussed to construct multiple objectives suitable for agri-environmental decision making.

Keywords: Agricultural systems, sustainability, multicriteria analysis, selection problems, life cycle impact assessment (LCIA), category indicators

1 INTRODUCTION

The interactions between agriculture and the environment are causing major public concern over the appropriateness of modern agricultural practices (Pretty, 2002). For example, intensive agricultural practices that make excessive use of chemical fertilizers and manure have negative impacts on the environment and are therefore recognized as the cause of agricultural nonpoint pollution. Moreover, modern intensive agriculture is threatening wild biodiversity. Since agriculture is now spreading to the remotest parts of the world in destructive forms, wild biodiversity has been reducing, thus undermining the sustainability of the food production system (McNeely and Scherr, 2003).

As a result, operations research for natural resource management has become important in assisting farmers and extension specialists decide whether to introduce alternative practices to make agriculture more sustainable and in helping policy makers judge the appropriateness of policies designed to remedy agri-environmental issues. Multicriteria analysis can be a typical methodology used for these purposes.

However, in most case studies, attention is centered on the relative performance of several agricultural systems and the selection of an appropriate agricultural system. This means that the problems pertaining to how evaluation criteria (attributes used for multicriteria analysis) should be constructed have not been sufficiently discussed, although the consideration of these problems is indispensable in the effective performance of multicriteria analysis. This is because the result of the analysis may be inadequate to ensure advantageous behavior unless the problem is structured appropriately.

Therefore, this chapter reviews the multicriteria analysis applied to analyze the relative performance of agricultural systems. Special attention is paid to the environmental impacts of agricultural practices. Life cycle impact assessment (LCIA) is also included in the following survey and is discussed from the perspective of multicriteria analysis. Considering the appropriateness of multicriteria analysis and related methods for resolving agri-environmental problems will offer food for thought on problem structuring in decision analysis for agriculture, through clarifying the issues on how to set appropriate multiple objectives in agricultural decision making.

Section 2 reviews the multicriteria analysis applied to the problems in selecting agricultural systems, in which the environmental impacts of agricultural systems are evaluated, and discusses the problems with weighting. In Section 3, the applications of LCIA to agricultural production are surveyed from the perspective of multicriteria analysis, after presenting a tripartite classification of methodology, which consists of the direct application of multicriteria analysis, multicriteria analysis using the midpoint approach to LCIA, and multicriteria analysis using the end point approach to LCIA. In the subsequent section, some related topics are presented to better understand the applicability of multicriteria analysis to natural resource management.

2 MULTICRITERIA EVALUATION OF AGRICULTURAL SYSTEMS

This section describes the selection problems in agricultural systems, and agricultural systems are defined as discrete alternatives and are selected (or ranked) with respect to multiple attributes. Although multicriteria analysis contains multiobjective planning (multiobjective mathematical programming, including goal programming), this chapter focuses on the methods for selecting discrete alternatives. (See Hayashi (2000) and Romero and Rehman (2003) for information on multicriteria analysis, including multiobjective mathematical programming, and see Hardaker *et al.* (2004) for information on decision analysis in agriculture.)

Table 1 illustrates the models used for evaluating agricultural systems by considering their impacts on the environment. In these applications, both compensatory and noncompensatory methods are used. The former methods use multiattribute value functions or compromise programming, and the latter methods apply the concept of outranking (see, e.g., DETR, 2000). Since it is difficult to elicit attribute weights from decision makers (the farmers or experts), only ranking information is used in the applications of the former methods. For example, to obtain the best and worst overall values, Yakowitz *et al.* (1993) define the following mathematical program:

$$\text{minimize or maximize } v(a_i) = \sum_{j=1}^{n} w_j v_j(x_{ij})$$

subject to

$$\sum_{j=1}^{n} w_j = 1$$

$$w_1 \geq w_2 \geq w_3 \ldots \geq w_n \geq 0$$

where $v(a_i)$ is the overall value of alternative a_i, w_j is the weight assigned to the jth attribute, $v_j(\cdot)$ is the value function for the jth attribute, x_{ij} is the jth attribute level for alternative a_i, and n is the number of attributes.

Table 2 lists the attributes (criteria) used for analyzing the selection problems in agricultural systems. Since these are examples of evaluating agricultural systems on the basis of environmental impacts, attention is paid to the trade-offs between economic objectives and environmental objectives, with the exception of Arondel and Girardin (2000). This implies that the problems can be depicted graphically as value trees (objectives hierarchies) that contain profitability and the environmental quality of soil and water.

Conducting multicriteria analysis for the selection problems provides us with an integrated perspective on the interaction between agriculture and the environment; thus, multicriteria analysis supports farmers' and policy makers'

Table 1. Models used for selecting an agricultural system.

Authors[a]	Model	Weighting method	Decision maker(s)
Yakowitz *et al.* (1993)	MAVF[b]	Ranking[c]	A farmer[d]
Foltz *et al.* (1995)	MAVF[b]	Pairwise ranking[e]	A farmer
Heilman *et al.* (1997)	MAVF[b]	Ranking[c]	Experts
Lawrence *et al.* (1997)	MAVF[b]	Ranking[c]	Experts
Tiwari *et al.* (1999)	Compromise programming	Verbal pairwise comparison (AHP)	Farmers and project officials
Arondel and Girardin (2000)	Outranking (ELECTRE TRI)	The revised "weighting with cards" method	Experts
Strassert and Prato (2002)	Outranking (balancing and ranking)	n.a.	n.a.

[a] References are restricted to referred journals in English.
[b] The multiattribute value function.
[c] The same method is applied.
[d] The explicit explanation of the decision maker is not provided.
[e] Kirkwood and Sarin (1985)

decisions on whether to introduce alternative agricultural systems. However, there are two difficulties in applying this methodology. One difficulty is the problem of weighting. Since the meaning of weights is dependent on models, weight parameters may have widely differing interpretations for different methodologies and different decision contexts (Belton and Stewart, 2002). In multiattribute value (utility) functions, which have clear algebraic meanings of attribute weights as compared with other methodologies such as outranking methods, weight elicitation methods that do not rely on attribute ranges might lead to biased weights (Von Nitzsch and Weber, 1993; Fischer, 1995; Belton and Stewart, 2002). Nevertheless, several of the applications listed in Table 2 elicit attribute weights without referring to attribute ranges. This difficulty with weighting is a common and serious mistake in the application of multicriteria analysis in various research fields, and this has already been realized in some application areas; for example, in the integration of geographical data by Geographical Information Systems (GIS), the range problem just mentioned has been recognized as a common source of error (Malczewski, 1999).

The other difficulty is related to problem structuring, the importance of which has been stressed recently. Since, for example, the trade-offs between nitrate concentrations in surface water and atrazine concentrations in percolation are difficult to understand for decision makers and even for experts,

Table 2. Attributes used for selecting an agricultural system.

Authors[a]	Economic	Environmental		
		Fertilizer	**Pesticide**	**Others**
Yakowitz *et al.* (1993)	Net income	N (percolation) N (surface) P (surface)	Atrazine (surface) Atrazine (percolation) Serin (surface) Carbofuran (surface)	Sediment yield
Foltz *et al.* (1995)	Net returns	N (surface)[b] N (percolation)[b]	Atrazine (surface)[c] Alachlor (surface)[c]	Soil loss[d]
Heilman *et al.* (1997)	Net returns	N (runoff) NO_3–N (percolation)	Atrazine (runoff) Atrazine (sediment) All other pesticides in surface or groundwater	Soil detachment Sediment yield
Lawrence *et al.* (1997)	Above ground net primary production			Range condition Channel erosion Annual runoff Annual maximum peak runoff rate Quail and javalina [NRCS[e] wildlife habitat index]
Tiwari *et al.* (1999)	Farmers' NPV Governmental NPV Societal NPV			Land suitability Energy output/input Water requirement Environmental cost
Arondel and Girardin (2000)		N management (amount, balance, date, splitting up, improving techniques)	Pesticide management (amount, half-life, mobility, toxicity, location, date)	Water management (hydric balance, amount)
Strassert and Prato (2002)	Net returns SD of net returns [as an economic risk]	NO_3–N (runoff) [as aquatic ecosystems]	Atrazine (application) [as drinking water quality]	Soil erosion

[a] See Table 1.
[b] Estimated by the Erosion Productivity Impact Calculator (EPIC)
[c] Estimated by Groundwater Loading Effects of Agricultural Management Systems (GLEAMS)
[d] Calculated by the Universal Soil Loss Equation (USLE)
[e] The USDA Natural Resources Conservation Service

it is necessary to introduce a methodology for transforming the evaluation data into other values to make the meaning easily comprehensible. This procedure can be considered as impact assessment. Although LCIA was developed in a research field different from decision analysis, it can be recognized as a type of multicriteria analysis. Hence, the subsequent section reviews the LCIA applied to agricultural production.

3 LCIA FROM THE PERSPECTIVE OF MULTICRITERIA ANALYSIS

As a preparation for discussing the earlier applications of LCIA in agriculture, this section first describes the trichotomy that consists of (1) the direct application of multicriteria analysis, (2) multicriteria analysis using midpoint approaches to LCIA, and (3) multicriteria analysis using end point approaches to LCIA. In the first category (Fig. 1), the inventory data are directly transformed into environmental indicators (overall values) by weighting. Although the system boundary of LCIA in general includes fertilizer and pesticide production processes, these figures depict only the direct impacts of agricultural practices. In addition, although the figure lacks the economic criteria that the applications in the previous section have, it is possible to add the criteria that measure economic performance.

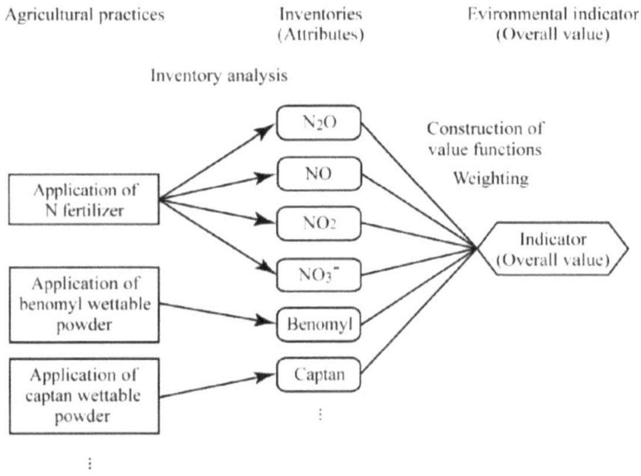

Figure 1. The direct application of multicriteria analysis. A rectangular node depicts an agricultural practice, which can be considered as a decision because the decision maker can control it directly. A rounded rectangular node means any intermediate concept or variable. A hexagonal node depicts an overall value or indicator, which can be used to evaluate the relative desirability of agricultural practices or agricultural systems. (Adapted from Hayashi and Kawashima, 2004.)

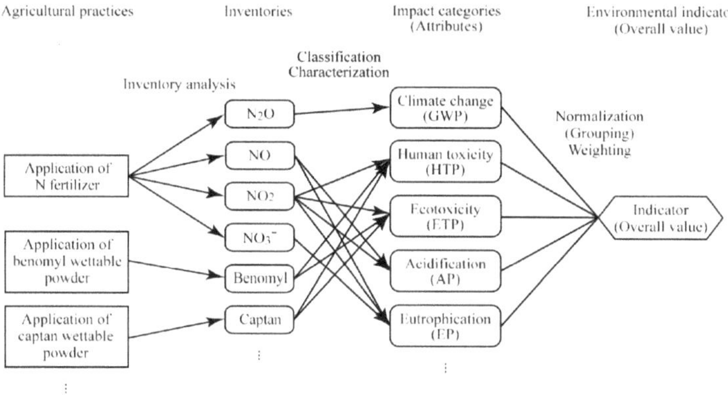

Figure 2. Multicriteria analysis using a midpoint approach to LCIA. The classification is based on Guineé (2002). See Fig. 1 for the notation. Adapted from Hayashi and Kawashima (2004).

The second category (Fig. 2) introduces the category indicators at the midpoint level, such as global warming potential (GWP), which correspond to the attributes in multicriteria analysis. In other words, inventory data are transformed into impact categories using, for example, GWP. Although the term "normalization" is used in the figure as is often the case with LCIA, it is recognized as the construction of value functions from the multiattribute value function framework.

In the third category (Fig. 3), inventory data are integrated into more comprehensive concepts such as human health and ecosystem quality, which are the category indicators at the end point level. Fate analysis, exposure and effect analysis, and damage analysis are used to derive the category indicators at this level. (For a discussion on the relationship between decision analysis and life cycle assessment (LCA), see Miettinen and Hämäläinen, 1997; Hertwich and Hammitt, 2001a, b; and Seppälä *et al.*, 2002).

Table 3 illustrates the applications of LCA to the assessment of the environmental impacts of agricultural systems. Examples include arable farming, milk production, and animal production. Most of the applications define the decision alternatives, although the problems in those applications are not explicitly formulated as decision problems. As for the functional unit, there are the unit weights of products and the unit area of production.

Table 3. Examples of LCA in agriculture (definition of the problem).

Authors[a]	**Issue**	**Alternatives**	**Functional Units**
Hanegraaf *et al.* (1998)	Energy crop production in the Netherlands	Route+Crop (GAP)	1 GJ and 1 ha
Cederberg and Mattsson (2000)	Milk production in Germany	Conventional and organic farming	1000 kg ECM (energy corrected milk)
Haas *et al.* (2001)	Grassland far-ming in Germany	Intensive, extensive, and organic farming	1 ha and 1 t milk
Brentrup *et al.* (2001a, b)	Sugar beet production in Germany	Sugar beet production with calcium ammonium nitrate (solid fertilizer), urea (solid fertilizer), and urea ammonium nitrate solution (liquid fertilizer)	1 t of extractable sugar
Eide (2002)	Industrial milk production in Norway	Small, middle-sized, and large dairy	1000 L of drinking milk brought to the consumers
Bennet *et al.* (2004)	GM sugar beet production in the UK and Germany	Conventional and GM-herbicide-tolerant sugar beet	50,000 kg fresh weight of sugar beet
Brentrup *et al.* (2004)	Winter wheat production in the UK	(Nitrogen fertilizer rate)	1 t of grain
Basset-Mens and van der Werf (2005)	Pig production in France	Conventional GAP, a French quality label (red label), and organic agriculture	1 ha and 1 kg of pig
Antón *et al.* (2005)	Greenhouse tomato produc-tion in Spain	Soil cultivation, open, and closed hydroponic systems (+3 waste management scenarios)	1 kg of tomatoes

Source: Hayashi et al. 2006

[a] The references are restricted to referred journals in English. Papers that analyze only food processing are not included. Web of Science was used to search the papers.

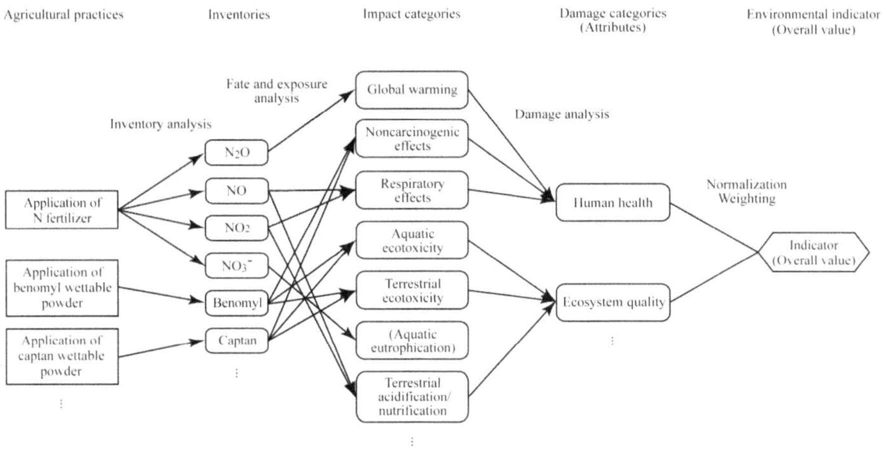

Figure 3. Multicriteria analysis using an end point approach to LCIA. Cause and effect relationships in fate analysis, exposure and effect analysis, and damage analysis are based on Jolliet *et al.* (2003), although the relationships regarding global warming are based on Itsubo and Inaba (2003). (Adapted from Hayashi and Kawashima, 2004.)

Table 4 summarizes the impact categories used in the applications to agriculture. The commonly used categories are climate change, human toxicity, acidification, and eutrophication. These categories can be recognized as attributes in multiattribute value functions, although weighting is not performed except in the work of Brentrup *et al.* (2001) in which Eco-indicator 95 is used.

Table 4. Examples of LCA in agriculture (impact categories).

Authors[a]	(1)	(2)	(3)	(4)	(5)	(6)	(7)	(8)	(9)	(10)
Hanegraaf *et al.* (1998)	x		x			x		x		x[b]
Cederberg and Mattsson (2000)	x	x	x	x		x	x	x	x	
Haas *et al.* (2001)	x	x	x	x	x					x[c]
Brentrup *et al.* (2001)			x			x	x		x	
Eide (2002)			x	x	x	x	x	x	x	
Bennet *et al.* (2004)	x		x	x	x	x	x	x	x	
Brentrup *et al.* (2004)	x	x	x	x	x	x	x			
Basset-Mens and van der Werf (2005)	x	x	x		x	x	x			x[d]
Antón *et al.* (2005)	x		x	x	x	x	x	x	x	x[e]

Note: The column headings are as follows: (1) resource use, (2) land use, (3) climate change, (4) human toxicity, (5) ecotoxicity, (6) acidification, (7) eutrophication, (8) stratospheric ozone depletion, (9) photo-oxidant formation and (10) others.

[a] The references are restricted to referred journals in English. Papers that analyze only food processing are not included.

[b] Minerals to soil and water, pesticides, soil erosion, groundwater depletion, waste production and utilization, contribution to biodiversity, contribution to landscape values

[c] Soil functions, water quality, and biodiversity

[d] Pesticide use

[e] Water resource

Although the applicability of the end point approaches to LCIA, such as Eco-indicator 99 (Goedkoop and Spriensma, 2001), to the evaluation of agricultural production has already been discussed (Jungbluth and Frischknecht, 2001), the impacts of fertilizer use and pesticide application, for example, are difficult to evaluate because of the difficulties in estimating related categorization factors. This may be the reason the case studies in Tables 3 and 4 applied the midpoint approaches.

4 WHAT TYPES OF ATTRIBUTES/CATEGORIES SHOULD BE USED?

This section discusses the problem of what types of attributes/categories should be used. First, the agri-environmental indicators developed by the OECD are presented, because they can be considered as attributes for evaluating the environmental impacts of agricultural systems and also because they can serve as a reference point for the ensuing discussion. Table 5 illustrates some of the indicators; it presents issues related to the environmental impacts of agriculture. This table contains a list of categories that are useful in establishing the midpoint categories (Issues in the table) suitable for assessing agricultural production. For example, Haas *et al.* (2001), used eight midpoint categories, of which four can be considered as impact categories that are particularly necessary for evaluating agricultural production. They are soil function/strain, biodiversity, landscape image (aesthetics), and animal husbandry (appropriate animal welfare). Except for the last category, all the other categories can be found in Table 5. Although "loss of biodiversity" is included in the general midpoint approach to LCIA as a study-specific impact category (Guineé, 2002), it appears that the approach using agri-environmental indicators as midpoint categories is useful in dealing with the problem of how to construct impact categories specific to agricultural production.

Since establishing the appropriate categories for LCIA is equivalent to constructing the appropriate attributes for multicriteria analysis, it is necessary to remark on certain aspects that are related to the discussion in the previous section. First, although radar charts are used to illustrate the results of LCIA, they do not provide adequate support for decision making; this is because dominance concepts can eliminate dominated alternatives but cannot determine the most preferable one. This is the reason weighting is necessary for supporting decisions. Although weighting is still difficult in the midpoint approaches to LCIA, discussing the general superiority of one

Table 5. List of OECD agri-environmental indicators (environmental impacts of agriculture).

Issues	**Indicators**
Soil quality	Risk of soil erosion by water
	Risk of soil erosion by wind
Water quality	Water quality risk indicator
	Water quality state indicator
Land conservation	Water-retaining capacity
	Off-farm sediment flow
Greenhouse gases	Gross agricultural greenhouse gas emissions
Biodiversity	Genetic diversity
	Species diversity
	Ecosystem diversity
Wildlife habitats	Intensively farmed agricultural habitats
	Semi-natural agricultural habitats
	Uncultivated natural habitats
	Habitat matrix
Landscape	The structure of landscape
	Landscape management
	Landscape costs and benefits

Source: OECD (2001)

approach over the other approach may be difficult; this is because it is pointed out that in LCIA, midpoint approaches yield results that are relatively certain but less environmentally relevant, whereas end point approaches yield results that are expressed in relevant terms but are relatively uncertain (Udo de Haes *et al.*, 2002). Thus, it may be fruitful to construct a framework that integrates midpoint and end point approaches.

Second, there is a possibility that establishing a set of categories that are suitable for assessing agricultural production could be complicated. That is, constructing a set of attributes that are nonredundant, which is one of the properties that fundamental objectives should satisfy (Keeney, 1992), would be difficult because of the confusion between general midpoint categories and agriculture-specific categories that are derived from agri-environmental indicators.

The earlier discussion indicates the necessity of problem structuring. In this case, graphical models for decision making, including influence diagrams and Bayesian networks, will be useful. In reality, graphical representation techniques have already been used in the explanation of LCIA, although terms such as the cause–impact network (Udo de Haes *et al.*, 2002), the impact web (Hertwich and Hammitt, 2001a), and the impact chain (Hertwich and Hammitt, 2001b) are used. In the modeling, it is important to be explicit regarding the complementarity between value-focused thinking (preference modeling) and causality-focused thinking (physical modeling).

Thus far, the discussion in this section has been based on the understanding that it is possible to calculate category indicators at the end point as well as the midpoint levels. In concluding this section, the other research directions have to be mentioned because it is unfair not to touch on them. The first example is the environmental accounting framework, which is used for comparing organic, integrated, and conventional farming systems (Pacini *et al.*, 2003). In the analysis, the compliance of farming systems with environmental sustainability thresholds is monitored. The second example is the reference point methodology (Wierzbicki *et al.*, 2000). Although this is originally a multiobjective programming method, it can be applied to the selection problems as in the case of compromise programming. In the methodology, in general, indicators and their thresholds at the inventory level are used and the solution is refined interactively, although category indicators can be introduced into mathematical programming.

5 CONCLUDING REMARKS

The topic contained in this final section is related to the reason we have discussed the problem of what types of attributes/categories should be used. Table 6 provides the explanation.

Table 6. How to deal with objectives?

Approach	Research question	How do researchers deal with objectives?
Normative	How should people behave?	Mathematical theory
Descriptive	How do people behave?	Surveys and experiments
Prescriptive	How to support people	Decision analysis and support

Normative approaches use mathematical theory to study how people should behave in a theoretical situation. Descriptive approaches attempt to find the explanation for people's actual behavior. The purpose of prescriptive approaches is to support people's decision making. Here, we find a difference in the presumptions between LCA and decision analysis. For example, in LCIA, the weights for end points (areas of protection) are measured by conjoint analysis and the assessment factors are determined on the basis of the weights. However, applying the weights to the assessment is quite different from the procedure of decision analysis. In order to support a decision maker, the preference of the decision maker has to be used. Furthermore, this table provides an explanation of the relationship between the earlier analysis on farmers' goals and behaviors (see, for example, Gasson, 1973) and the discussion in this chapter. In other words, the former is classified as a descriptive approach. Thus, there are differences among the objectives presented in the research on farmers' descriptive behavior and the objectives discussed in this chapter.

ACKNOWLEDGMENTS

This research was partially supported by the Japan Society for the Promotion of Science, Grant-in-Aid for Scientific Research (C), 16580194, 2004–2006.

6 REFERENCES

Antón, A., Montero, J.I. and Muñoz, P. (2005). LCA and tomato production in Mediterranean greenhouses. *International Journal of Agricultural Resources Governance and Ecology*, 4:102–112.

Arondel, C. and Girardin, P. (2000). Sorting cropping systems on the basis of their impact on groundwater quality. *European Journal of Operational Research*, 127:467–482.

Basset-Mens, C. and van der Werf, H.M.G. (2005). Scenario-based environmental assessment of farming systems: The case of pig production in France. *Agriculture, Ecosystems and Environment*, 105:127–144.

Belton, V. and Stewart, T.J. (2002). *Multiple Criteria Decision Analysis: An Integrated Approach*. Kluwer Academic Publishers, Dordrecht, The Netherlands, 372 pp.

Bennett, R., Phipps, R., Strange, A. and Grey, P. (2004). Environmental and human health impacts of growing genetically modified herbicide-tolerant sugar beet: A life-cycle assessment. *Plant Biotechnology Journal*, 2:273–278.

Brentrup, F., Küsters, J., Kuhlmann, H. and Lammel, J. (2001). Application of the Life Cycle Assessment methodology to agricultural production: An example of sugar beet production with different forms of nitrogen fertilisers. *European Journal Agronomy*, 14:221–233.

Brentrup, F., Küsters, J., Lammel, J., Barraclough, P. and Kuhlmann, H. (2004). Environmental impact assessment of agricultural production systems using the life cycle assessment (LCA) methodology II. The application to N fertilizer use in winter wheat production systems. *European Journal of Agronomy*, 20:265–279.

Cederberg, C. and Mattsson, B. (2000). Life cycle assessment of milk production: A comparison of conventional and organic farming. *Journal of Cleaner Production*, 8:49–60.

Department of the Environment, Transport and the Regions (DETR) (2000). *Multi-Criteria Analysis: A Manual*. London, 158 pp.

Eide, M.H. (2002). Life cycle assessment (LCA) of industrial milk production. The *International Journal of Life Cycle Assessment*, 7:115–126.

Fischer, G.W. (1995). Range sensitivity of attribute weights in multiattribute value models. *Organizational Behavior and Human Decision Processes*, 62:252–266.

Foltz, J.C., Lee, J.G., Martin, M.A. and Preckel, P.V. (1995). Multiattribute assessment of alternative cropping systems. *American Journal of Agricultural Economics*, 77:408–420.

Gasson, R. (1973). Goals and values of farmers. *Journal of Agricultural Economics*, 24: 521–542.

Goedkoop, M. and Spriensma, R. (2001). *The Eco-Indicator 99: A Damage Oriented Method for Life Cycle Impact Assessment, Methodology Report*, 3rd edition. PRé Consultants, Amersfoort, the Netherlands.

Guineé , J.B. (Editor) (2002). *Handbook on Life Cycle Assessment: Operational Guide to the ISO Standards*. Kluwer Academic Publishers, Dordrecht, the Netherlands, 692 pp.

Haas, G., Wetterrich, F. and Köpke, U. (2001). Comparing intensive, extensified and organic grassland farming in southern Germany by process life cycle assessment. *Agriculture, Ecosystems and Environment*, 83:43–53.

Hanegraaf, M.C., Biewinga, E.E. and van der Bijl, G. (1998). Assessing the ecological and economic sustainability of energy crops. *Biomass and Bioenergy*, 15:345–355.

Hardaker, J.B., Huirne, R.B.M., Anderson, J.R. and Lien, G. (2004). *Coping with Risk in Agriculture*. CABI Publishing, Wallingford, UK, 332 pp.

Hayashi, K. (2000). Multicriteria analysis for agricultural resource management: A critical survey and future perspectives. *European Journal of Operational Research*, 122:486–500.

Hayashi, K., Gaillard, G. and Nemecek, T. (2006). Life cycle assessment of agricultural production systems: Current issues and future perspectives. In: *Proceedings of International Seminar on Technology Development for Good Agricultural Practice in Asia and Oceania*, 127–143.

Hayashi, K. and Kawashima, H. (2004). Integrated evaluation of greenhouse vegetable production: Toward sustainable management. *Acta Horticulturae*, 655:489–496.

Heilman, P., Yakowitz, D.S. and Lane, L.J. (1997). Targeting farms to improve water quality. *Applied Mathematics and Computation*, 83:173–194.

Hertwich, E.G. and Hammitt, J.K. (2001a). A decision-analytic framework for impact assessment: Part 1: LCA and decision analysis. *The International Journal of Life Cycle Assessment*, 6:5–12.

Hertwich, E.G. and Hammitt, J.K. (2001b). A decision-analytic framework for impact assessment: Part 2: Midpoints, endpoints, and criteria for method development. *The International Journal of Life Cycle Assessment*, 6:265–272.

Itsubo, N. and Inaba, A. (2003). A new LCIA method: LIME has been completed. International Journal of Life Cycle Assessment, 8:305.

Jolliet, O., Margni, M., Charles, R., Humbert, S., Payet, J., Rebitzer, G. and Rosenbaum, R. (2003). IMPACT 2002+: A new life cycle impact assessment methodology. *International Journal of Life Cycle Assessment*, 8:324–330.

Jungbluth, N. and Frischknecht, R. (2001). The use of Eco-indicator 99 for the impact assessment of food products. In: *Proceedings: International Conference on LCA in Foods*, Gothenburg, Sweden, April 26–27, 2001. The Swedish Institute for Food and Biotechnology, SIK-Document, 143:1–6.

Keeney, R.L. (1992). *Value-Focused Thinking: A Path to Creative Decision Making*. Harvard University Press, Cambridge, MA., 416 pp.

Kirkwood, C.W. and Sarin, R.K. (1985). Ranking with partial information: A method and an application. *Operations Research*, 33:38–48.

Lawrence, P.A., Stone, J.J., Heilman, P. and Lane, L.J. (1997). Using measured data and expert opinion in a multiple objective decision support system for semiarid rangelands. *Transactions in the ASAE*, 40:1589–1597.

Malczewski, J. (1999). *GIS and Multicriteria Decision Analysis*. John Wiley & Sons, New York, 408 pp.

McNeely, J.A. and Scherr, S.J. (2003). *Ecoagriculture: Strategies to Feed the World and Save Biodiversity*. Island Press, London, 315 pp.

Miettinen, P. and Hämäläinen, R.P. (1997). How to benefit from decision analysis in environmental life cycle assessment (LCA). *European Journal of Operational Research*, 102:279–294.

OECD. (2001). *Environmental Indicators for Agriculture Volume 3, Methods and Results*. OECD, Paris, 409 pp.

Pacini, C., Wossink, A., Giesen, G., Vazzana, C. and Huirne, R. (2003). Evaluation of sustainability of organic, integrated and conventional farming systems: A farm and field-scale analysis. *Agriculture, Ecosystems and Environment*, 95:273–288.

Pretty, J. (2002). *Agri-Culture: Reconnecting People, Land and Nature*. Earthscan Publications Ltd, London, 261 pp.

Romero, C. and Rehman, R. (2003). *Multiple Criteria Analysis for Agricultural Decisions*, 2nd edition. Elsevier, Amsterdam, The Netherlands, 186 pp.

Seppälä , J., Basson, L. and Norris, G.A. (2002). Decision analysis frameworks for life-cycle impact assessment. *Journal of Industrial Ecology*, 5:45–68.

Strassert, G. and Prato, T. (2002). Selecting farming systems using a new multiple criteria decision model: The balancing and ranking method. *Ecological Economics*, 40:269–277.

Tiwari, D.N., Loof, R. and Paudyal, G.N. (1999). Environmental-economic decision-making in lowland irrigated agriculture using multi-criteria analysis techniques. *Agricultural Systems*, 60:99–112.

Udo de Haes, H.A., Finnveden, G., Goedkoop, M., Hauschild, M., Hertwich, E.G., Hofstetter, P., Jolliet, O., Klöpffer, W., Krewitt, W., Lindeijer, E., Müeller-Wenk, R., Olson, S.I., Pennington, D.W., Potting, J. and Steen, B. (Editors) (2002). *Life Cycle Impact Assessment: Striving Towards Best Practice. The Society of Environmental Toxicology and Chemistry (SETAC)*. Pensacola, FL., 249 pp.

Von Nitzsch, R. and Weber, M. (1993). The effect of attribute ranges on weights in multiattribute utility measurements. *Management Science*, 39:937–943.

Wierzbicki, A.P., Makowski, M. and Wessels J. (2000). *Model-Based Decision Support Methodology with Environmental Applications*. Kluwer Academic Publishers, Dordrecht, The Netherlands, 475 pp.

Yakowitz, D.S., Lane, L.J. and Szidarovszky, F. (1993). Multi-attribute decision making: Dominance with respect to an importance order of the attributes. *Applied Mathematics and Computation*, 54:167–181.

Chapter 3

MODELING MULTIFUNCTIONAL AGROFORESTRY SYSTEMS WITH ENVIRONMENTAL VALUES: DEHESA IN SPAIN AND WOODLAND RANCHES IN CALIFORNIA

Pablo Campos[1], Alejandro Caparrós[1], Emilio Cerdá[2], Lynn Huntsinger[3], and Richard B. Standiford[3]
[1]Institute of Economics and Geography (IEG), Spanish Council for Scientific Research (CSIC); [2]University Complutense, [3]College of Natural Resources University of California

Abstract The high environmental and amenity values of Mediterranean oak woodlands influence the response of the public and landowners to market forces and to public policies for the management of oak woodland areas. In California and in Spain, woodlands with a Quercus overstory open enough to allow the development of a significant grassy or shrubby understory harbor exceptional levels of biodiversity, provide watershed and habitat, sequester carbon, offer historically meaningful landscapes, and are pleasing to the eye. For historic reasons, and because of the social and environmental values of the woodlands for their owners, large private holdings based on sylvopastoral enterprises have and will have a crucial role in the future of the woodlands. Simple financial models for predicting landowner behavior based on response to market forces do not explain landowner retention of oaks without incorporation of landowner consumption of environmental and amenity values from the property, because landowner utility for oaks is not fully accounted for. By the same token, predicting the best afforestation approach considering carbon sequestration alone without consideration of the biodiversity and amenity values of native oaks risks an overvaluation of planting alien species that could have negative environmental and social consequences. Reforestation models for carbon sequestration that do not incorporate biodiversity and public amenity values might favor plantings of alien species such as eucalyptus; however, this does not take into account the high public and private consumption values of native oaks.

Keywords: Oak woodlands, optimization model, carbon sequestration, firewood, optimal control

1 INTRODUCTION

Mediterranean oak woodlands have high environmental and amenity values. Woodlands with a Quercus overstory open enough to allow the development of a significant grassy or shrubby understory harbor exceptional levels of biodiversity, provide watershed and habitat, sequester carbon, offer historically meaningful landscapes, and are pleasing to the eye. Such woodlands are important throughout the Mediterranean area, and also in California, where the climate and vegetation formations are similar (Fig. 1). Traditional sylvopastoral uses have proven to be essential in many cases and in others at least reasonably compatible with the continuance of these woodlands. For historic reasons, and because of the social and environmental values of the woodlands for their owners, large private holdings have had a crucial role. As a result, today the decisions of Spanish and Californian landowners will determine the fate of much of these woodlands. In this chapter we examine different economic models for predicting landowner response to various market forces, assessing the role of environmental and other social values for the landowner in explaining landowner decisions. We review the historical background of landownership and management; then examine the development of an optimal control model for explaining and predicting landowner stewardship of oaks; and finally examine a model assessing the potential future impact of carbon sequestration incentives for afforestation and reforestation, comparing approaches that do and do not internalize biodiversity and other values.

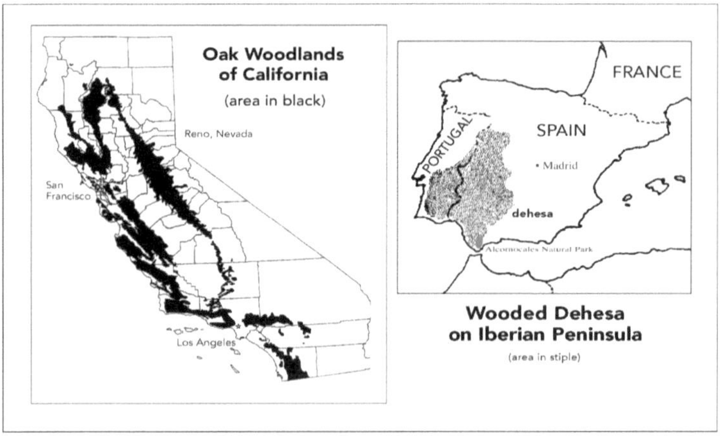

Figure 1. Californian oak woodlands and Spanish dehesa.

2 SPANISH WOODLAND DEHESA OWNERSHIP AND MANAGEMENT HISTORY

Known as *dehesa* in Spain and *montado* in Portugal, sylvopastoral oak woodlands extend for more than 3.2 million hectares in the west and south-western Iberian Peninsula. The typical dehesa is a private property larger than 500 ha (Campos, 1984) (Table 1). Dehesa is most common in areas where powerful knights and Castillian nobility were rewarded with the woodlands of reconquered Muslim areas beginning in the 11th century and ending with the surrender of Granada to the Catholic kings in the 15th century (Hernández and Pulido, 2004).

Despite evidence that dehesa was extant in the first millennium AD when land was divided among retired Roman legionnaires near Merida, in Extremadura (Cerrillo, 1984), widespread development of dehesa is relatively recent. The historically sparse Castillian population could convert closed forest to dehesa only slowly, and manual clearing of the vigorous shrubby woodland understory to create dehesa pasture lasts only a few years without intensive grazing followed by renewed clearing or understory cereal cultivation (Díaz *et al.*, 1997). Dehesa extent peaked only in the early twentieth century (Linares and Zapata, 2003).

During the Middle Ages high-quality merino wool was increasingly marketed to the textile industries of England, Flanders, and Genova. Livestock owners from the mountainous regions of Castilla and Leon found that the opened up Arab lands offered a solution for the problem of limited feed and winter forage. Transhumance, moving livestock to the south and southwest for the winter, began. The ease of collecting taxes and fees of all types as livestock threaded across bridges and through gates along the stockways meant that the transhumance was favored by the Crown. King Alfonso X the Wise sponsored a new association of transumant Castillian livestock producers under the powerful denomination of "Mesta" in 1273 (Klein, 1920).

The Mesta was for a long time one of the principal financial institutions of the Crown of Castille, and the sylvopastoralist enterprises originating from the allocation of woodlands among those considered responsible for the re-taking of the Muslim lands benefited. The leadership of the reconquest of Castillian territory south of the Duero River and the creation of the vast majority of the dehesa is thus tightly linked. Such powerful interests were able to maintain large herds and extensive grazing lands for the wool production, limiting expansion of the subsistence crops of local rural populations.

Only close to the towns are the lands mostly treeless and divided into small farms for intensive crop and pasture use.

In the nineteenth century, most church and municipal lands, and lands of the knighted orders, were expropriated by the state and sold at public auction. Some municipal common grazing lands (Dehesa Boyal), watersheds, and forests were excluded, and some nobility, able to weather the disentailment fever, retained their dehesa. The process of disentailing the dehesa increased the number of rich landowners, in a way that tended to benefit landowners from distant Madrid, Barcelona, or the Basque provinces more than local landowners.

This shared political history means that Iberian countries developed the large private ownerships that are today's dehesa and montado. They are the only countries on the Mediterranean that maintain livestock production integrated with oak and crop production. In continental Europe oaks were eliminated quickly from the ubiquitous medium and small properties to allow for crop production and grass pasture, just as oaks were eliminated by similar classes of Castillian landowners. The recent North African population explosion has led to an annual reduction in oak woodland cover of more than 1% per year (Campos, 2004). However, in the Iberian peninsula the woodlands are stable in extent in large part because of owners of large properties who value the environmental and other benefits they get from owning the land. Families with a history of dehesa and montado ownership consider them part of family identity and distinction. There is no doubt that newer landowners find dehesa and montado a means of achieving a higher social status. Dehesa and montado have persisted because owners have not responded to market signals that should have led them to clear oaks for cultivation. Instead they have kept their woodlands, profiting from woodland earnings, but profiting perhaps even more from family meaning, a second home, recreation, and the social status of rancher, a genteel status not enjoyed by other kinds of rich agriculturalists in Spain.

3 CALIFORNIA OAK WOODLAND RANCH OWNERSHIP AND MANAGEMENT

Oak woodlands with a developed understory cover more than 2 million hectares in California's Mediterranean climate zone, mostly in the rolling hills of the coast ranges or the Sierra Nevada foothills (CDF-FRAP 2003) (Table 1). Inhabited by more than 300 vertebrate species (Jensen *et al.* 1990), they are perhaps the most significant of the state's wildlife habitat when extent is considered. An "oak woodland ranch" is a livestock enterprise based on grazing the comparatively stable grass understory that, unlike in dehesa, most often persists without ongoing human intervention.

Before European settlement, California was home to an indigenous population of several hundred thousand with a long history of oak woodland management. What is known about the interaction of native management using fire and other methods, and the oak woodlands, is limited because of the widespread and rapid destruction of the indigenous way of life with the coming of Europeans (Keeley 2003), though California tribes today are making an effort to restore native management to some areas. The displacement and depopulation of native California opened up large areas to settlement, and as in Spain, the original land allocations were often made on the basis of service to the Kingdom.

It is California's Spanish and Mexican history that is largely responsible for the creation of large oak woodland ranch properties. California's coastal areas were settled starting in 1769 with missions, presidios, pueblos, and large land grants, called *ranchos*, used for livestock production. The foundation of the colonial economy, livestock hides and tallow were traded to Europe. About 30 rancho grants of thousands of hectares of expropriated lands were made, mostly to retired soldiers. When California became part of Mexico in 1821, the new government broadened and accelerated the granting of lands in large parcels, with more than 770 grants to individuals, especially following the secularization and sale of mission lands in 1834 (Perez, 1982). In surrounding states, the U.S. government allocated lands to private holders in much smaller parcels, resulting in most forest, woodland, and desert remaining in public ownership.

With the end of the Mexican-American War in 1848, California became a territory of the United States. The ranchos were largely broken up because of legal disputes or owner impoverishment, but the ranch properties derived from these breakups, from the sale and granting of mission lands, and from various kinds of reclamation programs under the U.S. government, are relatively large, averaging 800–960 hectares in size (Table 1). The few original ranchos that remain are generally of thousands of hectares. Today, 82% of oak woodlands are privately owned (CDF-FRAP 2003).

The 1849 Gold Rush stimulated a huge short-term population increase, as gold seekers flooded in and then left. Already reduced by more than half under Spanish and Mexican governance, native populations continued a precipitous decline caused by disease, poverty, warfare, and genocide, all now extended to the mountainous mining regions. In the post-Gold Rush vacuum, transhumance into the mountains developed in the 1860s. Cattle and sheep were driven to montane summer range, helping to compensate for the loss of watered lowlands to crop production. Though management of California oak woodland ranches is far less intensive than management of the dehesa because of a general lack of aggressive shrub growth as well as rural labor, ranchers do have some history of oak thinning, brush clearing, and seeding of improved forage. A single clearing, followed by grazing, will

Table 1. Summary of California oak woodland and Spanish dehesa characteristics.

	Californian oak woodland	**Spanish wooded dehesa**
Extent	More than 2 million hectare total oak woodlands and grasslands (CDF-FRAP 2003).	2.2 oak wooded dehesa out of 7 million hectares of dehesa woodlands, shrublands, and grasslands (Díaz, Campos, and Pulido 1997; Campos 1984).
Most common oak	Blue oak (*Quercus douglasii*)	Holm oak (*Quercus ilex*)
Ownership	82% + private (CDF-FRAP 2003). Public woodlands are part of large federal land holdings, utility corridors and watersheds, or regional, county, state and local parks. Sometimes they are leased for grazing.	75–80% + private based on study of a representative area in Extremadura (Campos-Palacín 1984). “Public dehesas, dehesa boyal and other municipal woodlands, are those maintained for community use and are less then 20% of the woodlands”.
Average ranch size	800–960 ha (Huntsinger, Buttolph, and Hopkinson 1997; Sulak and Huntsinger, 2002).	500 ha+ (Campos, 1984).
Amenity and investment ownership	Increasing owner self-consumption of environmental services	Increasing owner self-consumption of environmental services
Land use	Extensive sylvopastoral ranching over more than 60% of the woodland	Agrosylvopastoral complex, “Dehesa”
Stocking rate of livestock (does not meet total animal demand)	5–10 ha/A.U./year (Ewing and others 1988).	4 ha/A.U./year in Extremadura (Campos 1997)
Large stock	92% of animal demand is cattle (California Agricultural Statistics, 1990–2001)	42% of animal demand is cattle (Campos 1997).
Commodity products	Beef, lamb, wool, firewood, game, grazing resources.	Beef, Iberian pig, lamb, acorns, firewood, hay, cereal grain, grazing resources, wool, cabrito, goat milk, game, trufa, charcoal, cheese, fodder, honey, cork.

often last indefinitely. The removal of oaks for more intensive livestock production was subsidized by the state government in the 1940s through 1960s, as part of an effort to increase commodity production without consideration of environmental costs. Despite this, and an oak firewood market supported by the state's growing population, foothill and coastal ranches remain mostly oak woodland today, and studies have shown that ranchers do not often thin oaks or attempt to control them unless canopy cover becomes quite dense (Huntsinger *et al.*, 1997). More than 80% of ranchers live on the ranch with their families and manage the enterprise themselves, with few, if any, employees. Some are multiple-generation family owners, while others are wealthy individuals seeking a part- or full-time way of life that is widely admired by Americans. Numerous studies have shown that ranchers are highly motivated by lifestyle values, willing to accept considerable opportunity costs to remain in ranching, and often take off-ranch jobs or use off-ranch income to support the ranching operation (Liffmann *et al.*, 2000).

4 GOALS OF THIS ANALYSIS

Dehesa and California's oak woodland ranches are the result of particular natural conditions, but also of a particular history of human intervention (Table 1). Thus, models aiming at the understanding of the evolution of these ecosystems need to pay particular attention to the implications of human intervention. The rest of the chapter presents two models developed to understand: (i) the current behavior of landowners in stewarding their oak woodlands in response to market forces and private environmental and social values (Model A), and (ii) the possible future development of these ecosystems under the development of carbon sequestration markets and policies (Model B).

Model A is an optimal control model developed for ranches in California. The basic model was found to severely overestimate the cutting of oaks, so a positive mathematical programming (PMP) approach (Howitt, 1995) was used to derive missing elements of the true costs and returns of oak harvest that were omitted from the original, normative model. As shown below, this permits estimate of the environmental values consumed by the owners themselves. There are environmental values, not internalized in markets as flows (although they are indeed internalized in the price of land), which explain the actual behavior observed in California of not clearing oaks. In Spanish dehesa this "owner autoconsumption" of environmental services is also important, as work by Campos and Mariscal (2003) has shown.

In Spain, however this value was estimated using contingent valuation techniques, and not modeling tools.

Model B is a normative model proposed to evaluate the impact of carbon sequestration markets and policies in a model for two types of reforestations: cork oak (a native species with high environmental values, *Q. suber*) and eucalyptus (an alien species used in the past in Spain and California, *Eucalyptus globulus*). Although data for a full calibration are not available yet, the theoretical model is discussed to identify potential conflicts that may arise from internalizing only carbon sequestration, and not biodiversity and other values. Internalizing carbon sequestration only may imply an incentive to plant fast-growing alien species and a lower incentive to maintain or increase oak woodlands, which may have negative impacts on biodiversity and/or on scenic and other public and landowner values. Results from a contingent valuation study estimating the impact of these two different kinds of reforestations in Spain are presented to illustrate this concern.

5 MODELING LANDOWNER INVESTMENT IN ENVIRONMENTAL VALUES (MODEL A)

Models of likely silvopastoral management decisions must incorporate landowner values (utility), including landowner valuation of the environmental services from their lands. Poorly specified models based only on commodity production understate a manager's own consumption of amenity and environmental services and lead to erroneous conclusions about likely management behavior and appropriate public policies.

Standiford and Howitt (1992) developed a normative dynamic oak woodland optimization model including cattle, firewood, and hunting. The basic structure of the model is as follows:

$$\max \mathrm{NPV} = \int_{t=0} e^{rt} \{ \mathrm{WR}_t(\mathrm{WDSEL}_t) + \mathrm{HR}_t(\mathrm{WD}_t, \mathrm{HRD}_t, exog.) + [\mathrm{LR}_t\ \mathrm{HRD}_t, \mathrm{CS}_t, \mathrm{FOR}_t,](\mathrm{WD}_t, exog.)\}$$

s.t.

$$\dot{\mathrm{WD}} = \mathrm{F}(\mathrm{WD}_t,\ \mathrm{exog.}) - \mathrm{WDSEL}_t \text{ (Equation of motion for oaks),}$$

$$\dot{\mathrm{HRD}} = \mathrm{H}(\mathrm{HRD}_t, \mathrm{exog.}) - \mathrm{CS}_t \text{ (Equation of motion for livestock),}$$

the initial conditions

$$WD_0 = INITWD,$$

$$HRD_0 = INITHRD,$$

and the nonnegativity constraint:

$$WDSEL > 0,$$

where: WD and HRD are the stock of wood and livestock (cows); WR, HR, and LR are the net revenues of firewood, hunting, and livestock, respectively; WDSEL is the volume of firewood sold and CS a vector of the different classes of livestock sold; and FOR gives the number of forage quantity available.

This model was calibrated for the early nineties and concluded that markets at that time would lead landowners to clear their oaks to increase forage yield for livestock production (Standiford and Howitt, 1992). Although common in the 1940s to 1970s, this behavior was actually rare in the nineties, contradicting the prediction of the model (Standiford *et al.*, 1996). The model's shortcomings were due to failure to accurately account for a landowner's desire to keep oaks for their amenity value. A PMP approach (Howitt, 1995) was used to derive missing elements of the true costs and returns of oak harvest that were not in the original normative model. The dynamic optimization model was constrained by actual landowner behavior to derive these missing values. The shadow prices from the behavior constraint represent the marginal benefit of retaining trees as it differs from what might otherwise be predicted.

The firewood net revenue developed from market information and the hedonic pricing model calibrated from the actual behavior of oak woodland owners result in two curves (Fig. 2). The difference between them is the environmental self-consumption value of retaining trees — the value of oak trees to the landowner, which cannot be explained by the price of wood or other commodity values. This specification incorporates actual landowner behavior, giving a more realistic assessment of landowner behavior than a model which omits the value of trees to the landowner (Fig. 3) (Standiford and Howitt, 1992).

This optimization model, incorporating landowner utility, is used to evaluate oak cover, firewood harvest, and cattle grazing under different risk and land productivity conditions (Standiford and Howitt, 1992). Three major

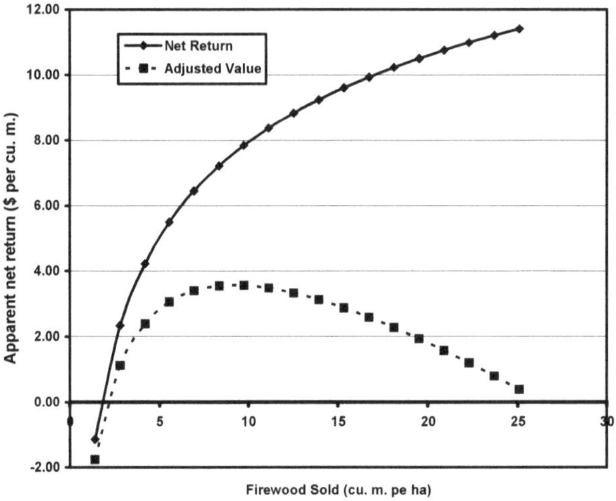

Figure 2. Net firewood return per cubic meter as a function of amount of wood harvested. Source: Standiford and Howitt (1992).

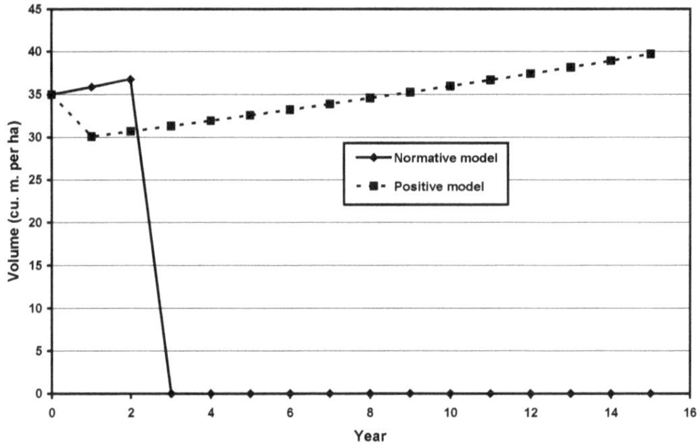

Figure 3. Oak volume levels in California oak woodlands under normative and positive modeling approaches (Standiford and Howitt 1992).

commercial enterprises typically contribute to total net present value of California oak woodlands (Fig. 4). With an initial oak volume of 50 m^3/ha (Standiford and Howitt, 1993), cattle production on average has a positive economic value. Fee hunting can be an important enterprise, contributing from 40% (on good range sites) to 70% (on poor range sites) of the total silvopastoral value. The economic contribution of wood harvest is low.

The model showed that diversification of silvopastoral enterprises reduced tree harvesting and cattle grazing. The marginal value of retaining oaks for wildlife habitat for hunt clubs exceeded the marginal value of the extra forage or firewood harvest (Standiford and Howitt, 1992). Wood harvest is used in years with poor forage production or low livestock prices. The capital value of the trees is a hedge against years with low livestock profitability. Inclusion of a risk term shows that firewood harvest and livestock grazing intensity both increase. Policies reducing landowner risk, such as a subsidized loan program during poor forage production or low livestock price years, might reduce the need to cut the trees for an infusion of capital.

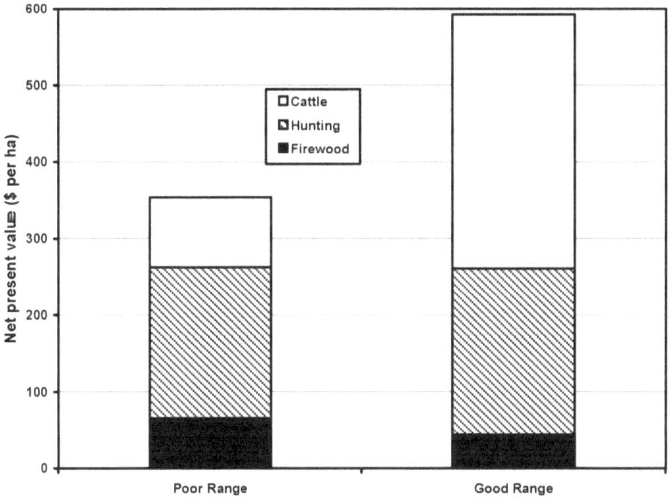

Figure 4. Net present value of California oak woodlands from various commercial enterprises. Initial oak volume is 50 cubic meters per ha (Standiford and Howitt 1993).

6 INCLUDING CARBON SEQUESTRATION AND BIODIVERSITY OR SCENIC VALUES IN THE ANALYSIS (MODEL B)

Countries ratifying the Kyoto Protocol, a development of the United Nations Framework Convention on Climate Change, will need to reduce their greenhouse gas emissions to an overall 5% below 1990 levels by 2012, though specific targets vary by country. One of the alternatives included in the Kyoto Protocol is to plant trees, since trees sequester carbon from the atmosphere by growing and thereby reduce carbon dioxide concentrations. This is 'afforestation and reforestation' in the terminology of the Kyoto

Protocol and the Marrakech Accords, an agreement that completes the Protocol.

According to the Marrakech Accords, parties can issue credits through 'afforestation and reforestation' by means of article 3.3 of the Kyoto Protocol if the land is located in an Annex I country (OECD countries and former economies in transition) that ratifies the Protocol (or eventually via article 6 and Joint Implementation), and by means of article 12 (Clean Development Mechanism) if the land is located in any Non-Annex I Party. Thus, incentives will probably be created to make carbon sequestration a forest management goal. The incentive scheme for afforestation and reforestation undertaken in an Annex-I country will probably associate payments with the actual carbon budget (since only the national budget is relevant) while for credits earned by CDM projects two methods have finally been accepted: the t-CERs and the l-CERs. The main difference between the two crediting procedures is the lifetime of the credit, 5 years (renewable) in the case of the t-CER and up to 30 years with the l-CERs. Therefore, three different crediting mechanisms are possible. However, for simplicity, the model presented here uses annualized values and a single framework, ensuring that the investment incentives are not changed, that is, the model includes the constant annual income that would equalize the actual future stream of incomes generated for each value over the entire reforestation cycle. As a result the model presented is general enough to be applied to any of the three crediting mechanisms and focuses on the additional income generated by reforestation with one or another species if carbon sequestration and/or biodiversity-scenic values are internalized.

It is usually accepted that biodiversity increases when degraded and agricultural lands are converted into forests (IPCC, 2000). However, this is only true of indigenous forests and not when the 'reforestation' is plantations of rapidly growing alien species like eucalyptus. It is also not true where existing land uses have high biodiversity values (IPCC, 2000). Matthews *et al.* (2002) have quantified bird biodiversity associated with reforestations in the United States and have found further evidence of the potential negative impacts of reforestation. As indicated in Jacquemont and Caparrós (2002), the 'afforestation and reforestation' alternative may conflict with the goal of the Convention on Biodiversity, since incentives to increase carbon sequestration may be negative for biodiversity under some conditions.

Van Kooten (2000) proposed an optimal control model to evaluate carbon sequestration via single species 'afforestation and reforestation', without taking into account biodiversity or scenic values. This model was extended in Caparrós and Jacquemont (2003) to include two species and biodiversity values. Nevertheless, since this paper focused on the legal and economic implications of the Protocol the model was not completely solved (only first-order conditions were used) and not applied. In Caparrós *et al.* (2005) the model is discussed in depth from a theoretical point of view, and applications are currently being made in Spain and California. In what remains of this chapter we summarize the main theoretical findings of the model in Caparrós *et al.* (2005) and present some preliminary results of a contingent valuations study done to identify the willingness to pay (WTP) to favor a reforestation with oak trees and to avoid a reforestation with eucalyptus.

Moons *et al.* (2004) also deal, using a GIS-based model, with the establishment of new forests for carbon sequestration purposes, including recreation and other values in the analysis. Their model is solved numerically and highlights the empirical importance of taking into account recreational values. Thus, we will analyze not only impacts on biodiversity values but also potential impact on scenic values, since they are relatively similar from a modeling point of view (Caparrós *et al.*, 2003).

Following Caparrós *et al.* (2005) we assume that the agent can choose between two types of forest, and that type 1 has greater biodiversity-scenic values while type 2 has greater carbon sequestration potential. A typical example of this situation is when reforestation with a natural indigenous species alternative (forest type 1) is compared with fast-growing alien species (forest type 2). In Spanish dehesa or Californian oak woodland ranches, we could see this model as comparing a reforestation program with oak trees (type 1) and with eucalyptus (type 2), a fast-growing alien species used in the past in Spain as well as in California.

Define: L= total land available; $f_0(t)$ = pasture land at time t; $f_i(t)$ = land of forest type i (i=1,2). To simplify, we can eliminate $f_0(t)$ from the model by setting $f_0(t) = L - f_1(t) - f_2(t)$ and leave $f_1(t)$ and $f_2(t)$ as state variables. Obviously, f_i cannot have negative values. Nevertheless, for simplicity, Caparrós *et al.* (2005) analyze the problem without explicitly incorporating this restriction and check afterwards the results for non-negativity. Define further: r = discount rate, $u_i(t)$ = total area reforested at time t of forest type i (i=1,2) (control variables), and $K_i(u_i)$ = reforestation cost for forest type i (i=1,2), a function of the amount of land reforested in a given year. The control variable $u_i(t)$ refers only to the amount of new land devoted to forest (or deforested) and not to the reforestation or natural regeneration needed to maintain the current forest surface. Assume $K'_i(u_i) > 0$ and $K''_i(u_i) > 0$ (e.g. as specialized labor becomes scarce, salaries increase). Finally, define $F_i(f_i)$ (i=0,1,2) as space-related functions showing the annual net capital income values for pasture land (i=0) or forest land of type i (i=1,2) and assume $F_i'(f_i) > 0$ and $F_i''(f_i) < 0$. These functions are supposed to have three terms: $F_i(f_i) = W_i(f_i) + C_i(f_i) + B_i(f_i)$, where $W_i(f_i) > 0$, $C_i(f_i) > 0$, and $B_i(f_i) > 0$ represent annual net capital income associated with commercial uses (timber, cork, fire-wood, livestock breeding, etc.), carbon sequestration, and biodiversity-scenic values, respectively. Note that forest-related data are sometimes strongly time-related but, for modeling reasons, it is interesting to annualize them, ensuring that investment incentives are not changed (Van Kooten, 2000). In the case of the Spanish *dehesa* this may be important, although in the case of Californian oak woodlands most of the data are already annualized.

The objective function is

$$\text{Max}V = \int_0^{\infty} \Pi(t) e^{-rt} dt$$

$$\Pi(t) = F_1(f_{1t}) - K_1(u_{1t}) + F_2(f_{2t}) - K_2(f_{2t}) + F_0(L - f_{1t} - f_{2t})$$

s.t.

$$\dot{f}_1 = u_1$$

$$\dot{f}_2 = u_2$$

And initial conditions: $f_1(0) = f_1^0; \quad f_2^0 = f_2^0$

Π is a concave function, since it is the sum of concave functions and convex functions (with a negative sign). In addition, the equations of motion for the state variables are linear in the control variables. Thus, the Mangasarian sufficient conditions will hold. Using the current-value Hamiltonian and the Pontryagin maximum principle, the following first-order conditions can be obtained for the steady state (Caparrós *et al.*, 2005):

$$\frac{F_1(f_1^*)}{r} - K_1'(0) = \frac{F_0(L - f_1^* - f_2^*)}{r} \tag{1}$$

$$\frac{F_2(f_1^*)}{r} - K_2'(0) = \frac{F_0(L - f_1^* - f_2^*)}{r} \tag{2}$$

Taking (1) and (2) together, and writing them out:

$$\frac{W_1(x)+C_1(x)+B_1(x)}{r} - K_1'(0) = \frac{W_2(x)+C_2(x)+B_2(x)}{r} - K_1'(0) \tag{3}$$
$$= \frac{F_0(L-f_1^*-f_2^*)}{r}$$

The interpretation of equation (3) follows conventional lines. In the steady-state equilibrium the stream of net revenues associated with the reforestation of one additional hectare of forest type 1 has to be equal to the revenues associated to one additional hectare reforested with forest type 2 and to the revenues associated to the use of that hectare as pasture.

Caparrós *et al.* (2005) show, after setting $F_0' = \alpha \geq 0, \forall x$ (i.e., the marginal value of pasture land is constant), that in the long-term equilibrium the amount of forest type i is: $f_1^*(rK_i(0)+\alpha) > 0$. They also show that this equilibrium is a saddle point and that (i) if the initial amount of forest type i is lower than the optimal amount f_i^* the optimal approach is to reforest forest type i (a positive u_i), and that (ii) if the initial amount of forest type i is higher than f_i^* the optimal approach is to reduce the amount of forest type i (a negative u_i). The optimal approach never implies reforesting first and deforesting afterwards, so that the annualization of the revenues as described above does not change investment incentives.

So far we have discussed the system focusing on the overall valuation function (F). Now we will discuss the impact of different values for conventional commercial uses (timber, cork, firewood), carbon sequestration (a value that might become a market value in the future), and biodiversity values. To make things interesting, Caparrós *et al.* (2005) assume $B_1' > C_2'$ and $C_1' < C_2'\ \forall x$ (i.e. species 1 has higher marginal values for biodiversity and species 2 has higher marginal values for carbon sequestration). Recalling the additive form of the valuation function assumed, we can compare the optimal amount of space devoted to each species in the equilibrium considering different values. We will call $\left(f_i^*\right)_X$ the amount of species i in equilibrium considering only the values indicated in the sub-index of the bracket (where X can be any combination of the three values defined above: W, C, and B). In an arbitrary situation where commercial values are supposed to be equal for species 1 and 2, and carbon values are higher for species 2, biodiversity values for species 2, eucalyptus, are supposed to be negative (Fig. 5). This is a reasonable assumption, as will be shown below.

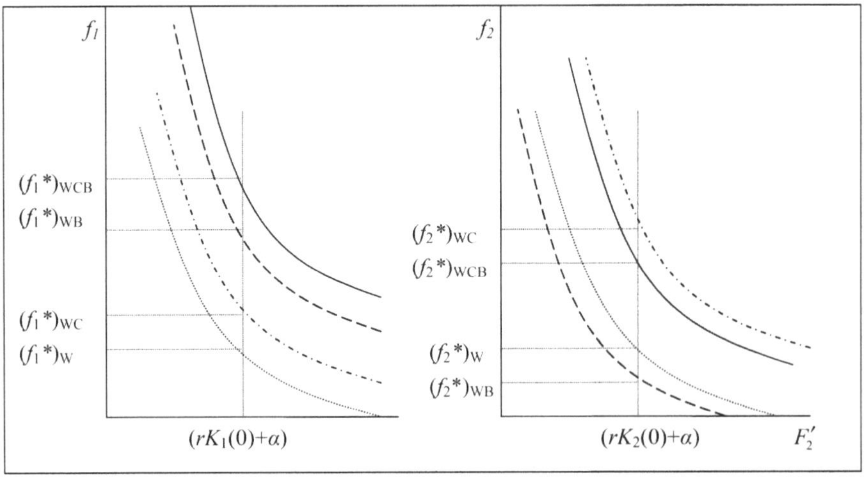

Figure 5. Equilibrium reforestation amounts for species 1 and 2. The functions shown are the marginal value functions for timber (*dotted line*), timber and carbon sequestration (*dashed-dotted line*), timber and biodiversity-scenic (*dashed line*) and timber, carbon, and biodiversity-scenic (*solid line*) (Caparrós *et al.* (2005).

For example, we might have a situation where future market forces (timber plus carbon) favor species 2, $(f_1^*)_{WC} < (f_2^*)_{WC}$, while present market forces equalize the amounts of both species, $(f_1^*)_W = (f_2^*)_W$, and social benefits (timber, carbon sequestration, and biodiversity-scenic values) would favor species 1: $(f_1^*)_{WCB} > (f_2^*)_{WCB}$. If only timber and biodiversity-scenic values are taken into account (probably the social values currently considered), the relative amount of species 1 in equilibrium should even be bigger. In addition, these values (especially scenic values) are local by their nature while carbon sequestration benefits are global. Thus, implementing an incentive for carbon sequestration might, in this particular case, be counter to local benefits.

The discussion so far does not allow us to say if this situation is relevant to the real world. Our current research is focused on applications to multiple-use forests in Spain and California. Data for a complete calibration of this model are not available yet; however, preliminary data for scenic values in Spain suggest that cork oak reforestations are seen as highly positive by visitors while reforestations with eucalyptus are seen as negative (Table 2 and 3). In a contingent valuation study with 900 interviews undertaken in the *Alcornocales* Natural Park (southwest Spain), half of the interviewees were asked about a reforestation with cork oak trees (showing them the evolution of this kind of reforestation in a booklet) and half were asked about reforestation with eucalyptus (giving them a similar booklet describing a reforestation with eucalyptus) (Table 2). The interviewees were then asked about their "willingness to pay" (WTP) to ensure a reforestation with cork oaks and about their WTP to avoid a reforestation with eucalyptus (Table 3).

Table 2. Subjective valuation of a reforestation with different species in the Alcornocales Natural Park (ANP) $n = 900$ (Caparrós *et al.*, 2005).

	What is your opinion about a reforestation in the Natural Park with ....?	
	cork oaks (%)	**Eucalyptus (%)**
Very negative	0.7	58.9
Negative	2.5	31.0
Indifferent	2.2	2.5
Positive	42.9	6.1
Very positive	51.8	1.6

Table 3. Willingness to pay to ensure reforestation with cork oaks and to avoid reforestation with eucalyptus ($n = 900$) (Caparrós *et al.*, 2005).

	Reforestation to maintain current forest surface (compensate deforestation)		**Reforestation to increase 20% of the current forest surface**	
	WTP to *ensure* this reforestation with *cork oaks (euros)*	**WTP to *avoid* this reforestation with *eucalyptus***	**WTP to *ensure* this reforestation with *cork oaks***	**WTP to *avoid* this reforestation with *eucalyptus***
Total answers	450	450	450	450
Valid answers	425	408	425	408
Mean (€)	26.96	24.21	30.49	29.68
Median (€)	12.00	10.00	12.00	10.00
Std. deviation	58.43	61.73	60.60	88.65

7 CONCLUSION

Mediterranean forests in Spain and California have in common climate, Spanish historical influence, ownership structure, and management. Thus, Spanish dehesa and California ranches are similar systems and present similar modeling challenges. Two optimal control models designed to incurporate environmental and social values into analysis of management options for Mediterranean forests were presented and discussed. The first model reveals that including the environmental goods valued as amenities by the landowner can better explain the fact that California landowners keep their oaks even if a simple financial model would suggest that the optimum action is to cut them down to maximize grazing resources. The second model includes carbon sequestration and biodiversity values in the analysis of reforestation alternatives for oak woodlands. The simple model suggests that fast-growing alien species are best for carbon sequestration. However, although data currently available are not enough for a full calibration of the model, the high biodiversity values of cork oak woodlands, and public preference for cork oaks compared to species such as eucalyptus, increase the benefit of cork oak reforestation. Care has to be taken not to promote aggressive incentives for carbon sequestration favoring alien species at the expense of oak woodlands. We find that at both the landowner and landscape scales the values of landowners and the public render models that to do not incorporate what have been shown to be high amenity and other social and environmental values to be potentially misleading for policy development, and of limited explanatory value. However, further research in this area is needed.

8 REFERENCES

Campos, P., 1984. *Economía y energía en la dehesa extremeña*. Instituto de Estudios Agrarios, Pesqueros y Alimentarios/MAPA, Madrid.

Campos, P., 1997. Análisis de la rentabilidad económica de la dehesa. *Situación. Serie estudios regionales: Extremadura*: 111–140.

Campos, P., 2004. Towards a sustainable global economy for Mediterranean agro forestry systems. In: S. Schnabel and A. Gonçalves (Editors), *Advances in GeoEcology* 37: *Sustainability of Agro-silvo-pastoral Systems* –Dehesas & Montados–. Catena Verlag, Reiskirchen, Germany, 13–28.

Campos, P. and Mariscal, P., 2003. Preferencias de los propietarios e intervención pública: el caso de las dehesas de la comarca de Monfragüe. *Investigación Agraria: Sistemas y Recursos Forestales* 12(3): 87–102.

Caparrós, A. and Jacquemont, F., 2003. Conflicts between biodiversity and carbon offset programs: economic and legal implications. *Ecol. Econ.* 46: 143–157.

Caparrós, A., Campos, A., and Martín, D., 2003. Influence of carbon dioxide abatement and recreational services on optimal forest rotation. *Int. J. Sustainable Dev.* 6(3): 1–14.

Caparrós, A., Cerdá, E., and Campos, P., 2007. Carbon sequestration with reforestations and and biodiversity-scenic values. FEEM Working Paper No. 28, 2007 Available at SSRN: http://ssrn.com/abstract=976402

Cerrillo, E., 1984. *La vida rural romana en Extremadura*. Universidad de Extremadura, Cáceres.

CDF-FRRAP, 2003. California Department of Forestry and Fire Protection, Fire and Resource Assessment Program, (2003). *The Changing California: Forest and Range 2003 Assessment. Sacramento*, CA. http://www.frap.cdf.ca.gov/assessment2003/index.html.

Díaz, M., Campos, P., and Pulido, F.J., 1997. The Spanish dehesas: a diversity in land-use and wildlife. In: D.J. Pain and W. Pienkowski (Editors), *Farming and Birds in Europe*. Academic Press, Londres, 178–209.

Ewing, R., Tosta, N., Tuazon, R., Huntsinger, L., Marose, R., Nielsen, K., and Motroni, R., 1988. *California's Forest and Range Resources: Growing Conflicts Over Changing Uses*. The 1988 Forest and Rangeland Resources Assessment. Anchor Press, California Department of Forestry and Fire Protection, Sacramento, California, 278 pp.

Hernández, J.A. and Pulido, F., 2004. Aproximación a la historia de la agricultura en Extremadura (II). De la Reconquista a los Austrias. In: *La agricultura y la ganadería extremeñas 2003*. Caja de Badajoz, Badajoz: 197–215.

Howitt, R., 1995. Positive mathematical programming. *Am. J. Agric. Econ.* 77: 329–342.

Huntsinger, L., Buttoloph, L., and Hopkinson, P., 1997. Ownership and management changes on California hardwood rangelands: 1985 to 1992.? *J. Range Manage*. 50(4): 423–430.

Jacquemont, F. and Caparrós, A., 2002. The Convention on Biological Diversity and the Climate Change Convention ten years after Rio: towards a synergy of the two regimes? *Rev. Eur. Community Int. Environ. Law* 11(2): 169–180.

Kan-Rice, P. and Sokolow, A., 2003. *A National View of Agricultural Easement Programs: Profiles and Maps – Report 1*. American Farmland Trust, Center for Agriculture in the Environment. http://www.aftresearch.org/PDRdatabase/NAPidx.htm.

Keeley, J.E., 2002. Native American impacts on fire regimes of the California coastal ranges. *J. Biogeogr*. 29: 303–320.

Klein, J., 1920. *The Mesta: a Study in Spanish Economic History, 1273–1836*. Cambridge, Harvard University Press, 444p.

Liffmann, R.H. Huntsinger, L., and Forero, L.C., 2000. To ranch or not to ranch: home on the urban range? *J. Range Manage*. 53: 362–370.

Linares, A.M. and Zapata, S., 2003. Una panorámica de ocho siglos. In: F. Pulido, P. Campos, and G. Montero (Editors), *La gestión forestal de la dehesa: historia, ecología, selvicultura y economía*. Iprocor, Mérida: 13–25.

Maqueda, A., Jiménez, J.L., and Mordillo, A., 2004. Las vías pecuarias de Extremadura. In: *La agricultura y la ganadería extremeñas 2003*. Caja de Badajoz, Badajoz: 165–179.

Matthews, S., O'Connor, R., and Plantinga, A.J., 2002. Quantifying the impacts on biodiversity of policies for carbon sequestration in forests. *Eco. Econ.* 40(1): 71–87.

Moons, E., Proost, S., Saveyn B., and Hermy, M., 2004. Optimal location of new forests in a suburban area. KU Leuven Working Paper.

Perez, C., 1982. *Grants of land in California made by Spanish or Mexican authorities*. Boundary Determination Office, State Lands Commission, Boundary Investigation Unit, August 23. Extracts from: http://www.lib.berkeley.edu/EART/rancho.html.

Standiford, R.B. and Howitt, R.E., 1992. Solving empirical bioeconomic models: a rangeland management application. *Am. J. Agric. Econ.* 74: 421–433.

Standiford, R.B. and Howitt, R.E., 1993. Multiple use management of California's hardwood rangelands. *J. Range Manage*. 46: 176–181.

Standiford, R.B., McCreary, D., Gaertner, S., and Forero, L., 1996. Impact of firewood harvesting on hardwood rangelands varies with region. *Calif. Agric.* 50(2): 7–12.

Standiford, R.B. and Scott, T.A., 2001. Value of oak woodlands and open space on private property values in Southern California. Special issue – *Investigación Agraria: Sistemas y Recursos Forestales* – Towards the New Forestlands Commercial and Environmental Benefits Accounting. Theories and Applications. 1: 137–152.

Standiford, R.B. and Tinnin, P., 1996. *Guidelines for managing California's hardwood rangelands*. University of California Division of Agriculture and Natural Resources Leaflet no. 3368, 180 pp.

Sulak, A. and Huntsinger, L., 2002. *Sierra Nevada grazing in transition: The role of Forest Service grazing in the foothill ranches of California*. Sierra Nevada Alliance, (http://www.sierranevadaalliance.org/publications/)
Van Kooten, G.C., 2000. Economic Dynamics of Tree Planting for Carbon Uptake on Marginal Agricultural Lands. *Can. J. Agric. Econ.* 48: 51–65.

Chapter 4

ENVIRONMENTAL CRITERIA IN PIG DIET FORMULATION WITH MULTI-OBJECTIVE FRACTIONAL PROGRAMMING

Teresa Peña[1], Carmen Castrodeza[1], and Pablo Lara[2]
[1]*Department of Applied Economics (Mathematics), Facutly of Economics, University of Valladolid, Avda. de Valle Esgueva, 6, 47011 Valladolid, Spain* [2]*Department of Animal Production, ETSIAM, University of Córdoba, Apdo. 3048, 14080 Córdoba, Spain*

Abstract The objective of this chapter is to show how the techniques of Multi-objective Fractional Programming (MFP) can enhance the process of animal diet formulation. We propose a model with three objectives and explain how animal diet formulation can be tackled using the method of Interactive Multiple Goal Programming (IMGP). This method allows us to treat the presence of fractional objectives, simplifying the problem from a computational point of view.

Keywords: Fractional multi-objective programming, Diets, Interactive methods

1 INTRODUCTION

Diet and/or ration formulation for livestock is the problem of finding a mix of feeds satisfying some nutritional considerations. The application of Operations Research techniques to this problem has rendered fruitful results. A Linear Programming (LP) model minimising the cost of the mix is a tool in the daily routine of compound feed manufacturers throughout the world farmers use management software extensively including facilities for optimising on-farm diets, and researchers have tackled with success the task of extending the model. One of the milestones in this extension was the recognition of Multiple Criteria Decision Making (MCDM) techniques as a more convenient paradigm than LP (Romero and Rehman, 1984).

Today, nutritional and environmental concerns have made the relation between the formulation problem and MCDM even fitter. First, in the least cost LP model the restrictions of the problem were the requirements of the animals in terms of energy, protein, amino acids, vitamins and minerals. No relations between these items are included in the model. However, nutritional advanced concepts aiming at an adequate understanding of the metabolisation of the diet and relating several of these items have been developed and they could be incorporated in the formulation model. For example, the most important amino acid for pig growth is lysine, and not only a precise requirement of it is needed, but the lysine requirement is better stated in relation to energy content of the mix. The lysine/energy ratio is used as a better indicator of the quality of a compound feed. Another important concept is the ideal protein balance, which relates the proportions to be maintained between the different amino acids in the diet.

Second, as the requirements in the LP model are set mainly as minimum requirements, the solution adjusts the most limiting and perhaps other nutrients and provides another nutrients in excess of the requirements. These excesses are eliminated by the animals and this waste contributes to environmental pollution. Particularly nitrogen and phosphorus excretions are of particular concern throughout the world. The previous nutritional concepts are also of interest here for a precise formulation because nitrogen is the main component of amino acids and proteins. Therefore, amino acids in excess of the ideal balance in the protein are a source of environmental pollution. The waste problem is aggravated by the use of "safety margins" in LP models. Safety margins increase the nutritional requirements with the aim of assuring acceptable levels of probability that a diet will be adequate.

The objective of this chapter is to show how a Multiobjective Fractional Programming (MFP) model can deal with the new demands of the formulation problem. An earlier work applying MFP to livestock ration formulation was made by Lara (1993). The present model has three criteria, the cost, the value of the ratio lysine/energy and the maximum deviation with regard to the ideal values of the percentage content of amino acids in the protein. In Section 2 we state the model and show how it can be solved using the Interactive Multiple Goal Programming (IMGP) method developed by Spronk (1981). Section 3 is an application of the model to a real formulation of a feed mix for growing pigs. A brief discussion closes the chapter.

2 THE MODEL

Our aim is to find a set of feed ingredients allowing high production results (weight gain, lean tissue deposit..) at a low cost and with the minimum nitrogen discharge to the environment.

Let $x = (x_1, x_2, .., x_n)$ where x_j, $j = 1, ..., n$ denotes the proportion of ingredient j in the diet and n is the number of available ingredients.

The possible combinations of ingredients are bounded by two types of constraints:

1. Nutritional requirements constraints, assuring that the levels of nutrients (protein, minerals, etc.) are kept between recommended figures. Usually, only a minimum required level is set in the LP framework, but it is also a good practice to include a *maximum:*

$$\underline{b_i} \le \sum_{j=1}^{n} a_{ij} x_j \le \overline{b_i}, \qquad i = 1, .., k,$$

where k is the number of nutrients considered, a_{ij} is the amount of nutrient i in ingredient j and $\underline{b_i}$ and $\overline{b_i}$ is the lower and upper bounds of nutrient i in the diet.

2. Constraints on the possible proportion of certain ingredients in the diet:

$$x_j \le s_j \quad j = 1, ..., n,$$

where s_j is the maximum proportion of ingredient j in the diet.

An additional constraint is usually introduced, requiring the sum of all the ingredients to be equal to the unit, so the solution appears in percentage terms.

For evaluating a diet, we consider three criteria:

1. The cost of the feed:

$$f_1 = \sum_{j=1}^{n} c_j x_j,$$

where c_j is the unit price of ingredient j.

2. As an indicator of the quality of the mix of ingredients, the lysine/energy ratio is employed as the second criterion:

$$f_2 = \frac{\sum_{j=1}^{n} l_j x_j}{\sum_{j=1}^{n} e_j x_j},$$

where l_j is the amount of lysine in ingredient j and e_j is the amount of energy in ingredient j.

3. Finally, as the amino acid composition of the diet must be as near as possible to the balance required by the animals, the third criterion will be the maximum deviation, with regard to the balanced values, of the percentage content of the main amino acids in pig feeding, methionine + cystine, threonine and tryptophan in comparison with lysine content,

$$f_3 = \max\left(\left|\frac{\sum_{j=1}^{n} m_j x_j}{\sum_{j=1}^{n} l_j x_j} - 0.5\right|, \left|\frac{\sum_{j=1}^{n} t_j x_j}{\sum_{j=1}^{n} l_j x_j} - 0.66\right|, \left|\frac{\sum_{j=1}^{n} tp_j x_j}{\sum_{j=1}^{n} l_j x_j} - 0.18\right|\right),$$

where m_j is the amount of methionine + cystine in ingredient j, t_j is the amount of threonine in ingredient j, tp_j is the amount of tryptophan in ingredient j, and 50, 66 and 18 are the ideal percentages of methionine + cystine, threonine and tryptophan, respectively with regard to lysine (Cole and Van Lunen, 1994).

The formulation of the problem is as follows:

$$\begin{aligned}
&(\min\ f_1,\ \max\ f_2,\ \min\ f_3)\\
&\text{s.t.}\ \ \underline{b_i} \le \sum_{j=1}^{n} a_{ij} x_j \le \overline{b_i}, \qquad i = 1,..,k\\
&\qquad 0 \le x_j \le s_j, \qquad\qquad\quad j = 1,...,n\\
&\qquad \sum_{j=1}^{n} x_j = 1,
\end{aligned}$$

where f_1 is a linear function, but f_2 and f_3 are non-linear, fractional and minmax fractional, respectively. This characteristic adds a complication to the operative treatment of the model. Methods aiming at finding the set of efficient solutions (see, for example, Romero and Rehman, 2003) are not effective here. However, the problem can be tackled using the IMGP method developed by Spronk (1981). The essential condition of IMGP is that in each iteration the objectives are optimised individually acting on the other ones as constraints. Therefore, the computational treatment of non-linear objectives is simplified. The objective of IMGP is to elicit the aspiration levels or targets of the decision maker interactively. The algorithm proceeds as follows:

First, a potency matrix is calculated. The dimension of this matrix is 2 × 3. The first row contains the best or ideal values computed when each criterion is optimised individually. The second row contains the worst or anti-ideal values obtained for each criterion in these three problems:

$$P_1 = \begin{pmatrix} f_1^* & f_2^* & f_3^* \\ \hat{f}_1 & \hat{f}_2 & \hat{f}_3 \end{pmatrix}.$$

Each column of the potency matrix represents the range over which it is possible to define the goals of the different criteria. Following this the pessimistic solution $S_1 = \begin{pmatrix} \hat{f}_1 & \hat{f}_2 & \hat{f}_3 \end{pmatrix}$[1] is established as the initial solution of the model. This solution, with the potency matrix P_1 is presented to the decision maker. If he/she considers that the solution is satisfactory, the process ends. If not, he/she must indicate which of the values $\hat{f}_i$, $i = 1,2,3$ he/she prefers to be improved first. A new solution S_2 is generated introducing in S_1 the new desired value for that objective.

The next step is to re-compute a new potency matrix, P_2. This matrix is calculated again optimising each of the objectives individually with added restrictions to assure that none of the objectives reach a worse value than in S_2. Once P_2 has been computed, the decision maker has to evaluate if the improvement of the $\hat{f}_i$ worth the changes in the values f_i^* of the remaining objectives. If this is not the case, the decision maker can adjust his/her desire over the value of the function $\hat{f}_i$. If the new value of $\hat{f}_i$ is accepted, the decision maker can go on with the process of improving another objective or even try if a better improvement is possible for the same objective. A constraint is updated in each iteration. Thus, the set of feasible solutions is reduced in each iteration. The iterative process continues until the decision maker finds a satisfactory solution or until the two rows of the potency matrix are the same.

[1] Or another solution worse than $\begin{pmatrix} \hat{f}_1 & \hat{f}_2 & \hat{f}_3 \end{pmatrix}$.

Now, we show the three problems that must be solved in each iteration.

The first problem is

$$
\begin{aligned}
&\text{Min } f_1 = \sum_{j=1}^{n} c_j x_j \\
&\text{s.t. } \left(\sum_{j=1}^{n} l_j x_j \Big/ \sum_{j=1}^{n} e_j x_j \right) \geq \tilde{f}_2 \\
&\quad \max\left(\left| \frac{\sum_{j=1}^{n} m_j x_j}{\sum_{j=1}^{n} l_j x_j} - 0.5 \right|, \left| \frac{\sum_{j=1}^{n} t_j x_j}{\sum_{j=1}^{n} l_j x_j} - 0.66 \right|, \left| \frac{\sum_{j=1}^{n} tp_j x_j}{\sum_{j=1}^{n} l_j x_j} - 0.18 \right| \right) \leq \tilde{f}_3, \\
&\quad \underline{b}_i \leq \sum_{j=1}^{n} a_{ij} x_j \leq \overline{b}_i, \qquad i = 1,..,k \\
&\quad 0 \leq x_j \leq s_j, \qquad j = 1,...,n \\
&\quad \sum_{j=1}^{n} x_j = 1,
\end{aligned}
$$

where $\tilde{f}_2$ denotes the minimally required value for the criteria, f_2 and $\tilde{f}_3$ indicate the maximum allowed values for f_3 in the current iteration.

An equivalent formulation is

$$
\begin{aligned}
&\text{Min } f_1 = \sum_{j=1}^{n} c_j x_j \\
&\text{s.t. } \left(\sum_{j=1}^{n} l_j x_j \Big/ \sum_{j=1}^{n} e_j x_j \right) \geq \tilde{f}_2 \\
&\quad -\tilde{f}_3 \leq \left(\sum_{j=1}^{n} m_j x_j \Big/ \sum_{j=1}^{n} l_j x_j \right) - 0.5 \leq \tilde{f}_3 \\
&\quad -\tilde{f}_3 \leq \left(\sum_{j=1}^{n} t_j x_j \Big/ \sum_{j=1}^{n} l_j x_j \right) - 0.66 \leq \tilde{f}_3 \\
&\quad -\tilde{f}_3 \leq \left(\sum_{j=1}^{n} tp_j x_j \Big/ \sum_{j=1}^{n} l_j x_j \right) - 0.18 \leq \tilde{f}_3 \\
&\quad \underline{b}_i \leq \sum_{j=1}^{n} a_{ij} x_j \leq \overline{b}_i, \qquad i = 1,..,k \\
&\quad 0 \leq x_j \leq s_j, \qquad j = 1,...,n \\
&\quad \sum_{j=1}^{n} x_j = 1.
\end{aligned}
$$

A linear form is obtained by cross multiplication of ratio inequalities.

The second problem is

$$
\begin{aligned}
&\text{Max } f_2 = \left(\sum_{j=1}^{n} l_j x_j \Big/ \sum_{j=1}^{n} e_j x_j \right) \\
&\text{s.t. } \sum_{j=1}^{n} c_j x_j \le \tilde{f}_1 \\
&\quad -\tilde{f}_3 \le \left(\sum_{j=1}^{n} m_j x_j \Big/ \sum_{j=1}^{n} l_j x_j \right) - 0.5 \le \tilde{f}_3 \\
&\quad -\tilde{f}_3 \le \left(\sum_{j=1}^{n} t_j x_j \Big/ \sum_{j=1}^{n} l_j x_j \right) - 0.66 \le \tilde{f}_3 \\
&\quad -\tilde{f}_3 \le \left(\sum_{j=1}^{n} tp_j x_j \Big/ \sum_{j=1}^{n} l_j x_j \right) - 0.18 \le \tilde{f}_3 \\
&\quad \underline{b}_i \le \sum_{j=1}^{n} a_{ij} x_j \le \overline{b}_i, \qquad i = 1,..,k \\
&\quad 0 \le x_j \le s_j, \qquad j = 1,...,n \\
&\quad \sum_{j=1}^{n} x_j = 1,
\end{aligned}
$$

where $\tilde{f}_1$ and $\tilde{f}_3$ are the maximum allowable values for f_1 y f_3 respectively in the current iteration.

Using the variable change of Charnes and Cooper (1962):

$$
\mathrm{y} = t\mathrm{x}, \; t = \left(1 \Big/ \sum_{j=1}^{n} e_j x_j \right),
$$

this single fractional problem can be transformed in the following linear problem:

$$
\begin{aligned}
&\text{Max } \sum_{j=1}^{n} l_j y_j \\
&\text{s.t. } \sum_{j=1}^{n} c_j y_j - \tilde{f}_1 t \le 0 \\
&-\tilde{f}_3 \sum_{j=1}^{n} l_j y_j \le \sum_{j=1}^{n} m_j y_j - 0.5 \sum_{j=1}^{n} l_j y_j \le \tilde{f}_3 \sum_{j=1}^{n} l_j y_j \\
&-\tilde{f}_3 \sum_{j=1}^{n} l_j y_j \le \sum_{j=1}^{n} t_j y_j - 0.66 \sum_{j=1}^{n} l_j y_j \le \tilde{f}_3 \sum_{j=1}^{n} l_j y_j \\
&-\tilde{f}_3 \sum_{j=1}^{n} l_j y_j \le \sum_{j=1}^{n} tp_j y_j - 0.18 \sum_{j=1}^{n} l_j y_j \le \tilde{f}_3 \sum_{j=1}^{n} l_j y_j \\
&\quad \underline{b}_i t \le \sum_{j=1}^{n} a_{ij} y_j \le \overline{b}_i t, \quad i = 1,..,k \\
&\quad 0 \le y_j \le s_j t, \quad j = 1,..,n \\
&\quad \sum_{j=1}^{n} y_j - t = 0, \\
&\quad \sum_{j=1}^{n} e_j y_j = 1, \\
&\quad t \ge 0.
\end{aligned}
$$

If (y^*, t^*) is an optimum solution of the linear transformed problem, $x^* = y^* / t^*$ is an optimum solution of the fractional problem (Charnes and Cooper, 1962; Schaible, 1976; Stancu-Minasian, 1997).

The third problem is

$$
\begin{aligned}
&\text{Min max} \left(\left| \frac{\sum_{j=1}^{n} mc_j x_j}{\sum_{j=1}^{n} l_j x_j} - 0.5 \right| , \left| \frac{\sum_{j=1}^{n} t_j x_j}{\sum_{j=1}^{n} l_j x_j} - 0.66 \right| , \left| \frac{\sum_{j=1}^{n} tp_j x_j}{\sum_{j=1}^{n} l_j x_j} - 0.18 \right| \right) \\
&\text{s.t.} \qquad \left(\sum_{j=1}^{n} l_j x_j \Big/ \sum_{j=1}^{n} e_j x_j \right) \ge \tilde{f}_2 \\
&\qquad\quad \sum_{j=1}^{n} c_j x_j \le \tilde{f}_1, \\
&\qquad\quad \underline{b}_i \le \sum_{j=1}^{n} a_{ij} x_j \le \overline{b}_i, \quad i = 1,..,k \\
&\qquad\quad 0 \le x_j \le s_j, \quad j = 1,..,n \\
&\qquad\quad \sum_{j=1}^{n} x_j = 1,
\end{aligned}
\tag{1}
$$

where $\tilde{f}_2$ is the minimally required value for the criterion f_2 and $\tilde{f}_1$ is the maximum allowed value for f_1 in the current iteration.

Formulation (1) is a non-smooth optimisation problem but can be replaced by the following equivalent smooth non-linear problem:

$$
\begin{aligned}
\text{Min} \quad & d \\
\text{s.t.} \quad & -d\sum_{j=1}^{n} l_j x_j \le \sum_{j=1}^{n} mc_j x_j - 0.5\sum_{j=1}^{n} l_j x_j \le d\sum_{j=1}^{n} l_j x_j \\
& -d\sum_{j=1}^{n} l_j x_j \le \sum_{j=1}^{n} t_j x_j - 0.66\sum_{j=1}^{n} l_j x_j \le d\sum_{j=1}^{n} l_j x_j \\
& -d\sum_{j=1}^{n} l_j x_j \le \sum_{j=1}^{n} tp_j x_j - 0.18\sum_{j=1}^{n} l_j x_j \le d\sum_{j=1}^{n} l_j x_j \\
& \left(\sum_{j=1}^{n} l_j x_j \Big/ \sum_{j=1}^{n} e_j x_j\right) \ge \tilde{f}_2 \\
& \sum_{j=1}^{n} c_j x_j \le \tilde{f}_1, \\
& \underline{b_i} \le \sum_{j=1}^{n} a_{ij} x_j \le \overline{b_i}, \quad i = 1,..,k \\
& 0 \le x_j \le s_j, \quad j = 1,..,n \\
& \sum_{j=1}^{n} x_j = 1,
\end{aligned}
\tag{2}
$$

where d is an auxiliary variable that represents the maximum deviation. Employing the change of variable:

$$u = d\sum_{j=1}^{n} l_j x_j ,$$

the previous problem becomes a linear fractional problem.

$$
\begin{aligned}
\text{Min} \quad & \frac{u}{\sum_{j=1}^{n} l_j x_j} \\
\text{s.t.} \quad & -u \le \sum_{j=1}^{n} mc_j x_j - 0.5\sum_{j=1}^{n} l_j x_j \le u \\
& -u \le \sum_{j=1}^{n} t_j x_j - 0.66\sum_{j=1}^{n} l_j x_j \le u \\
& -u \le \sum_{j=1}^{n} tp_j x_j - 0.18\sum_{j=1}^{n} l_j x_j \le u \\
& \left(\sum_{j=1}^{n} l_j x_j \Big/ \sum_{j=1}^{n} e_j x_j\right) \ge \tilde{f}_2 \\
& \sum_{j=1}^{n} c_j x_j \le \tilde{f}_1, \\
& \underline{b_i} \le \sum_{j=1}^{n} a_{ij} x_j \le \overline{b_i}, \quad i = 1,..,k \\
& 0 \le x_j \le s_j, \quad j = 1,..,n \\
& \sum_{j=1}^{n} x_j = 1.
\end{aligned}
\tag{3}
$$

And applying the transformation of variables proposed by Charnes and Cooper (1962) to this problem, we obtain the following linear programming problem:

$$\begin{aligned}
&\text{Min } d \\
&\text{s.t. } -d \le \sum_{j=1}^{n} mc_j y_j - 0.5\sum_{j=1}^{n} l_j y_j \le d \\
&\quad -d \le \sum_{j=1}^{n} t_j y_j - 0.66\sum_{j=1}^{n} l_j y_j \le d \\
&\quad -d \le \sum_{j=1}^{n} tp_j y_j - 0.18\sum_{j=1}^{n} l_j y_j \le d, \\
&\sum_{j=1}^{n} l_j y_j \ge \tilde{f}_2 \sum_{j=1}^{n} e_j y_j, \\
&\sum_{j=1}^{n} c_j y_j \le \tilde{f}_1 t, \\
&\underline{b}_i t \le \sum_{j=1}^{n} a_{ij} y_j \le \overline{b}_i t, \quad i = 1,..,k \\
&0 \le y_j \le t s_j, \quad j = 1,..,n \\
&\sum_{j=1}^{n} y_j - t = 0, \\
&\sum_{j=1}^{n} l_j y_j = 1, \\
&t \ge 0.
\end{aligned} \tag{4}$$

If *(y^*, t^*, d^*)* is an optimum solution of the problem (4), then *($x^*=y^*/t^*$, d^*)* will be an optimum solution of the problem (2).

Problem (4) is more tractable from a computational point of view than problem (2), since it can be solved easily by means of standard linear programming software.

3 APPLICATION AND RESULTS

To illustrate the use of the model we apply it in a real formulation context for growing pigs. Thirteen ingredients are considered: barley, wheat, corn, alfalfa, cassava meal, soybean meal, fish meal, gluten feed, mineral compound, pure lysine 78%, sunflower meal, animal fat and sugarbeet pulp. The nutrient contents of the ingredients were derived from the feed composition tables published by FEDNA (acronym in Spanish for "Spanish Foundation for the Development of Animal Nutrition"), and they are shown in Table 1. In this table the all the nutrient contents except digestible energy (DE) are expressed in percentage. DE is in megajoules/kg. Table 2 contains the limits for inclusion of feeds in the diet and the unit cost of each ingredient. Finally, Table 3 displays the nutritional requirements of the pigs according to the National Research Council (NRC).

Table 1. Nutrient content of feeds.

Feeds	CF	MC	Tp	T	Ca	P	DM	CP	L	DE
Barley	4.5	0.43	0.13	0.37	0.06	0.36	90.2	11.3	0.4	13.25
Wheat	2.8	0.46	0.13	0.34	0.04	0.35	89.4	11.6	0.33	14.33
Corn	2.5	0.33	0.06	0.27	0.02	0.27	86.3	7.7	0.22	14.42
Alfalfa	24.7	0.45	0.31	0.7	1.75	0.3	91.2	16.7	0.73	7.81
Cassava meal	6.1	0.06	0.02	0.07	0.24	0.1	88.8	2.5	0.09	13.29
Soybean meal	5.6	1.28	0.59	1.75	0.29	0.61	88	44	2.88	13.79
Fish meal	1.0	2.36	0.65	2.65	4.5	2.77	92	62.4	4.75	15.25
Gluten feed	8.0	4.36	0.68	3.89	0.16	0.8	88.6	19	3.26	11.36
Mineral com.					28.0	21				
Lysine 78%							98.5	95	78	20.56
Sunflower meal	22.5	1.25	0.43	1.06	0.35	1.00	89.3	30.5	1.06	9.19
Fat										34.06
Beet pulp	17.8	0.22	0.1	0.47	0.98	0.11	89.7	10.1	0.59	11.28

CF: crude fibre; MC: methionine + cystine; Tp: tryptophan; T: threonine; Ca: calcium; P: phosphorus; DM: dry matter; CP: crude protein; L: lysine; DE: digestible energy
Source: FEDNA (1999).

Table 2. Limits on ingredient contents and unitary costs.

Ingredients	Bounds (%)	Cost (€/Ton)
Barley		122.01
Wheat	40.00	126.21
Corn	40.00	134.63
Alfalfa	5.00	132.22
Cassava meal	22.00	151.45
Soybean meal 44		172.49
Fish meal	40.00	408.68
Gluten feed	80.00	122.61
Mineral compound		309.51
Lysine 78%	0.65	2434.04
Sunflower meal	6.00	118.40
Fat	4.00	384.69
Beet pulp	5.00	152.66

Source: FEDNA (1999) for bounds in ingredients. Actual prices recorded in several markets.

Table 3. Nutritional requirements.

Nutrients	Lower bounds (%)	Upper bounds (%)
Fibre		6.00
Calcium	1.07	1.47
Phosphorus	0.565	0.85
Dry matter	87.00	95.00
Crude protein	16.00	20.00

Source: NRC (1998).

The problem has been solved with LINDO and the spreadsheet Excel 7.0. A member of the Department of Animal Production at the University of Córdoba acted as the decision maker. The following is a sketch of the interactive process.

- *Iteration 1*
- First, each objective was optimised individually and the ideal and the anti-ideal values of the first potency matrix were obtained.
- $P_1 = \begin{pmatrix} 160.08 & 1.28 & 0.02357 \\ 202.12 & 0.67 & 0.2118 \end{pmatrix}$
- S_1 = (202.12 0.67 0.2118) was roposed as the first solution and it was presented with P_1 to the decision maker. Obviously, the solution was not a satisfactory one.
- *Iteration 2*
- The decision maker considered that the maximum deviation from the balanced composition of amino acids was too high and he advanced 0.06 as an upper desired bound. Thus, S_2 = (202.12 0.67 0.06) was proposed as the new solution and the new potency matrix P_2 was computed:
- $P_2 = \begin{pmatrix} 162.69 & 0.991 & 0.02357 \\ 190.58 & 0.736 & 0.06 \end{pmatrix}$.
- The decision maker considers that the reduction in the maximum deviation of the amino acids makes up for the potential worsening of the remaining objectives.
- *Iteration 3*
- However, the decision maker does not find S_2 as a satisfactory solution because the level of the lysine/energy ratio must be greater than 0.67 to optimise the growing potential of the pigs. In addition, the new solution S_3 = (202.12 0.95 0.06) is proposed. The corresponding potency matrix is P_3:
- $P_3 = \begin{pmatrix} 170.32 & 0.991 & 0.03310 \\ 190.58 & 0.95 & 0.06 \end{pmatrix}$.
- Again, the decision maker accepted the changes of the potency matrix.
- *Iteration 4*
- Now, S_3 is not an attractive solution from the economic perspective because the cost is high. The cost is modified to give the new solution S_4 = (174.29 0.95 0.06) with the associated potency matrix P_4:

- $P_4 = \begin{pmatrix} 170.32 & 0.9883 & 0.03313 \\ 174.29 & 0.95 & 0.06 \end{pmatrix}$

- *Iteration 5*
- Finally, the decision maker considered the value of the cost and the maximum deviation in the anterior solution to be acceptable and chose to raise the lysine/energy ratio to the maximum possible value. A new solution S_5 = (174.29 0.9883 0.06) is proposed. The corresponding potency matrix is P_5:

- $P_5 = \begin{pmatrix} 174.29 & 0.9883 & 0.06 \\ 174.29 & 0.9883 & 0.06 \end{pmatrix}$

- The iterative process ends because the two rows of the potency matrix are equal and S_5 = (174.29 0.9883 0.06) is the solution chosen.

The rows of the last potency matrix are the aspiration levels elicited interactively from the decision maker. The values of the feed ingredients in the solution are reported in the Table 4.

Table 4. Feeds with non-zero value in the preferred diet.

Feeds	Value (%)
Barley	19.47
Corn	17.90
Alfalfa	5.00
Cassava meal	22.00
Soybean meal	20.00
Fish meal	9.00
Mineral compound	1.51
Lysine 78%	0.13
Beet pulp	5.00

This is the most preferred composition of the diet from the point of view of the decision maker.

4 DISCUSSION

Our model adds complexity to previous approaches reported in the literature about the formulation problem in an MCDM context. The incorporation of economic and environmental objectives, together with nutritional advanced concepts encompassing production objectives allows to treat the formulation of a feed for growing pigs with a more precise model. An improved diet and precision feeding have been identified by pig production experts as a strategy for reducing nitrogen excretion in farm animals (Rotz, 2004). By the use of the lysine/energy ratio and the ideal protein balance concept we are improving the efficient use of nitrogen. In this section, we compare our model with two previous attempts to include environmental concerns in formulation of models and we make some comments about the results obtained.

Jean dit Bailleul *et al.* (2001) used a model to minimise simultaneously the cost and the excess of nitrogen over the requirements in the rations of growing and finishing pigs. They adopted an easy way to formalise and solve this bi-objective model, in the context of parametric programming as an extension of LP. They added the weighted excesses of dietary amino acid to the cost function. Accordingly, their model requires the decision maker to assign a value to the weight of the excess nitrogen.

An alternative approach was adopted by Tozer and Stokes (2001). Their model incorporates empirical functions for nutrient excretion, nitrogen and phosphorus. The non-linear character of these functions makes the model a non-linear multi-objective one with three objectives: to minimise the cost and the excretion of nitrogen and phosphorus. They applied the model to a dairy cow ration formulation and solved it, as in the case of Jean dit Bailleul *et al.* (2001), through a weighting procedure, demanding a huge amount of information from the decision maker.

Tackling these models through weighting procedures can be convenient in some circumstances, particularly when researchers confront the analysis of the effects of different diets on nutritional and environmental parameters. A sensitivity analysis of the weights is a proper way to proceed. In addition the applications reported in both papers are of this kind. However, for industry formulation by nutritionist experts, or even for on-farm formulation by farmers and advisers, a procedure demanding less information would be welcome. The model presented here does not require any a priori information from the decision maker, eliciting his preferences through the interactive process. We have extended our model (Castrodeza *et al.*, 2005) to include also the excretion of phosphorus as an objective. Although the addition of a fourth objective can raise difficulties in the mathematical understanding by model users, the algorithm performs efficiently.

We, and Jean dit Bailleul *et al.* (2001), have incorporated environmental concerns with an approach aimed at a more precise formulation. Instead, Tozer and Stokes (2001) used directly empirical impact functions. A different and interesting approach would be the incorporation in the coefficient matrix of impact indicators associated with the use of each feed. For example, the unitary contribution of, say, wheat, barley, etc., to such issues as eutrophication, climate change, acidification, or energy use. Although much research needs to be done by environmentalists to quantify these items with the required precision to be incorporated in the formulations with a minimum guarantee, some steps have been made (Van der Werf *et al.*, 2005).

To evaluate the results obtained, we must refer to the nutritional characteristics of the diet obtained and how they are traded off with the cost. Although this chapter is not the proper place for a deep analysis of the nutritional status of the diet, we would like to make a brief comment.

The value chosen for the lysine/energy ratio depends on the objectives of pig production. For growing pigs, the usual objective is to achieve the potential growth of the pigs. This potential varies not only with the breed of the pig, but also with the specific genetic line and the sex. Therefore, the proper value of the ratio for particular pigs must be chosen taking into account the empirical knowledge about the possibilities of transforming feed into growth of the pigs rationed. The potential growth use is stated as g/day of protein. The Agricultural Research Council (ARC, 1981) proposed a ratio of 0.675 g/MJ to maximise protein deposition in pigs with a growth potential of 100–130 g/day of protein. Subsequent studies have proposed values ranging from 0.73 to 1. For example, Van Lunen and Cole (1996) proposed 0.95 to 1 for selected males with a high potential of 150–190 g/day of protein. The value chosen by the decision maker in our problem, 0.9883 g lysine/MJ of energy is in the range of these figures of high growing animals.

REFERENCES

Agricultural Research Council (ARC), 1981, *The Nutrient Requirements of Pigs*, Commonwealth Agricultural Bureau, Slough.

Castrodeza, C., Lara, P. and Peña, T., 2005, Multicriteria fractional model for feed formulation: economic, nutritional and environmental criteria, *Agric. Syst.*, **86**: 76–96.

Charnes, A. and Cooper, W.W., 1962, Programming with linear fractionals functionals, *Naval Res. Log. Quart.*, **9**: 181–186.

Cole, D.J.A. and Van Lunen, T.A., 1994, Ideal amino acid patterns, in: *Amino Acids in Farm Animal Nutrition*, J.D.F. D'Mello, ed., CAB International, UK, pp. 99–112.

Fundación Española para el Desarrollo de la Nutrición Animal (FEDNA), 1999, *Normas FEDNA para la Formulación de Piensos Compuestos*, FEDNA, Madrid.

Jean dit Bailleul, P., Rivest, J., Dubeeu, F. and Pomar, C., 2001, Reducing nitrogen excretion in pigs by modifying the traditional least-cost formulation algorithm, *Livest. Prod. Sci.*, **72**: 199–211.

Lara, P., 1993, Multiple objective fractional programming and livestock ration formulation: a case study for dairy cow diets in Spain, *Agric. Syst.*, **41**: 321–334.
National Research Council (NRC), 1998, *Nutrient Requirements of Swine*, 10th ed., National Academic Press, Washington DC.
Rehman,T. and Romero, C., 1984, Multiple-criteria decision-making techniques and their role in livestock ration formulation, *Agric. Syst.*, **15**(1): 23–49.
Romero, C. and Rehman, T., 2003, *Multiple Criteria Analysis for Agricultural Decisions*, 2nd ed., Elsevier, Amsterdam.
Rotz, C.A., 2004, Management to reduce nitrogen losses in animal production, *J. Anim. Sci.*, **82**(E. Suppl.): E119–E137.
Schaible, S., 1976, Fractional programming I. Duality, *Manage. Sci.*, **22**: 858–867.
Spronk, J., 1981, *Interactive Multiple Goal Programming: Applications to Financial Planning*, Martinus Nijhoff, Boston.
Stancu-Minasian, I., 1997, *Fractional Programming: Theory, Methods and Applications*, Kluwer Academic Publishers, Dordrecht.
Tozer, P.R. and Stokes, J.R., 2001, A multi-objective programming approach to feed ration balancing and nutrient management, *Agric. Syst.*, **67**: 201–215.
Van Lunen, T.A. and Cole, D.J.A., 1996, The effect of lysine/digestible energy ratio on growth performance and nitrogen deposition on hybrid boards, gilts and castrated male pigs, *Anim. Sci.*, **63**: 1575–1583.
Van der Werf, H.M.G., Petit, J. and Sanders, J., 2005, The environmental impacts of the production of concentrated feed: the case of pig feed in Bretagne, *Agric. Syst.*, **83**: 153–177.

Chapter 5

MODELING THE INTERACTIONS BETWEEN AGRICULTURE AND THE ENVIRONMENT

Slim Zekri and Houcine Boughanmi

Department of Agricultural Economics and Rural Studies, College of Agricultural and Marine Sciences, Sultan Qaboos University, Muscat, Sultanate of Oman

Abstract Modeling agricultural systems that recognize the environmental dimensions of agriculture has evolved during the two past decades. Multi-objective mathematical models encompassed the diversity of objectives inherent in agricultural activities as a result of externalities and replaced single objective models. It has been observed that recent modeling efforts at farm level combined several simulation models at a time (crop simulation, weed simulation, hydrologic model, erosion) with a multi-criteria model. Many of the studies reviewed have not thoroughly considered the policy instruments to internalize the pollution problems and some studies have not considered any policy instrument. The multicriteria techniques used range from distance based approach, utility theory, generating methods, interactive methods, fractional programming and fuzzy programming. The later method is called for to deal with the inexact information generated with geographical information systems GIS or simulation models. Most applications of coupled GIS and decision models dealt with watershed management and soil erosion. Spatial GIS/multicriteria models that involve stakeholders are considered as a form of institutional reorganization which will help change the hierarchical mode of decision making. Stakeholder involvement in decision making has brought the modeling effort to include group and multiple decision makers. It is expected that future models will integrate several simulation models, GIS, and several stakeholders and would be applied at regional and national levels. Thus, multicriteria modelers will have to deal with uncertain and inexact information as well as asymmetric information.

Keywords: Multiple criteria decision making, geographical information systems, public participation, policy instruments

1 INTRODUCTION

The most common environmental pollutants from agriculture are sediment run-off, nutrients, manure, and production chemicals including herbicides and pesticides. Most of these pollutants are non-point pollution. Non-point pollution is characterized by the high cost of monitoring and control besides enforcement difficulties due to legal uncertainties. Consequently, it is almost impossible to make off-farm costs of pollution show up on farmers' profits. These characteristics as well as the atomicity of the farms restrict the policy options that could be considered in any mathematical model to tackle pollution reduction or prevention. Government's programs are mainly based on best management practices and land retirements coupled with subsidies and are voluntary instead of being based on the polluter pays principle. Additionally, the idea of paying the polluter is justified on the basis that farmers are providing positive externalities and contributing to rural development.

Modeling agricultural systems that recognize the environmental dimensions of agriculture has evolved during the last decade. Multi-objective mathematical models encompassed the diversity of objectives inherent in agricultural activities as a result of externalities and replaced single objective models. This was made possible because of the development of crop simulation models, which helped establish quantitative relationships between production and the environment. Most of the agricultural-pollution problems are heterogeneous, site-specific, and concern wide areas like water-sheds, river basins, or aquifers. Geographical Information Systems (GIS) have become a required tool to account for the spatial differences of pollution. The use of simulation models, GIS, and mathematical models made the integration of these tools into spatial decision support systems (SDSS), a necessity to create a friendly environment both for analysts and decision makers (DMs). Finally, the new governance ideas regarding the involvement of stakeholders coupled with the development of the Web made possible the emergence of a new generation of Web-SDSS, allowing for interaction of different stakeholders. Involvement of stakeholders has created in turn the need group decision for making techniques that consider several decision makers and not only farmer or farmers group.

The agricultural environmental linkages are multi-objective in nature. Romero and Rehman (1987) reviewed the use of multiple criteria decision methods in natural resource management. Recently Hayashi (2000) reviewed the application of multiple criteria to agriculture. The author reviewed more than 80 papers/books published during the period 1977–1998. He dealt with discrete and continuous methods and the difficulties encountered by practitioners in applying the multicriteria analysis to agriculture. The present chapter reviews, in a non exhaustive way, the articles published within the period 1998–2004 and dealing with the application of multiple criteria decision making (MCDM) techniques to agri-environmental problems. This chapter is organized as follows. Section 2 considers the agricultural and environmental linkages and the criteria adopted in the models. Section 3 addresses the choice of policy instruments at farm-level focusing on non-point pollution. Section 4 deals with the advantages and difficulties of linking GIS to MCDM techniques. Section 5 deals with the new governance principle and public participation and its implications for modeling agri-environmental problems. Sections 6 and 7 report the results of mathematical modeling at regional and nation–sector levels. The chapter ends with future research trends and application of multicriteria methods.

2 AGRICULTURE AND ENVIRONMENTAL LINKAGES

An increasing number of studies are considering the linkages between economic and environmental spheres. The papers are not only published in economic or environmental journals but also in agronomical, ecological, and policy journals, which is a positive sign of the adoption of the operations research (OR) techniques by a wide array of scientists. A large number of objectives are reported in the literature consequently, MCDM techniques are the most used OR technique. Besides the classical objective of profit maximization the most recurrent criteria are minimization of erosion or sediment delivery, minimization of agri–chemical inputs, minimization of ground water pumping, and minimization of nutrients as well as pesticide contaminants in surface and/or ground water (nitrogen, phosphate, atrazine, sevin, carbofuran, etc.) resulting from either chemical or manure fertilizers. Column 2 of Table 1 shows the objectives considered in different farm–level studies. The number of the objectives differs from a minimum of two to a maximum of ten criteria. In some cases, the number of objectives is artificially high as some of the objectives are redundant. This is the case for instance in Agrell *et al.* (2004) where at least four objectives are redundant. In two of the articles (Psychoudakis *et al.*, 2002, Stokes and Tozer, 2002), all the objectives are expressed in monetary units. Quantifying the objectives in monetary terms is neither necessary nor desirable because all the environmental standards are expressed in units of the pollutants and not in monetary units.

Several papers are based on the combination of a crop simulation model, which allows the quantification of the environmental externalities such as soil erosion or nitrate leach up, with the MCDM model. Others combine several simulation models at a time (crop simulation, weed simulation, hydrologic model, erosion, etc.) with an MCDM model (Pacini *et al.*, 2004). Usually the outputs of simulation packages are used as input data in the MCDM model manually without any interface. Simulation models have the advantage of synthesizing several characteristics of a pollution problem in one single unit, that is nitrogen pollution depends among other factors on the applied amount, balance of nitrogen, date of application, and splitting up. Instead of considering each one of these characteristics as a criteria (Arondel and Girardin, 2000), the simulation model allows a reliable estimation of the potential impact of the combined characteristics reducing considerably the complexity of the problem and interpretation of the solutions both for the analyst and DMs. Simulation models implicitly identify the scale of analysis and integrate targets to which the impacts should be compared. Once the scale is known, it is easier to generate the evaluation criteria, to compute the impact scores, and to identify polluters and victims (Munda, 2004). The use of simulation models circumvents the problem mentioned by Hayashi (2003) regarding the multiplicity of impacts of one single agricultural pollutant and the need to build indexes or indicators. Each environmental risk or health risk (OECD, 2001, p. 139) could be considered a goal in itself if targets are set by the simulation model. Hayashi argues that this may raise the number of objectives considerably and hence the difficulty of the DM in assigning weights. This problem could be solved in two complementary ways. First in an MCDM model the computation of the pay-off matrix helps reduce the complexity of the problem by dropping the complementary objectives and keeping only the conflictive objectives. Second, as far as models involve several decision makers each one of the DMs will have to express his/her preferences for a limited number of objectives only and not for the totality of them.

The most frequent MCDM techniques used, as reported in Table 1 column 3, are the constraint method and compromise programming. The weighting method and building of utility functions are also frequent with the conflicting aims of exploring the space of solutions and the convergence to a reduced set of alternatives. In fact in most cases where the objective is to compare the current situation with a future solution, methods that converge to one solution are preferred to make easier comparison. The weights used to generate solutions are either elicited and then parameterized or revealed through past choices. In other cases the objectives are simply given equal weights, which is not coherent with real–life situations as pollution is an externality and farmers never attribute to it the same weight as profit. These practices reflect the difficulties encountered by practitioners and the absence of consensual methods to account for the weights to attribute to the different objectives. Only two papers included the risk criteria (Ridier and Jacquet 2002; Gomez-Limon *et al.*, 2003). Although there is a consensus that the inclusion of risk criteria is of primary importance to improve the model's performances; usually lack of data makes it very difficult to account for such criteria besides environmental objectives.

Table 1. Modeling agricultural–environmental linkages at farm level.

Author (year)	Objectives	MCDM Technique	Environmental policy instrument	Simulation models	Country
Annetts and Audsley (2002)	Revenue (max) Nitrate pollution (min) Herbicide (min)	Eff (w) Weights elicited than parameterized	Best management practices Herbicide tax Legal nitrogen constraints	Machinery model Nitrogen model Weed model	UK
Loyce *et al.* (2002)	Gross margin (max) Cost of production (min) Technical operations (min) Nitrogen after harvest (min) Pesticides (min) Energy balance (max) By-product protein content (max)	Discrete optimize BETHA weight parameterized	Best management practices	Agronomic model Leaching model for nitrogen	France
Giasson *et al.* (2002)	Cost of manure management (min) Phosphorus pollution from manure (min) Risk phosphorus pollution (min)	Compromise programming Equal weights to all objectives	Best management practices	Phosphorus-index model	USA New York
Meyer-Aurich *et al.* (2001)	Gross Margin (max) Soil erosion (min) Nitrogen (tie equation) Global warming potential (tie equation) Energy input (tie equation)	Constraint method	Best management practices	Agronomic model Ecological model Erosion model	Germany
Ridier and Jacquet (2002)	Income (max) MOTAD (Financial risk) Target-MOTAD (Financial risk)	Multiobjective multiperiodic model Weights revealed	Animal stock density Decoupled subsidies	Production functions	France
Campus *et al.* (1999)	Gross margin (max) Nitrogen (min) Pesticides use (min)	Constraint method	Best management practices Trade-offs	Experiments	Italy

Author (year)	Objectives	MCDM Technique	Environmental policy instrument	Simulation models	Country
Hoag *et al.* (1999)	Profit (max) Financial risk (min) Nitrate leaching (min) Soil erosion (min)	Analytical hierarchy process	Irrigation management choices		USA Colorado
Arondel and Girardin, (2000)	Nitrogen Pesticide Irrigation No optimization,	ELECTRE TRI sorting process of farmers practices on ex-post basis	Management practices (subsidy)	None, recorded data by farmers, expert information	France
Hajkowicz and Prato (1998)	Net return (max) Risk (min) Soil erosion (min) Nitrate-nitrogen (min) Atrazine (min)	Analytical hierarchy process and utility function	Subsidy and best management practices	Watershed simulation model	USA Missouri
Van Huylenbroeck (2000)	Gross margin (max) Nitrate (min) Land use (min)	Compromise programming (CP)	Nitrogen constraint	Agronomic simulation model	Belgium
Jones and Barnes, 2000	Profitability (max) Nitrogen (min) Water (low, medium, high)	Fuzzy composite programming	Best management practices	Agronomic simulation model, GIS	USA Arizona
Falconer and Hodge (2001)	Profit (max) Pesticide index (min)	Constraint method	Tax on pesticides several indicators	Farm trials and expert information	UK
Jean dit Bailleul *et al.* (2001)	Cost (min) Excess nitrogen (min)	Eff (w) Weight of objective 2 parameterized	None	Farm trials	Canada and France
Gomez-Limon *et al.* (2003)	Gross margin (max) Risk (min) Labor (min) Working capital (min)	Multi-attribute utility function Weights revealed	Marginal utility of water	None	Spain
Psychoudakis *et al.* (2002)	Income (max) Fertilizers (min) Herbicides (min) Fungicides (min) Insecticides (min) (all expressed in monetary units)	Constraint method CP	Best management practices Subsidies	None	Greece
Stokes and Tozer (2002)	Cost (min) Phosphorus (min) Nitrogen (min) All expressed in monetary units	Constraint method NISE and CP Weights parameterized	Preventive approach Tax on feed inputs	Econometric functions	USA Pennsylvania
Castrodeza *et al.* (2004)	Cost (min) Lysine/energy ratio (max) Nutritional–ecological criteria (min) Phosphorus (min)	Multiobjective fractional programming	Preventive approach No policy instruments	Engineering functions	Spain

Author (year)	**Objectives**	**MCDM Technique**	**Environmental policy instrument**	**Simulation models**	**Country**
Latacz-Lohmann (2004)	Income Nitrate Crop diversity Labor Capital	Data envelopment analysis	Ex-post policy Efficiency analysis	None	Hypothetical farms
Pacini *et al.* (2004)	Gross margin (max) Nitrogen leaching and nitrogen run-off Soil erosion Ground and surface water balance Environmental potential pesticide risks Herbaceous plant biodiversity Hedge length Manure and slurry surplus Drainage system length	Constraint method Thresholds for the environmental criteria considered as constraints	Different kinds of subsidies for organic farming	Models for: Erosion Water Pesticides Plant biodiversity	Italy
Agrell *et al.* (2004)	Food production (max) Food production during several conditions(max) Self-sufficiency ratio (max) Value of production (max) Cost of production (min) Net revenue (max) Arable land (min) Harvested land (min) Total erosion (min) Erosion maximum level per cell (min)	Interactive weighted Chebychev programming	None	GIS Agro-ecological zoning	Kenya A whole district, Bungoma, is considered as a single farm

3 FARM LEVEL–MODELING AND POLICY INSTRUMENTS

Most agri-environmental problems tackled non-point pollution. Some of the studies considered the preventive approach through the exploration of BMPs and others considered taxes or subsidies to internalize the negative externalities caused by farming.

One of the serious problems facing modelers of agri-environmental systems is the consideration of the right environmental policy. Empirically accumulated experiences regarding environmental problems at farm level shows that the price-tax policy instrument to internalize externalities has uneven effects on farmers and crops. For instance while a tax on nitrogen price will have beneficial effects on nitrate pollution abatement for growers of winter cereals, sunflower and summer crop growers however will have to pay the tax without causing much environmental harm. Nitrogen is a common input for a large number of crops whether high nitrogen-efficient crops or low nitrogen-efficient ones. The problem is even more complicated within a single country with high differences in rainfall regimes among regions such as the Mediterranean countries. Nitrogen–risk pollution depends partly on rainfall; the higher the rainfall the higher the risk of nitrogen leaching and pollution. Nitrogen pollution is also site–specific and depends on soil characteristics and land slope. Consequently, non targeted environmental policy instruments, such a tax on nitrogen, could not be recommended, and this strong economic instrument is usually discarded due to the inequity effects which will result in a loss of competitiveness at international markets for countries adopting it. In an era of globalization, there is a need to implement specific and local policy instruments based on farm types, crops, and agro–ecological zoning for the sake of both efficiency and equity. Unfortunately, the polluter pays the principle or price-tax instrument is hardly feasible under such conditions. To generalize, we can say that the price-tax instrument can be used only for some pesticides or herbicides which are specific to a given type of crop and in no case for agri–chemicals that could be used for a wide range of crops or agricultural activities. However, Falconer and Hodge (2001) mentioned that unlike nitrogen, for pesticides there are no clear targets against which environmental quality improvements might be measured. Besides, pesticide hazard affects several different ecological and human health dimensions. They used a subjective hazard–aggregated index, which accounted for the hazards related to each pesticide. The index and other environmental indicators for pesticides were combined with a tax instrument. The authors were innovative in considering the side-effects of a pesticides-tax on nitrogen, which revealed a complementarity since the same pesticide-tax reduced nitrogen use.

Subsidy or paying the polluter is the most used instrument in the developed countries to enhance environmentally sound–farming practices. Currently direct payments are widely used in the European and US agricultural policies (investment aid for environmentally sound production methods, beef extensification premium, set-aside, buffer strips, or to enhance provision of positive externalities of agriculture). These payments are made possible under the argument of multi-functionality of agriculture and rural development. The first concept highlights the role of agriculture in providing amenities to society jointly produced with agricultural outputs, which are considered as positive externalities that should be paid for by tax payers. Subsidies are usually estimated using the MCDM techniques through the calculus of trade-off between economic and environmental objectives and are supposed to compensate farmers for the loss in profit they incur as a consequence of adoption of conservation measures or reduced use of agrichemicals. However, the subsidy (decoupled payment) instrument may become illegal in the future according to WTO conflict settlement (MacKenzie, 2004) and due to economists' reserves on the adverse effects of subsidy in the long term. This is not an assessment of whether subsidy is good or bad, but analysts should keep their eyes open to the feasibility of the instruments they are testing in their models.

The property right approach is seldom practiced in the case of negative externalities of farming. However, SDSS software might enhance negotiations between polluters and victims of pollution provided that property rights are explicitly defined and some sort of measurement of the externalities is feasible, agreed on, and at a cost that does not preclude negotiations. The property right approach may not be well adapted to non-point pollution problems, the most frequent type of pollution in agriculture. Even if negotiations may take place as a result of SDSS implementation, the situation may lead to the problem of moral hazard. The basis of moral hazard problem is imperfect information about farmers' actual compliance due to the difficulties in measuring the state of the environment and quantification of changes (Latacz-Lohmann, 2004) and the conversion of the problem from a non-point pollution to one of identified source pollutions, where it is possible to monitor the source, quantity, and quality of the toxic effluents. Woodward (2003) mentioned that even where effluent trading has been authorized, very few transactions have actually been carried out. The main physical and economic reasons enumerated to explain the absence of trade are water pollutants flow down-hill and concentrations change over time; predictions of pollution loads are likely imprecise, which may give rise to legal conflicts; water pollution problems are confined to a watershed. Consequently, resulting markets are "thin," prices may be manipulated by a few traders, and monitoring and enforcement costs are high. Nonetheless, Latacz-Lohmann reported examples of successful experiences where auctions for environmental public funded contracts is thought to be more cost efficient than a fixed lump sum payment.

The voluntary approach through the adoption of BMPs is favored by policy makers and researchers, as could be seen in Table-1 column 4, though its real impact on agricultural pollution abatement is small (Horan *et al.*, 2001). In fact with current technology any reduction in negative externalities is done at the expense of farmers' profit though it could be argued that the profit loss is small. Farmers are unable to pass on the extra environmental costs to consumers even though food demand is inelastic, which means that there are other market failures that should be addressed simultaneously and not just the pollution externalities.

Innovations and new technologies are considered as another instrument for pollution abatement. Precision agriculture is one of these technologies the aim of which is to deploy information and electronic technologies as GIS, global positioning system (GPS), and variable rate application equipments at farm level to apply precise quantities of agri–chemicals. The technology has been designed to increase productivity and reduce financial risk in addition to preventing pollution (Zeng *et al.*, 2004). Mathematical programming plays an important role since optimization of input use spatially and within time is a major component of precision agriculture. Gandonou and Dillon (2003) used a discrete stochastic sequential programming model to analyze the profitability of an investment in precision agriculture under financial risk (mean–variance). The non adoption of precision agriculture is mainly due to the high cost of equipment. Environmental objectives are considered secondary in this study and were not dealt with ignoring the potentialities of precision agriculture to reduce non-point pollution. Jones and Barnes (2000) presented a fuzzy composite programming model, a distance based model, for cotton precise management considering profitability, environment and sustainability as the three major criteria.

Given the continuous decline of electronic equipment's prices it is expected that precision agriculture may become more profitable and become one of the alternatives to prevent non-point pollution in countries with high farmer skills. MCDM can be coupled to stochastic programming to model the potential of pollution prevention. Besides, since computers are used in precision agriculture they can play the role of "big brother" at farm level. Then monitoring of non-point pollution control could be cheaply and easily accomplished as all inputs are recorded whenever used. However as pointed out by Horan *et al.* (2001) such a positive result could be obtained only if the demand for agricultural products in question is not too price-elastic. Otherwise the long-term effects may lead to an increased use of inputs as a consequence of output price decline and demand shift upward even though input usage per unit of output would fall. Consequently, analysts should consider coupling precision agriculture with other policy instruments to avoid the mentioned drawbacks.

Goal programming (GP) is progressively replacing linear programming and GP is being integrated in decision support systems (DSS) accessible to producers for ration formulation (Vickner and Hoag, 1998). Jean dit Bailleul *et al.* (2001) proposed a bi-criteria model, minimizing the cost of the diet and minimizing the total reducible fraction of nitrogen excretion in pig diets. In that study, there is a confusion as the weights attached to objectives are taken as a tax. Stokes and Tozer (2002) use MCDM techniques in ration formulation by including the objective of reducing phosphorus pollution from livestock manure. The links between phosphorus pollution and phosphorus intake and nitrogen pollution and dry matter intake and net energy lactation are done through two quadratic and linear functions respectively. These two functions are then minimized along with the classical objective of cost minimization. The earlier papers provided an excellent idea of the way the preventive approach can be considered. Feed ration formulation is at the root of environmental problems and farmers can make decisions on what they feed their livestock to reduce the negative externalities from excess nutrients leaching. Castrodeza *et al.* (2004) presented a multiobjective fractional programming model solved using the interactive multiple goal programming method. The model is really sound since besides cost minimization they considered two environmental criteria, nitrogen and phosphorus minimization, and a fourth criteria related to pig-meat quality. Results showed again that pollution reduction will increase feed cost. However, the authors have not estimated the trade-off neither any policy instrument on how to induce farmers to adopt such feed ration was proposed. The idea could be further enhanced by considering penalty functions in a goal programming model, which will allow more flexibility for the search of solutions (Romero and Rehman, 2003). Another issue which merits further investigation consists in the introduction of reproduction risk in animals as a consequence of phosphorus minimization.

Agricultural point pollution caused by manure is easier to manage than non-point pollution due to the low cost of monitoring and availability of information regarding the quantity of pollutants. Stonehouse *et al.* (2002) used a mixed integer programming model maximizing profit under ammonia, nitrogen and phosphorus constraints. Manure disposal alternatives were considered at farm level linked to crop production mix and hog ration. Lauwers *et al.* (1998) applied a linear programming model at farm level coupled with an aggregated module and a regional model to analyze the impacts of environmental regulations on manure disposal. They considered manure export from farms with excess manure to farms with shortage of manure counting transportation cost between regions. The farm model also analyzed the impact of environmental regulation on on-farm cost abatement alternatives. Manure monitoring is undertaken by the regional manure disposal coordination system.

4 GIS AND AGRI-ENVIRONMENTAL MODELING

The use of GIS in agri-environmental problems is justified by the fact that policies shape the environment. Most major land-based decisions are made at parcel level, and policy makers felt the need for geographic information to analyze ex-ante the impacts of their decisions. The major role of GIS is to capture, analyze and display spatial data. Most integration of GIS with decision models falls in the loose coupling category through the exchange of data files. The deep coupling approach links GIS and decision models with a common user interface (Wang *et al.*, 2004). Deep coupling of GIS with decision models is a difficult and cumbersome process. Among the barriers, some of the criteria are not easily translated into the sorts of spatial analysis for which a GIS is employed such as the category of farm and family criteria; lack of high–quality parcel mapping and updated data; and lack of clearly stated goals to be achieved by a program (Tulloch *et al.*, 2003). Table 2 shows the type of coupling and the multicriteria approach used for decision.

One field of application where SDSS has been largely applied is soil erosion and watershed management (see Table 2 column 6). GIS offers an enhanced way of targeting farmers based on several weather, soil, and topographic characteristics. Aggregation of farmers' decisions helps estimate the effects at a watershed level. GIS helps in reviewing and proposal of zoning changes and selection of parcels for conservation. In several cases, farmers are paid by the government to keep their property in permanent preservation. This subsidy instrument could be highly efficient if spatially differentiated to reflect critical sources or areas. Moreover, in the case of soil erosion the subsidy is conditioned on the implementation of visible techniques easily controlled and monitored (contour building, small check dams) unlike the agri–chemicals pollution problem, which means that the information problem of the principal is reduced. Targeting particular areas and/or conservation techniques helps considerably reduce the conservation costs compared with non targeting (Westra *et al.*, 2001).

5 PUBLIC PARTICIPATION

Although the modeling of linkages between agriculture and environment dates back to the early 1980s, the new features of the recent investigations are the consideration of DSS that seek to assist DMs and the integration of GIS technology. DSS that involve stakeholders are considered as a form of institutional reorganization that will help change the hierarchical mode of decision making and may lead to more environmental friendlier behavior (Arzt, 2003). Rinner (2003) reports that a common motivation for making SDSS accessible online (Web-SDSS) is to support group decision making, increase the number of potential users, and promote virtual democracy by increasing public participation which was confined to the right to object. The author stresses the fact that the quasi–absence of user studies cast some doubts on the utility of Web-SDSS. However, Giupponi *et al.* (2002) report that European Union legislation regarding WFD requests member states to encourage the active involvement of all interested parties. Participation of stakeholders might include public forums, focus groups, as well as software for decision making. Stagl (2003) discussed the advantages and limits related to public participation and addressed the questions on should participate and how. Munda (2004) stressed the fact that public participation results in a creative process as local people can imagine solutions never thought of by accredited experts.

Jones *et al.* (1998) stressed the fact that direct involvement of users creates a unique learning environment. When direct involvement is not possible, users may be indirectly involved through surveys, workshops, and public hearing. It should be stressed that such Web-SDSS should not be used for voting purposes only but rather for feeding back the models and providing new solutions. At the same time, the involvement of stakeholders moved the OR problem from one single decision to multiple decision makers.

Multiple criteria analysts' and GIS specialists' contribution to the effort of stakeholders involvement is strongly needed. Several questions still have to be closely analyzed by MCDM specialists such as the aggregation procedures of preferences, spatial aggregation, and addressing the environmental risk and uncertainties. Recent application has shown that consideration of interval goal programming instead of strict constraints as well as inexact fuzzy multiobjective programming helps to deal with the uncertainties and inexact values of the constraints' right-hand sides (Wang *et al.*, 2004).

Table 2. Multiobjective programming and SDSS.

Authors (year)	Objectives/ Goals	SDSS Web-SDSS	Multiobjective technique	Participative approach	Purpose
Janssen *et al.* (2005)	12 criteria	Models are not integrated in a DSS	Discrete method DEFINITE	Non-participative	Wet land management
Morari *et al.* (2004)	3 criteria	Models are not integrated in a DSS	Discrete method Concordance/ discordance	Participative: Farmer, environmentalist, and politician	River basin management Water quality
Giupponi *et al.* (2002)	5 criteria	SDSS	AHP and weighting method Discrete options	Interactive single DM	Water quality management
Jensen *et al.* (1997)	Farm budget chemicals optimization	Web-SDSS	Simulation model	Interactive single DM	Extension for chemicals and irrigation
McMaster *et al.* (2002)	Not explicated	SDSS	Not explicated	Interactive	Assisting farmers in selecting BMPs
Yaldir and Rehman (2002)	5 criteria	SDSS	Heurestic search	Interactive	Land consolidation
Veith (2002)	2 criteria	SDSS	Multiobjective programming Genetic algorithm	Nonparticipative	Watershed management
Wang *et al.* (2004)	4 criteria	SDSS	Inexact-fuzzy multiobjective programming	Nonparticipative	Watershed management

However, public participation showed the insufficiency of considering revealed or elicited preferences of parties involved in the decision-making process. Values and preferences of participants are constructed during the process and not just elicited (O'Connor, 2000, cited in Stagl, 2003). There is an important learning process that usually leads to adjustments of the weights attributed to the different objectives before the decision is made. Hayashi (2000) highlighted that attribute weights should be allowed to vary according to the goal range or attribute level. Analysts have then to find the best way to determine the stakeholders' preferences and either an iterative process should be engaged to build the aggregated social function or a weighted interval goal programming to account for the "goal range" requisite. Given the importance of the learning process, the use of exploratory techniques to provide participants with a wide set of alternatives is of utmost importance as a first stage. Only in later stages, convergence techniques should be used to find a compromise or an aggregated social function. In this sense, Linares and Romero (2002) proposed a method based on coupling the analytical hierarchy process and goal programming to aggregate stakeholders' preferences in a cardinal way.

From what has been discussed here, appears that several MCDM techniques might be needed in SDSS. However, recent theoretical investigations showed the linkages between several MCDM techniques. Thus for instance, a number of MCDM techniques (weighted GP, lexicographic GP, minimax GP, LMGP, extended GP, ELGP, weighted interval GP, reference point method, CP, interactive weighted Chebyshev procedure) are special cases of the general optimizing structure called extended linear interval GP (ELIGP) (Romero, 2001, 2004). The immediate impact of these unifying approaches is the building of a general mathematical model, without the requirement to create a software, which encompasses several techniques. This will improve communication among scientists across disciplines and with stakeholders and prevent the need to choose one MCDM technique among the techniques available.

The construction of an aggregated welfare social function helps to determine the social optimum. Unlike industrial activities where the social optimum could be relatively easily enforced as the company needs an approval of future plans by regulating bodies, farming does not obey to such a decision scheme. The farmer is the unique decision maker when it comes to crop mix and input uses. The social optimum is thus a simple reference and farmers cannot be forced to implement its result even though they may have participated in the decision-making process. As a consequence, the results of the social optima have to be used as a reference point to be reached by farmers. The development of farm models based on environmental policies to induce farmers to adopt the socially optimal alternative is still a necessity.

Incentives to farmers for co-operation and participation in the novel democratic institutions are not clear. Nevertheless, farmers' participation might enhance their learning and their efficiency may be considered as an incentive in itself. However, more than a decade of research has shown that farmers' collaboration based on a voluntary approach or BMPs has limits. The reasons may be the fact that MCDM practitioners have not considered the full implications of such BMPs at farm level, such as the yield risks and labor reorganization. Hayashi (2000) pointed out that methods for arranging work schedules may be needed more than just determining crop mix. In this sense, Annets and Audsley (2002) presented a farm multiobjective model, which captures the crop mix with the timing of operations and levels of labor and machinery. Yield risks due to reduction in chemical uses for instance have not been correctly addressed. We are heading towards more integrated models which will need higher human skills, multidisciplinarity, and larger data bases.

6 REGIONAL MODELING

One specificity of farm models is that they consider farmers as price takers. However in many circumstances, the proposed agri-environmental scenarios involve drastic changes in the cropping pattern, which in turn might affect prices received by farmers if such changes take place in large areas. Consequently, the farm models have to be integrated into a sector model to capture the possible changes in output prices and to get more reasonable policy alternatives assessments or at least to use crop price elasticity on an ad hoc basis for scenario analysis if it is not possible to build a sector model.

Studies on regional and nation-sector modeling are based on the premise that the behavior of the sector as a whole, following a policy change, is the aggregate behavior of the individual producers of that sector.

In regional modeling, "representative" farm models are chosen to jointly reflect the component of regional production relevant to the study. In addition, features reflecting regional factor markets (i.e., labor and land) and product markets are also included (McCarl,1992). At this level, producers (in a small region) are still assumed price takers, not large enough to influence prices of traded products or factors.

Environment issues in regional modeling in agriculture are considered in the form of added constraints imposed on agricultural production or are implicitly incorporated in the objective function. The objective function may therefore reflect the views of policy makers; some may be sensitive to the environment. An example of such a model is a study conducted by Dillon (1992). They studied the allocation of the Edward aquifer water in central Texas, where water is used by agricultural, municipal, and Industrial interests while feeding springs which support endangered species and recreation. The objective function maximizes expected regional welfare across the recharge distribution and includes terms of net farm income, municipal water consumers' surplus, municipal water supply cost, industrial water consumers' surplus, and industrial pumping cost. The model was used to examine optimal water allocation, potential management and property schemes, spring flow limits, usage limits, and drought management.

Environment concerns at the regional level can also be analyzed by simulating the effects on regional welfare of various scenarios representing different environment standards. Feinerman *et al.* (2004) used a spatial equilibrium model to evaluate the welfare costs of regulatory standards for animal manure application to crops. Because of the rapid growth in confined animal and poultry production, nitrogen (N) and phosphorus (P) water pollution from manure applications became a real environmental issue in many US states. Three scenarios of no standard, N standard, and P standard are evaluated to maximize the welfare, which equals the gain in net returns from using manure alone or in combination with commercial fertilizers. The author's results suggest that regulatory standards from manure application achieve large reduction in excess nitrogen and phosphorus with a welfare loss of 5–15% not including the non market environment valuation.

7 NATION-SECTORAL MODELING

Sectoral modeling investigates the behavior of the agricultural sector nationwide. It differs from farm or regional modeling in terms of pricing and details of representative farms (McCarl, 1992). Product and factor prices are determined by supply and demand considerations rather than being fixed as in the farm or regional models.

Mathematical programming in its various forms is the privileged technique among researchers to analyze the sector-wide effects of agricultural policies. In the structure of mathematical programming, the decision problem at the sector level can be viewed as a two-level process (Hazell and Norton, 1986). At the macrolevel, the policy maker is trying to find how best to allocate budgetary resource to achieve multiple objectives (macrogoals) in an environment where it is uncertain how farmers will respond to a change in policies. At the microlevel, producers are trying to maximize their own objectives, given the new policy environment and their resource constraints.

However, sector modeling confronts the facts that policy-makers' objectives are usually different from the objective of thousands of independent producers in agriculture, which makes the two-level planning problem specified earlier difficult to solve. To cope with this complexity, the two-level planning problem is transformed into a one-level, and the problem is usually transformed into the more practical one level approach of simulating the sector response to possible policy changes, where the policy goals are now included as variable in the model (Mc Carl and Spreen, 1980). The model can thus be used to simulate the effects of different kinds of policies and choose the outcome that best conforms with the decision-makers' preferences. In this transformation, the objective function becomes one of maximizing the joint consumer and producer surplus subject to technical and resource constraints of producers and market clearance conditions.

Interest in sector modeling dates back to the 1960s where they are used by governments and international aid donors as a tool to capture the multimarket and multiregional effects of policy changes. A known example of the sectoral model solved as a mathematical programming problem is the Egyptian model, which addressed the issues of water shortages of the Nile River and irrigation water productivity (Hazell and Norton, 1986).

More recent models put greater emphasis on the environment and explicitly include in the model the environment concerns that are raised by intensive agricultural production. Environment issues range from soil erosion and water salinity to ozone control and climate change. Two large models, which encompass many states (USA) or many countries (EU) and which addresses these environment issues arepresented later.

The US agricultural sector and greenhouse gas mitigation model (ASMGHG) (Schneider and McCarl, 2000) is developed to analyze the potential of US agriculture to mitigate greenhouse gas emissions. It is based on an agricultural sector model (Chang and McCarl, 1991), augmented by an envionmental component to analyze greenhouse gas emission mitigation. The environment component is linked to the agricultural component by estimating the relationship between agricultural crop and livestock management and associated levels of gas emissions. The emission data for livestock techologies are based on Environmental Protection Agency (EPA) and Inter Panel of Climate Change estimates whereas the emission data for crop production activities are based on crop growth simulation models such as Spatial Erosion Productivity Impact Calculator (EPIC). ASMGHG depicts production, consumption, and international trade in 63 US regions of 22 traditional crops, 29 animal products, and more than 60 processed agricultural products. Environmental criteria include emissions or absorption of carbon dioxide, methane, and nitrous oxide, groundwater pollution, and soil erosion. The model solves for commodity and factor prices, levels of production, exports and imports, greenhouse gas emissions management strategy adoption, resource usage, and environmental impact indicator. The model is used to simulate the effects of carbon prices on the role of each greenhouse gas mitigation strategy. The crop strategies concern changes in the crop mix, tillage, fertilization, irrigation, afforestation, and biofuel. The livestock management options involve herd size, liquid manure system alteration on hog and dairy farms, and stocker/feedlot production system adoption.

In Europe, a EU-wide economic modeling system, the Common Agricultural Policy Regional Impact (CAPRI) project was developed to analyze the regional economical and environmental impacts of the common agricultural policy (Heckelei and Britz, 2001). The modeling system is set to simultaneously analyze the effects of commodity market and policy developments on agriculture in the individual regions as well as the feedback from the regions to EU and world markets. The model is split into a supply and market demand. The supply module consists of individual programming models of 200 regions of the EU, whereas the market module follows the tradition of multicommodity models. Based on aggregate supply quantities from the regional models, the market model returns market-clearing prices. An iterative process between the supply and market component achieves a comparative static equilibrium. The supply module uses a combination of the Positive Mathematical Programming and the Maximum Enthropy approach to calibrate the model to observed data and produce empirically validated and plausible supply responses. The data set comprises a set of environmental indicators, which represent a direct link to the agricultural production systems. The main environmental issues considered are the nutrient balances and gas emissions for global climate changes in all regions in the system.

8 FUTURE RESEARCH TRENDS

The most important improvements observed in relation to agri-environmental modeling are the theory-based approach combined with detailed analytical applications. This is undertaken by linking the MCDM approach to utility theory and unifying the approach in terms of techniques. Farm models are more detailed and integrate labor and equipments management aspects. One of the serious problems still facing policy makers is the asymmetric information. For any policy to be efficient, thus there is a need to tackle this issue. Future models will be more data demanding and need more skilled mathematical modelers. This tendency is to be accelerated in the future due to the need to integrate environmental uncertainty and inexact information. The integration of environmental uncertainity would imply the employment of methods such inexact fuzzy multi-objective programming and interval goal programming. If long term environmental impacts have to be considered, stochastic recursive and dynamic MCDM models will be needed. Consequently, the level of human skills required has to be adjusted to fit the future demand.

The implementation of Web-SDSS for water management and mainly for water markets and water transfer between regions will be of great importance in the future. This is a group decision-making problem, which involves sellers, buyers, and administration as well as third parties who may be injured by the water transfer and have the right to oppose it. New research trends point to the need for considering the spatial distribution of such benefits as well as the impacts on local economies and on social groups. Weights should be attributed to third parties, that is, social groups and stakeholders in local economies. The Web-based approach has the advantage of keeping all parties informed and updated. Computer scientists, on the other hand, have to make it easy to understand how the input of multiple users is stored, combined, and discussed (Rinner, 2003).

The extension and advice activities are becoming one of the public and private services, where Spatial interactive DSS will be increasingly demanded in the future. Multiple criteria practitioners are called on to participate in this effort to ensure a larger dissemination of their knowledge. SDSS may be combined with weather forecast jointly with stochastic sequential mathematical models to help improve chemicals and water management at farm level and account for labor and equipment reorganization. One of the important requirements for a better future of agri-environmental modeling is the focus on the feasibility of policy instruments. Otherwise, models are built for the sake of modeling and will not have any impact on real-life problems. Cost effectiveness of pollution control is also a major economic objective which should be considered in agri-environmental modeling (Veith, 2002).

REFERENCES

Agrell, P.J., Stam, A., Fisher, G.W. 2004. Interactive multiobjective agro-ecological land use planning: The Bungoma Region in Kenya. *Eur. J. Oper. Res.* 158: 194–217.

Annetts, J.E., Audsley, E. 2002. Multiple objective linear programming for environmental farm planning. *J. Oper. Res. Soc.* 53: 933–943.

Arondel, C., Girardin, P. 2000. Sorting cropping systems on the basis of their impact on groundwater quality. *Eur. J. Oper. Res.* 127: 467–482.

Arzt, K. 2003. How to cope with agricultural–environmental problems? “Round tables” and their potentials to solve and avoid environmental problems in the field of agriculture. In *Frontiers 2: European Applications in Ecological Economics*. Tenerife, Spain, February 12–15, 2003.

Campus, F., Tellarini, V., Andreoli, E., Silvestri, N., Mazzoncini, M. 1999. Economic and environmental impact of input reduction in biennial rotations. In *Agriculture Beyond Production*. Maria Andreoli (Editor). University of Pisa. Edizioni II Borghetto, Pisa, Italy.

Castrodeza, C., Lara, P., Pena, T. 2005. Multicriteria fractional model for feed formulation: Economic, nutritional and environmental criteria. *Agr. Syst.* 86(1): 76–96.

Chang, C., McCarl, B. 1991. *Scope of ASM: The US Agricultural Sector Model.* Working paper. Texas Agricultural Experiment Station, Texas A&M University.

Dillon, C.R. 1992. Economic Analysis of Edwards Aquifers Water Management. Ph.D. Dissertation. Texas AM University. Texas, USA.

Falconer, K., Hodge, I. 2001. Pesticide taxation and multi-objective policy making: Farm modelling to evaluate profit/environment trade-offs. *Ecol. Econ.* 36: 263–279.

Feinerman, E.D., Bosh, D., Pease, J. 2004. Manure applications and nutrient standards. *Am. J. Agr. Econ.* 86(1): 14–25.

Gandonou, J.M., Dillon, C. 2003. *Precision Agriculture, Whole Field Farming and Irrigation Practices: A Financial Risk Analysis.* Paper selected for the American Agricultural Economics Association, Montreal, Canada, July 27–30, 2003. www. sweb.uky.edu/~jgand0/ Research.htm

Giasson, E., Bryant, R.B., Bills, N.L. 2002. Environmental and economic optimization of dairy manure management: A mathematical programming approach. *Agron. J.* 94: 757–766.

Giupponi, C., Cogan, V., La Jeunesse, I. 2002. *EU Water Policy: Research Developments and New Management Tools.* Working paper WP02-13. Center for International Food and Agriculture Policy, University of Minnesota.

Gomez-Limon, J.A., Riesgo, L. Arriaza, M. 2003. Multicriteria analysis of factors use level: The case of water for irrigation. Paper presented for the *25th International Conference of Agricutural Economists*, August 16–22, 2003, Durban, South Africa.

Hajkowicz, S. Prato, T. 1998. *Multiple Objective Decision Analysis of Farming Systems in Goodwater Creek Watershed.* University of Missouri, Columbia. CARES Research Report #24.

Hayashi, K. 2000. Multicriteria analysis for agricultural resource management: A critical survey and future perspectives. *Eur. J. Oper. Res.* 122: 486–500.

Hayashi, K. 2003. A multi-objective programming approach for evaluating agri-environmental policy. In *Multi-Objective Programming and Goal Programming: Theory and Applications.* Tanino, T., Tanaka, T., and Inuiguchi, M. (Editors). Springer-Verlag, Heidelberg, pp. 333–338.

Hazell, P., Norton, R. 1986. *Mathematical Programming for Economic Analysis in Agriculture*. The Johns Hopkins University Press, New York.

Heckelei, T., Britz, W. 2001. *Concept and Explorative Application of an EU-wide, Regional Agricultural Sector Model* (*Capri-Project*). Institute of Agricultural Policy, University of Bonn, Bonn.

Hoag, D., Page, S., Ascough, II, J. 1999. Using the analytical hierarchy process to determine economic and environmental tradeoffs on farms. Paper presented to the *2nd International Conference on Multiple Objective Decision Support Systems for Land, Water and Environmental Management*, MODSS'99.

Horan, R.D., Ribaudo, M., Abler, D.G. 2001. Voluntary and indirect approaches for reducing externalities and satisfying multiple objectives, Chapter 3. In *Environmental Policies for Agricultural Pollution Control.* Shortle, J.S. and Abler, D. (Editors). CAB International, UK.

Janssen, R., Goosen, H., Verhoeven, M.L., Verhoeven, J.T.A., Omtzigt, A.Q.A., Maltby, E. 2005. Decision support for integrated wetland management. *Environ. Modell. Softw.* 20(2): 215–229.

Jensen, A.L., Thysen, I., Boll, P.S., Hansen, J.G., Secher, B.J.M., Juhl, O. 1997. *Pl@nteInfo – Using the Internet for Custom Tailored Crop Information.* www.planteinfo.dk/information/ publikationer/efita97.

Jean dit Balleul, P., Rivest, J., Dubeau, C., Pomar, C. 2001. Reducing nitrogen excretion in pigs by modifying the traditional least-cost formulation algorithm. *Livest. Prod. Sci.* 72: 199–211.

Jones, A.C., El-Swaify, S.A., Graham, R., Stonehouse, D.P., Whitehouse, I. 1998. A synthesis of the state-of-the-art on multiple objective decision making for managing land, water and the environment. In *Multiple Objective Decision Making for Land, Water, and Environmental Management*. El-Swaify, S.A. and Yakowitz, D.S. (Editors). Lewis Publishers, Boca Raton, pp. 719–729.

Jones, D., Barnes, E.M. 2000. Fuzzy composite programming to combine remote sensing and crop models for decision support in precision crop management. *Agr. Syst.* 65: 137–158.

Latacz-Lohmann, U. 2004. Dealing with limited information in designing and evaluating agri-environmental policy. Paper presented in *90th Seminar of the European Association of Agricultural Economists*. Rennes, France, October 28–29, 2004.

Lauwers, L., Van Huylenbroeck, G., Martens, L. 1998. A systems approach to analyse the effects of flemish manure policy on structural changes and cost abatement in pig farming. *Agr. Syst.* 56(2): 167–183.

Linares, P., Romero, C. 2002. Aggregation of preferences in an environmental economics context: A goal programming approach. *Omega* 30: 89–95.

Loyce, C., Rellier, J.P., Meynard, J.M. 2002. Management planning for winter wheat with multiple objectives (2); ethanol–wheat production. *Agr. Syst.* 72: 33–57.

MacKenzie, D. 2004. *WTO Ruling may Spell End of Farmer's Subsidies*. New Scientist.Com www.newscientist.com/hottopics/pollution.

McCarl, B.A., Spreen, T.H. 1980. Price endogenous mathematical programming as a tool for sector analysis. *Am. J. Agr. Econ.* 62: 87–102.

McCarl, B.A. 1992. *Mathematical Programming for Resource Policy Appraisal under Multiple Objectives*. Working paper #6. Texas A&M University, USA.

McMaster, G.S., Ascough, J.C., Shaffer, M.J., Byrne, P.F., Haley, S.D., Neilsen, D.C., Andales, A.A., Dunn, G.H., Weltz, M.A., Ahuja, L.R. 2002. *Parameterizing GPFARM: An Agricultural Decision Support System for Integrating Science, Economics, Resource Use and Environment.* www.iemss.org/iemss2002/proceedings/pdf/volume20uno/293_mcmaster.pdf

Meyer-Aurich, A., Matthes, U., Osinki, E. 2001. Integrating sustainability in agriculture: Trade-offs and economic consequences demonstrated with a farm model. Paper presented to the *American Agricultural Economics Association Annual Meeting* 2001.

Morari, F., Lugato, E., Borin, M. 2004. An integrated non-point source model-GIS system for selecting criteria of best management practices in the Po Valley, North Italy. *Agr. Ecosyst. Environ.* 1002: 247–262.

Munda, G. 2004. Social multi-criteria evaluation: Methodological foundations and operational consequences. *Eur. J. Oper. Res.* 158: 662–667.

OECD. 2001. *Environmental Indicators for Agriculture. Volume 3. Methods and Results.* Paris.

O'Connor, M. 2000. Pathways for environmental evaluation: A walk in the (Hanging) Gardens of Babylon. *Ecol. Econ.* 34: 175–193.

Pacini, C., Giesen, G., Wossink, A., Omodei-Zorini, L., Huirne, R. 2004. The EU's Agenda 2000 reform and the sustainability of organic farming in Tuscany: Ecological-economic modelling at field and farm level. *Agr. Syst.* 80: 171–197.

Psychoudakis, A., Aggelopoulos, S., Dimitriadou, E. 2002. Agricultural land use in an environmentally sensitive area: An assessment of an agri-environmental policy measure. *Journal of Environmental Planning and Management*. 45(4): 481-491.

Ridier, A., Jacquet, F. 2002. Decoupling direct payments and the dynamics of decisions under price risk in cattle farms. *J. Agr. Econ.* 53(3), 549–565.

Rinner, C. 2003. Web-based spatial decision support: Status and research directions. *J. Geogr. Inf. Decision Anal.* 7(1), 14–31.

Romero, C., Rehman, T. 1987. Natural resource management and the use of multiple criteria decision-making techniques. *Eur. Rev. Agric. Econ.* 14(1), 76–89.

Romero, C. 2001. Extended lexicographic goal programming: A unifying approach. *Omega* 29: 63–71.

Romero, C., Rehman, T. 2003. *Multiple Criteria Analysis for Agricultural Decisions*, Second Edition, Elsevier, Amsterdam.

Romero, C. 2004. A general structure of achievement function for a goal programming model. *Eur. J. Oper. Res.* 153: 675–686.

Schneider, U., McCarl, B. 2000. *The Agricultural Sector and Greenhouse Gas Mitigation Model (ASMGHG)*. Unpublished paper, Department of Agricultural Economics, Texas A&M University.

Stagl, S. 2003. Multicriteria evaluation and public participation: In search for theoretical foundations. Paper presented in *Frontiers 2: European Applications in Ecological Economics*. Tenerife, Spain, February 12–15, 2003.

Stokes, J.R., Tozer, P.R. 2002. Cost minimization and managing soil nutrient loading: Conflict or compromise? *Can. J. Agr. Econ.* 50, 151–169.

Stonehouse, D.P., de Vos, G.W., Weersink, A. 2002. Livestock manure systems for swine finishing enterprises. *Agr. Syst.* 73: 279–296.

Tulloch, D.L., Myers, J.R., Hasse, J.E., Parks, J.P., Lathrop, R.G. 2003. Integrating GIS into Farmland Preservation Policy and decision making. *Landscape Urban Plan.* 63: 33–48.

Van Huylenbroeck, G., Jacobs, G., Vanrolleghem, P. 2000. A simulation model to evaluate the impact of environmental programmes on dairy farms. *Int. Trans. Oper. Res.* 7: 171–183.

Veith, T.L. 2002. *Agricultural BMP Placement for Cost-Effective Pollution Control at the Watershed Level.* Ph.D. Dissertation. Virginia Polytechnic Institute and State University. www.scholar.lib.vt.edu/theses/available/etd-04252002-132437/unrestricted/etd.PDF

Vickner, S., Hoag, D. 1998. Advances in ration formulation for beef cattle through multiple objective decision support systems. In *Multiple Objective Decision Making for Land, Water, and Environmental Management.* El-Swaify, S.A. and Yakowitz, D.S. (Editors). Lewis Publishers, Boca Raton, pp. 291–298.

Wang, X., Yu, S., Huang, G.H. 2004. Land allocation based on integrated GIS-optimization modeling at watershed level. *Landscape Urban Plan.* 66: 61–74.

Westra, J.V., Easter, K.W, Olson, K.D. 2002. Targeting Nonpoint Source Pollution Control: Phosphorus in the Minnesota River Basin. *Journal of the American Water Resources Association.* 38(2): 493–505.

Woodward, R. 2003. Lessons about effluent trading from a single trade. *Rev. Agric. Econ.* 25 (1): 235–245.

Yaldir, A.K., Rehman, T. 2002. A methodology for constructing multicriteria decision support systems for agricultural land consolidation using GIS and API: An illustration from Turkey. *Comput. Electron. Agr.* 36:55–78.

Zeng, G., Duan, S., Hu, H. 2004. On precision agriculture and the construction of agriculture sustainable-developing ability in China. Paper presented in the *AFITA/WCCA Joint Congress on Information Technology in Agriculture.* www.afitaandwcca2004.net/html/paper/2-On_Precision.pdf.

Chapter 6

MCDM FARM SYSTEM ANALYSIS FOR PUBLIC MANAGEMENT OF IRRIGATED AGRICULTURE

José A. Gómez-Limón[1], Julio Berbel[2], and Manuel Arriaza[3]
[1]*University of Valladolid, Spain;* [2]*University of Cordoba, Spain;* [3]*Instituto Andaluz de Investigación y Formación Agraria, Pesquera y Alimentaria, Spain*

Abstract In this chapter we present a methodology within the multi-criteria paradigm to assist policy decision-making on water management for irrigation. In order to predict farmers' response to policy changes a separate multi-attribute utility function for each homogeneous group, attained applying cluster analysis, is elicited. The results of several empirical applications of this methodology suggest an improvement of the ability to simulate farmers' decision-making process compared to other approaches. Once the utility functions are obtained the policy-maker can evaluate the differential impacts on each cluster and the overall impacts in the area of study (i.e. a river basin) by aggregation. On the empirical side, the authors present some studies for different policy instruments including water pricing, water markets, modernization of irrigation systems and a combination of them.

Keywords: Multi-Attribute Utility Theory, Water management, Irrigation, Policy analysis

1 INTRODUCTION: IRRIGATED AGRICULTURE AND THE MCDM PARADIGM

Since irrigated agriculture is simply a type of agriculture, the application of most of the literature on multicriteria decision-making (MCDM) for agricultural systems is straightforward (see Romero and Rehman, 2003). However, there are some special characteristics related to the farmers' decision-making processes:

1. The availability of water for irrigation allows farmers to obtain higher yields and the possibility of growing a larger amount of crops. Thus,

within this productive framework, the farmers' decision-making process in irrigated agriculture is more complex than that in rainfed farming.
2. Water is not merely an input of irrigated agricultural systems but also a scarce natural resource with alternative destinations: human consumption, the general environment, industry and agriculture. Therefore, water allocation policies are of decisive importance in terms of economic efficiency, territorial equilibrium and social equity.
3. Irrigated agriculture consumes much more inputs (labour, chemicals, machinery, etc.) than its rainfed counterpart. This results in the intensification of externalities, both positive and negative, which results in policy implementation conflicts between the farmers on one side (who primarily seek to maximize profit and reduce risk) and the public sector on the other (minimizing environmental impact, maximizing rural employment, etc.).

Furthermore, the qualitative importance of irrigation on agriculture is clearly reflected in its contribution to agricultural world production: although only 18% of the world agricultural land (250 million ha) is under irrigation, irrigated agriculture accounts for 80% of global water consumption (3000 km^3/year) and produces 43% of the world's food supply (more than 50% in monetary terms), according to official statistics (FAO, 2000).

All the above observations justify the proliferation in recent years of scientific papers by agricultural and resources economists, as well as civil engineering studies. Many experts in these fields have opted for MCDM as the methodological guide to analyse the agricultural systems. This is why a chapter in this book devoted to the application of MCDM to irrigation is also justified.

The aim of this chapter is thus to present a suitable methodology for guiding the decision-making process of the authorities regarding efficient water management for irrigation, subject to economic, environmental and social sustainability. To achieve this end, the authorities have a wide variety of policy instruments for agriculture (subsidies, tariffs, etc.) and water (pricing, markets, etc.). However, given the multidimensional implications of welfare optimization, more traditional approaches (e.g. cost–benefit analysis) may be overwhelmed by the complexity of the decision-making process. This work supports the use of multicriteria techniques to simulate policy scenarios, in particular, the integration of results in a multiattribute utility function that ranks all the alternatives according to the preferences of society, enabling us to determine in advance the suitability of individual policy instruments.

This chapter is organized into six sections. Section 2, following the introduction, presents a review of the literature on MCDM as applied to irrigated agriculture. Section 3 highlights some challenges faced by these simulation techniques to become a useful tool in policy decision-making. The methodological contribution required to meet this objective is outlined in Section 4. Section 5 analyses some empirical applications following this approach. Finally, we draw some conclusions in Section 6.

2 IRRIGATED SYSTEMS, DECISIONS, DECISION-MAKERS AND DECISION CRITERIA

2.1 Irrigated Systems and Decision-Making

As pointed out above, there are numerous empirical applications of MCDM techniques to analyse irrigated agricultural systems. We can classify these into three levels of aggregation: river basin, irrigated area and farm. For each level, the type of problem analysed and the approach selected have been different.

The first use of MCDM techniques, beginning in the 1970s, corresponded to the river basin level. In most cases, the problems analysed were related to water use planning: investment appraisal, water allocation to various economic sectors, and within the agricultural field, to irrigation areas and crops. More recently, as environmental regulations have become stricter, many studies have focused on conflicting environmental, economic and social criteria in these particular agricultural systems.

From a methodological point of view, most MCDM techniques are covered in these studies. It is also important to note that in all of them the basin authority is the only decision-maker and that it seeks to maximize the benefits to society as a whole through its decisions. In this sense, the public criteria can be categorized as follows:

- *Economic development*: Economic efficiency, national economic development (growth rates of national income, inflation) and regional economic development (direct income, territorial equity, market development).
- *Social welfare*: Social equity (employment level, income redistribution) and self-sufficiency in food production.
- *Environmental protection*: Water quality impacts control (nitrogen and phosphorus discharges, increasing biological oxygen demand load, groundwater level), control of soil quality impacts (salinization, erosion), other

ecological impacts (biodiversity, energy balance) and reservoir safety (sediment, flood impact on dams).

In a multi-objective analytical framework, the objective function to be considered in such problems is usually defined in such a way as to simultaneously maximize economic development and social welfare and to minimize environmental impacts, considering the institutional framework, and the social, physical, economic and environmental limitations included in the set of constraints.

At the second level of aggregation, the irrigation area in most cases we find again the public sector as the sole decision-maker. There are studies on water dosage and optimum crop distribution, irrigation technology, irrigation schedules and environmental problems. As in the studies at basin level, the public criteria in the irrigation areas include economic (profitability of crops, cost of irrigation systems, etc.), social (equity or rural employment) and environmental aspects (water volume, water quality after irrigation, land capability/suitability, efficiency of water usage, resistance to floods or droughts, energy balance, etc.), applied to relatively homogeneous geographical areas.

At farm level, most of the MCDM applications focus on crop-mix optimization, following, unlike in the other two levels of aggregation, private criteria: profits (level of income and costs), risk avoidance or farm labour (as a proxy for farmer's leisure time). For further details see Hayashi (2000).

2.2 Normative models versus descriptive models

Decision models of irrigated agricultural systems show a clear distinction between the most frequent normative and a very small number of descriptive models. Normative models, however, are less favoured for use as a centrally planned approach; nowadays economists place greater importance on the decisions made by private economic agents. In this context, therefore, normative optimum solutions are rarely achieved by any society. We cannot conclude from the previous statement that normative solutions are pointless. On the contrary, they show the potential of agricultural systems to satisfy the needs of society. It follows that policy instruments should be selected on the basis of inducing those farmers' responses, on an aggregated level, that are as close as possible to the normative solutions. Descriptive models, however, may help us to arrive at better explanations (backward use) and predictions (forward use) of farmers' responses to policy changes.

In order to develop descriptive models, neo-classical economic theory supports the single-objective maximization behaviour of economic agents. Nevertheless, it has frequently been observed that the optimum solution of models developed within this theoretical framework do not seem to adequately match the observed behaviour of producers, which suggests that there is a need for more complex models capable of providing more accurate results.

A number of studies have rejected the hypothesis that farmers seek to maximize profits only, arguing that producers seek to optimize a broader set of objectives such as the maximization of leisure time, the minimization of management complexity and working capital, and so on. In this context, we may mention recent studies by Willock *et al.* (1999), Costa and Rehman (1999), Solano *et al.* (2001) or Bergevoet *et al.* (2004). The implication is clear: when modelling farmers' decision-making processes (building models capable of simulating farmers' behaviour) it is essential to take more than one criterion into account.

Therefore, it is necessary to put forward more realistic hypotheses based on the psychology of decision-makers. One alternative, the one proposed in this work, tackles the MCDM decision-making problem via multi-attribute utility theory (MAUT). This advances a set of descriptive models that assume optimizing behaviour on the part of the farmer and present a mathematical formulation of his or her preferences in a multicriteria context; that is, a multi-attribute utility function (MAUF), as we explain in the following section.

3 FRAMEWORK FOR MODELLING IRRIGATED AGRICULTURE SYSTEMS

Most irrigated agriculture is located in economies that are characterized by growing demand for water, a limited long-term supply, increasing operating costs of storage and distribution, growing competition among regions for alternative uses and rising environmental problems (negative externalities). However, the whole question is more a problem of water management and inefficiencies than an input shortage (Randall, 1981).

In order to partially overcome these limitations, water policies have shifted from an exclusively supply-side approach towards a more integrated analysis that includes demand-side policies. In this context, water policies aim to allocate this natural resource according to socio-economic efficiency criteria via three main policy instruments: water pricing, implementation of water markets and subsidies to improve the technical efficiency of the distribution infrastructure.

In the search for water allocation efficiency, one of the first initiatives to be taken was the transfer to the producer of part of the total cost of providing water. The second instrument, the implementation of water markets, may help to improve this allocative efficiency in a decentralized manner, as well as reducing the effects of water scarcity. Finally, we have the provision of subsidies to modernize the distribution infrastructure and the irrigation systems in the farms. The suitability of each instrument of water policy depends on the social, economic and environmental impact upon the agricultural systems, via the farmers' responses. Therefore, if we aim to build functional simulation models for the regional or national authorities, all these three aspects need to be considered in the analysis.

Furthermore, it is worth noting that in addition to environmental (water) policy on irrigation there is another crucial policy to be considered: the agricultural policy. In this respect, agricultural policies have evolved to meet both internal (e.g. environmental concerns, budget limitations, bio-technological advances, etc.) and external demands (e.g. agricultural market liberalization). In any case, it is important to provide these types of models to assist policy-makers in their assessments of the adequacy of alternative instruments.

Within this social and political framework, the methodological framework proposed to achieve the goal set out in this chapter (modelling irrigated systems for policy decision support) is displayed graphically in Fig. 1. This methodological outline is based on five stages, which are further explained in the next section.

4 METHODOLOGICAL APPROACH

4.1 Farm Typology Definition

Modelling agricultural systems at any level other than that of the individual farm introduces the problem of aggregation bias. The introduction of a set of farms in a unique programming model overestimates the mobility of resources among production units, allowing combinations of resources that are not possible in the real world. The final result of these models is that the

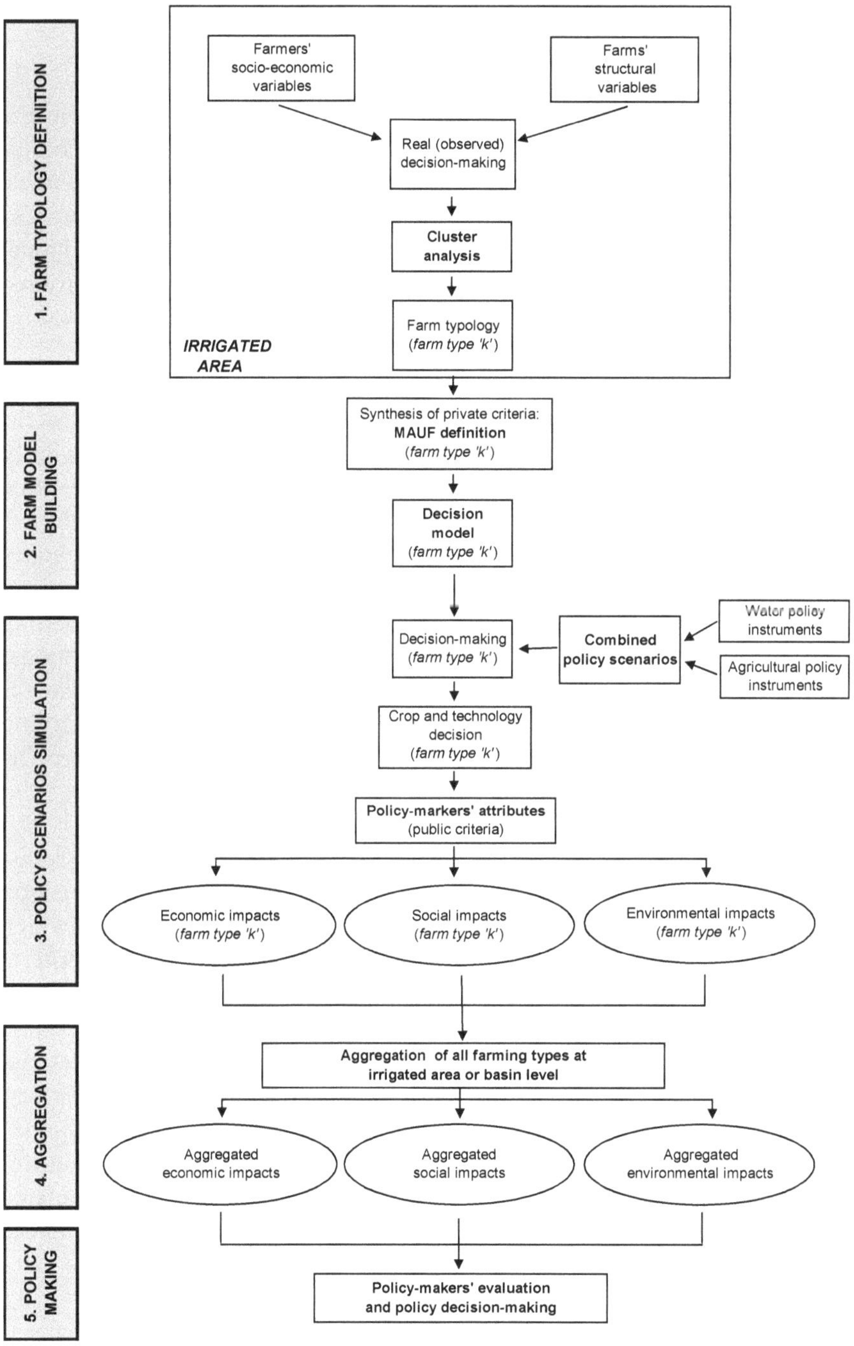

Figure 1. Methodology diagram.

value obtained for the objective function is biased upward and the values obtained for decision variables tend to be unachievable in real life.

This aggregation bias can only be avoided if the farms included in the models fulfil certain criteria regarding homogeneity (Day, 1963): technological homogeneity (same possibilities of production, same type of resources, same technological level and same management capacity), pecuniary proportionality (proportional profit expectations for each crop) and institutional proportionality (availability of resources to the individual farm proportional to average availability).

This requirement of homogeneity brings us to consider the irrigated unit as the basic element to be analysed. These units are relatively small areas (ranging usually from 1,000 to 20,000 ha) that can be regarded as fairly homogeneous in terms of soil quality and climate, and in which the same range of crops can be cultivated with similar yields. Furthermore, the set of farms that comprise each of these agricultural systems usually operates the same technology at a similar level of mechanization. Moreover, given efficient capital and labour markets, the constraints included in modelling this system have been limited to agronomic requirements (crop rotations) and the restrictions imposed by the agricultural policy, which are similar for all farms. All these facts allow us to assume that the requirements regarding technological homogeneity, pecuniary proportionality and institutional proportionality are basically fulfilled.

We might thus conclude that particular irrigated areas could be modelled by means of a unique linear programme with relatively small aggregation bias, but this is based upon the assumption that the sole criterion on which decisions are based is profit maximization. When we adopt a multi-criteria perspective, a new homogeneity requirement emerges if we wish to avoid aggregation bias; viz., *homogeneity related to choice criteria.* We assume that decision criteria are primarily based on the psychological characteristics of decision-makers, which differ significantly from farmer to farmer. According to this perspective, the differences in decision-making (crop mix) among farmers in the same production area must be primarily due to differences in farmers' objective functions, rather than other differences related to farm characteristics regarding either crop profits or disparities in resources requirements or endowments.

In order to avoid aggregation bias resulting from lumping together farmers with significantly different objective functions, a classification of farmers into homogeneous groups with similar decision-making behaviour (objective functions) is required. As Berbel and Rodríguez (1998) have pointed out, we can assume that in a homogeneous area (irrigated unit), differences in the crop mix are mainly caused by farmers' different management criteria (utility functions) rather than by other constraints such as land quality, capital, labour or water availability. Thus, following these authors, the observed area (as percentages) devoted to individual crops, considered as proxies for the real criteria, can be used as classification variables to group farmers using the cluster technique.

Finally, it is worth noting that the homogeneous groups obtained in this way can be regarded as 'fixed' in the short and medium terms. As noted above, the decision criteria are based on psychological features of the decision-makers, which is why they may be regarded as structural characteristics of producers. These psychological features, and thus the criteria, are unlikely to change in the near future. This means that the selection variables chosen allow farmers to be grouped into clusters that are robust to changes in the policy framework. In other words, once the homogeneous groups of producers have been defined for actual data (crop mix), we can assume that all elements (farmers) of each group will behave in a similar way if policy variables change.

4.2 Farm type Model-Building

As Fig. 1 shows, the first stage of the modelling process produces a set of homogeneous groups of farmers, whereas the second builds the mathematical models for each farm type, consisting of a specific multi-criteria model at farm-type level. This enables independent simulations to be run based on the decision-making behaviour of the various groups of farmers. For this purpose, the basic elements of any mathematical model; that is, decision variables, objective function and set of constraints need to be outlined.

While the choice of crop areas as a decision variable does not cause any problem (observing crop diversity in the area studied is sufficient), the objective function and constraints require more detailed analysis.

4.2.1 Multi-Attribute Utility Theory Approach

In view of the evidence on how farmers take decisions while trying to simultaneously optimize a range of conflicting objectives (see Section 2.2), we propose MAUT as the theoretical framework for the MCDM programming to be implemented. The aim of MAUT is to reduce a decision problem with multiple criteria to a cardinal function that ranks alternatives according to a single criterion (Keeney and Raiffa, 1993). Thus, the utilities of n attributes are captured in a quantitative way via a utility function, mathematically, $U = U(r_1, r_2, ..., r_n)$, where U is the MAUF and r_i are the attributes regarded by the decision-maker as relevant in the decision-making process.

In spite of the interest of developing the analysis from the above expression, the main drawback to this approach comes from the difficulty of eliciting the MAUF (Hardaker *et al.*, 1997, p. 162). In order to simplify this process, some assumptions need to be made about the mathematical features of the utility function.

Fishburn (1982) and Keeney and Raiffa (1993) have explained the mathematical requirements for the assumption of an *additive* utility function. From a practical point of view, the basic condition that needs to be satisfied is that the attributes considered r_i should be mutually utility-independent. Although this condition is somewhat restrictive, Edwards (1977), Farmer (1987), Huirne and Hardaker (1998) and Amador *et al.* (1998) have shown that the additive utility function yields extremely close approximations to the hypothetical true utility function even when these conditions are not satisfied. For this reason, additive utility functions for modelling farmers' behaviour have been widely employed.

Given this justification for the use of the additive utility function, we take the further step of assuming that the individual attribute utility functions are *linear*. Hence, the MAUF expression becomes:

$$U = \sum_{i=1}^{n} w_i r_i . \qquad (1)$$

This expression implies linear utility-indifferent curves (constant partial marginal utility), a rather strong assumption that can be regarded as a close enough approximation if the attributes vary within a narrow range (Edwards, 1977 and Hardaker *et al.*, 1997, p. 165). There is some evidence for this hypothesis in agriculture. Thus, Huirne and Hardaker (1998) have shown how the slope of the single-attribute utility function has little impact on the ranking of alternatives. Likewise, Amador *et al.* (1998) analysed how linear and quasi-concave functions yield almost the same results. We therefore adopt this simplification in the elicitation of the additive utility function. Thus, MAUFs with this shape may be regarded as objective functions for the different farm-type models.

4.2.2 The Objective Function: MAUF Elicitation Technique

To estimate the relative weightings w_i we select a methodology that avoids the necessity of interacting directly with farmers, and in which the utility function is elicited on the basis of the revealed preferences implicit in the real values of decision variables (i.e. the actual crop mix). The methodology adopted for the estimation of the additive MAUFs is based on the technique proposed by Sumpsi *et al.* (1997) and extended by Amador *et al.* (1998). It is based upon weighted goal programming. To avoid unnecessary repetition, we refer to works of these authors for details of all aspects of this multi-criteria technique. Here, we wish only to point out that the results obtained by this technique are the weights (w_i) that imply utility functions that are capable of reproducing farmers' observed behaviour. As Dyer (1977) demonstrates, these weights are consistent with the following separable and additive utility functions:

$$U = \sum_{i=1}^{q} \frac{w_i}{k_i} f_i(x), \tag{2}$$

where k_i is a normalizing factor.

Applying this technique to each farm-type enables us to estimate the different objective functions in each case.

4.2.3 Model Constraints

Finally, it is worth noting that the farm-types' decision-making models need to be completed with the constraints that must be satisfied. These constraints are mainly due to the structural characteristics (climate, soil fertility, market limits, agricultural policy requirements, etc.) of the farms which are similar for all farm types in a particular irrigated area. Only slight differences could be fixed by clusters (farm size, production quotas, etc.) according to the data obtained in the farm survey implemented for primary data gathering.

In sum, the descriptive decision model at farm level can be set out as follows:

$$\begin{aligned} &\text{Max}\, U = \sum_{i=1}^{q} \frac{w_i}{k_i} f_i(x) \qquad (3)\\ &\text{s.t.} \quad g_j(x) \le b_j \ \forall j, \end{aligned}$$

where $g_j(x) \leq b_j$ represents the set of constraints applied to each group (cluster) of farmers (land, rotational, market and agricultural policy constraints, etc.).

4.3 Simulation of Policy Scenarios

4.3.1 Definition of Policy Scenarios

The third stage of the methodology proposal simulates the policy scenarios. For this purpose the scenarios must already have been defined. Here it is essential to clearly identify the instruments to be implemented, both in a qualitative and quantitative sense. For example, in case of water pricing, the control method employed should be clarified: for example, per cubic metre, mixed (volume and irrigated hectare), by blocks, and so on, in addition to the price.

4.3.2 Simulation of Farm-Type Behaviour

Once we have established the policy scenario to be analysed, the farm-type models should modify the decision variables and parameters as appropriate. At this point, it is necessary to address certain issues.

It is worth pointing out that the estimates of the *utility functions* have been obtained by farm models that have been fed with data gathered for the current situation. In doing so we assume that the utility functions obtained at this point can be regarded as a structural feature of each cluster. As these objective weightings are the result of the farmers' own attitudes, it is reasonable to assume that they will remain constant in both the short and medium terms. This assumption is a key point of the methodology, since the estimated utility functions are assumed to be those that the farmers in each cluster will attempt to maximize in the future, for any scenario that they will be likely to face. This assumption is based upon the hypothesis that values reflected in the MAUF are stable characteristics of decision-makers.

Furthermore, in order to simulate the impacts of different scenarios, *decision variables* to be included in the tailored decision-making models should consider all the ways in which farmers are likely to adapt to any given policy scenario. Potential changes in the institutional framework should include at least the following:

1. Changes in the crop plans, allowing irrigated versus rain-fed (no irrigation) crops, and annual versus perennial crops. The fallow alternative (abandonment of agriculture) should also be considered.
2. Implementation of water stressing (deficit irrigation).
3. Changes in farming technology: irrigation technology (taking into account the substitution of surface irrigation by sprinklers or drip whenever possible), tillage technology, and so on.

Once the adapted models have been built, farm-type behaviour as a reaction to policy scenarios is simulated by simply running the models. This identifies the decisions likely to be taken (i.e. crop mixes and technology) by the different clusters of irrigators.

4.3.3 Policy-Makers' Attributes

The crop-technology plans obtained by the simulation models are intermediate tools for policy-makers, who are primarily interested in the values that result from the adoption of a series of public criteria (see Section 2.1). Nevertheless it is important to note that these policy criteria are attributes obtained in the simulated private decision-making process, but they are called attributes just because they do not belong to farmers' private objective function; neither are they considered as goals or constraints in simulation models.

Nowadays, the political paradigm in agriculture, as in any other industry, is to achieve *sustainability*. This global criterion may be decomposed into three main dimensions: economic, social and environmental sustainability. When modelling policy alternatives, the level of achievement of these criteria requires the use of *indicators*, as attributes obtained by the simulation models. Although there are many indicators of sustainability (Brouwer and Crabtree, 1998; OECD, 1999; Rigby *et al.* 2001), the selection among them depends greatly on the policy-makers' own preferences.

4.4 Aggregation

Albeit the particular study of the results by group of farms (differential assessment of impacts) is relevant, the policy choices are based on the aggregated analysis. Therefore we need to extend the conclusions to the area or river basin level, aggregating each weighted impact by its relative hectarage.

4.5 Policy-Makers' Decision-Making

Once the social utility function that includes all the relevant criteria has been defined, the methodology ends with the policy choices. Assessment of the alternative policy instruments is based on the value achieved in the utility function of society as a whole, in which all the public criteria considered (values reached by the selected indicators) are taken into account.

Although several MCDM techniques to attain this last step are available, the authors favour the analytical hierarchy process (AHP), developed by Saaty (1980). Its straightforwardness and the utility of the public criteria ranking are the reasons for our choice.

5 APPLICATIONS

This section presents some of the authors' empirical applications of this methodology. They include studies of water pricing, water markets and modernization of irrigation systems.

5.1 Water Pricing

One of the most ambitious applications of this methodology can be found in the European Project WADI (2000-2004). The ultimate objective of this research project is the design of a tool that can be used to assist policy-makers in pricing water, following the approval of the Water Framework Directive (2000), which obliges all State Members to use economic instruments and to recover the costs of providing water services (i.e. water pricing as a major instrument of water policy inside the EU). Specifically, the norm states that "Member States shall take account of the principle of recovery of the costs of water services, including environmental and resource costs".

As Fig. 1 explains, the models used in the WADI project are based on the definition of farming types by cluster analysis (or any other technique), based in each case on the particular farming model, and finding weights for MAUT objective functions. A first result worth pointing out is the wide differences that have been found among farmers' objective weightings. An example of these variations is shown in Table 1, which illustrates the results from some Spanish irrigated areas located in the Duero and Guadalquivir basins. In all cases crops and natural conditions are homogeneous within the irrigated unit, and variations in behaviour are mainly due to the socio-economic characteristics of individual farmers.

Table 1. MAUF weights for selected Spanish locations.

Irrigated area	**Farming model**	**Label**	**Weights**			
			Max. gross margin	**Min. variance (risk)**	**Min. total labour**	**Min. working capital**
Canales Bajo Carrión *(Duero)*	CBC1	Part-time farmers	0.33	0.67	0.00	0.00
	CBC2	Livestock Farmers	0.43	0.57	0.00	0.00
	CBC3	Small commercial farmers	0.71	0.07	0.00	0.22
	CBC4	Risk-averse farmers	0.66	0.34	0.00	0.00
Canal del Pisuerga *(Duero)*	CPI2	Risk diversification farmers	0.00	1.00	0.00	0.00
	CPI2	Young commercial farmers	0.30	0.70	0.00	0.00
	CPI3	Maize growers	0.58	0.42	0.00	0.00
Fuente Palmera *(Guadalquivir)*	FP1	Cotton growers	0.99	0.00	0.01	–
	FP2	Wheat growers	0.84	0.00	0.16	–
	FP3	Maize growers	0.96	0.00	0.04	–
	FP4	Groves growers	0.99	0.00	0.01	–

Source: WADI www.uco.es/grupos/wadi

The examples illustrated in the preceding table show variability in utility functions found by the MAUF elicitation technique, with the common feature of improving the predictive ability of models for each farm type.

An application of these models is the analysis of policy instruments. We have done this for the study of water pricing. A detailed analysis of all European case studies developed by the WADI project can be found in Berbel and Gutiérrez (2004).

In order to explain the main findings, we first comment as an example on the case study described by Gómez-Limón and Riesgo (2004), where the methodology proposed is applied to a single irrigated area in the Duero basin (Northern Spain) called '*Canal del Pisuerga*' (9,300 ha). Figure 2 shows the three different curves developed for the clusters representing different farming types in this particular area.

On the basis of these results, the authors conclude that the analysis of water pricing policy impacts clearly demonstrates that farmers display different behaviour patterns related to this natural resource. This diversity is shown by the different shapes of the demand curves for each of the clusters considered. The effects of irrigation water pricing thus vary significantly, depending on the group of farmers being considered.

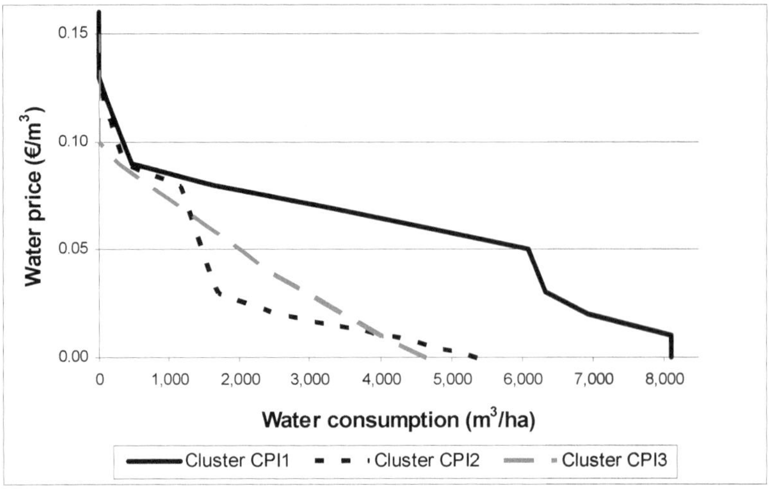

Figure 2. Demand curves for three farm types inside an irrigated area.

By aggregating the cluster results in a particular irrigated area by using the percentage of their respective agricultural areas, we can obtain the aggregated demand for the whole irrigated area. In a further step, when we aggregate different irrigated areas within a basin or region demand curve, we obtain the simulated demand curve at basin/regional level. Figure 3 shows three examples of demand curves obtained using this methodology at aggregate level in three European river basins.

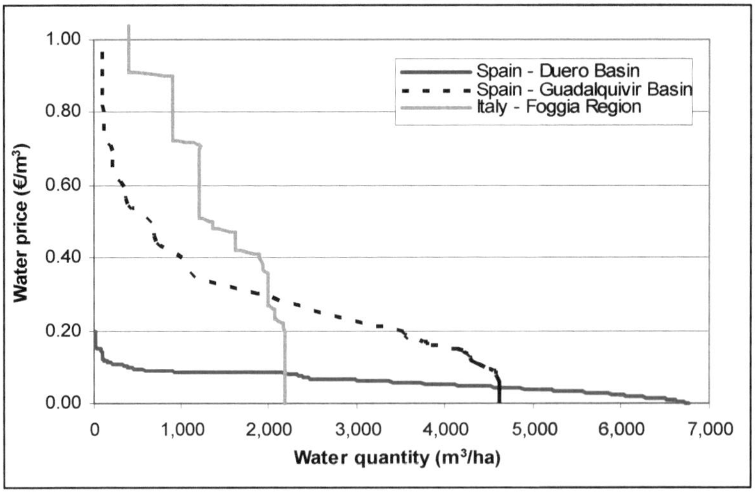

Figure 3. Aggregated demand curves for irrigation in three European river basins.

As we can see, there are large differences between basins, due to local natural conditions (soil, climate, etc.) and economic infrastructure (human resources, locational advantages, etc.), that limit farmers' decision variables and sets of constraints, and thus determine water productivity and farmers' behaviour (shape of MAUFs).

All the models were developed with the aim of deriving from farmers' production plans (crops and technology), the values of attributes that are of interest to policy-makers. These indicators used in connection with the MAUT models measure various impacts of water pricing at different levels of aggregation: economic balance (farm income, farm contribution to GDP and public support), social impact (farm employment and seasonality), landscape and biodiversity issues (genetic diversity and soil cover), water use (water consumption and irrigation technology) and nutrient and pollutant balances (nitrogen balance, pesticide risk and energy balance). For an analysis of indicator definitions and the results obtained, readers are referred to Berbel and Gutiérrez (2004).

5.2 Water Markets

The impact assessment of this water policy instrument from a multicriteria point of view is addressed in Arriaza *et al.* (2002). The authors examine an irrigated area of 54,000 ha located in Southern Spain and elicit three multi-attribute utility functions, with farm size the classifying variable used to define farm types. Two criteria are regarded as simulating the farmers' response to policy changes: the maximization of total gross margin (TGM) (as a proxy for short-term profit) and the minimization of risk (measured as the variance of the margins). The differences observed in the mathematical formulation of the utility functions support this approach to the problem by considering relatively homogeneous groups of farmers. The results achieved in this chapter show that for most price levels (water availability situations), small and medium farmers buy water from large farmers, because of the higher utility of water for the smaller farmers. Furthermore, the simulation implemented demonstrates that the volume of traded water is very small in comparison with the total amount in the market, and is less than neo-classical theory would suggest.

Assuming that it is necessary to analyse farmers' decision-making within the MCDM paradigm, it is evident that water use (allocation to different crops and/or its transfer in the market) depends on the *utility* that this input offers them (contribution to MAUF value: attaining the various objectives that farmers try to simultaneously optimize), and not only on its *productivity* (profit generation). In this respect, we believe that water market modelling is more realistic when we assume that water reallocations move this resource from the uses that generates a relatively low level of utility towards those that generate greater utility, until an equilibrium point is reached at which the marginal utilities provided by water to all users equal the market price. This utilitarian approach assumes an extension of neo-classical theory, which assumes, as a particular case, that profit maximization alone is taken into account as a unique management criterion, and that this defines market equilibrium when the value of the marginal product for all users is equal to the market price.

Dealing with public criteria achievement, the study by Arriaza *et al.* (2002) considers only two indicators. First, in order to assess the whole economic impact of the introduction of local water markets, the variations in aggregated gross margin due to selling/buying water, as a proxy of economic efficiency, were selected. In this respect, the results obtained contradict the traditional assumption of higher expected farmers' income at aggregated level following the implementation of water markets. The results show that at certain price levels there is a reduction in the economic efficiency of the system. The second indicator implemented is the use of farm labour. Here, the results suggest that the social impact of water market implementation is very limited. In fact, the total increase in farm employment under the water market scenario is insignificant.

Gómez-Limón and Martínez (2006) take a further step in this methodology, simulating a spot market for irrigation water for a whole basin. The case study analysed in this chapter is the Duero valley (78,000 km^2) in Northern Spain, where 555,582 ha are dedicated to irrigated agriculture. In the basin, a total of 21 farms in 7 irrigated areas were selected using to a cluster technique to capture the variability in farm types. Regarding the utility functions, three criteria were considered: maximization of TGM, minimization of risk (VAR) and minimization of the total labour input (TL).

In a further step towards the optimization of an individual farm type, the authors propose a mathematical programming model that simulates the market equilibrium for different scenarios of water availability, transaction costs and water prices, quantifying for each case the socio-economic impacts considered as public criteria (economic efficiency and labour demand).

the basis of the results, some interesting practical conclusions can also be drawn, the most important of which is the potential of water markets to act as a demand policy instrument to improve economic efficiency and agricultural labour demand in this basin-scale framework, particularly in periods of water scarcity. The results achieved confirm this positive impact from the economic and social points of view. These gains are due to transfers being made to those producers with more highly commercial profiles (greater weight devoted to the TGM attribute), and who enjoy greater competitive advantages (favourable soil and climate conditions) and better geographic locations (downstream).

In any case, one of the key aspects of the previous works lies in the application of a methodology that improves the ability to simulate the farmers' response to policy changes, as validation procedures suggest. Therefore, the case studies discussed here represent an interesting approach to a better understanding and modelling of water markets in the real world.

5.3 Modernization of the Irrigation Infrastructure

Regarding the modernization of the irrigation infrastructure, Riesgo and Gómez-Limón (2002) propose a similar methodology to estimate the farmers' willingness to pay (WTP) for the new irrigation technology. Within this context, WTP embeds not only the productivity increases due to the implementation of the new technology, as neo-classical theory states, but also the increase in farmers' utility. This approach has been put into practice in an irrigated area of 9,392 ha in Northern Spain (*Canal del Pisuerga*), with three relatively homogeneous groups obtained by cluster analysis, and group utility functions with four attributes: TGM, risk, farm labour and working capital.

Using the elicited utility functions, the authors obtain the water demand curves for each irrigation technology, and then the maximum farmers' WTP. For a WTP lower than the investment cost, the difference is considered as the minimum subsidy from the public sector that would be needed to adopt the new technology.

The study concludes that the WTP for water saving technologies is related to the shape of the farmers' MAUF, the technical efficiency of the technology and the water price. In particular, higher WTPs correspond to farmers who place greater weight on profit maximization. This result is consistent with an input valuation close to its marginal product value.

6 CONCLUDING REMARKS

Taking into account the evidence about how irrigators and policy-makers take their own decisions within a multi-criteria context (considering private and public criteria respectively), the first obvious conclusion is that any analysis focused on the management of irrigated agriculture ought to be developed within the MCDM paradigm.

This chapter has attempted to illustrate some aspects of the MCDM methodology as applied to the management of irrigated agriculture. The methodological approach proposed is initially descriptive, in that it tries to simulate farmers' responses to policy changes. For this purpose, in order to avoid aggregation biases, farmers are classified into homogeneous groups using cluster analysis, with the observed crop distribution being the classifying variable (proxies of farmers' utility functions). For each homogeneous group, a separate multi-attribute utility function is elicited on the basis of the weights that farmers attach to each individual objective. Next, once the objective functions have been fixed, the rest of the decision models (constraints sets) are built. Validations of the different empirical applications developed prove the worth of this modelling approach to simulating farmers' behaviour. In fact, we can affirm that the methodology proposed here improves the ability of traditional and other more recent MCDM models of simulating the farmers' responses to alternative policy instruments.

The low data requirement of this approach is also worth noticing: we need only the actual crop distribution and the mathematical formulation of each attribute with respect to the decision variables to develop these models. This is a critical point, because it allows the implementation of the methodology in the real word (excellent cost/benefit ratio: that is effort required/quality of results).

We have also proceeded to the simulation of the new water and agricultural policy scenarios in order to obtain the impacts on each cluster, and then the aggregated impacts on the area of study from an economic, social and environmental perspective. Thus, this approach allows policy-makers' decision-making to be fed with quality data regarding the multiple effects of the instruments that may potentially be implemented. We believe this is also a useful feature of the methodology proposed, offering efficient selection of policy instruments.

Although the results of the empirical applications, as well as the validation of the models, are promising in the MCDM field, there are several aspects that should be further analysed in future studies. First, the use of additive and linear MAUFs is based on rather restrictive assumptions. Thus, new developments are needed, which that will permit us to use other separable and non-separable functions, in each case without losing the simplicity and the low data requirement features of the approach presented here. Secondly, in order to support policy-makers' decisions, public decision-making models should be developed. Only in this way can the 'governance' (transparency and public debate in public decision-making processes) of these agricultural systems be improved. Thirdly, the use of descriptive model predictions, as currently proposed, is limited to short-term analysis, since we are assuming static models (no structural changes in the farms). However, there are certain prospects of overcoming this limitation using multi-period and dynamic programming, which would allow for possible developments (technological changes, farm sizes, etc.) in irrigated areas. In this respect, the study of discounting criteria other than profit (non-monetary criteria) is still an open field for research.

7 REFERENCES

Amador, F., Sumpsi, J. M., and Romero, C., 1998, A non-interactive methodology to assess farmers' utility functions: an application to large farms in Andalusia, Spain, *Eur. Rev. Agr. Econ.* 25(1):92–109.

Arriaza, M., Gómez-Limón, J. A., and Upton, M., 2002, Local water markets for irrigation in Southern Spain: a multicriteria approach, *Aust. J. Agr. Resour. Econ.* 46(1):21–43.

Berbel, J., and Gutiérrez, C., eds., 2004, *The Sustainability of European Irrigated Agriculture under Water Framework Directive and Agenda 2000,* Commission of the European Communities, Brussels.

Berbel, J., and Rodríguez, A., 1998, An MCDM approach to production analysis: an application to irrigated farms in Southern Spain, *Eur. J. Oper. Res.* 107(1):108–118.

Bergevoet, R. H. M., Ondersteijn, C. J. M., Saatkamp, H. W., Van Woerkum, C. M. J., and Huirne, R. B. M., 2004, Entrepreneurial behaviour of Dutch dairy farmers under a milk quota system: goals, objectives and attitudes, *Agr. Sys.* 80(1):1–21.

Brouwer, F., and Crabtree, B., 1998, *Environmental Indicators and Agricultural Policy*, CABI Publishing, Oxon.

Costa, F. P., and Rehman, T., 1999, Exploring the link between farmers' objectives and the phenomenon of pasture degradation in the beef production systems of Central Brazil, *Agr. Sys.* 61(1):135–146.

Day, R. H., 1963, On aggregating linear programming models of production, *J. Farm Econ.* 45:797–813.

Dyer, J. S., 1977, *On the relationship between Goal Programming and Multi-attribute Utility Theory*, Discussion paper 69, Management Study Centre, University of California, Los Angeles.

Edwards, W., 1977, Use of multi-attribute utility measurement for social decision making, in: D. E. Bell, R. L. Keeney, and H. Raiffa, eds., *Decisions*, John Wiley & Sons, Chichester.

FAO, 2000, *Anuario de producción*, Organización de las Naciones Unidas para la Agricultura, Roma.

Farmer, P. C., 1987, Testing the robustness of multi-attribute utility theory in an applied setting, *Decision Sci.* 18(2):178–193.

Fishburn, P. C., 1982, *The Foundations of Expected Utility*, Reidel Publishing Company, Dordrecht.

Gómez-Limón, J. A., and Riesgo, L., 2004, Irrigation water pricing: differential impacts on irrigated farms, *Agr. Econ.* 31(1):47–66.

Gómez-Limón, J. A., and Martínez, Y., 2006, Multi-criteria modelling of irrigation water market at basin level: a Spanish case study, *Eur. J. Oper. Res.* 173(1):313–336.

Hardaker, J. B., Huirne, R. B. M., and Anderson, J. R., 1997, Coping with Risk in Agriculture. CAB International, Oxon (UK).

Hayashi, K., 2000, Multicriteria analysis for agricultural resource management: a critical survey and future perspectives, *Eur. J. Oper. Res.* 122(3):486–500.

Huirne, R. B. M., and Hardaker, J. B., 1998, A multi-attribute model to optimise sow replacement decisions, *Eur. Rev. Agr. Econ.* 25(4):488–505.

Keeney, R. L., and Raiffa, H., 1993, *Decisions with Multiple Objectives: Preferences and Value Trade-Offs*, Cambridge University Press, Cambridge.

OECD, 1999, *Environmental indicators for agriculture. Issues and design*, OECD, Paris.

Randall, A., 1981, Property entitlements and pricing policies for a maturing water economy, *Aust. J. Agr. Econ.* 25(3):195–220.

Riesgo, L., and Gómez-Limón, J. A., 2002, Políticas de tarifación y de ahorro de agua en el regadío. Análisis de su aplicación conjunta, *Revista de Estudios Agrosociales y Pesqueros* 197:63–104.

Rigby, D., Woodhouse, P., Young, T., and Burton, M., 2001, Constructing a farm level indicator of sustainable agricultural practice, *Ecol. Econ.* 39(3):463–478.

Romero, C., and Rehman, T., 2003, *Multiple criteria analysis for agricultural decisions*, Elsevier Science Publishers, Amsterdam.

Saaty, T. L., 1980, The *Analytic Hierarchy Process*, McGraw, New York.

Solano, C., León, H., Pérez, E., and Herrero, M., 2001, Characterising objective profiles of Costa Rican dairy farmers, *Agr. Sys.* 67(1):153–179.

Sumpsi, J. M., Amador, F., and Romero, C., 1997, On Farmers' Objectives: A Multi-Criteria Approach, *Eur. J. Oper. Res.* 96(1), 64–71.

Willock, J., Deary, I. J., Edwards-Jones, G., Gibson, G. J., McGregor, M. J., Sutherland, A., Dent, J. B., Morgan, O., and Grieve, R., 1999, The role of attitudes and objectives in farmer decision-making: business and environmentally-oriented behaviour in Scotland, *J. Agr. Econ.* 50(2):286–303.

Chapter 7

WATER PUBLIC AGENCIES AGREEING TO A COVENANT FOR WATER TRANSFERS: HOW TO ARBITRATE PRICE–QUANTITY CLAUSES

Enrique Ballestero
Technical University of Valencia (Alcoy School)

Abstract This chapter deals with inter-basin water covenants guided by arbitration, which are feasible when the principle of interregional solidarity is politically accepted and implemented through public agencies sharing policies of common welfare and interregional efficiency. As a main objective, the stochastic arbitration model herein helps the agencies make joint decisions on quantity and price, namely, the commercial side of the agreement. The model is built by simulating a water demand and supply setting from information on random surplus in the donor basin and random deficit in the recipient, the required geographical/economic information being available in technical and statistical studies. Although environmental needs should be considered to estimate surplus in the donor basin, an efficient use of water in the recipient basin should be assured to estimate deficit in this area. Besides irrigation, the model is also applicable to urban supply projects. A numerical example highlights its easy application through tables.

Keywords: Arbitration, inter-basin water transfer, simulated demand and supply, stochastic models, water public agencies

1 INTRODUCTION

Intraregional and interregional inefficiencies in the use of irrigation water are observed not only in developing countries where about 500 million people experience water stress or scarcity but also in developed countries, the problem of improving water management and increasing efficiency being currently in the center of academic and public debate (World Bank, 1993). For even southern European regions, the survival of obsolete farming systems, the ongoing policies of allocating large water quantities among prior right farmers at low price, and above all, the limited presence of water markets for irrigation allow the continuity of water-wasting techniques such as furrow and flood irrigation, which means intraregional inefficiency. On the other hand, interregional inefficiency appears when a wet area with water surplus (after meeting all its needs and environmental constraints) does not transfer this surplus to a neighboring drought-exposed area despite the fact that the transfer project is technically, economically and ecologically viable. A significant aspect of inter-basin water transfer agreements concerns commercial clauses on price and water quantity. In some countries and environments, these commercial clauses are arranged after bargaining between the parties. However, in other environments, the parties can reach an arbitration-guided agreement on the commercial aspects.

This chapter especially deals with the latter problem within the following framework of decision:

1. European-like scenario, where the parties are public agencies/basin authorities, which are willing to enter into a covenant assuring that both the donor and the recipient regions will share the project benefits equitably. If the donor and the recipient basins are governed by two public agencies under the same jurisdiction within a country, then water transfers may be relatively trouble-free as the main responsible government sets common guidelines for the agreements. "As soon as multiple jurisdictions are involved, particularly in the case of transfers among different countries, the complexity of the issues is multiplied, exacerbated by the lack of jurisprudence" (UNESCO, 1999, summary statement, p. 216).
2. In the recipient area, water consumption should be saved by measures guaranteeing high efficiency in its use, namely, intraregional inefficiencies should be removed. If not, public reactions against the transfer project (not only from ecologists but also from economists and taxpayers) would be inevitable. To this end, nontransferable rights and average cost-based prices should be replaced by market prices, once the public agency in the recipient area has established a water market where right-holders can trade water with low transaction costs (Colby, 1990; Easter *et al.*, 1998; Archibald and Renwick, 1998). Indeed, establishing the market involves transition periods which are modeled elsewhere (Ballestero *et al.*, 2002). Inefficiency due to average cost-based prices is analyzed by Brill *et al.* (1997). See also Zekri and Romero (1993), and Gómez-Limón *et al.* (2002).
3. Water quantities to be transferred from the donor catchment basin must be surplus after satisfying the needs of the donor region in the reasonably foreseeable future and safeguarding all environmental require-

ments. Water rights in the donor area must be respected (Anderson, 1983; Michelsen, 1993). Two distinct types of compensation are envisaged: (i) price, namely, the donor charges the recipient an agreed amount of money per water unit, this characterizing the commercial side of the agreement; and (ii) political compensations, which are intended for economic development and welfare in the donor region, so that the impact of potential sociocultural losses can be offset and public disruption avoided. Quantity and price are decided by arbitration. To this end, an approach to arbitrated covenants is developed in detail in Section 2. This approach, easy to apply by an arbiter (expert), is obtained by simulating water supply and demand curves (Ballestero, 2004). Experts' reports use geographical/economic information on the catchment basins. "Professional analyses should support the decision making process and provide unbiased information to the decision makers and the general public" (UNESCO, 1999, summary statement, p. 216). Thus, the agreement between the public agencies does not involve the use of bilateral or multilateral water bargaining (Saleth *et al.*, 1991; Frisvold and Caswell, 1995; Colby, 1998).

Cases and perspectives of water transfers in Europe and other continents have been brought together by UNESCO (1999). This includes experiences in South Africa (Muller, 1999), India (Parashar, 1999), Bangladesh (Rahman, 1999), central Asia (Vasiliev and Voropaev, 1999), and Australia (Wright, 1999). Among the European projects, some involving large-scale investment, it is worth noting the Barcelona region transfer from the Llobregat and Ter rivers (Vilaró, 1999), the Languedoc-Catologne aqueduct (Blanc and Imbert, 1999), the Tajo-Segura aqueduct transferring most water for irrigation from a cold inland area to a dry coastal basin (Blasco *et al.*, 1999), water supply in Germany (Schumann, 1999; Scheuerlein, 1999), and the Helsinki water transfer project (Lemmela *et al.*, 1999).

Although the paper does not deal with water bargaining, some reference to this type of agreements is given in the following lines. They are typical in the USA and other countries where even public agencies act as private enterprises. Through these agreements, water is transferred not only for irrigation but often for urban consumption. A well-known example is California, where continuous drought over the period 1987–1991 led to a four-point government plan to alleviate the situation (Howitt *et al.*, 1992; Jercich, 1997). California's population is expected to double by 2040, and consequently, water transfers from wet areas are in the making. In addition in the USA, a number of water transfer agreements have been promoted in the past decade, such as the San Diego-Imperial Valley (1998) and the Orange-San Bernardino (1999). Regarding the San Diego agreement, a total volume up to 200,000 acre-feet of conserved agricultural water is transferred from the Imperial Irrigation District to the San Diego County Water Agency for at least 45 years, either agency being entitled to extend the contract by 30 years. Water prices are computed by a formula outlined in the agreement. The 1-year arrangement between the San Bernardino Valley Municipal Water District, Western Municipal Water District, and the Orange County Water District provides the latter with additional water sources from the San Bernardino Valley, where high levels of groundwater are even potentially damaging; because of this circumstance, water is supplied at low price. Water banks and markets in the USA are commented on in Griffin (1998). The proper approach to analyze transfer agreements from bargaining involves a game-theoretic model. To the author's knowledge, an analysis that combines bargaining and stochastic water supply/demand does not exist. This lack of previous results in the current literature makes more appealing the arbitration-based solution developed herein.This chapter is organized into six sections. Section 2, following the introduction, presents a review of the literature on MCDM as applied to irrigated agriculture. Section 3 highlights some challenges faced by these simulation techniques in order to become a useful tool in policy decision-making. The methodological contribution required to meet this objective is outlined in Section 4. Section 5 analyses some empirical applications following this approach. Finally, we draw some conclusions in Section 6.

2 ARBITRATING WATER TRANSFER COVENANTS: A METHODOLOGY

Advantages of this procedure are as follows: (i) if compared with a long bargaining process, the arbitration solution allows to save time and money; and (ii) when the commercial clauses are negotiated without an arbitration guideline, the parties often reach deadlock.

Suppose two nonprofit agencies A_r and A_d, which are responsible for the recipient and the donor basins, respectively, facing an inter-basin water transfer agreement. To undertake the approach to arbitration, a demand–supply equilibrium setting is simulated by the independent experts. The following notation is used

q = yearly transferable water flow to be provided for by the agreement.
p = water price to be agreed. This price does not include water production/ transportation costs, namely, the donor agency charges the recipient agency these costs, apart from price P.

P_d = origin water surplus value (per water unit), namely, the estimated value of water surplus in the donor basin per water unit taking into account alternative economic and environmental uses. For the sake of simplicity, parameter P_d is assumed to be constant, which is acceptable providing changes in surplus are not too large. Sometimes, surplus is evaluated at zero price, namely, $P_d = 0$.
P_r = water shortfall cost (per water unit), namely, the estimated cost of water deficit in the recipient region per water unit. For the sake of simplicity, parameter P_r is assumed to be constant, which is acceptable, provided changes in water deficit are not too large.

1. *Simulation approach to the demand curve.* Let us start with the equation:

$$E_r = p\int_0^{q_{r\max}} \xi f_r(\xi)d\xi + \int_0^q (p - p_d)(q-\xi)f_r(\xi)d\xi + \int_q^{q_{r\max}} (p_r - p)(\xi - q)f_r(\xi)d\xi \quad (1)$$

Where ξ is the random water quantity yearly demanded by the recipient agency through A_r the interbasin transfer. In other words, random ξ is the difference between the recipient basin water needs and the recipient basin random water production (availability) each year.

$f_r(\xi)$ is the probability distribution (density function) for random ξ, $q_{r\max}$ is the upper limit, namely, the probability distribution of random ξ is ranged over $(0, q_{r\max})$, and E_r is the expected costs borne by agency A_r as a result of its random demand.

Equation 1 is characterized saying that the recipient agency A_r does not purchase the random quantity ξ at price p exactly, since extra costs/compensatory payments are charged to this agency. In other words, agency A_r pays for water the amount just specified by the first term of Eq. 1, and in addition it should bear the extra costs/compensatory payments just specified by the second and third integrals. Indeed, the agency incurs these extra costs/compensatory payments due to any of the following random events:

1. Surplus. This means that agency A_r occasionally needs water transfers less than the agreed quantity q. Surplus randomly appears because of yearly variability in available water, that is, when water available in the recipient basin randomly increases. To make up for the other party's fall in sales due to the recipient basin random surplus, the model assumes a compensatory payment of $(p-p_d)$ $(q-\xi)$ monetary units to be made by agency A_r to agency A_d, this amount being the donor agency's sales margin.

2. Shortage. This means that agency A_r occasionally needs to purchase more water than the agreed quantity q, this event depending on random water production (availability) in the recipient basin. In times of shortage, the water shortfall ($\xi - q$) entails extra costs estimated at $(p_r - p)$ $(\xi - q)$ monetary units borne by the recipient region, namely, the model assumes that extra costs could be tentatively evaluated at $(p_r - p)$ times shortfall.

To simulate the demand curve, the independent experts consider every feasible price p (between the limits p_d and p_r) as a possible term of agreement. Then, the experts estimate from Eq. 1 what water quantity q the recipient agency would be willing to agree at the price under consideration. Plainly, the interest of agency A_r would be to minimize its expected costs E_r. Therefore, the model minimizes Eq. 1 with respect to the q quantity, as price p plays a parametric role. Hence, the experts should calculate the first derivative of (1) with respect to q and make this derivative equal to zero, which yields

$$\int_0^q f_r(\xi)d\xi = (p_r - p)/(p_r - p_d). \qquad (2)$$

Equation 2 is meaningful as representing the demand curve for water transfers. First, the quantity q to be demanded depends on the recipient's random needs as expressed by the $f_r(\xi)$ probability distribution. Second, remember that price p is ranged between the low p_d and the high p_r, namely, we have $p_d \leq p \leq p_r$. Therefore, the right-hand side in Eq. 2 is ranged over (0, 1). If p was fixed at its highest level p_r in the agreement, then agency A_r would claim a level equal to zero for quantity q such as Eq. 2 states. If p was set at its lowest level p_d, then agency A_r would claim the maximum value for quantity q. In broad terms, the recipient demands larger (smaller) quantities q as price p decreases (increases). In sum, the recipient would only accept p and q values satisfying the above law of downward sloping demand.

2. *Simulation approach to the demand curve.*

Let us start with the equation:

$$E_r = p\int_0^{q_{d\max}} \xi f_d(\xi)d\xi - \left[- \int_0^q (p_r - p_d)(q - \xi) f_d(\xi)d\xi \right.$$
$$\left. + \int_q^{q_{d\max}} (p - p_d)(\xi - q) f_d(\xi)d\xi - \right], \qquad (3)$$

where ζ is the random water quantity yearly available to be transferred from agency A_d, $f_d(\zeta)$ is the probability distribution (density function) for random ζ, $q_{d\,max}$ is the upper limit, namely, the probability distribution of random ζ is ranged over (0, $q_{d\,max}$), and E_d is the expected margin, namely, expected A_d revenue at price p minus costs/compensatory payments borne by this agency as a result of its random supply.

Equation 3 is characterized as follows. If the agreement provided for costs/ compensatory payments due to randomness, then the donor agency A_d would not exactly receive the expected revenue as given by the first term of Eq. 3. In other words, agency A_d would receive the amount just specified by the first term of Eq. 3 minus costs/compensatory payments just specified by the amount between square brackets. Indeed, the agency would incur these costs/compensatory payments due to any of the following random events:

3. Shortage. This now means that the donor agency cannot occasionally transfer the agreed quantity q, but only smaller flows. Shortage randomly appears owing to yearly variability in available water, that is, when surplus available in the donor basin randomly decreases. In this case, the recipient agency A_r would seek redress for the water quantity $(q-\zeta)$ that it does not receive. Consequently, the donor agency is required to make a compensatory payment of $(p_r - p)\,(q-\zeta)$ monetary units to the other party.
4. Surplus. When water available for transfer in the donor basin exceeds the agreed quantity q, the recipient agency is not obliged to purchase surplus at the agreed price p. Then, surplus should be allocated to alternative uses, this involving a cost of $(p - p_d)\ (\zeta - q)$ monetary units for the donor.

To simulate the supply curve, the independent experts consider every feasible price p (between the limits P_d and P_r) as a possible term of agreement. Then, the experts estimate from Eq. 3 what water quantity q the donor agency would be willing to agree at the price under consideration. Plainly, the interest of the donor agency would be to maximize its expected margin $\boldsymbol{E}_r$. Therefore, the model minimizes costs/compensatory payments, that is, the amount between square brackets in equation (3) with respect to the q quantity, as price p plays a parametric role. Hence, the experts should calculate the first derivative of this amount with respect to q and make this derivative equal to zero, which yields

$$\int_0^q f_d(\xi)d\xi = (p - p_d)/(p_r - p_d) \tag{4}$$

Equation 4 portrays the supply curve. First, the quantity q to be supplied depends on the donor's random surplus as expressed by the $f_{\mathrm{d}}(\zeta)$ probability distribution. Second, since the right-hand side in Eq. 4 is ranged over (0, 1), the donor agency would claim a level equal to zero for quantity q if p was set at its lowest level p_{d}, while it would claim the maximum value q for quantity p if it was fixed at its highest level p_{r}. More generally put, the donor is willing to supply larger (smaller) quantities q as price p increases (decreases). Indeed, the donor will only accept p and q values satisfying the above law of upwards sloping supply.

Hence, the solution advised by the experts will be to establish price p and quantity q at levels given by the intersection of the demand curve (2) and the supply curve (4), as the equilibrium position. Notice that p_r and p_{d} come from available information gathered in geographical and economic studies on the areas. Therefore, the experts with a good geographical knowledge can obtain the simulated solution by numerical tables approximating the frequency distributions $f_{\mathrm{r}}(\xi)$ in the recipient basin and $f_{\mathrm{d}}(\zeta)$ in the donor basin. In Section 3, the table is displayed and the numerical solution found through an illustrative example.

As a final remark, the approach overlooks optimistic conjectures about feasible agreements between the agencies to counterbalance surplus in a basin and shortfalls in the other. Simulation relying on compensatory payments is a mere methodology to recommend price and quantity, and consequently, the actual covenant does not need to include a clause about this kind of payments.

3 ILLUSTRATIVE EXAMPLE

As inspired from a real-world case elsewhere (Ballestero, 2004), the present make-believe example is only intended to highlight applications of the above model through numerical tables whose structure and content widely differ from the previous paper. Consider a potential donor area described as an inland region crisscrossed by large dammed rivers, and a drought-exposed maritime region near the Mediterranean Sea, the latter being a potential recipient basin.

In Table 1, historical series of geographical data on water resources in both the donor and the recipient basins from 1975–1976 to 1994–1995 hydrological years are shown with six columns displaying the following information. In column 2, the available water quantity to be supplied each

Table 1. Annual geographical information on the donor and recipient basins.

Donor basin			Recipient basin		
Hydrological year	**Surplus (supplied water), Mm^3**	**Cumulative frequencies**	**Hydrological year**	**Demanded water from transfers, Mm^3**	**Cumulative frequencies**
1994–95	7.50	0.05	1983–84	0	0.05
1983–84	16.98	0.10	1984–85	6.39	0.10
1992–93	26.23	0.15	1981–82	16.12	0.15
1982–83	27.84	0.20	1994–95	16.26	0.20
1993–94	29.46	0.25	1985–86	22.43	0.25
1991–92	37.89	0.30	1980–81	26.92	0.30
1981–82	60.08	0.35	1982–83	29.04	0.35
1978–79	63.55	0.40	1992–93	32.52	0.40
1977–78	71.64	0.45	1987–88	34.98	0.45
1990–91	71.64	0.50	1993–94	35.43	0.50
1987–88	74.87	0.55	1991–92	41.82	0.55
1989–90	75.10	0.60	1978–79	59.77	0.60
1988–89	76.26	0.65	1979–80	62.17	0.65
1975–76	76.58	0.70	1990–91	62.73	0.70
1986–87	80.88	0.75	1977–78	65.09	0.75
1984–85	83.77	0.80	1986–87	71.34	0.80
1976–77	84.13	0.85	1976–77	71.41	0.85
1985–86	84.69	0.90	1975–76	73.35	0.90
1980–81	113.24	0.95	1988–89	74.49	0.95
1979–80	160.85	1.00	1989–90	77.54	1.00

year is recorded. This is the random surplus obtained after subtracting foreseeable consumption and environmental needs in the donor basin. A part and parcel of consumption is irrigation water. In column 5, the water quantity demanded from external sources (inter-basin transfer) by the recipient basin each year is given. This demanded water is computed as available resources less foreseeable consumption, both yearly quantities given by geographical studies. Columns 1 and 3 are associated with column 2 to indicate, respectively, the hydrological year and the cumulative frequency corresponding to each supplied water quantity. Columns 4 and 6 play an analogous role with respect to column 5. Therefore, the probability distribution of random water quantity yearly available to be transferred from the donor agency is tabulated on the left side of the table, while on the right side the probability distribution of random water quantity yearly demanded by the recipient agency is also tabulated. In other words, the left side of Table 1 provides in numerical terms the cumulative distribution function for the donor basin, namely, the integral

$$F_{\mathrm{d}}(q) = \int_0^q f_{\mathrm{d}}(\zeta)d\zeta \tag{5}$$

whereas the right side of the table provides the analogous integral for the recipient basin

$$F_{\mathrm{r}}(q) = \int_0^q f_{\mathrm{r}}(\xi)d\xi. \tag{6}$$

Once Table 1 has been constructed only from available geographical studies, the arbiter should know parameters p_{r} and p_{d} appearing on the right-hand side of Eqs. 2 and 4. Concerning the estimated per unit cost p_{r} of water deficit in the recipient maritime region, this is equivalent to the market price of desalinated water, which would be used to cover the water shortfall. In Mediterranean environments, this price is about 0.60 euros/m^3. Now, only the origin water surplus value p_{d} remains to be estimated from its alternative economic and environmental uses, if any. As this surplus flows to the sea with no environmental impact (e.g., on fisheries), parameter p_{d} is assigned a zero value. Therefore, the right-hand sides of Eqs. 4 and 2 become

$$(p - p_{\mathrm{d}})/(p_{\mathrm{r}} - p_{\mathrm{d}}) = p/0.60, \tag{7}$$

$$(p_{\mathrm{r}} - p)/(p_{\mathrm{r}} - p_{\mathrm{d}}) = (0.60 - p)/0.60, \tag{8}$$

for the donor and the recipient areas, respectively.

Table 2 describes the arbitration process through which the simulated supply and demand curves are drawn. First, the arbiter proposes a sequence of possible prices between 0 and 0.60 euros/m^3, this sequence appearing in column 1. For each price p, the arbiter computes the cumulative frequencies (7) and (8), the respective sequences being brought to columns 2 and 4. From Table 1, each cumulative frequency is associated with the q value of water quantity. Thus, for example, price $p = 0.27$ in Table 2 involves a cumulative frequency of 0.45 relative to simulated supply, which leads to 71.64 Mm^3 from Table 1, columns 2–3. This quantity (71.64) now appears in Table 2, column 3, for price $p = 0.27$. Finally, in Table 2 the arbiter compares columns 3 and 5 to find the row in which both water quantities q approximately coincide. Check that this row corresponds to price $p = 0.21$, the water quantities being 60.08 Mm^3 (column 3) and 62.17 Mm^3 (column 5), which are very close to one another. This is the arbitrated solution in terms of price and quantity to the commercial clauses of the covenant.

Table 2. Simulated supply and demand curves in the arbitration process.

Price (euro/m³)	Simulated supply		Simulated demand	
	Cumulative frequencies	Water quantity (Mm³)	Cumulative frequencies	Water quantity (Mm³)
0.03	0.05	7.50	0.95	74.49
0.06	0.10	16.98	0.90	73.35
0.09	0.15	26.23	0.85	71.41
0.12	0.20	27.84	0.80	71.34
0.15	0.25	29.46	0.75	65.09
0.18	0.30	37.89	0.70	62.73
0.21	0.35	60.08	0.65	62.17
0.24	0.40	63.55	0.60	59.77
0.27	0.45	71.64	0.55	41.82
0.30	0.50	71.64	0.50	35.43
0.33	0.55	74.87	0.45	34.98
0.36	0.60	75.10	0.40	32.52
0.39	0.65	76.26	0.35	29.04
0.42	0.70	76.58	0.30	26.92
0.45	0.75	80.88	0.25	22.43
0.48	0.80	83.77	0.20	16.26
0.51	0.85	84.13	0.15	16.12
0.54	0.90	84.69	0.10	6.39
0.57	0.95	113.24	0.05	0
0.60	1.00	160.85	0.00	–

4 CONCLUDING REMARKS

Limitations and contributions of this chapter are as follows. First, the decision-making model in Section 2 focuses on a relevant but not unique aspect, namely, the commercial side of the agreement consisting in determining water quantity and price providing that some technical, environmental, and political constraints are met. Therefore, the model does not help agree the political compensations to the donor area but it only helps the parties agree what water volume should be transferred and what price should be paid for this resource without including costs of water production in the donor area and transportation along the aqueduct. A vital premise to the covenant is assuring efficiency in the recipient basin. A second limitation concerns the agencies' behavior. In fact, if the agencies behave as profitability seekers rather than non-profit public entities, they would have no reason to employ an arbiter instead of bargaining strategies like those used in aggressive negotiations. This is frequently the US case, where one can contend that arbiters are used to resolving labor and business contract disputes but not to negotiating water transactions. However, even for profitability-seeking agencies, each party may be convinced that the joint decision of agreeing to an arbitrated covenant is preferable to a long bargaining process that often comes to a standstill, especially when the parties cannot reasonably expect significant extra profit by negotiating. In European countries and other areas where public agencies (not profitability seekers) are responsible for interbasin water agreements, the proposed arbitration model provides a consistent solution as based on simulated supply and demand curves, whose parameters are obtained from available geographical studies, namely, without cumbersome information problems. As the numerical example in Section 3 has shown, the solution is easy to obtain through tables, this being a remarkable advantage over bargaining models often difficult to apply.

5 REFERENCES

Anderson, T.L. (1983). Water Rights. San Francisco, CA: Pacific Institute for Public Policy Research.

Archibald, S.O. and Renwick, M.E. (1998). Expected transaction costs and incentives for water market development. In K.W. Easter, M.W. Rosegrant, and A. Dinar, eds., *Markets for water, potential and performance*. Boston, MA: Kluwer, 95–117.

Ballestero, E. (2004). Inter-basin water transfer public agreements: A decision approach to quantity and price. *Water Resources Management*, 18: 75–88.

Ballestero, E., Alarcon, S., and Garcia-Bernabeu, A. (2002). Establishing politically feasible water markets: A multi-criteria approach. *Journal of Environmental Management*, 65: 411–429.

Blanc, J-L and Imbert, F. (1999). L'aqueduct Languedoc-Rousillon-Catalogne. In *Interbasin Water Transfer, Proceedings of the International Workshop*, UNESCO, Paris, pp. 33–44.

Blasco, J.A.L., Marquez, J.M.M., and Cerezo, J.T. (1999). El acueducto Tajo-Segura: experiencia desde 1980, in *Interbasin Water Transfer, Proceedings of the International Workshop*, UNESCO, Paris, pp. 45–60.

Brill, E., Hochman, E. and Zilberman, D. 1997. Allocation and pricing at the water district level. *American Journal of Agricultural Economics*, 79: 952–963.

Colby, B.G. (1990). Transaction costs and efficiency in western water allocation. *American Journal of Agricultural Economics*, 72: 1184–1192.

Colby, B.G. (1998) Negotiated transactions as conflict resolution mechanisms: water bargaining in the U.S. West. In K.W. Easter, M.W. Rosegrant, and A.Dinar, eds., *Markets for water, potential and performance*. Boston, MA: Kluwer, 77–94.

Easter, K.W., Dinar, A., and Rosegrant, M.W. (1998). Water markets: Transaction costs and institutional options. In K.W. Easter, M.W. Rosegrant, and A. Dinar, eds., *Markets for water, potential and performance*. Boston, MA: Kluwer, 1–18.

Frisvold, G.B. and Caswell, M.S. (1995). A bargaining model for water transfers including water quality. In Dinar and Loehman (eds.), *Water quantity/quality management and conflict resolution*, Chapter 30, Westport, CT: Praeger Publishers.

Geoff, W. (1999). Interbasin water transfer: The Australian experience with the Snowy Mountains Scheme. In *Interbasin Water Transfer, Proceedings of the International Workshop*, Unesco, Paris, 101–106.

Gómez-Limón, J.A., Arriaza, M., and Berbel, J. (2002). Conflicting implementation of agricultural and water policies in irrigated areas in the EU. *Journal of Agricultural Economics,* 53: 259–281.

Griffin, R.C. (1998). The application of water market doctrines in Texas. In K.W. Easter, M.W. Rosegrant, and A. Dinar, eds., *Markets for water, potential and performance.* Boston, MA: Kluwer, 51–63.

Howitt, R.E., Moore, N., and Smith, R.T. (1992). *A Retrospective on California's 1991 Emergency Drought Water Bank.* Sacramento, CA: California Department of Water Resources.

Jercich, S.A. (1997). California's 1995 Water Bank Program: Purchasing water supply options. *Journal of Water Resources Planning and Management,* 123: 59–65.

Lemmela, R., Hirsto, I., and Pekkarinen, M. (1999). Water transfer to Helsinki metropolitan area. In *Interbasin Water Transfer, Proceedings of the International Workshop*, UNESCO, Paris, 123–132.

Michelsen, A.M. and Young R.A. (1993). Optioning agricultural water rights for urban water supplies during drought. *American Journal of Agricultural Economics*, 75: 1010–1020.

Muller, M. (1999). Interbasin water sharing: A South African perspective. In *Interbasin Water Transfer, Proceedings of the International Workshop*, UNESCO, Paris, 61–70.

Parashar, R.K. (1999). Interbasin water transfer: Indian scenario. In *Interbasin Water Transfer, Proceedings of the International Workshop*, UNESCO, Paris, 73–80.

Rahman, K. (1999). Interbasin water transfer: Bangladesh perspective. In *Interbasin Water Transfer, Proceedings of the International Workshop*, UNESCO, Paris, 81–94.

Saleth, R.M., Braden, J.B., and Eheart, J.W. (1991). Bargaining rules for a thin spot water market. *Land Economics*, 67(3): 326–339.

Scheuerlein, H. (1999). Interbasin water transfer from the Danube to the Rhine Basin in Bavaria. In *Interbasin Water Transfer, Proceedings of the International Workshop*, UNESCO, Paris, pp. 107–114.

Schumann, H. (1999). Water transfer systems for fresh water supply in Germany. In *Interbasin Water Transfer, Proceedings of the International Workshop*, UNESCO, Paris, 115–122.

UNESCO (1999). *Interbasin water transfer, Proceedings of the International Workshop*, UNESCO, Paris, 25–27 April.

Vasiliev, O.F. and Voropaev, G.V. (1999). SIBARAL and other water transfer projects in the former USSR. In *Interbasin Water Transfer, Proceedings of the International Workshop*, UNESCO, Paris, 95–100.

Vilaró, F. (1999). Abastecimiento a la región de Barcelona. In *Interbasin Water Transfer, Proceedings of the International Workshop*, UNESCO, Paris, 17–32.

World Bank (1993). *Water Resources Management,* A World Bank Policy Paper. Washington D.C.: World Bank.

Zekri, S. and Romero, C. (1993). Public and private compromises in agricultural water management. *Journal of Environmental Management*, 37: 281–290.

Chapter 8

POSITIVE MATHEMATICAL PROGRAMMING FOR AGRICULTURAL AND ENVIRONMENTAL POLICY ANALYSIS: REVIEW AND PRACTICE

Bruno Henry de Frahan[1], Jeroen Buysse[2], Philippe Polomé[1], Bruno Fernagut[3], Olivier Harmignie[1], Ludwig Lauwers[3], Guido Van Huylenbroeck[2], and Jef Van Meensel[3]

[1]*Université catholique de Louvain;* [2]*Ghent University;* [3]*Institute for Agricultural and Fisheries Research, Merelbeke*

Abstract Positive mathematical programming (PMP) has renewed the interest in mathematical modelling of agricultural and environmental policies. This chapter explains first the main advantages and disadvantages of the PMP approach, followed by a presentation of an individual farm-based sector model, called SEPALE. The farm-based approach allows the introduction of differences in individual farm structures in the PMP modelling framework. Furthermore, a farm-level model gives the possibility of identifying the impacts according to various farm characteristics. Simulations of possible alternatives to the implementation of the Agenda 2000 mid-term review illustrate the value of such a model. This chapter concludes with some topics for further research to resolve some of the PMP limitations.

Keywords: Positive mathematical programming, Common Agricultural Policy, agro-environmental policy analysis

1 INTRODUCTION

There is a renewed interest in mathematical programming (MP) to model economic behaviour. This originates from a combination of factors. First, the emergence in the late 1980s of positive mathematical programming (PMP) has brought an appealing breath of positivism in the determination of the optimizing function parameters. This method formalized later by Howitt (1995a) makes it indeed possible to calibrate MP models exactly. Second, as a result of the former, PMP has provided a more flexible and realistic simulation behaviour of MP models avoiding unlikely abrupt discontinuities in the simulation solutions. Third, the increasing need to model and simulate behavioural functions under numerous technical, economic, policy and, more recently, environmental conditions has strengthened the recourse to MP models. Fourth, in an environment of often limited amount of adequate information and data to treat complex decisions, MP models are nevertheless able to handle decision problems which econometrics cannot. With the increasing number of available databases assembled from data collected at the regional, territorial, farm and even land plot levels, the construction of MP models is now possible at a more disaggregated level of decision making. This allows the analysis of agricultural, environmental and land use policy in accordance with local conditions. This renewed interest in MP modelling for analysing agricultural and environmental policies has generated numerous applications as well as extensions at different investigation levels of which several are reported in Heckelei and Britz (2005).

This chapter concentrates on PMP and its recent developments as a tool for policy analysis and on the practical elaboration in Belgium. It is organized as follows. Section 2 shows how PMP renovates the calibration of mathematical programming models. Section 3 explains how its weaknesses have generated various developments including extensions bridging mathematical programming and econometrics. Exploiting together the advantages of mathematical programming and econometric approaches leads to a new field of empirical investigation, which we would like to name 'econometric mathematical programming'. Section 4 shows how the Belgian regional mathematical programming model SEPALE tackles some of the PMP weaknesses and adopts a calibration method able to exploit the richness of the European Union's Farm Accountancy Data Network (FADN) in representing the economic behaviour of a collection of farmers. The application shows how current reforms of the Common Agricultural Policy (CAP) are treated and simulated in SEPALE. Section 5 discusses some issues left to PMP. The last section concludes with a summary of the advantages and limitations of PMP for agricultural and environmental policy analysis.

2 THE STANDARD PMP APPROACH

PMP is a method to calibrate mathematical programming models to observed behaviour during a reference period by using the information provided by the dual variables of the calibration constraints (Howitt, 1995a; Paris and Howitt, 1998). The dual information is used to calibrate a non-linear objective function such that the observed activity levels are reproduced for the reference period but without the calibration constraints. The term 'positive' that qualifies this method implies that, like in econometrics, the parameters of the non-linear objective function are derived from an economic behaviour assumed to be rational given all the observed and non-observed conditions that generates the observed activity levels. The main difference with econometrics is that PMP does not require a series of observations to reveal the economic behaviour, which as a drawback deprives PMP from inference and validation tests.

Formalized by Howitt (1995a), PMP follows a procedure in three steps. The first step consists in writing an MP model as usual but adding to the set of limiting resource constraints a set of calibration constraints that bound the activities to the observed levels of the reference period. Taking the case of maximizing gross margins with upper bounded calibration constraints, we write the initial model as in Paris and Howitt (1998):

$$\text{Max } Z = \mathbf{p}'\,\mathbf{x} - \mathbf{c}'\,\mathbf{x} \tag{1}$$

$$\text{s.t.:} \quad \mathbf{A}\,\mathbf{x} \le \mathbf{b}[\boldsymbol{\lambda}] \tag{1a}$$

$$\mathbf{x} \le \mathbf{x_0} + \varepsilon\,\lfloor\boldsymbol{\rho}\rfloor \tag{1b}$$

$$\mathbf{x} \le 0, \tag{1c}$$

where

Z is the scalar of the objective function value,

$\mathbf{p}$ $(n \times 1)$ the vector of product prices,

$\mathbf{x}$ $(n \times 1)$ the non-negative vector of production activity levels,

$\mathbf{c}$ $(n \times 1)$ the vector of accounting costs per unit of activity,

$\mathbf{A}$ $(m \times \text{n})$ the matrix of technical coefficients in resource constraints,

$\mathbf{b}$ $(m \times 1)$ the vector of available resource levels,

$\mathbf{x_0}$ $(n \times 1)$ the non-negative vector of observed activity levels,

$\boldsymbol{\varepsilon}$ $(n \times 1)$ the vector of small positive numbers for preventing linear dependency between the structural constraints (1a) and the calibration constraints (1b),

$\boldsymbol{\lambda}$ $(m \times 1)$ the vector of duals associated with the allocable resource constraints, and

$\boldsymbol{\rho}$ $(n \times 1)$ the vector of duals associated with the calibration constraints.

Assuming that all activity levels are strictly positive and all allocable resource constraints are binding at the optimal solution, the first-order conditions of model (1) provide the following dual values as in Heckelei and Wolff (2003):

$$\boldsymbol{\rho}^{\text{p}} = p^{\text{p}} - \mathbf{A}^{\text{p}'}\boldsymbol{\lambda}, \tag{2}$$

$$\boldsymbol{\rho}^{\mathbf{m}} = \mathbf{0}, \tag{3}$$

$$\boldsymbol{\lambda} = (\mathbf{A}^{\mathbf{m}'})^{-1} (\mathbf{p}^{\mathbf{m}} - \mathbf{c}^{\mathbf{m}}), \tag{4}$$

The vector $\mathbf{x}$ is partitioned into $[(n-m) \times 1]$ vector of preferable activities x^p constrained by the calibration constraints (1b) and $(m \times 1)$ vector of marginal activities x^m constrained by the allocable resource constraints (1a). The other vectors ρ, p and $\mathbf{c}$ and the matrix $\mathbf{A}$ are partitioned accordingly.

Howitt (1995a) and Paris and Howitt (1998) interpret the dual variable vector $\boldsymbol{\rho}$ associated with the calibration constraints as capturing any type of model mis-specification, data errors, aggregate bias, risk behaviour and price expectations. In the perspective of calibrating a non-linear decreasing yield function as in Howitt (1995a), this dual vector ρ represents the difference between the activity average and marginal value products. In the alternative perspective of calibrating a non-linear increasing cost function as in Paris and Howitt (1998), this dual vector ρ is interpreted as a differential marginal cost vector that together with the activity accounting cost vector $\mathbf{c}$ reveals the actual variable marginal cost of supplying the observed activity vector x_0.

The second step of PMP consists in using these duals to calibrate the parameters of the non-linear objective function. A usual case considers calibrating the parameters of a variable cost function C^v, which has the typical multi-output quadratic functional form, however, holding constant variable input prices at the observed market level as follows:

$$C^v(x) = d'\,x + x'\,\mathbf{Q}\,x\,/\,2 \tag{5}$$

where

d $(n \times 1)$ is the vector of parameters of the cost function,

Q $(n \times n)$ is the symmetric, positive (semi-) definite matrix with typical element q_{ij} for activities i and j.

Other functional forms are possible. The generalized Leontief and the weighted-entropy variable cost function (Paris and Howitt, 1998) and the constant elasticity of substitution (CES) production function (Howitt, 1995b) in addition to the constant elasticity of transformation production function (Graindorge *et al.*, 2001) have also been used.

The variable marginal cost vector $\mathbf{MC}^v$ of this typical cost function is set equal to the sum of the accounting cost vector $\mathbf{c}$ and the differential marginal cost vector $\boldsymbol{\rho}$ as follows:

$$MC^v = \nabla C^v(\mathbf{x})_{x0} = \mathbf{d} + \mathbf{Q}\,\mathbf{x}_0 = \mathbf{c} + \boldsymbol{\rho} \tag{6}$$

where
$\nabla C^v(\mathbf{x})_{x0}$ a $(1 \times n)$ gradient vector of first derivatives of $C^v(\mathbf{x})$ for $\mathbf{x} = \mathbf{x}_0$.

To solve this system of n equations for $[n + n(n + 1)/2]$ parameters and, thus, overcome the under-determination of the system, PMP modellers rely on various solutions that are reviewed in the next section.

The third step of PMP uses the calibrated non-linear objective function in a non-linear programming problem similar to the original one except for the calibration constraints. This calibrated non-linear model is consistent with the choice of the non-linear activity yield or cost function derived in the preceding step and exactly reproduces observed activity levels and original duals of the limiting resource constraints. The following PMP model is ready for simulation.

$$\text{Max } Z = p'x - \hat{\mathbf{d}}'x - x'\hat{\mathbf{Q}}\,x/2, \tag{7}$$

$$\text{s.t.} \qquad \mathbf{A}\,\mathbf{x} \le \mathbf{b}[\boldsymbol{\lambda}], \tag{7a}$$

$$\mathbf{x} \ge \mathbf{0}, \tag{7b}$$

where the vector $\hat{\mathbf{d}}$ and matrix $\hat{\mathbf{Q}}$ are the calibrated parameters of the non-linear objective function.

Assuming again that all optimal activity levels are strictly positive and allocable resource constraints are all binding at the optimal solution, the first-order conditions of model (7) provide the following dual values of the resource constraints as in Heckelei and Wolff (2003):

$$\boldsymbol{\lambda} = (\mathbf{A}\,\mathbf{Q}^{-1}\,\mathbf{A}')^{-1}\,(\mathbf{A}\,\hat{\mathbf{Q}}^{-1}\,(\mathbf{p} - \hat{\mathbf{d}})). \tag{8}$$

This calibration approach can be applied at the farm, regional and sector levels. When accounting data of a sample of F farms are available such as from the FADN, F PMP models can be defined for each farm of the sample. Simulations can then be performed on these individual PMP models and simulation results may be aggregated as shown in the application below.

3 FURTHER PMP DEVELOPMENTS

Althrough being an appealing method for calibration, PMP has shown shortcomings in model calibration that, in turn, motivated further developments. One of these shortcomings is the missing representation of economic behaviour with regard to activities of farms whose initial observed supply level is zero during the reference period. To overcome this self-selection problem during the calibration as well as during the simulation steps, Paris and Arfini (2000) add to the *F* PMP models a supplementary PMP model for the whole farm sample and calibrate a frontier cost function for all the activities included in the whole farm sample.

A second development of the PMP methodology concerns the integration of risk. For example, Paris (1997) uses a von Neumann–Morgenstern expected utility approach assuming a normal distribution of output prices and a constant absolute risk aversion.

A third development is the inclusion of greater competitiveness among close competitive activities whose requirements for limiting resources are more similar than with other activities. Rohm and Dabbert (2003) represent these close competitive activities as variant activities from their generic activities and add to the first PMP step calibration constraints for these variant activities that are less restrictive than their counterparts for their generic activities.

A fourth development to overcome criticism that has been raised against the use of a linear technology in limiting resources and the zero-marginal product for one of the calibrating constraints is the expansion of the PMP framework into a Symmetric Positive Equilibrium Problem (SPEP). Paris (2001) and Paris and Howitt (2001) express the first step of this new structure as an equilibrium problem consisting of symmetric primal and dual constraints and the third step as an equilibrium problem between demand and supply functions of inputs, on the one hand, and between marginal cost and marginal revenue of the output activities, on the other hand. For these authors, the key novelty of this new framework is rendering the availability of limiting inputs responsive to output levels and input price changes. Britz *et al.* (2003), however, address several conceptual concerns with respect to the SPEP methodology and question the economic interpretation of the final model ready for simulations.

Other shortcomings comprise the under-determination of the system, the unequal treatment of the marginal and preferable activities and the first phase estimation bias. They are treated in the following three subsections.

3.1 The Under-Determination Problem

To overcome the shortcoming of under-determination of the system in equation (6), an earlier *ad hoc* solution consists in assuming that the symmetric matrix **Q** is diagonal, implying that the change in the actual marginal cost of activity *i* with respect to the level of activity *j* ($i \neq j$) is null and, then, in relying on additional assumptions. Common additional assumptions consist in posing the vector **d** of the quadratic cost function to be either equal to zero, which leads to

$$q_{ii} = (c_i + p_i)/x_{io} \text{ and } d_i = 0 \text{ for all } i = 1,\dots n$$

or equal to the accounting cost vector **c**, which leads to:

$$q_{ii} = \rho_i / x_{io} \text{ and } d_i = c_i \text{ for all } i = 1,\dots n.$$

Another calibration rule called the average cost approach equates the accounting cost vector **c** to the average cost vector of the quadratic cost function, which leads to:

$$q_{ii} = 2\rho_i / X_{io} \text{ and } d_i = c_i - \rho_i \text{ for all } i = 1,\dots n.$$

Exogenous supply elasticities ε_{ii} are also used to derive the parameters of the quadratic cost function as in Helming *et al.* (2001):

$$q_{ii} = p_{io} / \varepsilon_{ii} x_{io} \text{ and } d_i = c_i + \rho_i - q_{ii} x_{io} \text{ for } i = 1,\dots n.$$

All these specifications would exactly calibrate the initial model as long as Eq. 6 is verified, but lead to different simulation responses to external changes.

A subsequent development from Paris and Howitt (1998) to calibrate the marginal cost function is to exploit the maximum entropy estimator to determine all the $[n + n(n + 1)/2]$ elements of the vector **d** and matrix **Q** using the Cholesky factorization of this matrix **Q** to guarantee that the calibrated matrix **Q** is actually symmetric positive semi-definite.[1] This estimator in combination with PMP enables to calibrate a quadratic variable cost function accommodating complementarity and competitiveness among

[1] In short, the maximum entropy approach consists in estimating parameters regarded as expected values of associated probability distributions defined over a set of *a priori* discrete supports (Golan *et al.*, 1996).

activities still based on a single observation but using *a priori* information on support bounds. Nevertheless, as argued in Heckelei and Britz (2000), the simulation behaviours of the resulting calibrated model would still be arbitrary because heavily dominated by the supports.

Heckelei and Britz (2000) exploit the suggestion from Paris and Howitt (1998) to use the maximum entropy estimator to determine these parameters on the basis of additional observations of the same farm or region with a view to collect information on second-order derivatives. They estimate the parameters of the vector **d** and matrix **Q** on the basis of cross-sectional vectors of marginal costs and the use of the Cholesky decomposition of the matrix of the second-order derivatives as additional constraints. They obtain a greater successful *expost* validation than using the standard "single observation" maximum entropy approach. This cross-sectional procedure is an interesting response to the lack of empirical validation for models that are calibrated on a single reference period. It is used to calibrate the cost functions of the regional activity supplies of the Common Agricultural Policy Regional Impact (CAPRI) modelling system (Heckelei and Britz, 2001).

3.2 The Unequal Treatment of Marginal and Preferable Activities

Another PMP shortcoming discussed in the literature is the unequal treatment of the marginal and preferable activities. Because the differential marginal costs of the marginal activities captured by the dual vector $\boldsymbol{\rho}$ are zero, the actual marginal costs of supplying these activities are independent of their levels while those of supplying the preferable activities are not under the average cost approach of calibration. For these marginal activities, calibrated marginal costs are equal to average costs and marginal profits are equal to average profits. Gohin and Chantreuil (1999) show that an exogenous shock on one preferable activity would uniquely modify the levels of this activity and the levels of the marginal activities, not those of the other preferable activities.

One *ad hoc* solution to obtain an increasing marginal cost function for these marginal activities consists in retrieving some share of one limiting resource dual value λ and adding it to the calibration dual vector $\boldsymbol{\rho}$ to obtain a modified calibration dual vector $\boldsymbol{\rho}_M$ (Rohm and Dabbert, 2003). A more severe solution consists in skipping the first step of PMP altogether. Judez *et al.* (2001) use this approach to represent the economic behaviour of different farm types based on farm accounting data from the Spanish part of the FADN.

3.3 The First Phase Estimation Bias

More fundamentally, Heckelei and Wolff (2003) recently explained that PMP is, however, not well suited to the estimation of programming models that use multiple cross-sectional or chronological observations. They show that the derived marginal cost conditions (6) prevent a consistent estimation of the parameters when the ultimate model (7) is seen as representing adequately the true data-generating process. Their argument goes as follows. On the one hand, the shadow price value vector $\boldsymbol{\lambda}$ implied by the ultimate model (7) is determined by the vectors $\mathbf{p}$, $\mathbf{d}$ and $\mathbf{b}$ and the matrices $\mathbf{A}$ and $\mathbf{Q}$ through the first-order condition derived expression (8). On the other hand, the various dual value vectors $\boldsymbol{\lambda}$ from the sample initial models (1) are solely determined by the vectors $\mathbf{p}$ and $\mathbf{c}$ and matrix $\mathbf{A}$ of only those marginal activities bounded by the resource constraints through the first-order derived expression (4). As a result, the various vectors $\boldsymbol{\lambda}$ of resource duals of the initial models are generally different from the vector $\boldsymbol{\lambda}$ of resource duals of the ultimate model. Since the first step simultaneously sets both the initial dual vectors $\boldsymbol{\rho}$ and $\boldsymbol{\lambda}$ and the second step uses the initial dual vector $\boldsymbol{\rho}$ to estimate the vector $\mathbf{MC}^v$, this latter vector must generally be inconsistent with the ultimate model (7). The derived marginal conditions (6) are, therefore, most likely to be biased estimating equations yielding inconsistent parameter estimates.[2]

Howitt (2005), however, disputes this inconsistency. First, he shows that, for any single observation, the dual value vector $\boldsymbol{\lambda}$ derived from the initial constrained linear model is numerically identical to the dual value vector $\boldsymbol{\lambda}$ that the ultimate quadratic model would generate for the same observation point. Second, he shows that, for a given set of observations, the dual value vector $\boldsymbol{\lambda}$ derived from each initial constrained linear model must be generally different from the dual value vector $\boldsymbol{\lambda}$ generated by the ultimate quadratic model unless each initial linear model uses numerical values of marginal cost that are identical to those generated by the ultimate quadratic model for each observation point. Different observations bring changes in costs that are not necessarily fully captured by the changes in their cost vector $\mathbf{c}$.

To avoid inconsistency between steps 1 and 3 as further exposed in Heckelei and Britz (2005), Heckelei and Wolff (2003) suggest to skip the first step altogether and employ directly the optimality conditions of the desired programming model to estimate, not calibrate, shadow prices and parameters

[2]In other words, the 'estimated' value of the dual vector λ cannot converge to the true dual vector λ as more observations are added because PMP always selects the highest possible value for the dual vector λ.

simultaneously. They illustrate this general alternative to the original PMP through three examples relying on the Generalized Maximum Entropy (GME) procedure for estimating the model parameters. Their examples deal with the estimation of the parameters of various optimization models that (1) incorporate a quadratic cost function and only one constraint on land availability, (2) allocate variable and fixed inputs to production activities represented by activity-specific production functions or (3) allocate fixed inputs to production activities represented by activity-specific profit functions.

As stated by their authors, this alternative approach to PMP has some theoretical advantage over the original PMP for the estimation of programming models. It also has some empirical advantage over standard econometric procedures of duality-based behavioural functions for the estimation of more complex models characterized by more flexible functional forms and more constraints as well as the incorporation of additional constraints relevant for simulation purpose. The application in the next section also skips the first step of PMP to use directly the optimality conditions of the desired programming model. Howitt (2005) does not dispute here the desirability and efficiency of the simultaneous estimation of dual shadow prices and parameters when sufficient observations are available. He, however, shows that this approach could result in the error and bias of the estimated cost coefficients that are sensitive to the error specification and the initial value assigned to the dual value vector $\boldsymbol{\lambda}$ when relying on the GME estimator. He shows that the use of dual value vectors $\boldsymbol{\lambda}$ from initial linear models can significantly increase the precision of the estimate of the dual value vector $\boldsymbol{\lambda}$ of the ultimate quadratic model with small losses in the precision of the estimates of the ultimate cost coefficients.

4 THE SEPALE MODEL AND APPLICATIONS

This section illustrates how the PMP concept can be applied into an agricultural model that can be used to simulate various policy scenarios. The agricultural model is composed of a collection of microeconomic mathematical programming models each representing the optimizing farmer's behaviour at the farm level. Parameters of each PMP model are calibrated on decision data observed during a reference period exploiting the optimality first-order conditions and the observed opportunity cost of limiting resources. Simulation results can be aggregated according to farm location, type and size.

Exploiting the richness of the FADN data, this model is part of an effort initially funded by the Belgian Federal Ministry of Agriculture to develop a decision support system for agricultural and environmental policy analysis. The model is known under the name of SEPALE and is developed by a group of agricultural economists based at the Université catholique de Louvain, the University of Ghent and the Institute for Agricultural and Fisheries Research of the Ministry of the Flemish Community. Since this model only predominantly uses FADN data, it is conceivably applicable to all the EU-15 58,000 representative commercial farms recorded in this database accessible by any national or regional administrative agencies.

Before presenting an application drawn from the recently agreed mid-term review (MTR) of Agenda 2000, the following subsection first presents how key parameters of the model are calibrated in the farm generic model and how animal feeding and quota constraints are added to the generic farm model.

4.1 Calibration of the Cost Parameters

The SEPALE model relies on a modified version of the standard PMP calibration method, which skips the first step of the standard approach for two reasons. First, following Heckelei and Wolff (2003), the first step of PMP provides duals of the resource constraints that are biased. Second, resources such as farmland are supposed to be not binding at the farm level and enter into the variable cost component on the premise that farms are able to acquire farmland from other farms. As a result, we directly start with the second step, which is the calibration of the cost function.

The model relies on a farm-level profit function using a quadratic functional form for its cost component. In matrix notation, this gives

$$Z_{\mathrm{f}} = \mathbf{p}_{\mathrm{f}}\,\mathbf{x}_{\mathrm{f}} + \mathbf{a}_{\mathrm{f}}\,\mathbf{S}_{\mathrm{f}}\mathbf{x}_{\mathrm{f}} - \mathbf{x}_{\mathrm{f}}\,\mathbf{Q}_{\mathrm{f}}\mathbf{x}_{\mathrm{f}}/2 - \mathbf{d}_{\mathrm{f}}\,\mathbf{x}_{\mathrm{f}}, \tag{15}$$

where

- Z is the scalar of the objective function value,
- $\mathbf{x}_{\mathrm{f}}$ $(n \times 1)$ the vector of production quantities with n production activities,
- $\mathbf{p}_{\mathrm{f}}$ $(n \times 1)$ the vector of output prices per unit of production quantity,
- $\mathbf{Q}_{\mathrm{f}}$ $(n \times \mathrm{n})$ the diagonal matrix of quadratic cost function parameters,
- $\mathbf{d}_{\mathrm{f}}$ $(n \times 1)$ the vector of linear cost function parameters,
- $\mathbf{a}_{\mathrm{f}}$ $(n \times 1)$ the vector of technical coefficients determining how much resource base (land or animal) is needed per production quantity $\mathbf{x}_{\mathrm{f}}$,

S_f $(n \times n)$ the diagonal matrix of direct payments per unit of resource base, and

f is the index for farms.

Two sets of equations calibrate the parameters of the matrix $\mathbf{Q_f}$ and the vector $\mathbf{d_f}$, relying on output prices $\mathbf{p_{fo}}$, direct payments $\mathbf{S_{fo}}$ and average variable production costs $\mathbf{c_{fo}}$ observed at the reference period. The first-order conditions of model (15) determine the first set of equations as follows:

$$\mathbf{p}_{\mathrm{fo}} + \mathbf{S}_{\mathrm{fo}}\,\mathbf{a}_{\mathrm{f}} = \mathbf{Q}_{\mathrm{f}}\mathbf{x}_{\mathrm{f}} + \mathbf{d}_{\mathrm{f}}. \tag{16}$$

The second set of equations equates the observed average costs $\mathbf{c_{fo}}$ to the average costs implied by model (15) as follows:

$$\mathbf{c}_{\mathrm{fo}} = \mathbf{Q}_{\mathrm{f}}\mathbf{x}_{\mathrm{fo}} / 2 + \mathbf{d}_{\mathrm{f}} \tag{17}$$

with $\mathbf{c_{fo}}$ the vector of observed average variable costs per unit of production quantity that include costs of seeds, fertilizers, pesticides, contract work and other costs gathered from the FADN for each farm *f* including farmland rental cost.

The following two sets of equations calibrate the diagonal matrix **Q** and the vector **d** for each farm *f* of the sample:

$$\mathbf{Q}_{\mathrm{f}} = 2\left(\mathbf{p}_{\mathrm{fo}}\,\mathbf{x}_{\mathrm{fo'}} + \mathbf{S}_{\mathrm{f}}\,\mathbf{a}_{\mathrm{f}}\mathbf{x}_{\mathrm{fo'}} - \mathbf{c}_{\mathrm{f}}\mathbf{x}_{\mathrm{fo'}}\right)\left(\mathbf{x}_{\mathrm{fo}}\,\mathbf{x}_{\mathrm{fo'}}\right)^{-1} \tag{18}$$

$$\mathbf{d}_{\mathrm{f}} = \mathbf{p}_{\mathrm{fo}} + \mathbf{S}_{\mathrm{f}}\mathbf{a}_{\mathrm{f}} - 2\left(\mathbf{p}_{\mathrm{fo}}\mathbf{x}_{\mathrm{fo'}} + \mathbf{S}_{\mathrm{f}}\mathbf{a}_{\mathrm{f}}\mathbf{x}_{\mathrm{fo'}} - \mathbf{c}_{\mathrm{f}}\mathbf{x}_{\mathrm{fo'}}\right)\left(\mathbf{x}_{\mathrm{fo}}\,\mathbf{x}_{\mathrm{fo'}}\right)^{-1}\mathbf{x}_{\mathrm{fo}}. \tag{19}$$

With these parameters, model (15) is exactly calibrated to the reference period and is ready for simulation applications.

The basic model is further extended with feeding and quota constraints. The feeding constraint uses a CES function that allows substitution between on-farm forage crops and off-farm feed, which is calibrated on feedings observed at the reference period. The A and B sugar quota constraint is included in the first-order conditions of model (15) by adding the dual of the sugar beet quota to the right side of Eq. 16. The gross margin differential between A and B sugar beets and the next best alternative crop that is observed at the reference period approximates this dual. As explained in Buysse *et al.* (2004), the supply of A and B sugar beets includes a precautionary C supply and a quota exchange mechanism allows for a quota redistribution among sugar beet farms within the sample.

4.2 Simulation of the Mid-term Review of Agenda 2000

The three main elements in the MTR of Agenda 2000 are direct payment decoupling, cross-compliance and modulation. First, the decoupling of direct payments implies that a single farm payment replaces the previous direct payments that were linked to activities. Second, the cross-compliance renders the single farm payment subject to farm compliance with rules related to food safety, animal health and welfare and good agricultural and environmental practices. Third, the modulation introduces a system of a 5% progressive reduction of the direct payments that are higher than a threshold of 5,000 € per farm. The savings on these direct payments are added to the financing of the rural development measures defined in the CAP. Within the transitory options offered by the MTR, the Belgian government chooses to decouple all direct payments except payments for suckler cows and veal slaughters. The following subsections show how the basic model is modified to incorporate the provisions of the new MTR policy instruments.

4.2.1 Activation of the Single Payment Entitlement

The MTR assigns a single farm payment entitlement per hectare for every farm. This per hectare single entitlement is the ratio of the amount of direct payments granted to the farm during a reference period over the farmland declared for requesting the direct payments during the same reference period, including farmland for cereals, oil yielding and protein (COP) and fodder crops, but not including farmland for potatoes, vegetables and sugar beets.

Farmland planted with the eligible crops, i.e., all crops except potatoes and vegetables in open air, can activate the per hectare single payment entitlement. Three situations could occur:

1. A farm that plants an area with eligible crops of the same size of the reference farmland is entitled to receive the same amount of direct payments as before the MTR.
2. A farm that increases its area planted with eligible crops is not entitled to additional direct payments.
3. A farm that reduces its area planted with eligible crops is entitled to proportionate lower direct payments than before the MTR.

To model the MTR single farm payment adequately, a new variable denoted by 'aa_f' is defined to represent the maximum eligible area that can activate the per hectare single payment entitlement. A first constraint prevents the total single payment from exceeding the reference amount of direct payments. A second constraint restricts the per hectare single payment entitlement to the eligible area.

$$aa_{\mathrm{f}} \leq a_{\mathrm{fo'}}\, L_{\mathrm{f}}\, x_{\mathrm{fo}} \tag{20}$$

$$aa_{\mathrm{f}} \leq a_{\mathrm{f'}}\, E_{\mathrm{f}}\, x_{\mathrm{f,}} \tag{21}$$

where

L

- $\mathbf{L_f}$ $(n \times \mathrm{n})$ is the diagonal matrix with unit elements indicating whether the activity i has been declared for obtaining direct payments during the reference period and zero elements for other activities,
- $\mathbf{E_f}$ $(n \times 1)$ the diagonal matrix with unit elements for eligible crops and zero elements for others,
- aa_f the scalar of the maximum eligible area for the per ha single payment entitlement.

The single payment extends the profit function as follows:

$$Z_{\mathrm{f}} = \mathbf{p}_{\mathrm{f'}}\,\mathbf{x}_{\mathrm{f}} + aa_{\mathrm{f}}\,\mathbf{a}_{\mathrm{fo'}}\,\mathbf{S}_{\mathrm{fo}}\,\mathbf{D}_{\mathrm{f}}\,\mathbf{x}_{\mathrm{fo}}\left(\mathbf{a}_{\mathrm{fo'}}\,\mathbf{x}_{\mathrm{fo}}\right)^{-1} + \mathbf{a}_{\mathrm{f'}}\,\mathbf{S}_{\mathrm{fo}}\left(U - \mathbf{D}_{\mathrm{f}}\right)\mathbf{x}_{\mathrm{f}} - \mathbf{x}_{\mathrm{f'}}\mathbf{Q}_{\mathrm{f}}\mathbf{x}_{\mathrm{f}}/2 - \mathbf{d}_{\mathrm{f'}}\mathbf{x}_{\mathrm{f}}, \tag{22}$$

where

- $\mathbf{D_f}$ $(n \times n)$ is the diagonal matrix with the production decoupling ratio of activity i and
- **U** $(n \times n)$ the unit matrix.

4.2.2 Modulation of Direct Payments

Modulation reduces all direct, coupled and non-coupled payments beyond 5,000 € per farm by a maximum of 5% in 2007. Farms with direct payments higher than the threshold of 5,000 € can, however, choose either to not activate their direct payment entitlements or to transfer their direct payment entitlements to farms with direct payments lower than the threshold of 5,000 €. This transfer mechanism is also included in the optimization process of the model.

The following constraint introduces modulation into the model:

$$md \leq a_{\mathrm{fo'}}\, Subs_{\mathrm{fo}}\, D_{\mathrm{f}} x_{\mathrm{fo}} (a_{\mathrm{fo'}}\, x_{\mathrm{fo}})^{-1} + a_{\mathrm{f'}} Subs_{\mathrm{fo}} (I - D_{\mathrm{f}}) x_{\mathrm{f}} - mt, \tag{23}$$

where

md is the scalar of the positive amount of direct payments subject to modulation and

mt the scalar of the amount of direct payments free from modulation.

Modulation extends the profit function as follows:

$$Z_f = \mathbf{p}_{f'}\mathbf{x}_f + aa_f\,\mathbf{a}_{fo'}Subs_{fo}\,\mathbf{D}_f\,\mathbf{x}_{fo}\left(\mathbf{a}_{fo'}\mathbf{x}_{fo}\right)^{-1} + \mathbf{a}_{f'}\,Subs_{fo}\left(I-\mathbf{D}_f\right)\mathbf{x}_f - \mathbf{x}_{f'}\mathbf{Q}_f x_f/2 - \mathbf{d}_{f'}\mathbf{x}_f - md\,mp \tag{24}$$

where

mp is the scalar of the modulation percentage.

Although the MTR modulation imposes an increase in the modulation percentage in three steps from 3% in the first year, 4% in the second year and 5% in the third year, the following analysis is restricted to the simulation of the final modulation percentage.

4.2.3 Transfers of Direct Payment Entitlements

Transfers of direct payment entitlements can occur both with and without transfer of land. A certain percentage of the entitlements that are transferred can, however, be withheld by the member state. For entitlement transfers with land, 10% of the entitlement can revert to the national reserve while, for sole transfers of direct payment entitlements, up to 30% of the entitlement can revert to national reserve. Seven additional constraints and seven additional variables that are not shown here for lack of space are used to model the transfers of direct payment entitlements leaving open the possibility to realize these transfers with and without land transfers. Unobserved transaction costs can play a major role in the decision to transfer direct payment entitlements but are not modelled here.

4.2.4 Cross-compliance

Currently, the model assumes that every farm satisfies the conditions imposed by the member state. The model further assumes that these conditions do not generate additional costs. This is a reasonable assumption given that most of these conditions were already compulsory before the MTR.

4.3 Impact Analysis

The model is calibrated and run for a FADN sub-sample of 159 arable and cattle farms for which data are available for the year 2002. Because of the non-representativeness of this sub-sample, one has to be careful in extrapolat the calibrated parameters and the simulation results to the whole sector. Being only indicative of the outcome of the MTR, the simulation results illustrate the various possibilities of the model in simulating differential effects of changes in the policy-controlled parameters.

The impact analysis focuses on the decoupling and modulation elements of the MTR. The following subsections show the effects of three policy-controlled parameters: the decoupling ratio, the modulation threshold and the modulation percentage on land allocation and gross margin according to farm size. Results are given in percentage changes with respect to the reference period of 2002.

4.3.1 Impact Analysis of the Decoupling Ratio

Figure 1 shows the effects of increasing the decoupling ratio from 0% to 100% on land allocation among different types of crops with a modulation threshold set at 5,000 € and percentage set at 5%. As the decoupling ratio increases to 100%, farms substitute crops that were not subsidized before the MTR for crops that were subsidized before the MTR. This substitution effect is larger for previously subsidized crops such as wheat and barley than for previously subsidized fodder crops such as fodder maize. For the former, the decline reaches 6% while, for the latter, the decline reaches 5% for the full decoupling scenario compared with the reference period of 2002. Substitution among fodder crops is tighter as a result of the feeding constraints and few alternative fodder crops. Effects of the MTR on allocation of non-eligible crops are minor because the simulation limits the activation of decoupled direct payments to the maximum amount granted during the reference period.

Figure 2 shows the effects of increasing the decoupling ratio from 0% to 100% on farm gross margins across farm sizes with a modulation threshold set at 5,000 € and percentage set at 5%. Effects of the MTR on farm gross margins are relatively smaller than effects on land allocation. As expected, a complete decoupling of the direct payments generates a positive effect on farm gross margins across all farm sizes. The larger positive effect in gross margin for farms of smaller size is due to the 5% modulation of direct payments above the threshold of 5,000 €.

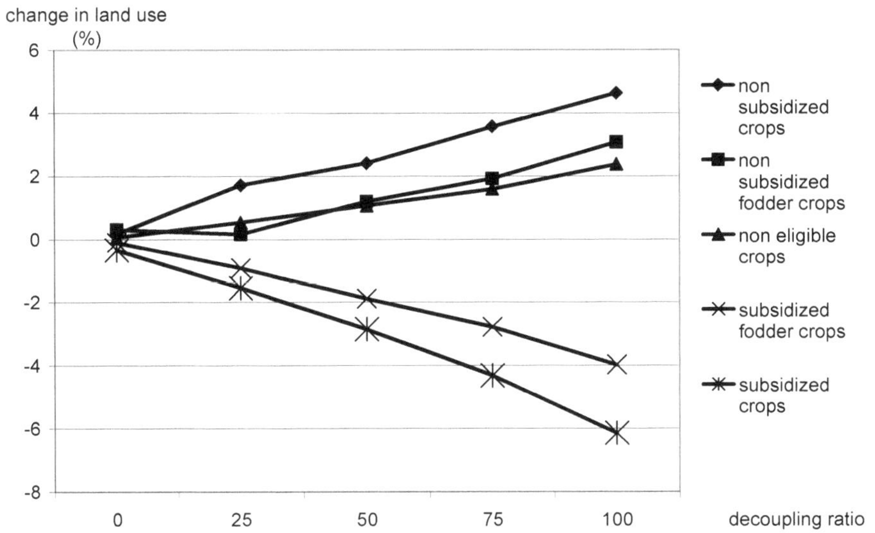

Figure 1. Changes in land allocation among crop categories with respect to the decoupling ratio.

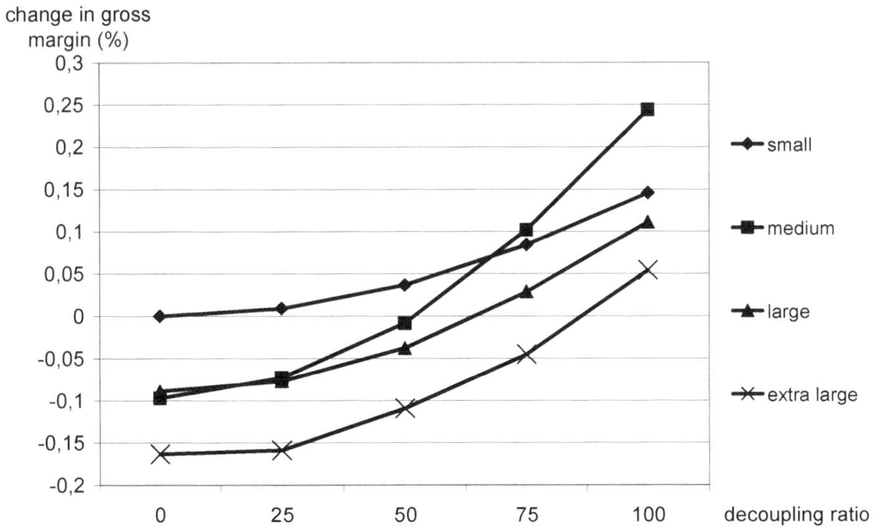

Figure 2. Changes in farm gross margin with respect to the decoupling ratio across farm sizes.

4.3.2 Impact Analysis of the Modulation

Figure 3 shows the effects of increasing the modulation percentage from 10% to 30% on farm gross margins across farm sizes with a modulation threshold set at 5,000 € and full decoupling. As expected, the effects of an increasing modulation percentage on farm gross margins are higher on farms of larger size. Since small farms with a farm gross margin lower than 9,600 € do not receive an amount of direct payments exceeding the threshold of 5,000 €, these farms are not affected by this simulation. The extra large farms with a farm gross margin higher than 48,000 € have the highest share of direct payments above the 5,000 € threshold and, therefore, see their farm gross margin reduced by almost 0.8% with a 30% modulation. The medium and large farms with a farm gross margin lower than 19,200 and 48,000 € respectively, see their farm gross margin reduced by about 0.3% with a 30% modulation.

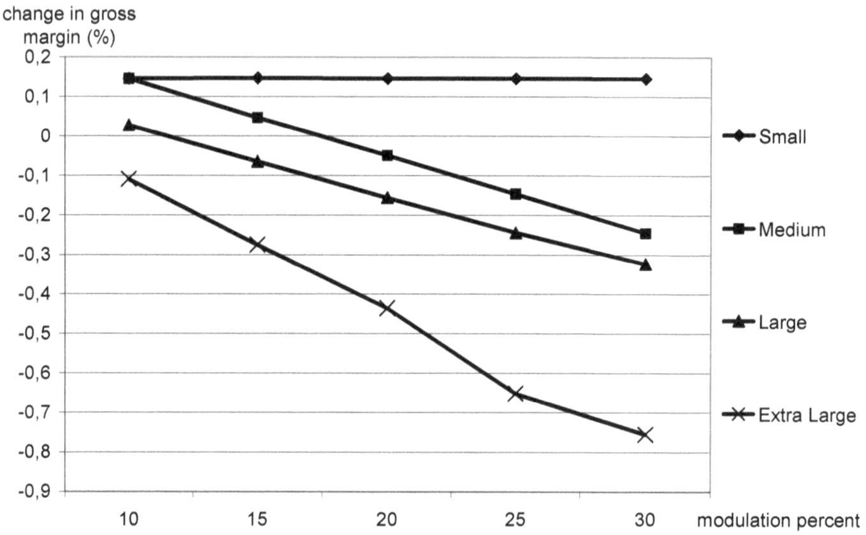

Figure 3. Changes in farm gross margin with respect to the modulation percentage across farm sizes at full decoupling.

Figure 4 shows the effects of decreasing the modulation threshold from 5,000 to 2,000 € on farm gross margins across farm sizes with a modulation percentage set at 5% and full decoupling. As expected, a lower modulation threshold leads to a decline in farm gross margin across all farm sizes. This decline is larger for farms of smaller size. A reduction of the modulation threshold combined with an increase in the modulation percentage results in even larger decline in farm gross margins.

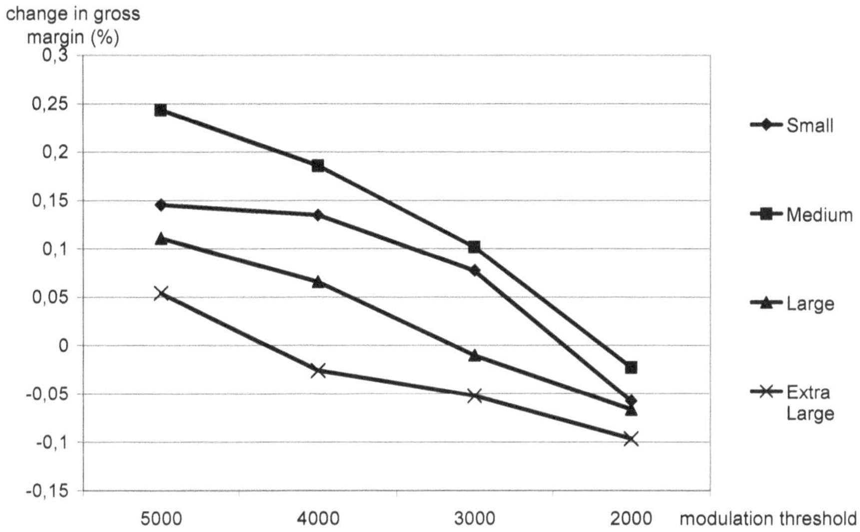

Figure 4. Impact of changes in the modulation threshold according to farm size.

4.3.3 Conclusions

In sum, the simulation results point out that the decoupling of direct payments decreases farmland allocated to crops that were subsidized in the reference period and increases farmland allocated to crops that were not subsidized in the reference period. In contrast, farmland allocated to crops that are not eligible for direct payments does not vary, a consequence of maximizing the activation of the single payment entitlement on available farmland. In addition, the simulation results confirm the positive but still minor impact of decoupling direct payments on the farm gross margin. They also show the negative but still minor impact of modulating direct payments on the gross margin of the farms with the largest size. Although these illustrative simulation results show the capacity of a farm-based PMP model to differentiate the results according to farm size, they can also be easily be differentiated according to other parameters available in the data base such as farm location and type.

5 PENDING PROBLEMS AND FURTHER DEVELOPMENT

This section discusses some of the possible extensions of PMP and some of the issues that still need to be addressed. PMP is a method that has been developed for situations in which the researcher has either very few information or faces a situation with a high heterogeneity in farms, but is willing to impose strong hypotheses on the functional form of the cost function. In PMP, one does not test economic theory but imposes it because there is not enough data to test it. PMP is often interpreted as an attempt to move from programming models to "mixed" models in which some inference from the data can be drawn (Just and Pope, 2001) and calibration of the coefficients of the cost or production function can be substituted by estimation. The difference between calibration and estimation is that in the former the researcher assigns some value to the coefficients on the basis of external information while, in the latter, the value of the coefficients is computed from a set of data using some econometric technique. PMP is therefore really in between calibration and estimation because, in its original formulation (Howitt, 1995), there are not enough data to estimate all the coefficients of the cost function and some additional hypotheses must be made. Subsections 3.1 and 3.2 give a set of such restrictions that guarantee that the cost function is regular in the sense that the marginal cost is constructed to be larger than the average cost.

However, a regular cost function does not guarantee that simulations are credible (see Heckelei and Britz, 2005). One of the problems of PMP is that it is not robust: with very little information, estimation and inference may be very unreliable. The credibility of the simulations relies mainly on the investigator's judgement. Without additional data, there is probably little improvement that can be achieved. However, as large samples such as the FADN become available, it becomes more and more useful to extend PMP and to prefer econometric estimation approaches to calibration approaches as they are less demanding in terms of hypotheses and more robust.

5.1 Application of PMP When More Data are Available

SPEP (Paris, 2001) is an example of extension of PMP to a full sample of farms sharing the same technology. In that case, the amount of information is considerably higher than in the typical single farm case of PMP. Yet the method is designed for only 1 year of data – a cross section. It is a strong hypothesis to assume that differences in output prices across farms in a cross section do indeed reveal the supply curve. More likely, differences in prices reveal differences in products, possibly local marketing conditions or differences in quality. Figure 5 shows a plot of price versus quantity produced across the year 2000 FADN sample of winter wheat producers in Belgium. If such a sample is used in SPEP to extract a cost function and the marginal cost is set equal to the price, the supply curve slope is negative.

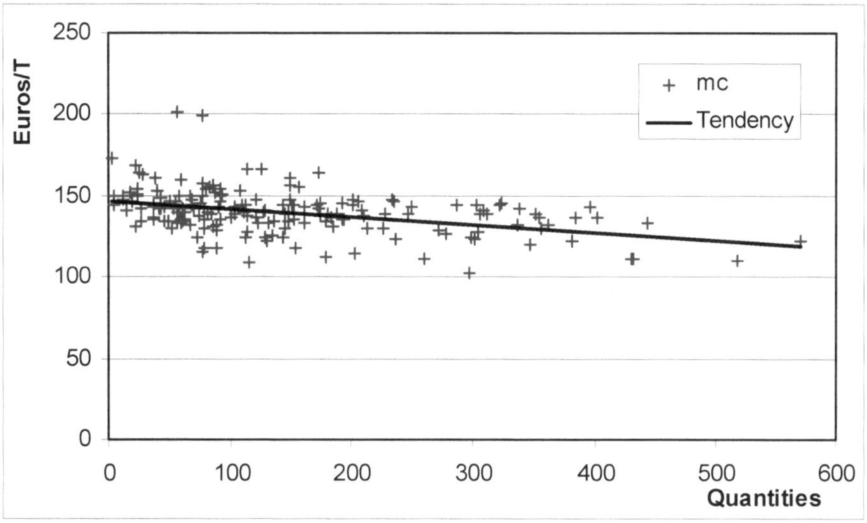

Figure 5. Price and output of wheat (including by-products) in Belgium (2000) and least-squares regression.

On the other hand, it is certainly true that when a producer expects the price of a crop to rise relatively to the other crops, he will increase production. In other words, observing the farm at different points in time seems important. This points out to panel data estimation, for which FADN data are suitable.

5.2 Constraints on Input and Output Quantities

PMP is designed to accommodate any number of constraints on input quantities. Those inputs are called binding resources. In many PMP applications at regional level, total land is a binding resource because the sum of the land for all the farms in a region cannot exceed the total agricultural land of that region. At farm level, however, that restriction does not hold anymore: from 1 year to the next, the farm can acquire any amount of land. Therefore, land is not a limiting resource; it is merely an expensive one. In that sense, inferring a shadow cost of land by means of a quantity constraint on available farmland might be questionable. It may be more reasonable to let land vary freely and obtain a proxy for its price from external sources.

Some inputs are nevertheless truly quantity-constrained at the farm level, for example family labour for obvious reasons and also pesticide or fertilizer uses because they may be limited by law.[3] Quotas, such as those existing in the milk and sugar sub-sectors, may also be binding at farm level, although it seems that in most member states, they can be traded. This means that, similar to applications of PMP at regional level, there may be limiting resources that affect estimation of the parameters of the cost function at farm level. When there is no price for additional units of one resource, the marginal cost of producing one output is not anymore equal to that output price and the shadow price of the limiting resource is used to modify the marginal cost as in Eq. 6. Therefore, to extend PMP to a sample of farms, constraints on input quantities are relevant for issues such as pesticide or fertilizer use and manure production, but these topics are yet to be addressed.

5.3 Functional Form

The original quadratic cost function of PMP, although quite simple, allows for simultaneous production of several outputs. This is a necessity in agricultural modelling, where most farms supply more than one product. Following Mundlak (2001), such diversification may have four causes: interdependence in production, fixed inputs, savings due to vertical integration, and risk management.

The simplicity of the PMP quadratic cost function is, however, obtained by suppressing all input prices from the cost function, leaving only the output quantities and some quantity-constrained inputs. All the inputs that are not quantity-constrained are implicitly used in fixed proportions to the quantity-constrained ones (most often, land).

Regarding the PMP applications in the EU, some specificities of the FADN sample with respect to inputs are noteworthy. First, data on land use and land price are available per farm and per output. To some extent, that is also true for fertilizers. For other inputs, such as pesticides, seeds and hired services, only the expenses per output are known not the quantity. Other inputs, such as capital, labour and machinery, are not allocated per output. Multi-product cost functions developed in the literature (e.g. Kumbhakar 1994) are designed only for the last type of inputs. Because the FADN farm-level data holds much richer information, there is scope and need to develop a cost function or, equivalently, a profit or production function that exploits this information fully.

[3]Family labor must be considered separately from hired labor because it is immune to moral hazard.

5.4 Aggregation Issues

Cost and production functions are defined at farm level. At an aggregate level, it is not clear what properties these functions should have. In particular, the interest for diversification may shift from risk at the farm level to trade costs at the aggregate level (Mundlak 2001), that is, a country may be diversified because importing is more expensive than producing locally, not only in pure transport costs, but also in marketing costs.

An aggregate farm results from summing all the farms in a sample. This aggregate farm is always more diversified than any farm in the sample. Therefore, the cost function that can be calibrated from such an aggregate farm bears little resemblance to the cost function that is extracted from the individual farms.

With farm-level data, there is a serious problem of heterogeneity in the sense that few farms produce the same products: this is the selection problem mentioned in Section 3. Selection causes zero production for some outputs leading to two problems. First, the cost function must accommodate true zeroes. Second, it is necessary for simulation that the parameters of the cost function are estimated for all the outputs for all the farms in the sample. Hence, some hypothesis must be made regarding the homogeneity of the sample: can we use for some farm parameter values that have been estimated on the basis of the production of other farms?

An additional aggregation issue is that in any sample, most farms are involved in a series of activities whose output levels are very limited. It is unclear whether those activities really belong to the core economic activities of the farm because they may be experimental or heavily regulated such as tobacco. The question is whether to remove such activities from the farms or to aggregate them. The former option may seem dramatic, but the total farm area and income in fact virtually do not change. The later option may appear more cautious, but induces a strong heterogeneity. Generally speaking, aggregating within a farm causes heterogeneity in the sample because an output that is seemingly identical across farms may appear with widely different prices and technical characteristics.

6 CONCLUSIONS

PMP has renewed the interest in mathematical modelling for agricultural and environmental policies for several reasons. The main advantages of the PMP approach are the simplicity of the modelling of bio-economic constraints or policy instruments, the smoothness of the model responses to policy changes and the possibility to make use of very few data to model agricultural policies. In this chapter, the focus has been on farm-level data.[4] The individual farm-based sector model SEPALE is an illustration of how PMP can be used with large farm-level samples. This model not only makes it possible to account for the individual farm structure, but also for the direct payment entitlement trade mechanisms. The results prove the relevance of the model for simulating possible alternatives to the implementation of the Agenda 2000 MTR, but this example is certainly not limitative as other applications of the model (Henry de Frahan *et al.*, 2003; Buysse *et al.*, 2004) have already shown. The possibility of distinguishing the effects according to farm size or other criteria such as region or farm type is one of the main advantages of the individual farm-base modelling. It also opens avenues to model structural changes of the sector.

Although already widely applied, as illustrated by the many references in this chapter and elsewhere, PMP is still developing and each new application raises new questions and challenges. In Section 5, some of the pending issues have been mentioned, but this is certainly not an exhaustive list. One can think, for example, about the inclusion of risk or other behavioural parameters in the model or about the extension of the model with environmental parameters.

At the farm level, strong hypotheses must be maintained for PMP to be operational. The basic shortcoming, when considering large farm-level samples such as the FADN, is that PMP only makes use of a single data point and imposes considerable structure on the technology as embodied in the cost function. It disregards all the information that is present when considering several years of data (time series) or when the data on several farms can be pooled together. As reminded by Heckelei and Britz (2005), one observation of activity level on one farm is not enough to estimate how that farm could respond to changing economic conditions. In addition, the quadratic cost function used in standard PMP is not flexible and may constrain the farm behaviour in various ways. In particular, it could be "too smooth" with respect to reasonable expectations, as shown by Röhm and

[4] Heckelei and Britz (2005) supply additional insights for applications of PMP with regional data.

Dabbert (2003). When large farm-level datasets are available, econometric estimation of general flexible functional form cost functions should solve these problems, but will pose others, especially regarding the regularity properties of those cost functions (see Wolff *et al.*, 2004). The challenge is to maintain the flexibility of the PMP approach, in particular for the modelling of bio-economic constraints, in an econometric model that can better capture the information contained in large panel datasets.

ACKNOWLEDGMENTS

This chapter is a result of a project financed from 2001 to 2003 by the Belgian Federal Ministry of Agriculture and from 2004 until now by the Flemish Institute for Science and Technology and the Ministry of Wallonia.

7 REFERENCES

Britz W., Heckelei, T., and Wolff, H., 2003, Symmetric positive equilibrium problem: a framework for rationalizing economic behavior with limited information: comment, *Am. J. Agr. Econ.* 85(4): 1078–81.

Buysse, J., Fernagut, B., Harmignie, O., Henry de Frahan, B., Lauwers, L., Polomé, P., Van Huylenbroeck, G., and Van Meensel, J., 2004, Modelling the impact of sugar reform on Belgian Agriculture, Selected paper presented at the International Conference on Policy Modelling, Paris, 30 June–2 July, 2004.

Gohin, A., and Chantreuil, F., 1999, La programmation mathématique positive dans le modèles d'exploitation agricole: principes et importance du calibrage, *Cahiers d'Economie et Sociologie Rurales* 52: 59–78.

Golan, A., Judge, G., and Miller, D., 1996, *Maximum Entropy Econometrics*, Wiley, Chichester UK.

Graindorge C., Henry de Frahan, B., and Howitt, R., 2001, Analysing the effects of agenda 2000 using a CES calibrated model of Belgian agriculture, in Heckelei T., Witzke, H.P., and Henrichsmeyer, W. (eds.) *Agricultural Sector Modelling and Policy Information Systems*, Vauk Verlag, Kiel.

Heckelei, T., and Britz, W., 2000, Positive mathematical programming with multiple data points: a cross-sectional estimation procedure, *Cahiers d'Economie et Sociologie Rurales* 57: 28–50.

Heckelei, T., and Britz., W., 2001, Concept and explorative application of an eu-wide, regional agricultural sector mode*l*, in Heckelei T., Witzke, H.P., and Henrichsmeyer, W. (eds.) *Agricultural Sector Modelling and Policy Information Systems*, Vauk Verlag, Kiel.

Heckelei, T., and Britz, W., 2005, Models based on positive mathematical programming: state of the art and further extensions, in Arfini, F. (ed.). *Modelling Agricultural Policies: State of the Art and new Challenges,* Monte Universita Parma, Parma.

Heckelei, T., and Wolff, H., 2003, Estimation of constrained optimisation models for agricultural supply analysis based on generalised Maximum Entropy, *Eur. Rev. Agric. Econ.* 30(1): 27–50.

Helming, J.F.M., Peeters, L., and Veendendaal. P.J.J., 2001, Assessing the consequences of environmental policy scenarios in Flemish agriculture, in Heckelei T., Witzke, H.P., and Henrichsmeyer, W. (eds.) *Agricultural Sector Modelling and Policy Information Systems*, Vauk Verlag, Kiel.

Henry de Frahan, B., Elouhichi, K., Harmignie, O., Polomé, P., Van Huylenbroeck, G., Buysse, J., Lauwers L., and Fermagut, B., 2004, La Politique Agricole Commune: une analyse de la réforme récente, *Regards économiques* 19: 2–13.

Howitt, R.E., 1995a, Positive mathematical programming, *Am. J. Agr. Econ.* 77(2): 329–342.

Howitt, R.E 1995b. A calibration method for agricultural economic production models, *J. Agr. Econ.* 46(2): 147–159.

Howitt, R.E., 2005, PMP Based Production Models - Development and Integration, Paper presented at the organised session "PMP, Extensions and Alternative methods" for the XIth EAAE Congress, The Future of Rural Europe in the Global Agri-Food System, Copenhagen, Denmark, August 23–27, 2005.

Judez, L., Chaya, C., Martinez, S., and Gonsalez, A.A., 2001, "Effects of the measures envisaged in 'Agenda 2000' on arable crop producers and beef and veal producers: an application of Positive Mathematical Programming to representative farms of a Spanish region" *Agr. Syst.* 67: 121–138.

Just, R.E., and Pope, R.D., 2001, The agricultural producer: theory and statistical measurement, in Gardner, B.L., and Rausser, G.C., (eds.), *Handbook of Agricultural Economics Volume 1A – Agricultural Production*, Elsevier, Amsterdam.

Kumbhakar, S. C., 1994, A multiproduct symmetric generalized McFadden cost function, *J. Prod. Ana.* 5: 349–57.

Mundlak, Y., 2001, Production and supply, in Gardner, B.L., and Rausser, G.C., (eds.), *Handbook of Agricultural Economics Volume 1A - Agricultural Production*, Elsevier, Amsterdam.

Paris, Q., 1997, CAPRI Meeting Follow-up, CAPRI Workshop in Reggio Emilia, Italy, May 1997.

Paris, Q., 2001, Symmetric positive equilibrium problem: a framework for rationalizing economic behavior with limited information, *Am. J. Agr. Econ.* 83(4): 1049–1061.

Paris, Q., and Arfini, F., 2000, Frontier Cost Functions, Self-Selection, Price Risk, PMP and Agenda 2000, Eurotools Working Papers Series, No. 20, May 2000.

Paris, Q., and Howitt, R.E., 1998, An analysis of ill-posed production problems using Maximum Entropy, *Am. J. Agr. Econ.* 80(1): 124–138.

Paris, Q., and Howitt, R.E., 2001, The multi-output and multi-input symmetric positive equilibrium problem, in Heckelei T., Witzke, H.P., and Henrichsmeyer, W. (eds.) *Agricultural Sector Modelling and Policy Information Systems*, Vauk Verlag, Kiel.

Röhm, O., and Dabbert, S., 2003, Integrating agri-environmental programs into regional production models: an extension of positive mathematical programming, *Am. J. Agr. Econ.* 85(1): 254–65.

Wolff, H., Heckelei, T., and Mittelhammer, R.C., 2004, Imposing Monotonicity and Curvature on Flexible Functional Forms, Selected Paper prepared for presentation at the American Agricultural Economics Association Annual Meeting, Denver, Colorado, August 1–4, 2004.

PART II

FISHERIES

OR/MS methodologies have found widespread use in fisheries and aquaculture over several decades. The fisheries section illustrates some important applications, although the list of topics presented is far from exhaustive. Nevertheless, the chapters show a broad range of areas where important contributions have been and are being made.

Ragnar Arnason gives an overview of major developments in fisheries management over the past few decades. OR/MS methodologies have always played an important role in this area.

Most fish stocks are shared by two or more nations, and some stocks are harvested on the high seas, outside the control of coastal states. These management issues are analysed by Trond Bjørndal and Gordon Munro.

Kaitala and Marko Lindroos review non-cooperative and cooperative game theoretic models applied to fisheries. They first compare two-player static and dynamic non-cooperative games, and second multi-player non-cooperative and cooperative coalition games.

Linda Nøstbakken and Jon Conrad review the literature dealing with uncertainty in bioeconomic modelling of fisheries. An overview of some of the main developments in the field since the introduction in the early 1970s is given.

Dan Lane presents applications in the use of multicriteria decision analysis in support of management in fisheries systems. The applications identify the multicriteria structures of the problems taken from Canadian fisheries and demonstrate how pairwise comparison feedback is integrated to compare alternative fisheries policies in ranked decision evaluations.

Sean Pascoe and Diana Tingley outline the main methods used for capacity and efficiency analysis in fisheries, and discuss data issues that are unique to fisheries when applying these methods. Several examples are presented to illustrate the potential and limitations of the methods.

Finally, the chapter by Frank Asche, Trond Bjørndal and Daniel Gordon is devoted to fish markets and looks at the demand structure for fish and seafood products. Various methodologies are discussed, and empirical results for different species presented.

Chapter 9

FISHERIES MANAGEMENT

Ragnar Arnason
Department of Economics, University of Iceland, Iceland

Abstract Fisheries worldwide are subject to economic mismanagement of major proportions. Although most commercial fish stocks are capable of yielding high net profits (rents), only a relatively few fisheries are actually profitable. Rough estimates suggest that on a global scale, the loss of net economic benefits due to mismanagement of fisheries may easily amount to 50% or more of the global landed value of some 100 billion US$ annually. The fisheries problem derives fundamentally from inappropriate social institutions controlling the fishing activity, the foremost of which is the common property arrangement. Fisheries management consists of replacing these institutions with more appropriate ones and, within that framework, setting the correct parameters over time. Global evidence suggests that the cost of fisheries management often constitutes a substantial fraction of the value of the harvest. A part of the fisheries management problem, therefore, is to strike the right balance between the efficacy of the fisheries management regime and the cost of operating it.

The problem of fisheries management is by its nature multi-disciplinary. It involves marine ecology and biology, mathematics, economics, game theory, political science and anthropology to name a few. The problem is, moreover, typically quite complex, requiring powerful modelling and calculation techniques. In many respects, this is the kind of problem operations research techniques are designed to deal with.

Keywords: Fisheries management, cost of fisheries management, efficient fisheries management, multi-objective fisheries management, operations research and fisheries management

1 INTRODUCTION

Fisheries worldwide are subject to economic mismanagement of major proportions.[1] Although most commercial fish stocks are capable of yielding high net profits (rents), only a relatively few fisheries actually manage to be profitable. In fact, seen as a whole, the world's fishing industry is heavily subsidized (Milazzo, 1998). Rough estimates suggest that on a global scale, the loss of economic rents (profits) due to mismanagement of fisheries may easily amount to 50% or more of the global landed value of some 100 billion US$ annually.[2] This loss, approximately 50 billion US$, is of a similar magnitude as the total annual development assistance from industrialized nations to the underdeveloped nations of the world.

While mismanagement characterizes the global fishery as a whole, it is important to realize that there are fisheries, sometimes quite sizeable fisheries, that do not adhere to this general pattern and are both biologically sustainable and highly profitable. These fisheries, which comprise such diverse marine conditions as those of New Zealand, Falkland Islands and Iceland, are in no way different from the other fisheries which exhibit declining stocks and negative profits. The only thing they have in common is good management. Generally, this management is based on high quality and well-enforced property rights.

Ocean fish stocks have traditionally been arranged as common property resources. This means that anyone, at least anyone belonging to a certain group (often a complete nation), is entitled to harvest from these resources. Thirty years ago, the common property arrangement was virtually universal. Today, at the beginning of the twenty-first century, it is still the most common arrangement of ocean fisheries.

Since the work of Gordon (1954) it has been known that common property resources are subject to fundamental economic problems of over-exploitation and economic waste. In fisheries, the common property problem manifests itself in:

---

[1] A global overview of the current state of fish stock exploitation can be found in FAO, 2004. A well-researched example of global fisheries mismanagement is North Atlantic cod see for example Hannesson, 1996.

[2] Global landings from ocean fisheries have in recent years been in the neighbourhood of 84 million metric tones. Average landed value may be close to US$ 1.20/kg. Various empirical studies of fisheries around the world typically suggest loss of potential profits of some 50% of the value of landings.

1. Excessive fishing fleets and effort.
2. Too small fish stocks.
3. Little or no profitability and unnecessarily low personal incomes.
4. Unnecessarily low contribution of the fishing industry to the GDP.
5. A threat to the sustainability of the fishery.
6. A threat to the sustainability of human habitation.

The essence of the fundamental problem is captured by the diagram in Fig. 1.

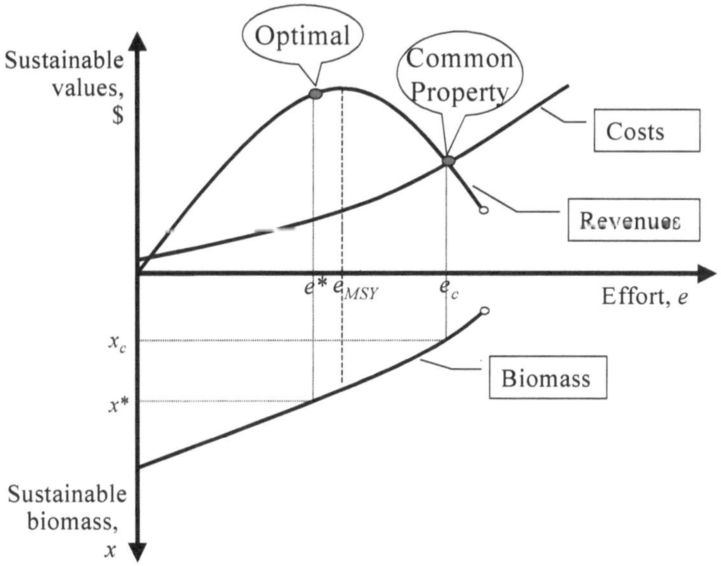

Figure 1. The sustainable fisheries model.

Figure 1 illustrates the revenue, biomass and cost curves of a typical fishery as a function of fishing effort. Fishing effort here may be regarded as the application of the fishing fleet to fishing. The revenue and biomass curves are sustainable in the sense that these are the revenues and biomass that would apply on average in the long run, if fishing effort was kept constant at the corresponding level.

The upper part of Fig. 1 is the well-known sustainable fisheries model initially forwarded by Gordon (1954). As illustrated, sustainable revenues initially increase with fishing effort but at a declining rate as the biomass is reduced. At a certain level of fishing effort, sustainable revenues are maximized. If fishing effort is increased beyond this point, sustainable revenues decline as the biomass level is reduced still further. Finally, at a certain level of fishing effort, the fishery is no longer sustainable. The stock collapses and there will be no sustainable revenues. As drawn in Fig. 1, costs, on the other hand, increase monotonically with fishing effort.

The lower part of Fig. 1 describes what happens to sustainable biomass as fishing effort is increased. Note that the level of biomass is measured in a downward direction so that the further down in the diagram the higher the biomass. The relationship between biomass and fishing effort, drawn in the diagram, shows that sustainable biomass is monotonically decreasing as fishing effort is increased. If, as illustrated in the diagram, fishing effort exceeds a certain level, the stock size becomes insufficient for regeneration – the fishery is no longer sustainable at that effort level – and the stock collapses.

Figure 1 reveals that the profit maximizing level of the fishery occurs at fishing effort level e^*. At this level of fishing effort, profits and consequently the contribution of the fisheries to GDP is maximized.[3] Note that the profit maximizing fishing effort e^* is less than the one corresponding to the maximum sustainable yield (MSY), e_{MSY}. Consequently, the profit maximizing sustainable stock level, x^*, is comparatively high as can be read from the lower part of Fig. 1. The profit maximizing fisheries policy, consequently, is biologically conservative. Indeed the risk of a serious stock decline is generally very low under the profit maximizing sustainable fisheries policy.

Under the common property arrangement, the fishing industry will find equilibrium at fishing effort level, e_c. At this level of fishing effort, costs equal revenues and there are no profits or rents in the industry. If, at the same time fishing labour is paid its reservation wage the net contribution of the fishery to the GDP is approximately zero.[4] In other words, the competitive fishery contributes virtually no net benefits to the economy. Note that this is the equilibrium outcome in any common property fishery irrespective of the size and productivity of the underlying natural resource.

[3]This assumes that all the relevant prices correctly reflect economic valuables.

[4]The assumption that labour receives its reservation wage is, in the market economy, equivalent to assuming that the labour market is in equilibrium, including no involuntary unemployment. In a situation of unemployment, that is excess supply of labour in the fishing industry or generally, even a common property fishery will generate some contribution to the GDP.

Compared to the net benefits obtainable by the profit-maximizing fishery, the common property arrangement is highly wasteful. Not only does it generate little or no net economic benefits, it also implies a much smaller biomass level. Indeed, as can easily be verified from inspection of Fig. 1, the common property fishery may easily imply the exhaustion of the biomass altogether.

It is important to realize that individual fishermen subject to the common property arrangement can do nothing to avoid this wasteful outcome. When many fishermen share ownership in a common fish stock, each has every reason to grasp as large a share of the potential yield and as fast as possible. Prudent harvesting by one fisherman to maintain the stocks will, for the most part, only benefit the other more aggressive fishermen without preventing the ultimate decline of the stocks. Thus, each fisherman, acting in isolation, is powerless to alter the course of the fishery. His best strategy is to try to grasp as large a share in the fishery as possible while the biomass is still large enough to yield some profits (Bjorndal and Scott, 1988). This in a nutshell is what has been called the tragedy of commons (Hardin, 1968). The common property arrangement in fisheries basically forces the fishermen to overexploit the fish resources, even against their own better judgment. As a result, the potential benefits of these resources, no matter how great, become wasted under the onslaught of a multitude of users.

The extreme economic waste associated with the common property problem is the reason why management of the fisheries is needed. The fundamental purpose of this management is to induce the fishery to operate at the socially most beneficial point and to do so at the least possible management cost.

It is of considerable importance to be clear about what fisheries management is. Fisheries management is not the identification of a fisheries policy that achieves some objective. That is essentially a technical exercise involving a number of different sciences. There is a great number of studies attempting to do this for particular fisheries. Fisheries management, by contrast, is people management. It is the science and art of inducing fishers to act in accordance with the social objective of the fisheries. Sometimes this social objective has been translated into a fisheries policy, sometimes it has not. Irrespective of this, fisheries management generally involves setting up and operating the appropriate institutional framework for the fishery. Designing this framework is as much art as a science. That may be the reason there are so few operation research (OR)-based or even empirically based studies on the topic. Fisheries have not, in my opinion, been mismanaged around the world because of lack of studies on the appropriate fisheries policy. Shortage of well-founded studies on the appropriate fisheries management regimes, how to set them up and operate them, may, on the other hand, have something to do with it.

In what follows, we will outline the main principles of fisheries management to achieve the above purpose. Most commercial fisheries take place in a multi-species or, more precisely, ecosystem framework and are rarely found in a sustainable state or equilibrium. Nevertheless, our analysis will for the most part take place in the context of single species, sustainable fisheries models. It is important to realize that this is only for presentational simplicity. The same basic management principles apply equally in disequilibrium as in equilibrium (see e.g. Arnason, 1990) and with only minor modifications in the multi-species or ecosystem context (Arnason, 1998).

2 OBJECTIVES OF FISHERIES MANAGEMENT

Well-defined objectives are obviously a prerequisite for sensible management of fisheries. Many objectives for fisheries have been suggested (see e.g. Charles, 2001 and references therein). Among the most frequently mentioned objectives are: (i) maximum employment, (ii) maintaining regional habitation, (iii) MSY, (iv) conservation of fish stocks and the environment, (v) generation of exports and foreign exchange (vi) economic efficiency, that is maximum economic rents and (vii) social equity. Clearly, not all of these objectives are independent.

The observation that several different objectives have been proposed for fisheries suggests that fisheries management may be seen as a problem in multi-objective maximization. To solve problems of this kind, techniques have been developed (see e.g. Coello, 1999; Kalyanmoy, 2001). Several studies of fisheries employing the multi-objective approach have been conducted (see e.g. Criddle and Streletski, 2000; Mardle *et al.*, 2000).

While the multi-objective approach is undoubtedly useful in many contexts, it appears that it is fundamentally superfluous. The basic reason is that when it comes to actually making decisions, there can only be one objective or, more precisely, one objective function. A simple argument supported by a well-known theorem in optimization theory (Afriat, 1967) is sufficient to establish this.

First, note that management implies choices. In fisheries management for instance the basic choices are: (i) which management system to adopt and (ii) what management measures to select. Logic dictates that it is either possible to make a choice or it is not. In the latter case, management will not be possible so there is no reason to waste valuable resources on research. The multi-objective approach or, for that matter, any other approach to the problem is fundamentally superfluous.

The first case, where choice is possible, is more interesting. In this case, a simple application of a theorem by Afriat (1967) shows that if a choice can be made, there must exist a function that maps the possible choices into a real number.

For our purposes, we may state Afriat's theorem as follows:

Theorem (Afriat, 1967)
If a choice, x, out of a possibility set X, say, is possible, then there must exist a continuous, concave function mapping the attributes of each possible choice into a real number.

The proof of the theorem is too long and involved to be recounted here. A relatively straight-forward proof can be found in Varian (1982).

The essence of Afriat's theorem, the existence of an objective function, is quite intuitive. If a decision can be made, the decision maker (an individual or a group of individuals) is able to select an option out of the available set. This means that the decision maker has a preference ordering over the options. He can at least compare the option selected with all other options. It follows that he must have a way to assign at least ordinal values to these options. This preference ordering defines our objective function. This objective function, of course, is what economists would refer to as a utility function or, in the case of a group, a social welfare function. Since this function appropriately represents the various objectives involved, its existence basically renders the multi-objective approach unnecessary.

There is an important practical qualification to this result. It is often quite difficult for the researcher to determine the form and content of the objective function. Sometimes, even the decision-making unit is unable to do so. This difficulty or lack of knowledge appears to be the basic justification for the multi-objective approach. When the decision makers cannot reach a decision, that is because they are uncertain about their objective function, it may help to lay-out the consequences of choices over a range of different objectives.

The social welfare function is one of those objective functions whose precise form is poorly determined. To counter this, there are certain principles of long standing in economic theory – the Pareto principle and the Hicks-Kaldor principle (see e.g. Varian, 1992) – that greatly simplify the issues. Basically, if the Hicks-Kaldor principle is adopted, the fisheries problem is reduced to maximizing the net economic benefits from the fishery. This is generally referred to as economic efficiency. If prices are correct and complete (in the sense that all valuables are priced), economic efficiency is equivalent to maximizing profits in the fishery.

It is important to appreciate that economic efficiency or profit maximization achieves many of the fishery objectives discussed at the outset of this chapter. More importantly, if prices are correct and complete, it achieves these objectives to the appropriate extent and ignores objectives that do not contribute to individual utility.[5] Thus, for instance, profit maximization implies conservation of fish stocks and marine environment to the extent that this conservation contributes to harvesting profits and conservation in itself is valuable. The same essentially applies to employment, regional habitation and the other objectives.

For the rest of this chapter we will proceed as if the social objective is to maximize profits in the fishery. This, of course, is only appropriate to the extent that prices are actually correct and complete. In the real world, they are not. Hopefully, however, they are sufficiently correct and complete for this to be a reasonable approximation. At any rate, this is the usual assumption in the design of optimal policies in economics. More, importantly, if other objectives are thought to be more appropriate, it is possible to derive the corresponding optimal fisheries management employing an approach similar to that below. In fact, in many cases, it is sufficient to simply modify some of the input and output prices and then proceed in exactly the same way. Thus, if profit maximization is thought to give insufficient weight to employment, this can be easily remedied by reducing the cost of labour in the profit calculation. Similarly, increased conservation will come out of profit maximization, if the price of landed fish is reduced in the profit calculations.

3 THE FISHERIES MANAGEMENT REGIME

All fisheries, irrespective of whether they are explicitly managed or not, are subject to an overall framework of social institutions. We refer to this institutional framework as the fisheries management regime. Essentially, the fisheries

[5]This is a simple consequence of the first welfare theorem see for example Varian, 1992.

management regime is a set of social prescriptions and procedures that control the fishing activity. All fisheries management regimes must logically comprise the following three basic components: (i) the fisheries management system (FMS), (ii) the monitoring, control and surveillance (MCS) system and (iii) the fisheries judicial system (FJS) (Fig. 2).

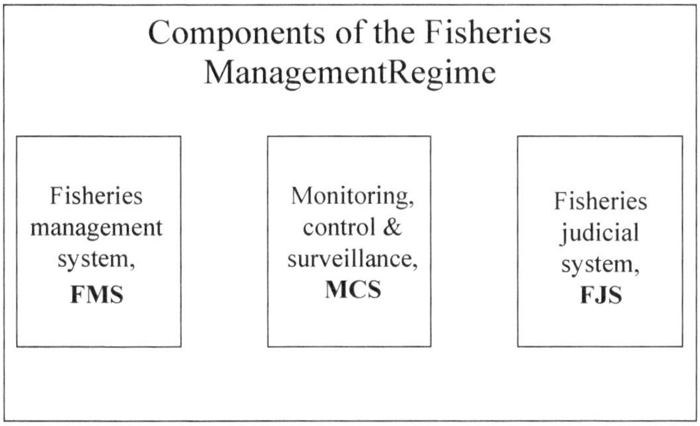

Figure 2. Components of the fisheries management regime.

FMS specifies the regulatory framework for the fishing activity. It consists of all the rules that the fishing activity must obey such as gear and area restrictions, fishing licences, catch quotas and so on.

The primary task of the *MCS* system is to observe the fishing industry's activities and to enforce its adherence to the rules of the FMS. Its secondary, but nevertheless very important, task is to collect data about the fishery that can be used to improve both the fisheries management and judicial systems as well as the MCS system itself.

FJS processes alleged violations of fisheries management rules and issues sanctions to those deemed to have violated the rules. The FJS thus complements the MCS activity in enforcing the fisheries management rules.

To achieve full benefits from fisheries management, all three components of the fisheries management regime must be appropriately designed, fully functional and well coordinated. The importance of this cannot be overemphasized. These three components of the fisheries management regimes are like links in the same chain. If any of them fails, the other components, however well designed and implemented, will generate precious little, if any, benefits.

3.1 Fisheries Management Systems

To solve the fisheries problem, a great number of different FMSs have been suggested. Most of these, however, may be conveniently grouped into two broad classes: (i) biological fisheries management and (ii) economic fisheries management. Economic fisheries management may be further divided into (i) direct restrictions and (ii) indirect economic management. The difference between these two categories is that direct restrictions impose explicit constraints on the activity of the fishermen, whereas indirect management merely changes the incentives facing the fishermen. Finally, indirect economic management may be divided into two categories: taxes and property rights. This classification is illustrated in Fig. 3.

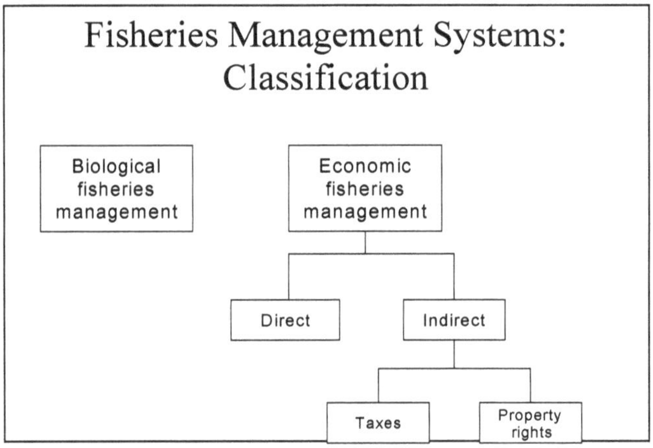

Figure 3. Fisheries management systems: classification.

To make a long story short, the only FMSs that are theoretically and appear empirically to be capable of generating net economic benefits (profits) are indirect economic ones.

Biological fisheries management, such as mesh size regulations, total allowable catch, area closures, nursery ground protection and so on, may conserve and even enhance the fish stocks. They, however, fail to generate net economic benefits because they do not remove the common property nature of the fishery that is at the root of the fisheries problem.

To see this, it is useful to take a moment to think about the effect of total allowable catches (TACs). To fix ideas, let us imagine that fishery is initially at a competitive equilibrium where there are no profits and the fishing effort is at e_c in Fig. 1. Let TAC restriction be imposed at this point and enforced by effort limitations.[6] Then, at this reduced level of fishing effort, the fishery will become profitable. As a result, each company will have an economic incentive to increase fishing effort to get more profits. Consequently, there will be a movement to improve fishing vessels, build new ones and so on to partake in these profits. Thus, to maintain the TAC restriction, the operating time of the fleet will have to be curtailed further and so on. At the end, when this process has worked itself out and a new equilibrium been established, sustainable catches may have increased and the fish stocks improved. However, what really counts, the profits, that is net economic benefits, from the fishery, will be the same as before, namely zero.

Very much the same applies to *direct economic restrictions*. Such restrictions take various forms. There are limitations on days at sea, fishing time, number of vessels, holding capacity of the vessels, engine size, and so on. Just as biological fisheries management, these methods fail to generate economic rents because they do not remove the common property nature of the fishery. As a result, the fishermen are still forced to compete with each other for a share in the catch until all net economic benefits have been wasted through expansion of the fisheries inputs that are not controlled.

In addition to this rather negative outcome, it is important to realize that setting and enforcing biological and economic fisheries restrictions is invariably costly. Usually, these costs are quite substantial.[7] Since, as we have seen, biological and economic restrictions do not generate any economic benefits, at least not in the long run, these management costs represent a net economic loss. Consequently, we are driven to the somewhat distressing conclusion that these fisheries management methods – biological fisheries management and direct economic restrictions – may be worse than nothing.

Indirect economic fisheries management may be divided into taxation and various types of private property rights as depicted in Fig. 3.

The appropriate *taxation of the fishing industry* can in principle induce the industry to operate in the social optimal way. Taxes can do this by reducing revenues, as illustrated in Fig. 4 (e.g. tax on landings), or increasing the costs of fishing (e.g. tax on fishery inputs). In practice, however, there are severe technical and social problems with using taxes as a fisheries management tool (Arnason, 1990). For this reason, fisheries management by means of taxes has not been used in any significant ocean fishery so far.

---

[6]For example, limited fishing days.

[7]According to Arnason *et al.*, 2000, fisheries management costs typically range from 3%–20% of the gross value of the harvest.

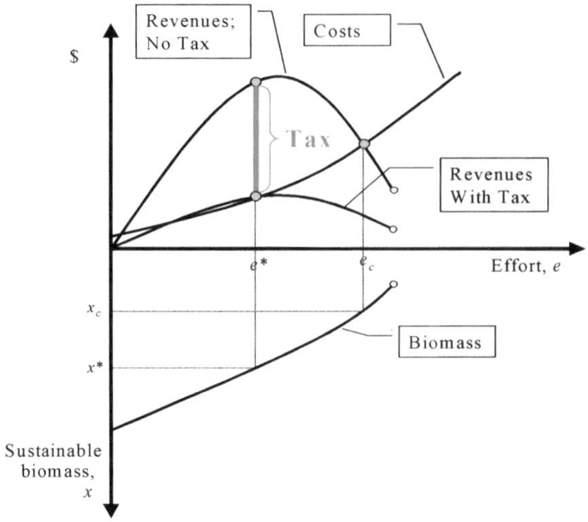

Figure 4. The effect of tax on landings.

Property rights-based regimes, on the other hand, especially ITQ systems, have been widely applied and met with a fair degree of success.

Property rights-based approaches to fisheries management attempt to eliminate the common property problem by establishing private property rights over the fish stocks. Since the source of the economic problems in fisheries is the absence of property rights, this approach should in principle be successful in securing full economic benefits from the fishery.

Several types of property rights regimes have been employed to alleviate the fisheries problem including: (i) fishing licences, (ii) sole ownership, (iii) territorial use rights in fisheries (TURFs), (iv) individual catch quotas and (v) community fishing rights (see Fig. 5). Here, we briefly discuss each of these.

Fishing licences, that is the right to conduct fishing, constitute a property right. However, this property right is quite far removed from the source of the common property problem, namely the fish stocks and harvest from them. As a result, licence holders will still be forced to compete for shares in the catch with the resulting use of excessive fishing effort and capital. Therefore, fishing licences, even when their issue is very restrictive, are not capable of significantly remedying the common property problem.

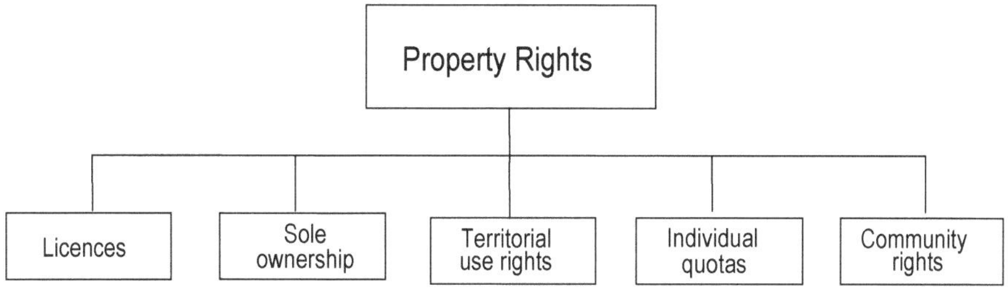

Figure 5. Types of property rights regimes.

Under *sole ownership*, one agent (individual or a firm) is awarded exclusive ownership over the resource or part thereof. Therefore, virtually by definition, sole ownership eliminates the common property problem. Consequently, the fishery should reach full efficiency (Scott, 1955).

TURFs consist of the allocation of a certain area of the ocean and the associated seabed to a single owner (user). Being similar to a property right in a farm, the owner will have every incentive to husband his TURF efficiently. Consequently, this arrangement works very well for relatively sedentary fish stocks, that is those that remain within the confines of the TURF. In fact, under those circumstances, TURFs are virtually the same as sole ownership and should lead to full economic efficiency. Empirical studies seem to confirm this prediction (see e.g. Panayoutou (1984). For relatively migratory stocks, that is stocks that periodically migrate in and out of the TURF-area, the effectiveness of TURFs is much reduced. Indeed, the indications are that the stock in question does not have to spend much time outside the TURF for its beneficial effects to be virtually nullified (Arnason *et al.*, 2000). This, obviously, greatly reduces the applicability of this method.

Individual quotas have been widely applied around the world with a fair degree of success. Transferable and perfectly divisible catch quotas are usually referred to as individual transferable quotas or ITQs. If the ITQs are also permanent they constitute a complete property right just like a building or a piece of land. In that case, standard economic theory should apply and, barring market imperfections, the fishery should automatically reach point of maximum profits (i.e. net economic benefits).

ITQs do this essentially in two ways. First, secure rights to a certain quantity of harvest[8] allow the holder to take this harvest in the economically most efficient way. The second way by which ITQs further economic efficiency is

[8]Actually defined as a share in whatever TAC is set.

by quota trades. Given quota tradability, there will be a tendency for only the most efficient fishing firms to operate in the fishery. The less efficient firms will simply find it to their advantage to sell their quota and leave the fishery. Thus, under an ITQ system, there will be a convergence to the optimal use of overall fishing capital and fishing effort and to the most efficient fishing firms operating in the fishery. This prediction has been verified in numerous empirical studies of actual ITQ fisheries (see e.g. Shotton, 2000 and the references therein).

It is important to realize however, that unlike sole ownership and well-designed TURFs, the ITQ system will not automatically lead to full efficiency in fisheries. For that to happen, the path of TACs over time must also be optimal. Under ITQs a central authority, currently usually the government, sets the TACs. For this authority to find the optimal of TAC, every season is major problem (Arnason, 1990).

One of the most visible outcomes of a quota system is the quota price, that is the price by which quotas are traded in the market. This price, just as any other market price, represents the value of the marginal fish to society as a whole. At the same time, it represents a cost to the user of the quota. After all the quota used for harvesting fish cannot be sold in the market. Thus, the quota price, acts as a deterrent to harvesting very much like the tax on landings discussed earlier. A significant difference, however, is that the tax reverts to the government whereas the quota price stays with the members of the fishing industry. This effect of the ITQ system is illustrated in Fig. 6 later.

Now, it can be shown (Arnason, 1990) that the quota price, more precisely the price of permanent quota shares, is maximized along the profit maximizing path of the fishery. Thus, the quota price contains crucial and very visible information about the optimality of the TAC or lack thereof. Thus, if the quota price rises in response to a particular TAC setting, this indicates that the TAC was in the right direction and vice versa. In this way, the quota price can serve as a guidance to the TAC setting authority (Arnason, 2000).

Under *community fishing rights*, exclusive harvesting rights are given to a community for example group of fishermen, village, municipality and so on. These exclusive rights may apply to the whole fishery or a certain share of it, for example in the form of a community fishing quota. With this exclusive asset in hand, the hope is that the group will somehow find a way to

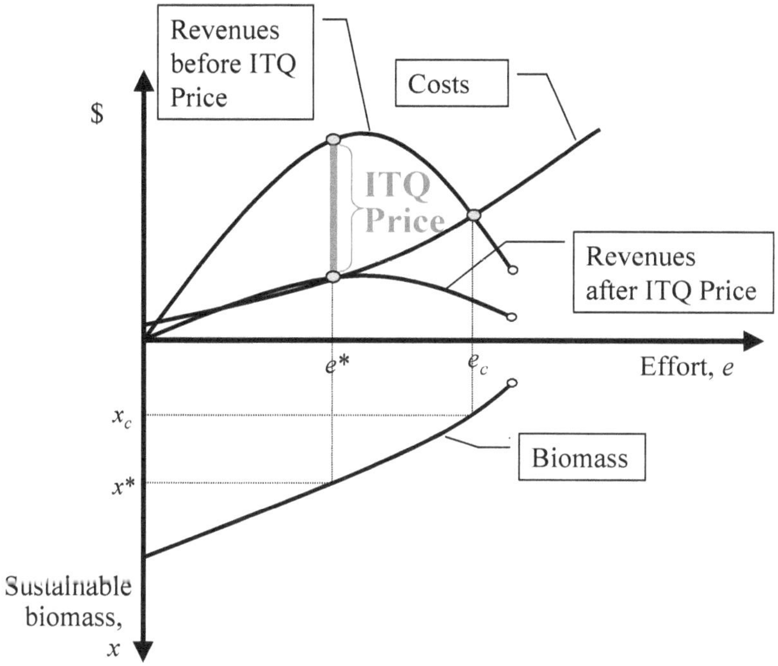

Figure 6. The effect of the ITQ system.

manage it efficiently. Fundamentally, this belief rests on a well-know argument by Coase (1960). The argument is that if property rights are clearly defined and bargaining and enforcement costs sufficiently low, then there is good reason to expect the parties involved to come to a mutually beneficial conclusion. In the case of the fishery, the mutually beneficial conclusion would be the introduction of efficient fisheries management.

The main disadvantage of communal fishing rights as a way towards good fisheries management is that this simply may not happen. It is important to realize that community fishing rights do not constitute an FMS. They merely represent devolution of the fisheries management authority from a higher level to a lower level. The community will still have to deal with the problem of designing and implementing a good FMS.

The fisheries management coming out of the community depends on various factors including the decision-making process, group dynamics and coherence. The management system adopted can easily be just as inefficient as the one preceding the community rights. Therefore to increase the probability of success, it is imperative that (i) the rights allocated to the communities should be as high quality as possible, (ii) the communities have the ability to exclude new members, (iii) the communities should consist of as homogenous group of fishermen as possible and the communities should, if at all possible, be set up so that each member's pay-off is an increasing function of the aggregate pay-off. It can be shown that if these four conditions are met, there is a high probability that the fisheries community will manage its rights in an efficient manner.

Other advantages of communal fishing rights is that they are often socially acceptable and facilitate effective enforcement of fisheries management rules on the basis of social and physical proximity and social group pressure. Obviously for this latter advantage to be effective, the group or community in question has to be reasonably small and socially coherent.

3.2 The MCS Activity and the Cost of Management

The two main functions of the MCS activity are to enforce the rules of the FMS and to generate information for use in fisheries management. The first function is much like policing. The second consists of data collection and research.

The necessary enforcement depends on the nature of the FMS. For instance under limited fishing days and closed areas, it is necessary to monitor the fishing vessels' actual fishing days and their location when out fishing. Under a system of harvest quotas, harvesting volumes have to be monitored. Much of this enforcement, not to mention the research, is quite labour and equipment demanding. This applies not least to the part of it that has to take place at sea.

Not surprisingly, it turns out that the MCS activity is quite costly. Thus, in a detailed study by Arnason *et al.* (2000) of the fisheries management costs in Iceland, Newfoundland and Norway during the 1990s, it was found that management costs, as percentage of the value of landings, ranged between 3% and 28%. In a recent study covering 26 OECD countries, Wallis and Flaaten (2003) found that fisheries management costs averaged some 6% of the landed value of fish. Other studies (see Schrank *et al.*, 2003) come up with similar results. Thus, it appears that fisheries management costs, far from being negligible, in fact amount to a substantial fraction of the maximum attainable economic rents. This suggests that in designing rent maximizing fisheries policies it is necessary to take full account of the costs necessary to implement such policies.

In an attempt to model this, let ec−e, represents the deviation of actual fishing effort, e, from the one corresponding to no management, ec. It stands to reason that the higher this deviation the higher the enforcement costs. Correspondingly, write the MCS cost function as:

$$MC(ec - e, P),$$

where, the marginal cost of reducing fishing effort is positive, that is $MCe > 0$. The variable *P* is a measure of the effectiveness of the FJS to be further discussed in Section 3.3. Obviously, the more effective the FJS, the lower the cost of enforcement. Hence, we take it that $MCP < 0$.

Under the above specifications, it should be obvious that taking account of the costs of management, the profit maximizing fishing effort would normally be greater than the one when management costs are ignored (Arnason, 2003). This will be explicitly shown in Section 3.4.

3.3 The FJS and its Cost

FJS is an often forgotten but, nevertheless, crucial part of the FMR. Its purpose is to issue sanctions to violators. Obviously, if this did not occur, there would be little or no reason for fishing firms to adhere to fisheries management prescriptions. Therefore, without an effective FJS, fisheries management will work badly, if at all.

According to the standard theory of crimes (Becker, 1968), rational agents will break rules (or at least tend to) if the expected value of violations exceeds the expected costs. Let us briefly explore the implications of this basic principle.

The expected cost of violation may be written as $\pi \cdot P$, where *P* represents the penalty for a violation and π the probability that a violation will result in a penalty. For most enforcement systems, and fisheries enforcement is no exception, π is quite small. To see this note that π is the multiple of conditional sub-probabilities each one of which must be less than unity. For instance, let π_1 be the probability that a violation is observed. Let π_2 be the probability that an observed violation leads to conviction. Note that his probability is itself a multiple of sub-probabilities, namely the probability that an observed violation is prosecuted, the probability that if prosecuted the violator will actually be found guilty and the probability that if found guilty, he will assessed a penalty. Finally, let π_3 be the probability that the penalty will actually be paid if the violator is found guilty. Now, obviously,

$\pi = \pi_1 \cdot \pi_2 \cdot \pi_3$. However, all of these probabilities must be less than unity and some would normally be very small. This applies in particular to π_1, the probability of a violator being observed. This probability is typically very low. In most fisheries, π_1 would be well under 0.1. π_2 would similarly rarely be much above 0.5. Finally, π_3 would normally be significantly less than unity. Thus the overall probability of a violation resulting in a penalty, that is π, would normally be quite small, less than 0.05 in most realistic cases. It follows that, to provide a cost effective deterrence, the penalty, P, has to be high. For instance, if $\pi = 0.05$, P has to be at least 20 times the expected gain from the violation to have the intended impact.

Now, to substantially increase π, it is necessary to expand the enforcement activity. In fisheries, however, this is generally very expensive, as we have seen in Section 3.2, especially if the fishery is generating profits. Increasing the effectiveness of the judicial system, that is probabilities π_2 and π_3, is also costly and has, moreover, limited impact if $\cdot\pi_1$ remains small. This suggests that high penalties are the cost effective way to fisheries management. It is important to realize, however, that to increase penalties also costly. It is generally more costly to administer heavier penalties. This added cost is not only in terms of collection (or imprisonment if that is necessary) but also in terms of the social costs involved.

Letting P represent the effectiveness of the FJS (including penalties) we may express the cost of the FJS with the increasing function $FC(P)$. For mathematical convenience, we assume this function is convex as well.

3.4 Optimal Enforcement Activity

As discussed in Section 3.2, the cost of enforcing FMSs is quite substantial compared to the attainable rents. Therefore, it is an important part of the FMR to operate the enforcement process in the optimal way. We can use the results of Sections 3.2 and 3.3 to lay out the elements of optimal enforcement activity.

To focus on the enforcement aspect, let us first assume that an efficient FMS (taxes or property rights) is in place. Also in the interest of simplicity, let us restrict the analysis to equilibrium, that is the sustainable fisheries model depicted in Fig. 1. The fully dynamic analysis is analytically more complex without adding anything of substance.

Let Π(*e*), where *e* stands for fishing effort represent the sustainable fisheries profit function. This function we take to be concave with maximum at some level of fishing effort (see Fig. 1).

From Section 3.2 we have the MCS cost function as $MC(e_0-e,P)$, where e_0 denotes the open access fishing effort and *P* the effectiveness of the FJS. This function is taken to be convex in both arguments.

From Section 3.3 we have the FJS cost function as *FC*(*P*), an increasing, convex function in *P*.

Given these specifications, we may express the fisheries management (or enforcement) problem as

$$\underset{e,P}{\text{Max}}\ V(e,P) = \Pi(e) - MC(e_0 - e, P) - FC(P).$$

The first order conditions for solution are

$$\Pi_e + MC_e = 0,$$

$$FC_p = -MC_p.$$

The first condition simply says that fishing effort should be increased until the total marginal benefits (note enforcement costs fall as effort increases) are zero. As indicated in Section 3.2, this implies greater optimal fishing effort than that obtained when enforcement costs are ignored. This solution is illustrated in Fig. 7.

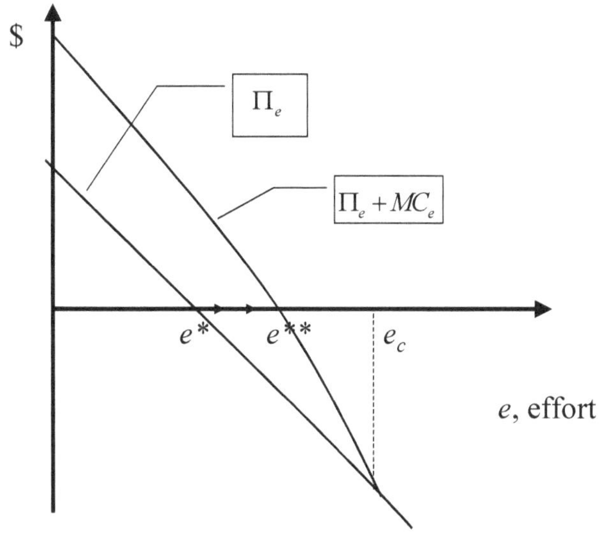

Figure 7. Optimal fishing effort with costly enforcement.

The second condition says that the FJS should be made more efficient (P increased) until the marginal benefits in terms of less MCS costs, that is $-MC_p$, equal the marginal costs, FC_p. The nature of this solution is illustrated in Fig. 8.

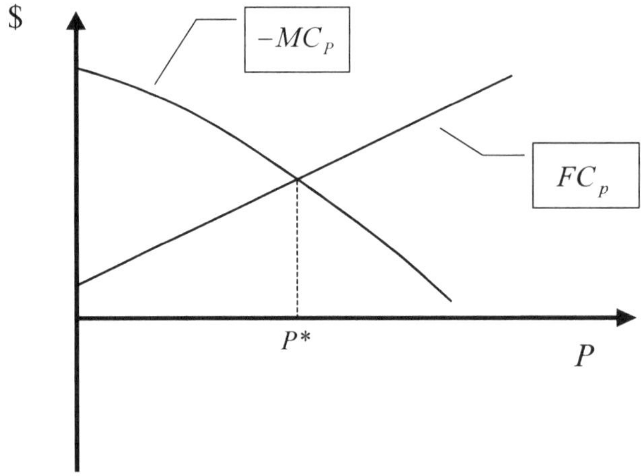

Figure 8. Optimal efficiency of the fisheries judicial system.

These solutions to the enforcement problem are conceptually simple. To apply them is much more difficult. For that, the relevant functions have to be estimated. Thus, not only must the fisheries profit function be estimated, an acknowledged major task, but also the MCS and FJS cost functions. These functions, of course, vary across fisheries and societies. Therefore to apply the theory, a great deal of difficult empirical work has to be conducted.

4 ROLE OF OPERATIONS RESEARCH IN FISHERIES MANAGEMENT

Fisheries management is really a multi-disciplinary task involving biology, ecology, economics, sociology, anthropology and political science. Biology and ecology are needed to understand the processes that determine fish availability and its response to harvesting and other human activities. Economics is needed to convert fish availability into potential for human gain and to design ways to realize this potential. Sociology and anthropology offer help in understanding the social relationships pertaining to institutional change and facilitate the assessment of social valuables that are not adequately represented by market prices. Finally, political science throws light on the power structures related to the fisheries management regime and its modification.

Many aspects of the application of the earlier sciences to fisheries management are technically demanding. It generally requires extensive biological, economic and social modelling. Given the complexity of the underlying reality – intricate, multidimensional relationships, complicated dynamics and pervasive stochasticity – these models can be very complex. To design, construct and manipulate these models requires mathematics and statistics, often of a high order. In addition, to solve them, not to mention identifying optimal paths for control variables over time, is generally a formidable task, requiring substantial numerical computations.

Therefore, the fisheries management problem, being multi-disciplinary, modelling-oriented and technically demanding, appears to be precisely the kind of problem OR is designed to deal with. This, however, does not mean that OR is equally applicable to all aspects of the problem. First note that the biology and fisheries economics involved are already subject to fairly advanced scientific inquiry by specialists and are for the most part well developed. The same applies to FMSs, which for some time have been extensively explored by fisheries economists with many actually being applied to numerous fisheries around the world. The role for OR approach in these areas is therefore limited, largely constrained to bringing already existing knowledge to particular tasks.

When it comes to the MCS and FJSs, however, the situation is different. These crucial components of the fisheries management regime are still poorly chartered territory. While FMSs have been studied for decades, fisheries and institutional economists have only recently started to explore the conditions and constraints relevant to the MCS and the FJSs and lay-out the principles for the best possible design of these parts of the fisheries management regime. It follows that in these areas a good deal of fairly basic, although highly applied research, needs to be carried out. No doubt, not the least due to the nature of these systems, OR can contribute in this respect.

Apart from this, the role of OR in fisheries management would primarily be to assist in solving the fisheries management problem in particular fisheries. This involves applying existing knowledge to model building, model manipulation and solution, selection of the most appropriate FMS, the design of the appropriate MCS system and the FJS and, even, to facilitate the emergence of a political consensus concerning fisheries management.

Numerous OR-oriented studies of fisheries problems have been conducted (see e.g. Anderson and Ben-Israel, 1981; Rodrigues, 1990; Wallace and Ólafsson, 1994). Most of these are more concerned with identifying the optimal fisheries policy than designing management systems. More recently, however, the focus has shifted somewhat to the latter (see e.g. Anon., 2001; IREPA, 2003; BEMMFISH, 2004; Bjorndal *et al.*, 2004). Nevertheless, in this field of practical fisheries economics a great deal of research work remains to be done.

5 REFERENCES

Afriat, S. 1967. The construction of a utility function from expenditure data. *International Economic Review* 8:67–77.

Anderson, L. G. and Ben-Isreal. 1981. Modelling and simulation of independent fisheries: An optimal effort allocation using mathermatical programming. In K.B. Haley (ed.) *Applied Operations Research in Fisheries.*

Anon. 2001. *The Significance of Economic Incentives in Fisheries Management under the CFP.* Fodevareokonomisk Institut, rapport n. 127. Statens Jordbrugs - og Fiskeriokonomiske Institut. Copenhagen.

Arnason, R. 1990. Minimum Information Management in Fisheries. *Canadian Journal of Economics* 23:630–653.

Arnason, R. 1998. Ecological fisheries management using individual transferable share quotas. *Ecological Applications* 8(1) Supplement:151–159.

Arnason, R. 2000. Economic instruments to achieve ecosystem objectives in fisheries management. *ICES Journal of Marine Science* 57(3):742–751.

Arnason, R. 2003. Fisheries management costs: some theoretical implications. In: W.E. Schrank, R. Arnason, R. Hannesson (eds.) *The Cost of Fisheries Management.* Ashgate Aldershot, UK.

Arnason, R., R. Hannesson, and W.E. Schrank. 2000. Costs of fisheries management: the cases of Iceland, Norway and Newfoundland. *Marine Policy* 24:233–243.

Arnason, R., Magnusson, G., and Agnarsson, S., 2000. Herring Fishery Games. *Marine Resource Economics* 15(4):293–319.

Becker, G.S. 1968. Crime and punishment: an economic approach. *Journal of Political Economy* 76:169–217.

BEMMFISH. 2004. *Final Report.* Bio-Economic Modelling of Mediterranean Fisheries. EU Research Program. Q5RS-2001–01533.

Bjorndal, T., D. Lane, and A. Weintraub. 2004. Operational research models and the management of fisheries and aquaculture: a review. *European Journal of operations Research* 156:533–540.

Bjorndal, T. and A.D. Scott. 1988. Does the Prisoners' Dilemma Apply to a Fishery. Institute of Fisheries Economics. *Discussion paper series* 5/1988. Norwegian School of Economics.

Charles, A.T. 2001. *Sustainable Fishery Systems.* Fish and Aquatic Resources Series 5. Blackwell Science. Oxford.

Coase, R.H. 1960. The problem of social cost. *Journal of Law and Economics* 3:1–44.

Coello, C. 1999. A comprehensive survey of evolutionary-based multiobjective optimisation techniques. *Knowledge and Information Systems. An International Journal* 1:269–308.

Criddle, K.R. and A.Y. Streletski. 2000. Multiple criterion management of a sequential fishery. *Annals of Operations Research* 94:259–273.

FAO. 2004. *The State of World Fisheries and Aquaculture 2004.* Food and Agricultural Organization of the United Nations. Rome.

Gordon, H.S. 1954. Economic theory of a common property resource: The fishery. *Journal of Political Economy* 62:124–142.

Hannesson, R. 1996. *Fisheries Mismanagement: The Case of the North Atlantic Cod.* Fishing News Books. Oxford.

Hardin, G. 1968. The tragedy of the commons. *Science* 162:1243–1247.

IREPA. 2003. *Fishery Regulation and the Economic Responses of Fishermen: Perceptions and Compliance.* Final Report. European Union Fifth Framework Programme: Quality of Life and Management of Living Resources. Project QLK5-CT1999-01405. Bruxelles.

Kalyanmoy D. 2001. Nonlinear goal programming using multi-objective genetic algorithms. *Journal of the Operational Research Society* 52:291–302.

Milazzo, M. 1998. *Subsidies in World Fisheries: A Re-examination.* World Bank Technical Papers no. 406. Washington DC.

Panayoutou, T. 1984. *Territorial Use Rights in Fisheries.* FAO Fisheries Report No 289, suppl. 2. Food and Agricultural Organization of the United Nations. Rome.

Rodrigues, G. (ed.). 1990. *Operations Research in Fisheries and Fisheries Management.* Kluwer. Dordrecht.

Schrank, W.E., R. Arnason, and R. Hannesson. 2003. *The Cost of Fisheries Management.* Ashgate Aldershot, UK.

Scott, A.D. 1955. The fishery: The objectives of sole ownership. *Journal of Political Economy* 63:727–56.

Shotton, R. (ed.). 2000. Use of property rights in fisheries management. *FAO Fisheries Technical Paper* 401/1. Food and Agriculture Organization of the United Nations. Rome.

Mardle, S., S. Pascoe, and M. Tamiz. 2000. An investigation of genetic algorithms for the optimization of multi-objective fisheries bioeconomic models. *International Transactions of Operations Research* 7:33–49.

Varian, H. 1982. The nonparametric approach to demand analysis. *Econometrica* 50:945–973.

Varian, H. 1992. *Microeconomic Analysis.* 3rd ed. W.W. Norton & Company. New York.

Wallace, S.W. and S. Ólafsson (ed.). 1994. The Nordic fisheries management model: description and experience. *Nord* 1994:4. Nordic Council of Ministers, Denmark.

Wallis, P. and O. Flaaten. 2003. Fisheries management costs: Concepts and studies. In W.E. Schrank, R. Arnason, R. Hannesson (eds.). *The Cost of Fisheries Management.* Ashgate Aldershot, UK.

Chapter 10

SHARED FISH STOCKS AND HIGH SEAS ISSUES

Trond Bjørndal[1] and Gordon Munro[2]
[1]*Centre for Fisheries Economics, SNF, Bergen, Norway;* [2]*CEMARE, University of Portsmouth, Portsmouth, UK*

Abstract One of the most significant fishery resource management problems to have arisen under the New Law of the Sea consists of the management of internationally shared fish stocks. A game theoretic analysis is given of the management of the two key classes of such stocks, those shared between and among neighbouring coastal states – transboundary stocks – and those crossing the boundary of the coastal state exclusive economic zone into the adjacent high seas – highly migratory and straddling stocks. The management of highly migratory and straddling stocks raises particularly difficult management issues, several of which remain unresolved. The analysis is accompanied by two case studies: Norwegian spring-spawning herring, a straddling stock and Northeast Atlantic bluefin tuna, a highly migratory stock.

Keywords: Transboundary fish stocks, highly migratory fish stocks, straddling fish stocks, non-cooperative and cooperative games, unregulated fishing

1 INTRODUCTION

The management of world capture fishery resources, which currently yield annual harvests in the order of 85 million tonnes (Normura, 2004), has been plagued by the fact that, historically, the resources have been "common pool" in nature, that is open to all. It is easy to demonstrate that, when commercially valuable fishery resources have the characteristic of being "common pool," the consequences are overexploitation of the resources, from society's point of view, and economic waste (Bjørndal and Munro, 1998).

An international attempt to mitigate the "common pool" aspect of world capture fishery resources was undertaken through the 1973–1982 UN Third

Conference on the Law of the Sea, which brought forth the UN Convention on the Law of the Sea (UN, 1982), which did, in turn, come into force in 1994. Prior to the 1982 UN Convention, the fisheries jurisdiction of coastal states (states with significant marine coastlines) extended out to a maximum of 12 miles (19 km) from shore. Under the 1982 UN Convention, coastal states have been enabled to establish exclusive economic zones (EEZs) out to 200 nautical miles (370 km) from shore. The coastal states do, to all intents and purposes, have property rights to the fishery resources contained within their respective EEZs (McRae and Munro, 1989). Vast amounts of hitherto international "common pool" fishery resources became coastal state property, as a consequence. It has been estimated that as much as 90% of the world's capture fishery resources are encompassed by EEZs (Eckert, 1979).

The EEZ regime has, however, brought with it its own set of fishery resource management problems, one of the most important of which arises from the mobility of fish encompassed by the EEZs. The typical coastal state, upon establishing an EEZ, found that some of the fishery resources, encompassed by the EEZ, crossed the EEZ boundary into neighbouring EEZs, the adjacent high seas, or both, where they were subject to exploitation by other states. Such fishery resources are deemed by the FAO to be "shared," that is, subject to exploitation by two or more states. It is estimated that as much as one third of the world capture fishery harvests are based upon shared fishery resources (Munro *et al.*, 2004). This chapter is concerned with the economic management of such shared capture fishery resources.

The FAO sets out four non-mutually exclusive categories of shared fish stocks, these being:

1. Transboundary stocks – fishery resources moving from one EEZ to one, or more neighbouring EEZs.
2. Highly migratory stocks (tuna primarily), which because of their nature cross the EEZ boundary into the adjacent high seas, where they become subject to exploitation by so-called distant water fishing states (DWFSs). A DWFS is a fishing nation, some of whose fishing fleets operate far beyond that state's home waters,
3. Straddling stocks – all other fish stocks crossing the EEZ boundary into the adjacent high seas.
4. Discrete high seas stocks – those few stocks remaining wholly in the high seas (Munro *et al.*, 2004).

We shall, in this chapter, have nothing to say about Category (4), which, currently, are of minor economic importance. Categories (2) and (3) can, for our purposes, be safely merged. We shall, from hereon in, refer to the merged categories, simply as straddling fish stocks.

In light of the small percentage of capture fish stocks estimated to lie outside of the EEZs, it might seem reasonable to suppose that the management of straddling fish stocks would present only a minor resource management problem. Such was the view at the close of the UN Third Conference on the Law of the Sea in 1982 (Kaitala and Munro, 1993). This view proved to be wholly unfounded, however. So serious a problem did the management of these resources become that the UN found it necessary to mount an international conference devoted solely to the management of these stocks, a conference popularly referred to as the UN Fish Stocks Conference, 1993–1995. The Conference produced an agreement, popularly referred to as the UN Fish Stocks Conference,[1] which came into force in 2001 (UN, 1995). The 1995 UN Agreement serves, not to supplant any part of the 1982 UN Convention, but rather to supplement and buttress the Convention (Munro *et al.*, ibid.).

In proceeding to examine the economics of the management of shared fish stocks, we shall first consider the relatively simple case of transboundary fish stocks, and then deal with the more complex case of straddling fish stocks. We conclude with a case study of the cooperative management of a major straddling fish stock, namely the Norwegian spring spawning herring stock.

The economics of the management of transboundary and straddling fish stock consists of a blend of the dynamic economic model of fishery resources confined to a single EEZ – unshared fish stocks – and the theory of games. To set the stage, then, let us first review the economics of the management of unshared fish stocks:

2 THE BASIC ECONOMICS OF THE MANAGEMENT OF UNSHARED FISH STOCKS: A REVIEW[2]

In this review, we look first at the "ideal", the management of an unshared fish resource under an all-powerful social manager, in which all "common pool" aspects of the resource are eliminated. We then turn to the polar

[1]The full names of the Conference and the Agreement are: United Nations Conference on Straddling Fish Stocks and Highly Migratory Fish Stocks, 1993–1995; Agreement for the Implementation of the Provisions of the United Nations Convention on the Law of the Sea of 10 December 1982 Relating to the Conservation and Management of Straddling Fish Stocks and Highly Migratory Fish Stocks (UN, 1995).

[2] Sections 2 and 3 of this chapter draw heavily upon Bjørndal *et al.* (2000).

opposite of a pure open access fishery, in which the "common pool" aspects of the resource are unchecked.

2.1 The All-Powerful Social Manager

We assume a deterministic world and suppose that the relevant underlying biological model is the general production model of M.B. Schaefer (see, Clark, 1990). The resource dynamics are described by the following differential equation model.

$$\frac{\mathrm{d}x}{\mathrm{d}t} = F(x) - h(t), \quad x(0) = x_0, \tag{1}$$

$$h(t) = qE(t)x(t), \tag{2}$$

where $x(t)$ is the non-negative state variable representing the biomass at time t, $F(x)$ is the growth function of the biomass, $h(t)$ is the harvest rate, q, a constant, is the catchability coefficient, hereafter assumed to equal 1, and $E(t)$ is the fishing effort, defined as the flow of labor and capital serviced devoted to harvesting fish. The usual assumptions made are that $F(x) > 0$ for $0 < x < K$. The biomass K is often referred to as the carrying capacity. In the analyses here, it is assumed that E is of the feedback form, such that $E(t) = E(x(t))$. Furthermore, we assume that $0 < E(x) < E^{\max}$.

Now let us assume that the demand for harvested fish and the supply of fishing effort are both perfectly elastic. Let p and a, both constants, denote the price of harvested fish and the unit cost of fishing effort, respectively. The net revenue from the fishery, or resource rent, at time t is given by:

$$\pi = (px - a)E \tag{3}$$

or alternatively

$$\pi = (px - c(x))h, \tag{4}$$

where $c(x)$, the unit cost of harvesting, is given by $c(x) = a/x$ (recall our assumption that $q = 1$).

The objective of management, from society's point of view, is to maximize the present value of the resource rent:

$$\max J(x_0, E) = \int_0^{\infty} e^{-\delta t}[px(t) - a]E(t)dt \tag{5}$$

such that Eq. 1 holds, and where δ is the social rate of discount.

Consider next the optimal strategy for the all-powerful social manager, in managing the resource. The problem has a unique optimal solution in which the harvesting strategy *E(x)* is discontinuous in the state variable x. There is an optimal steady state x^* (that is, an optimal resource stock level x^*) that is determined by the following equation (see, e.g. Clark and Munro, 1975):

$$F'(x^*) + \frac{\partial \pi / \partial x^*}{\partial \pi / \partial h\big|_{h=F(x^*)}} = \delta \tag{6}$$

The second term on the LHS of Eq. 6 is the so-called marginal stock effect, which reflects the impact of marginal investment in the resource on harvesting costs.

Given that the capital employed in harvesting is perfectly malleable, the optimal approach path is the most rapid one. Denoting the optimal fishing effort by $E^*(x)$, we have the following most-rapid approach rule:

$$E^*(x) = \begin{cases} E^{\max} & \text{for } x(t) > x^* \\ F(x^*)/x^* & \text{for } x(t) = x^* \\ 0 & \text{for } x(t) < x^* \end{cases} \tag{7}$$

In other words, when the stock level exceeds the optimal steady-state level x^* then the maximum fishing effort is applied until the stock level reaches the level x^*. On the other hand, if the stock level is sub-optimal, then $E(x)$ is set equal to zero, and the stock is allowed to recover at the most rapid rate, until the level x^* is reached. If it should have been the case that our model was non-linear, for example, if we had $p = p(h)$, or that the capital employed in harvesting was not perfectly malleable, the most rapid approach path would not have been optimal (see Clark, 1990).

2.2 "Common Pool" Fishery à Outrance

The polar opposite to the case of an all-powerful social manager is that, in which, the fishery is a pure open-access one, subject to no government regulations of any form. In this case, the biomass will be driven below the social optimum, x^*, to a point where the resource rent is fully dissipated. The founder of modern fisheries economics, H. Scott Gordon, referred to the resultant equilibrium as bionomic equilibrium (Gordon, 1954), which can be seen as a benchmark of fisheries exploitation undesirability. Denoting the bionomic equilibrium biomass as x^∞, the equation is given by:

$$p - c(x^{\infty}) = 0 \tag{8}$$

It can be shown that $x^* = x^{\infty}$, if and only if, $\delta = \infty$ (Clark and Munro, 1975). The implication is that, under pure open access, fishermen are given the incentive to discount wholly all future returns from the fishery.

2.3 Management of Transboundary Fish Stocks

Now let it be supposed that the resource in question, rather than being confined to the waters of a single EEZ, crosses the EEZ boundary into the EEZs of one or more neighbouring coastal states, and, as such, is a shared stock. Let us suppose, initially, that the resource is shared by two neighbouring coastal states only, which we shall designate as coastal states I and II.

2.3.1 The Two Coastal State Case

Under the terms of the 1982 UN Convention on the Law of the Sea, the two coastal states I and II, are called upon to enter into negotiations over the cooperative management of the resource (UN, 1982, Article 63(1)). However, the two states are not required to establish a cooperative resource management arrangement. If the two states do not establish a cooperative arrangement, then each is to manage the segment of the resource within its EEZ in accordance with the other provisions of the 1982 UN Convention (Munro *et al.*, 2004). Let us refer to this as the default position.

The default position, non-cooperative management, is modeled theoretically as an infinite horizon non-cooperative game (Kaitala, 1986; Mesterton-Gibbons, 1993). We shall employ Nash's model of a two player non-cooperative game (Nash, 1951). It is assumed that there is strategic interaction between the coastal states, in the sense that I's harvesting of the resource will have an impact on II's harvesting activities, and vice versa. If there was no strategic interaction between the two, there would be no basis for a non-cooperative, or a cooperative, game, and considerations of cooperative management would largely be beside the point.

Let us now assume that the fishing effort costs of the two coastal states differ, such that $a_{\mathrm{I}} < a_{\mathrm{II}}$, that is, I is a more efficient harvester than II. It can be easily shown, in the context of our model, that: $x_{\mathrm{I}}^* < x_{\mathrm{II}}^*$; $x_{\mathrm{I}}^{\infty} < x_{\mathrm{II}}^{\infty}$. We shall assume that $x_{\mathrm{II}}^{\infty} < x_{\mathrm{I}}^*$.[3]

---

[3]If $x_{\mathrm{I}}^* < x_{\mathrm{II}}^{\infty}$, the case would be uninteresting. Coastal state I would drive II out of the fishery and manage it optimally as an unshared fishery.

We know from game theory analysis, that when the two coastal states act non-cooperatively, the Nash non-cooperative feedback equilibrium solution (Nash, ibid.) is such that the resource will be depleted in a most rapid approach manner until x_{II}^{∞} has been reached (Clark, 1980). Then, the non-cooperative feedback strategies of the two coastal states can be defined as:

$$E_{\mathrm{I}}^{N}(x) = \begin{cases} E_{\mathrm{I}}^{\max} & \text{for } x > \min(x_{\mathrm{I}}^{*}, x_{\mathrm{II}}^{\infty}) \\ F(x)/x & \text{for } x = \min(x_{\mathrm{I}}^{*}, x_{\mathrm{II}}^{\infty}), \\ 0 & \text{for } x < \min(x_{\mathrm{I}}^{*}, x_{\mathrm{II}}^{\infty}) \end{cases} \tag{9a}$$

$$E_{\mathrm{II}}^{N}(x) = \begin{cases} E_{\mathrm{II}}^{\max} & \text{for } x > x_{\mathrm{II}}^{\infty} \\ 0 & \text{for } x < x_{\mathrm{II}}^{\infty} \end{cases} \tag{9b}$$

Given our assumption that $x_{\mathrm{I}}^{*} < x_{\mathrm{II}}^{\infty}$, the transboundary resource will be subject to overexploitation. Clark (1980) demonstrates that, if $a_{\mathrm{I}} = a_{\mathrm{II}}$, that is, the two players are symmetrical, the resource will be driven down to the common bionomic equilibrium. Thus, other than in exceptional circumstances, cooperation does indeed matter.

The payoffs to I and II in the non-cooperative game can be seen as the present value of the net economic return, or resource rent, accruing to I and II, respectively. Let us denote the payoffs to I and II, arising from the solution to the non-cooperative game as: $J_{\mathrm{I}}(x(0), E_{\mathrm{I}}^{\mathrm{N}}, E_{\mathrm{II}}^{\mathrm{N}})$ and $J_{\mathrm{II}}(x(0), E_{\mathrm{I}}^{\mathrm{N}}, E_{\mathrm{II}}^{\mathrm{N}})$. In our following discussion of a cooperative fisheries game, these payoffs are seen to constitute the threat point payoffs.

Having determined the consequences of non-cooperation, we investigate opportunities for cooperative management. In doing so, we seek to measure the economic benefits of cooperation, and to address the question of the sharing these cooperative benefits in an equitable manner. We turn to Nash's two player model of a cooperative game for guidance (Nash, 1950, 1953).

The two minimum conditions, which must be met if the solution to the cooperative fisheries game is to prove to be stable, are as follows. First the solution must be Pareto optimal. Secondly, the individual rationality constant must be satisfied, in that each player must be assured a payoff at least as great as its threat point payoff, which we have defined as arising from the solution to a non-cooperative game.

Now let us examine the details of cooperative resource management arrangements of the transboundary stock by the two coastal states. We assume that cooperative arrangements, upon being achieved, are binding (but see Kaitala and Pohjola, 1988), and that side payments between the players are a feasible policy option. Kaitala and Munro (1993) show that in this particular game setting I will buy out II. Furthermore, the cooperative arrangement will be focused on the sharing of the total net returns from the fishery between the two.

Let $\omega(x(0))$ denote the present value of the global net economic returns from the fishery, commencing at $x = x(0)$, following the optimal harvest strategy of I. Let $\omega_{\text{I}}(x(0))$ and $\omega_{\text{II}}(x(0))$ denote the share of I and II respectively of the global net economic returns from the fishery under a cooperative resource management arrangement. We have:

$$\omega(x(0)) = \omega_{\text{I}}(x(0)) + \omega_{\text{II}}(x(0)). \tag{10}$$

The shares, defined by Eq. 10, are Pareto optimal. If I(II) receives additional benefits from the fishery, it can only do so at the expense of II(I).

Given that the individual rationality constraint is satisfied, we can define the cooperative surplus, $e(x(0))$ as the difference between $\omega(x(0))$ and the sum of the threat point payoffs. Thus, we have:

$$e(x(0)) = \omega(x(0)) - \left[\sum_{i=\text{I}}^{\text{II}} J_{\text{i}}(x(0), E_{\text{I}}^{\text{N}}, E_{\text{II}}^{\text{N}}\right]. \tag{11}$$

An application of the bargaining scheme of Nash (1950) leads to the outcome that, under the side payments regime, the cooperative surplus will be divided evenly between I and II (Kaitala and Munro, 1997). The cooperative solution payoffs to I and II can be expressed as follows:

$$\omega_i(x(0)) = e(x(0))/2 + J_{\text{i}}(x(0), E_{\text{I}}^{\text{N}}, E_{\text{II}}^{\text{N}}), \quad i = \text{I, II}. \tag{12}$$

This result will hold true, even though I and II are, in economic terms, quite different. The rationale in applying the Nash bargaining scheme is that upon joining the agreement each coastal state can be seen to make equal contribution to reaching the agreement and to generating the subsequent economic benefits.

A recent empirical application of the model described is to be found in the article of Bjørndal and Lindroos (2004) on the North Sea herring resource. The transboundary resource is managed cooperatively by a coastal state, Norway, and by what we might call a coastal state entity, the EU. The authors' best estimates show Norway's fishing effort costs to be significantly below those of the EU. Political considerations prevent Norway from "buying out" the EU in this case. Even though the *optimum optimorum* is not achievable, the cooperative surplus is substantial, as a consequence of the fact that the resource is highly vulnerable to overexploitation (Bjørndal and Lindroos, ibid.).

2.3.2 The Case of Three or More Coastal States

Now let it be supposed that the resource, rather than being shared between two neighbouring coastal states, is shared among three such states, I, II and III. When three, or more, players are involved in the cooperative fishery game, the possibility of sub-coalitions arising between, or among, the players must be recognized. Therefore, an approach, which explicitly recognizes the existence of coalitions, is preferred. We shall review the results obtained by Kaitala and Lindroos (1998), when using a coalitional bargaining approach, namely the characteristic function game approach. Our discussion will be restricted to the Shapley value (Shapley, 1988).

The characteristic function game (c-game) approach (Mesterton-Gibbons, 1993) assumes a rather different perspective from the Nash bargaining approach, in that the coastal states are seen as having no bargaining power on their own. It is the coalitions, which the coastal states can form with one another that define their contribution in the cooperative agreement, and consequently their bargaining strengths. Thus, it is natural that the result of the two-player game coincides with the Nash bargaining solution. In our three-player game, we assume that there is only one two-player coalition that has bargaining power during the negotiations, and that the value of this coalition determines the sharing of total benefits from cooperation for all three players. In addition, we continue to assume transferable utility, that is we allow for side payments.

Let it be supposed, as before, that the coastal states differ only in terms of fishing effort costs, and that we have

$$a_{\text{I}} < a_{\text{III}} < a_{\text{II}} \tag{13}$$

and

$$x_{\text{III}}^{\infty} < x_{\text{II}}^{\infty} < x_{\text{I}}^{*}. \tag{14}$$

The two-player coalition, having the bargaining power, obviously consists of I and III, the two most efficient coastal states. The two most efficient players will always have veto power in any cooperative fisheries game, involving more than two players (see, eg. Arin and Feltkamp, 1997), since their presence is necessary for any coalition to have positive bargaining strength. The second most efficient player has the ability, if it chooses to exercise it, of harvesting the resource down to the non-cooperative level.

In our three-player game, the Shapley value gives I and III each more than one third of the cooperative surplus (unlike the Nash bargaining solution).

Let $v(\{I,III\})$ denote the value of the I–III coalition, when playing non-cooperatively against II. Let us normalize the cooperative surplus, $e(x(0))$ to 1, and assume that $v(\{I,III\}) < 1$. Let z_i^s, $i = I, II, III$, denote the cooperative game payoffs dictated by the Shapley value. We have

$$z_{I}^{s} = z_{III}^{s} = \frac{v(\{I,III\})}{6} + \frac{1}{3} \tag{15}$$

$$z_{II}^{s} = \frac{1}{3} - \frac{v(\{I,III\})}{3}. \tag{16}$$

The fairness of the Shapley value arises from the equal treatment of coastal states in the coalition formation process, as well as the difference of the bargaining strengths with respect to the coalitions of which a given coastal state is a member, and those of which it is not. While our example is of a three-player game, the results extend to any cooperative games with the number of players, n, greater than two. The veto players always receive equal shares, whereas all others receive shares, which vary according to their relative efficiency, but which are always less than the shares of the veto players.

In the $n > 2$ player games described, the players are distinctly asymmetrical. Lindroos, in a paper written a number of years after his 1998 paper with Kaitala, warns that, if the players are symmetric, the number of players, which full cooperation – the grand coalition – can support, in the absence of strong legal constraints, is small, maybe no more than two (Lindroos, 2002). It becomes too attractive for individual players to attempt to defect and to enjoy the cooperative benefits of the other players.

The probability of a stable full cooperative solution, with $n > 2$ players, is much enhanced, if the players are asymmetrical – given that side payments are feasible (Lindroos, ibid.). Fortunately, asymmetry among states sharing fishery resources appears to be the rule, not the exception.

Nonetheless, Lindroos' concerns about the stability of cooperative fisheries arrangements in the face of large numbers is of great relevance when one considers the management of straddling fish stocks, to which we now turn.

3 THE MANAGEMENT OF STRADDLING FISH STOCKS

Straddling type of fish stocks, those to be found both within the EEZ and the adjacent high seas, are subject to exploitation by both coastal states and DWFSs. The 1982 UN Convention is vague and imprecise, regarding the rights and duties of coastal states, on the one hand, and those of DWFSs on the other, with respect to the high seas portions of straddling stocks (Bjørndal and Munro, 2003). As a consequence, in the years following the close of the UN Third Conference on the Law of the Sea, non-cooperative management of the resources was all but guaranteed. The economic model of non-cooperative management of transboundary stocks applies, without modification, to straddling stocks. As this model would have predicated, case after case of straddling stock overexploitation occurred in the mid to late 1980s, and early 1990s. The UN responded by convening the UN Fish Stocks Conference, 1993–1995.

The 1995 UN Fish Stocks Agreement, which emerged from the Conference (UN, 1995), calls for the management of straddling-type stocks to be undertaken on a region basis by region basis, through Regional Fisheries Management Organizations (RFOs), the members of which are to consist of the relevant coastal states and DWFSs. The Northwest Atlantic Fisheries Organization (NAFO) and the newly emerging Western Central Pacific Fisheries Convention (WCPFC) are examples of RFOs (Munro *et al.*, 2004).

In analysing the economics of cooperative management of straddling fish stocks through RFOs, economists employ the economics of the cooperative management of transboundary fish stocks and introduce modifications where required. The cooperative management of straddling fish stocks differs from that of transboundary fish stocks in two respects. The first is in terms of the number of participants, or "players". While examples of transboundary stock cooperative arrangements involving large numbers can be found, these are the exceptions. RFOs, since they include both coastal states and distant water fishing nations (DWFNs) can involve very large numbers indeed. Admittedly, however, this is a difference in degree, rather than in kind.

There is nothing particularly new in the analysis required here. The cooperative transboundary fish stock game models, involving $n > 2$ players, apply, essentially without modification. We have the usual problem of the threat of non-compliance – defections – steadily increasing with the number of players (Lindroos, 2002).

The second difference, which we shall refer to as the New Member problem, is a difference in kind. In the cooperative management of transboundary fish stocks, the players, both in nature and number, can be expected to be invariant over time, in other than exceptional circumstances. Such is not the case in the cooperative management of straddling stocks. Some of the participants will be DWFSs, the fleets of which are nothing, if not mobile.

Conceivably, an initial, or "charter", DWFS member of a RFO might withdraw. More importantly DWFSs, which were not among the founders of a RFO, may demand admission to the "club". The 1995 UN Agreement maintains explicitly that any state wishing to exploit a straddling stock, under RFO management, must become a member of the RFO, or agree to abide by the RFO's management provisions. In an effort to be fair to latecomers, the Agreement also explicitly states that "charter" members of a RFO cannot bar would be new members, or entrants, outright (UN, 1995; Munro *et al.*, 2004).

The New Member provision carries with it definite risks. Kaitala and Munro (1997) demonstrate that, if all New Members agree faithfully to abide by the management provisions of the RFO, but demand full pro rata shares of the allowed harvest, "free of charge" as it were, the RFO could be undermined. "Charter" members, anticipating a swarm of New Members, could calculate that their expected payoffs from cooperation would be less than their threat point payoffs (Kaitala and Munro, ibid.), and the RFO would be stillborn.

If prospective New Members are offered less than full pro rata shares, "free of charge", however, they may be strongly tempted to ignore the provisions of the 1995 UN Agreement, by refusing to join the RFO, and then by becoming free riders in the adjacent high seas. Obviously, a RFO faced with rampant, and uncontrollable, free riding would cease to be stable.

This points to an outstanding legal issue, which must be resolved, if the RFO regime is to prosper. Under current international law, vessels of a state, not party to a RFO, which are found to be operating without authorization in the EEZ of a coastal state member of the RFO, are deemed to be engaged in illegal fishing. The coastal state can take vigorous action to repel the vessels. If the same vessels are, however, found to be exploiting the high seas portions of the straddling stock(s) being managed by the RFO, contrary to RFO management provisions, the vessels are deemed to be engaging in *unregulated* fishing. It is much less clear what measures RFO members can take to deal with unregulated fishing.

The FAO currently has underway a plan of action to address the problem of illegal, unreported and unregulated (IUU) fishing (FAO, 2001). What is clearly required is that the FAO initiative must succeed, and that customary international law – state practice – should evolve in such a manner that unregulated fishing achieves the status – de facto if not de jure – of illegal fishing.

4 CASE STUDIES

4.1 The Norwegian Spring-Spawning Herring Fishery

In the 1950s and the 1960s, Norwegian spring-spawning herring (*Clupea harengus* L.) was a major commercial species, harvested by vessels from Norway, Iceland, Faroe Islands and the former Soviet Union. During the 1950s, the fishable component of the Norwegian spring-spawning herring stock measured about 10 million tonnes. However, the stock was subjected to heavy exploitation by the parties mentioned here, employing new and substantially more effective fishing technology. The annual harvest peaked at 2 million tonnes in 1966. By this time, however, the stock was in serious decline and a complete stock collapse occurred by the end of the decade.

Prior to stock depletion, the species was a straddling stock migrating through several coastal states and the high seas. The migratory pattern and number of components to the stock changed between 1950 and 1970. In its depleted state, however, the adult population ceased migration and while adults remained in Norwegian waters year round, their offspring also were distributed in the Barents Sea.[4]

---

[4] The issue of migration is controversial Patterson argues that the causes of migration include changes in water temperature and availability of zooplankton. However, there are studies that suggest that migration may be genetically linked and it is possible that in a small non-migratory stock of herring there is a risk that the migratory genes may disappear and migration would stop.

Recruitment remained weak throughout the 1970s and it was not until the strong year class of 1983 joined the adult population in 1986 that the stock began to recover. In 1993–1994 after spawning along the coast of Norway, the adult herring of this growing stock began a westerly migration into the international waters called the “ocean loop” and occasionally into the EEZs of the European Union, Faroe Islands and Iceland, on their way to the summer feeding area near Jan Mayen Island.

The new migration pattern (as from 1993–1994) of the Norwegian spring-spawning herring takes on importance since, as a straddling stock the herring are exposed to territorial and possibly distant water fleets with strong incentives to harvest the population before it moves elsewhere (Bjørndal *et al.*, 1998). If a cooperative management policy, with an equitable distribution of harvest, cannot be agreed upon, Norway, Iceland, Faroe Islands, countries of the European Union, Russia and possibly distant water vessels fishing in the ocean loop, may resort to ‘strategic over fishing’ that could jeopardise continued sustainability of the stock.

During the first years of the new migration pattern, the situation was quite chaotic. There was no comprehensive regional agreement about the utilisation of the stock. It followed that Norway, Russia, Iceland and Faroe Islands were able to harvest the stock at will within their own jurisdictions. Moreover, in international waters the stock could be harvested legally by any interested fishing nation.

In 1995, the Advisory Committee on Fishery Management (ACFM) of the International Council for the Exploration of the Sea (ICES) recommended a total allowable catch (TAC) for the Norwegian spring-spawning herring of 513,000 t. Norway ignored the recommendation and announced an individual TAC of 650,000 t of which 100,000 t would be allocated to Russian vessels. Iceland and Faroe Islands followed suit and announced their own combined TAC of 250,000 t. In total, the collective harvest of Norway, Russia, Iceland, Faroe Island and the EU was approximately 902,000 t of herring, almost twice the quantity recommended by ACFM (Bjørndal *et al.*, 1998). Nevertheless, in spite of these high catch levels, the herring spawning stock continued to increase, due to high recruitment in 1991 and 1992.

There was, however, some progress towards cooperation. In 1996, Norway, Russia, Iceland and Faroe Islands reached an agreement for a combined TAC. The agreement was reached by increasing the quota levels for each country and setting a total maximum limit of 1,267,000 t. Nevertheless, the European Union did not take part in a TAC commitment and continued fishing at near capacity. In 1997, the EU became a signatory to an agreement, limiting the maximum total catch to 1,498,000 t. The significance of this agreement is that the EU in a commitment to international fisheries cooperation agreed to reduce their total catch levels from the previous period, whereas the four other countries again increased individual TACs (Bjørndal *et al.*, 1998). Notwithstanding, the stock of spring-spawning herring was robust and continued to increase.[5]

The countries involved agreed to continue cooperation and in 1998 the total TAC was set at 1.3 million tonnes. The new quotas for 1998 were allocated between the parties with the same key as in 1997. As a part of the agreement, bilateral access quotas were granted. For example, for fishing spring-spawning herring Russia, the EU, Iceland and Faroe Islands are all granted limited access to Norwegian fishing waters and vice versa. For 1999, the TAC was 1.3 million tonnes and for 2000 the TAC was set at 1.25 million tonnes.

The five-party cooperative agreement broke down in the autumn of 2002, so that as of the 2003 season there has been no agreement. The main reason for the breakdown is that the Norwegian claims, much based on zonal attachment of the herring to the Norwegian EEZ, is much higher than the Norwegian quota of 57%. As a consequence, Norway has demanded a higher share of the TAC, a demand that has not been met by the other parties. For 2005, Norway has unilaterally set a national quota that represents a Norwegian share of 65% of the TAC recommended by ICES, an increase of its national quota by 14%. Iceland has similarly increased its national quota with 14%. If all parties increase their quotas, the sum of the nationally determined quotas will exceed the TAC recommended by ICES. To what degree total catches in 2005 will be higher than the TAC recommended by ICES, cannot be said until the end of the fishing season. As a consequence of the breakdown of the agreement, the other parties, with the exception of Russia, no longer have access to the Norwegian EEZ and the fishery zone around Jan Mayen, which is under Norwegian jurisdiction.

According to the UN Fish Stocks Agreement, the management of straddling and highly migratory fish stocks is to be carried out through Regional Fisheries Management Organisations (RFMOs). For Norwegian spring-spawning herring, which is classified as a straddling stock, management takes place through the North East Atlantic Fishery Commission.

---

[5]The continued recovery of the herring stock even under heavy fishing pressure was due to good growth conditions, conceivably partly due to the reduced stocks of predatory fish species.

The recovery of the Norwegian spring-spawning stock offers the opportunity for substantial annual harvests on a sustainable basis for the benefit of all nations involved. It is clear that if, as a consequence of the breakdown of the co-operative arrangement among the countries, there is a return to competitive harvesting and open access condition, this will result in increased international competition for harvest shares that will be biologically, economically and politically damaging. Eventually, this could threaten a new stock collapse for the fishery and result in substantial economic damage for all nations concerned in terms of lost revenue and employment as catch levels decline.

4.2 The Northeast Atlantic Bluefin Tuna Fishery

The Northeastern bluefin tuna stock is distributed from the east of the Canary Islands to Norway, in the North Sea, in Ireland, in the whole of the Mediterranean and in the south of the Black Sea. Occasionally, it goes to Iceland and Murmansk. The bluefin tuna moves according to food abundance and water temperature. Spawning is located in the warm waters (around 24ºC) of the Mediterranean around the Balearic Islands and in the south of the Tyrrhenian Sea, starting in June and continuing until July. In the beginning of this season, a great flow of bluefin tunas can be observed. Afterwards, some specimens remain in the Mediterranean throughout the year, and others, either young or adult, leave these waters and go to Morocco, the Viscaya Gulf, the Canary Islands and the Madeira Islands. The larger bluefin tuna can be found in the North Sea and along the Norwegian coast, since they are more resistant to colder waters. In the winter, they return to the tempered waters of the African coast.

Bluefin tuna is the most valuable fish in the ocean. High-quality tuna fetch a price premium in the Japanese sushi market, where a single fish can command a price of up to US $ 100,000. Moreover, the price has been increasing in recent years due to a world wide decline in catches of high-quality tuna.

The bluefin tuna fisheries are characterised by a variety of vessel types and fishing gears operating from many countries. Different circumstances – economic, biological, geographical, political as well as traditional – dictate the actual gear choice. The most important fishing gears in the Northeast Atlantic are the purse seine, the long line, the trap and the bait boat.

Throughout the years, the importance of each gear has changed. Certain fisheries, such as trap, go back to ancient times. Other gears, such as the long line and the Mediterranean purse seine, reached full development in the mid-1970s. The spatial distribution of the different gears has changed through the years. The most important change in this respect has been the relocation of the long-line fishery to latitudes above 40° and longitudes between 20° and 50° west, that is to fishing grounds on the high seas outside coastal state 200 mile EEZs.

Historically, more than 50 countries have participated in the fishery for bluefin tuna; currently, 25–30 participate. European countries such as Italy, France and Spain use bait boat, long line, purse seine and trap. DWFNs such as Japan come to the high seas of the North Atlantic to catch bluefin tuna using long line. The large number of countries harvesting bluefin tuna imposes a severe pressure on the stock. In the 1970s, annual catches varied between 10,500 t in 1970 and 22,300 t in 1976. Subsequently, catches increased to a maximum of about 53,000 t in 1997. Thereafter, there has been a decrease to 28,000 t in 2000, mainly due to lower stock levels.

The lower number of participants in the fishery is primarily due to reduced stock levels as compared with historical figures. This has been compounded by the fact that as the stock declines, the distribution area of the stock is reduced. This explains why countries like Norway, Iceland and Russia are not currently active in the fishery. Nevertheless, the situation points to a potential threat to the stock: if and when the stock recovers, there are many potential entrants to the fishery. This is compounded by the high value of the fish.

Bluefin tuna is classified as a highly migratory fish stock. According to the 1995 UN Fish Stocks Agreement, both coastal states and high seas fishing states are required to cooperate directly or through the establishment of sub-regional or RFMOs to this end.

The management of the Northern Atlantic bluefin tuna falls under the aegis of the International Commission for the Conservation of Atlantic Tunas (ICCAT). ICCAT was established in 1969 with two main functions: to provide scientific assessments of Atlantic tunas and tuna-like fish and to give management recommendations that will permit a sustainable fishery. At present, there are 23 contracting parties to ICCAT. These include coastal states in Europe and Africa as well as DWFNs such as Korea and Japan.

As early as 1974, ICCAT recommended limiting the bluefin tuna catch in both the Atlantic and the Mediterranean. In spite of the recommendations being officially implemented in 1975, they had no or little impact, as they were not respected. Present regulations include catch limits (quotas for each member country), prohibition of juvenile landings and closed seasons. So far, the regulations have proved to be rather ineffective. This is due to the inability of ICCAT to monitor and enforce its regulations, which is compounded by the large number of participants in the fishery, members as well as non-members of ICCAT. Therefore, to a large extent many of the characteristics of an open access fishery still prevail.

Stock size decreased from 210,000 t in 1971 to 133,000 t in 1981. Thereafter, the stock remained fairly stable, experiencing a slight increase in 1993–1994 to about 150,000 t, which was also the stock level in 2000. As noted, in the 1990s catches have remained at fairly high levels, especially in the Mediterranean, causing a decline in stock size.

The situation is very grave. If the current trend is maintained, a complete stock collapse is expected within a few years (Brasão *et al.*, 2001). On the other hand, according to Bjørndal and Brãsao (2005), a cooperatively managed fishery bears the promise of generating very substantial rents.

5 REFERENCES

Arin, J. and V. Feltkamp, 1997. The nucleous and kernel of veto-rich transferable utility games, *International Journal of Game Theory*, 26, 61–73.

Bjørndal, T. and A. Brãsao, 2005. The Northeast Atlantic Bluefin Tuna Fisheries: Stock Collapse or Recovery? Centre for Fisheries Economics SNF-Working paper No. xx, 2005.

Bjørndal, T., Hole, A. S., Slinde, W. M. and F. Asche, 1998. Norwegian Spring-Spawning Herring – Some Biological and Economic Issues, Centre for Fisheries Economics SNF-Working paper No. 46/1998.

Bjørndal, T., Kaitala, V., Lindroos, M. and G. Munro, 2000. The management of high seas fisheries, *Annals of Operation Research*, 94, 183–196.

Bjørndal, T. and M. Lindroos, 2004. International management of North-Sea herring, *Environmental and Resource Economics*, 29 (1), 83–96.

Bjørndal, T. and G. Munro, 1998. The economics of fisheries management: A survey, in T. Tietenberg and H. Folmer (eds.), *The International Yearbook of Environmental and Resource Economics 1998/1999: A Survey of Current Issues*, Edward Elgar, Cheltenham, pp. 153–188.

Bjørndal T. and G. Munro, 2003. The management of high seas fisheries resources and the implementation of the UN fish stocks agreement of 1995, in T. Tietenberg and H. Folmer (eds.), *The International Yearbook of Environmental and Resource Economics 2003/2004: A Survey of Current Issues*, Edward Elgar, Cheltenham, pp. 1–35.

Clark, C., 1980. Restricted access to common-property resources: A game theoretic analysis, in P. Liu (ed.), *Dynamic Optimization and Mathematical Economics*, Wiley, New York, pp. 117–132.

Clark, C., 1990. *Mathematical Bioeconomics: The Optimal Management of Renewable Resources*, Second Edition, Wiley Science, New York.

Clark, C. and G. Munro, 1975. The economics of fishing and modern capital theory: A simplified approach, *Journal of Environmental and Resource Economics*, 2, 92–106.

Eckert, R., 1979. *The Enclosure of the Oceans: Economics and the Law of the Sea*, Hoover Institution Press, Stanford.

Food and Agriculture Organization of the UN, 2001. *International Plan of Action to Prevent, Deter and Eliminate Illegal, Unreported and Unregulated Fishing*, Rome.

Gordon, H., 1954. The economic theory of a common property resource: The fishery, *Journal of Political Economy*, 62, 124–142.

Kaitala, V., 1986. Game theory models in fisheries management – A survey, in T. Basar (ed.), *Dynamic Games and Application in Economics*, Lecture Notes in Economics and Mathematical Systems, Springer-Verlag, Berlin, pp. 252–266.

Kaitala, V. and M. Lindroos, 1998. Sharing the benefits of cooperation in high seas fisheries: A characteristic function game approach, *Natural Resource Modeling*, 11, 275–299.

Kaitala, V. and G. Munro, 1993. The management of high seas fisheries, *Marine Resource Economics*, 8, 313–329.

Kaitala V. and G. Munro, 1997. The conservation and management of high sea fishery resources under the New Law of the Sea, *Natural Resource Modeling*, 10, 87–108.

Kaitala, V. and M. Pohjola, 1988. Optimal recovery of a shared resource stock: A differential game model with efficient memory equilibria, *Natural Resource Modeling*, 3, 91–119.

Lindroos, M., 2002. *Coalitions in Fisheries*, Helsinki School of Economics, Working Papers W-321.

McRae and G. Munro, 1989. Coastal state "rights" within the 200 mile exclusive economic zone, in P. Neher, R. Arnason and N. Mollet (eds.), *Rights Based Fishing*, Kluwer, Dordrecht, pp. 97–122.

Mesterton-Gibbons, M., 1993. Game-theoretic resource modeling, *Natural Resource Modeling*, 7, 93–147.

Munro, G., Van Houtte, A. and R. Willmann, 2004. *The Conservation and Management of Shared Fish Stocks: Legal and Economic Aspects*, FAO Fisheries, Technical Paper 465, Rome.

Nash, J., 1950. The bargaining problem, *Econometrica*, 18, 155–162.

Nash, J., 1951. Noncooperative games, *Annals of Mathematics*, 54, 289–295.

Nash, J., 1953. Two-person cooperative games, *Econometrica*, 21, 128–140.

Nomura, I., 2004. State of World Fisheries and Future Sustainability Issues, *Proceedings of the Twelfth Biennial Conference of the International Institute of Fisheries Economics and Trade: What are Responsible Fisheries?*, International Institute of Fisheries Economics and Trade, Corvallis.

Shapley, L., 1988. A value for N-person games, in A. Roth (ed.), *The Shapley Value –Essays in Honor of Lloyd S. Shapley*, Cambridge Press, Cambridge, pp. 31–40.

United Nations, 1982. United Nations Convention on the Law of the Sea. UN Doc. A/Conf.61/122.

United Nations, 1995. United Nations Conference on Straddling and Highly Migratory Fish Stocks. Agreement for the Implementation of the United Nations Convention on the Law of the Sea of 10 December 1982 Relating to the Conservation and Management of Straddling and Highly Migratory Fish Stocks. UN Doc. A/Conf.164/37.

Chapter 11

GAME THEORETIC APPLICATIONS TO FISHERIES

Veijo Kaitala and Marko Lindroos
University of Helsinki

Abstract This chapter reviews non-cooperative and cooperative game theoretic models applied to the economics and management of multi-agent fisheries. We first compare two-player static and dynamic non-cooperative games, and then multi-player non-cooperative and cooperative coalition games. In both comparisons we will make use of same type of models in order to facilitate the comparison of analyses and conclusions.

Keywords: Static games, dynamic games, non-cooperative games, cooperative games, coalitions

1 INTRODUCTION

Marine fish resources are typically overexploited – there is simply not enough fish for every nation. This is due to an overcapacity of fishing gear and vessels in international fisheries where countries build large fishing fleets to be able to rapidly exhaust profitable fish stocks. As a result, fishing pressure is biologically damaging: stocks are decreasing. Consequently, extinctions and collapses of resources have become increasingly common. In addition, as an economic consequence, fishermen's revenue decreases as the stocks decrease.

The following questions arise inevitably: How can we explain this seemingly irrational behaviour in international fisheries? Why is it so difficult to prevent overexploitation of international marine resources? Are there any ways to overcome the problems?

The present study discusses how game theory can be used in fisheries economics applications to predict, explain and resolve the issues that are of major importance for many countries and regions. Whenever there is more than one interest group present in the fishery it is possible to make use of game theory. The current chapter concentrates on international fisheries problems where countries negotiate on common fisheries policies and fisheries agreements.

We review the central contributions in the literature of game theoretic applications to fisheries by making distinction between static and dynamic non-cooperative approaches. In addition, we shall make a difference between cooperative and non-cooperative coalition games applied to international fisheries problems.

Section 2 discusses non-cooperative two-player games, both dynamic and static. Section 3 introduces applications of non-cooperative and cooperative multi-player games to fisheries. What becomes apparent is that the non-cooperative and cooperative approaches are many times difficult to completely separate from each other. We discuss the recent progress in coalition games that is essentially a merger between cooperative and non-cooperative fisheries games. Finally, Section 4 concludes and discusses possible future research needs.

2 NON-COOPERATIVE FISHERIES MODELS

We explore and compare dynamic and static game theoretic approaches to international fisheries management. We begin by considering dynamic games and then continue with static games. This is due to the chronological development of the game theory applications in the literature.

2.1 Dynamic Non-Cooperative Games

Assume that there are two countries exploiting a common fishery. This may well be a realistic assumption in the case of so-called shared fish stocks that are found within the Exclusive Economic Zones (EEZ) of two coastal countries. The two countries have a set of feasible strategies that they may use. From this set of strategies they will choose a fishing effort level that will maximize their own benefits from the fishery taking into account the strategic choices of the other country.

The result of the strategic interaction between the two countries will be a non-cooperative Nash equilibrium. In equilibrium, it is not profitable for any player to unilaterally deviate from the chosen strategy. Thus, non-cooperative Nash equilibrium is a safe strategy for each player since it is not vulnerable to cheating. However, typically a bilateral deviation from the equilibrium strategies, which is usually required in full mutual cooperation, would be beneficial to both countries.

Let us use a model developed by Clark (1980). Assume that the two countries are asymmetric with respect to their unit effort costs

$$c_1 < c_2. \tag{1}$$

Both countries maximize their net benefits over an infinite time horizon subject to the stock dynamics as follows

$$\max\ J_i = \int_0^{\infty} e^{-rt}\left[p - \frac{c_i}{x}\right] h(t)\mathrm{d}t, \tag{2}$$

$$\text{s.t. } \dot{x}_t = F(x) - h_1(t) - h_2(t),$$

where r is the discount rate, p is the constant price, x is the fish stock, $F(x)$ is biological growth of the stock. Finally, $h(t) = E(t)x(t)$ is the production function, that is harvest.

Assume further, that the zero-profit stock level of country 2 is smaller than the optimal stock level of country 1:

$$x_2^{\infty} < x_1^{*}. \tag{3}$$

The non-cooperative feedback Nash equilibrium of the game is given as follows

$$E_1^N(x) = \begin{matrix} E_1^{\max}, & x > \min(x_1^*, x_2^{\infty}) = x_2^{\infty} \\ F(x)/x, & x = \min(x_1^*, x_2^{\infty}) = x_2^{\infty}. \\ 0, & x < \min(x_1^*, x_2^{\infty}) = x_2^{\infty} \end{matrix} \tag{4}$$

$$E_2^N(x) = \begin{matrix} E_2^{\max}, & x > x_2^{\infty} \\ 0, & x \le x_2^{\infty} \end{matrix} \tag{5}$$

In equilibrium, the less efficient country 2 is driven out of the fishery. Country 2 has no other option than to harvest with maximum effort until its profits are diminished to zero. The more efficient country 1 also uses maximum effort until country 2 is driven out. From that point onwards country 1 harvests the stock sustainably so that the stock level will stays unchanged at the zero-profit level of country 2.

Munro (1979) was the first to present a full economic analysis of this game theoretic model for a two-agent fishery. In his seminal article he studied several important issues in international fisheries management such as sharing of quotas, consumer preferences and discount rate differences. Although Munro's paper is mainly dealing with cooperative games, it also includes elements from the theory of dynamic non-cooperative games. A similar model has also been analysed for example by Kaitala and Pohjola (1988). They showed how transfer payments and threat (trigger) strategies can be used to guarantee efficient use of the resource. Trigger strategies mean that players constantly monitor one another. When a deviation from the agreed strategy is observed it triggers the use of the non-cooperative (threat) strategy for the rest of the game.

2.2 Static Non-Cooperative Games

The development of static fisheries games has perhaps surprisingly been slower than dynamic games. Most of the development has taken place from 1993 onwards despite that Schaefer-Gordon model was developed almost forty years earlier. The advantage of static games is that analytical results are easier to derive and interpret. Next, we address the model analysed by Mesterton-Gibbons (1993).

Assume sustainable use of the fish stock by two asymmetric countries. Thus, the resource level remains unchanged:

$$\frac{dx}{dt} = F(x) - \sum_{i=1}^{2} h_i = 0. \tag{6}$$

The growth function is explicitly formulated as logistic growth, given as

$$F(x) = Rx(1 - x / K), \qquad (7),$$

where R is the intrinsic growth rate of fish and K is the carrying capacity of the ecosystem. The production function is assumed to be linear:

$$h_i = qE_i x. \tag{8}$$

Note that we assume that the catchability coefficient, q, is identical for the countries (for a game with asymmetric q, see Kronbak and Lindroos 2003).

It follows from (6) to (8) that the sustainable fish stock is given as

$$x = \frac{K}{R}(R - q\sum_{i=1}^{2} E_i). \tag{9}$$

Hence, the stock decreases linearly in effort.

The objective of each country is to maximize its profit:

$$\max \; ph_i - c_i E_i. \tag{10}$$

Note that here we continue to assume that the unit effort costs are asymmetric (condition (1)). Therefore, country 1 is still the most efficient country with lowest unit cost of effort.

The equilibrium efforts for the players in the game can be characterised as follows:

$$E_1 = \frac{2R}{3q}(1-b_1) - \frac{R}{3q}(1-b_2), \tag{11a}$$

and

$$E_2 = \frac{2R}{3q}(1-b_2) - \frac{R}{3q}(1-b_1), \qquad \text{(11b)},$$

where $b_i = c_i/pqK$, i=1,2, is described by Mesterton-Gibbons (1993) as an (inverse) efficiency parameter of countries such that decreasing value of b_i means increasing efficiency. We immediately note that the equilibrium fishing effort of country i increases with increasing efficiency. Further, the strategic interaction is characterized such that the effort of country 1 decreases as country 2 improves its efficiency, and vice versa.

Assuming that both countries remain active in the fishery the equilibrium fish stock level is given as:

$$x = \frac{K}{3}(1 + b_1 + b_2). \tag{12}$$

From Eq. (12) we see that the higher the efficiency of the countries the lower the equilibrium stock level. Note that only one of the biological parameters plays a role in determining the equilibrium stock level, that is the carrying capacity *K*. The remaining factors influencing the stock are price of fish, unit costs of effort and the catchability coefficient.

Ruseski (1998) and Quinn and Ruseski (2001) have used this static game model to analyse for example overcapacity and subsidies in international fisheries management. Their analyses illustrate how the simple static game can produce intuitive results that explain many of the problems in world's fisheries.

Given the dynamic and static game theoretic approaches described in Sections 2.1 and 2.2 the question arises how the results of these approaches differ? Another issue is what alternative approaches can we find in the literature and how these two approaches have been applied?

The main difference between dynamic and static approaches is that the dynamic game model predicts that one of the countries will eventually be driven out of the fishery. However, on that path the less efficient country will gain benefits from the fishery given that the game is started above the equilibrium stock level.

The static model predicts that both countries will be active in the fishery in the non-cooperative equilibrium. The less efficient country 2 will just modify its effort to a lower level than the most efficient country 1. Both countries are also making positive profits at the static equilibrium. This is a remarkable similarity between the approaches.

Both approaches can be useful in the analysis of international fisheries management. Which approach should be chosen depends on the nature of research questions one needs to answer. From a static model it is sometimes more straightforward to produce intuitive analytical results. However, when applying a static model several important issues arising on the path to the equilibrium and in the dynamic steady state are omitted.

2.3 Discussion of Dynamic Game Applications

This section considers other approaches in the dynamic non-cooperative games. Levhari and Mirman (1980) studied a discrete-time fish-war concentrating to dynamic externalities. They showed that in this setting overfishing occurs. Fischer and Mirman (1992) later developed the model to also consider predator-prey interactions as well as economic interactions in a multispecies fishery. In other words, they studied biological externalities in multispecies fisheries.

In the case of biological externalities, both overfishing and underfishing, can be observed depending on the type of biological interactions. Fischer and Mirman (1996) combined both types of externalities. They studied in particular the following types of biological interactions: symbiotic, negative (competition) and predator-prey. It appears that when both players harvest both species then the fishing strategies may change as compared to the case where each agent is harvesting only one species (Fischer and Mirman 1992). In addition, the efficiency of the outcomes from harvest games may depend on the types of biological interactions. Of course, adding market externalities could complicate the game analyses even more.

McKelvey *et al.* (2002) have been important contributors to straddling stocks issues in the high seas recently. They argued that the establishment of the Exclusive Economic Zones (EEZ) accelerated overfishing and depletion of world's many valuable resource stocks. In McKelvey *et al.* (2002) a new member problem is studied.

Laukkanen (2003) has developed game-theoretic models in sequential stochastic fisheries. She studied cooperative harvesting in a sequential fishery with stochastic shocks in recruitment. Cheating (deviations from cooperative harvesting) is deterred by the threat of harvesting at non-cooperative levels for a fixed number of periods (Kaitala, 1993) whenever the initial stock falls below a trigger level. Cooperation in harvesting provides considerable gains to the players. Based on this analysis she concludes that fish wars are often launched by shocks in recruitment, which trigger non-cooperative harvesting. This may be avoided, however, if the detection of cheating is based on the observations on fishing effort (Kaitala, 1993) rather than stock. Unfortunately, the theory in this matter is still underdeveloped.

Finally, Hannesson (1997) considered fishery as a repeated game. He showed that the number of countries is likely to be small in international fisheries agreements. The result is particularly strong in the case of cost heterogeneity. However, in the case of migratory fish stocks cooperation may be more likely.

3 COALITION MODELS IN FISHERIES

There are many examples where the assumption of two countries may not be very relevant. Many of such cases are included in straddling and highly migratory fish stocks. This section studies the coalition games applied to study multi-country fisheries negotiations. Section 3.1 extends the results of section 2.1 and section 3.2 extends the results of section 2.2 into multi-country coalition games.

3.1 Cooperative Coalition Games

Assume that we have three countries harvesting a common fishery resource. This is the simplest possible case where we can analyse the effect of coalitions that the countries can form with each other. Following Kaitala and Lindroos (1998) we have eight possible coalitions:

$$\{\{C, D_1, D_2\}, \{C, D_1\}, \{C, D_2\}, \{D_1, D_2\}, \{C\}, \{D_1\}, \{D_2\}, \varnothing\}. \quad (13)$$

Here C denotes a coastal state and D distant water fishing nations.

Assume that the objective of each of these three countries is given by Eq. 2: the countries aim at maximizing their net present values from the fishery. Assume further that they have successfully engaged in negotiations and an agreement is binding to each country. The remaining question is how to share benefits according some reasonable and fair cooperative solution so that each country will be satisfied.

Before proceeding to sharing rules a characteristic function must be constructed. The characteristic functions assign a value to each possible coalition. A crucial assumption here is that a coalition enters in a non-cooperative game with those remained outside of it. In the current game, the normalized characteristic function (before normalization the value of the coalition is characterized as the difference between coalition gain and the gain from non-cooperative game) is such that only one coalition in addition to grand coalition has a positive value:

$$v(\{C, D_2\}) > 0. \quad (14)$$

This follows from two assumptions

$$c_{\mathrm{C}} < c_{\mathrm{D}_2} < c_{\mathrm{D}_1} \tag{15}$$

and

$$x_{\mathrm{D}_2}^{\infty} < x_{\mathrm{D}_1}^{\infty} < x_{\mathrm{C}}^{*} \tag{16}$$

In this case, a coalition including the coastal state and the more efficient distant water fishing nation is able to maintain the stock level at a level higher than the non-cooperative stock level given in (4) and (5). However, a coalition of the coastal states, or a coalition of the coastal state and the less efficient distant water fishing nation 2 are not able to do that.

Employing the Shapley (1953) value yields a cooperative solution to each country:

$$z_{\mathrm{C}}^{\mathrm{S}} = z_{\mathrm{D}_2}^{\mathrm{S}} = v(\{\mathrm{C}, \mathrm{D}_2\})/6 + 1/3, \tag{17}$$

$$z_{\mathrm{D}_1}^{\mathrm{S}} = [1 - v(\{\mathrm{C}, \mathrm{D}_2\})]/3.$$

Thus, the two most efficient countries should receive a share higher than one third of cooperative benefits. This is due to their higher contribution to the overall coalitional values. No other two-player coalition is able to obtain a positive value.

The result can be compared to the Nash bargaining model of Kaitala and Munro (1995). Their approach suggests that the cooperative benefits would be shared equally between all members of the cooperative organisation.

Lindroos (2004) expanded the cooperative coalition game to the case where coalition formation is restricted to the coastal country group and to the distant water fishing nation group. It is shown that under these restrictions the less efficient countries may improve their bargaining positions.

3.2 Non-Cooperative Coalition Games

We next employ the static game model analysed in Section 2.2 to examine the issue of coalition formation. Assume that there are now three asymmetric countries with the fishing effort costs ordered as follows

$$c_1 < c_2 < c_3. \tag{18}$$

The coalition formation games can be thought of as a two-stage game. In stage one, the players decide which coalition to belong to. In the second stage, countries or coalitions play a non-cooperative game by choosing fishing efforts to maximize their profits (Eq. 10). Given the payoffs of the second stage, the equilibrium coalition structure is the Nash equilibrium of the first stage. Possible equilibrium coalition structures are full cooperation (grand coalition), partial cooperation (two-player coalition) and non-cooperation.

Non-cooperative equilibrium efforts are analogous to Eq. 11.

$$E_i^N = -\sum_{j,k} \frac{R}{4q}(1-b_k) + \frac{3R}{4q}(1-b_i), \quad i = 1,2,3 \tag{19}$$

Two-player coalition efforts are given as.

$$\begin{aligned} E_i^{\mathrm{PC}} &= -\frac{R}{3q}(1-b_k) + \frac{2R}{3q}(1-b_i) \\ E_j^{\mathrm{PC}} &= 0 \end{aligned}, \quad (i,j) = (1,2), (1,3), (2,3), \tag{20a}$$

This means that in coalition (i,j) only the most efficient country remains active as its marginal cost is always lower due to linearity (for non-linear costs and coalition games see Kronbak and Lindroos, 2005). Note that there are three possible two-player coalitions.

Country k that remains outside of cooperation chooses its strategy from its reaction curve, which is linear in this game:

$$E_k^F = -\frac{R}{3q}(1-b_i) + \frac{2R}{3q}(1-b_k), \qquad k \in \{1,2,3\} \setminus (\mathrm{i,j}). \tag{20b}$$

The resulting equilibrium stock levels are of similar form as in eq. (12)

$$x^{\mathrm{PC}} = \frac{K}{3}(1 + b_i + b_k). \tag{20c}$$

Finally, the cooperative effort where all three countries maximize their joint economic benefits by allowing country 1 to be the only harvester is

$$E_1^{\mathrm{C}} = \frac{R}{2q}(1-b_1). \tag{21a}$$

The corresponding stock is

$$x^{\mathrm{C}} = \frac{K}{2}(1+b_1). \tag{21b}$$

Let us then examine the conditions that affect stability of grand coalition (see Lindroos, 2002)

$$pqE_1^{\mathrm{C}}x^{\mathrm{C}} - c_1E_1^{\mathrm{C}} > \sum_{k=1}^{3}(pqE_k^{F}x^{\mathrm{PC}} - c_kE_k^{F}), \tag{22}$$

where F stands for Free Rider. Condition (22) says that grand coalition is stable if the cooperative benefits are larger than the sum of possible free-rider benefits. This means that there has to be enough cooperative benefits to be shared in order to guarantee that no single country will leave the grand coalition.

Stability is dependent on the unit effort costs in the following way.

$$c_3 < \frac{c_1}{2} + \frac{pqK}{2}, \tag{23}$$

$$c_2 < \frac{4c_1}{5} + \frac{pqK}{5}. \tag{24}$$

Conditions (23) and (24) have to be satisfied so that grand coalition can be stable. These results mean that if there is enough cost heterogeneity in the game, then cooperation can be an equilibrium. More cost heterogeneity means here that cooperation, where country 1 acts as the only harvester, becomes relatively more profitable.

3.3 Discussion of other Coalition Game Applications

Negotiating an international environmental agreement is complicated due to, among other aspects, the presence of different forms of externalities. In the fisheries, we usually have many complicating factors present: several and uncertain number of players, biological, economic and dynamic externalities, difficulties in observation, monitoring and implementing, and inherently stochastic nature of the fisheries. Thus, it is not surprising that asymmetries between the players put them in different positions within the framework of the agreement. In many cases, it may occur that some countries will accept agreement immediately after the negotiations have been concluded but some countries would like to delay signing several years.

Kaitala and Lindroos (2004) argue that timing of an agreement may include a strategic component. It is well-known that the value of the fishery depends on the initial value of the stock. Kaitala and Lindroos (2004) analyse the case where two countries are negotiating when playing a non-cooperative game. The purpose is to switch to a cooperative agreement. Thus, both countries attempt to optimize the timing of the agreement. If the optimal timing for the agreement differs between the countries, then a conflict arises. Kaitala and Lindroos (2004) show that several outcomes are possible: it is optimal for both players to initiate cooperation immediately, at least for one of the countries it is never optimal to cooperate, and one of the players wishes to delay the agreement. This leads to a complicated situation if there does not exist any "must" for the agreement. The theory is still immature in this matter.

Coalition games applied to fisheries problems have not merely stayed in the theoretical level as presented in the two previous sections. Coalition games have been applied to Norwegian spring-spawning herring (Arnason *et al.*, 2000, Lindroos and Kaitala, 2000), North-Eastern mackerel (Kennedy, 2003), and Northern Atlantic bluefin tuna (Brasao *et al.*, 2000, Pintassilgo and Duarte, 2000). Pintassilgo (2003) and Pham Do and Folmer (2003) are examples of applying the partition function game approach. Pintassilgo (2003) applies the framework to Northern Atlantic bluefin tuna, whereas Pham Do and Folmer (2003) show that the Shapley Value is an appropriate cooperative solution to share the benefits of cooperation in fisheries games. For a thorough review of coalition games in fisheries see Lindroos *et al.* (2007).

4 FUTURE RESEARCH ON GAME THEORY AND FISHERIES

We have reviewed selected game-theoretic models applied to fisheries. Game theory has lot of useful potential to be applied in international fisheries management and negotiations. International agreements should be made sustainable in order to guarantee economically and biologically efficient use of fish resources. If international management fails, national management, whether based on biology or economics, is a futile effort.

At least three issues need immediate attention in the near future of game theory and fisheries. Firstly, merging the cooperative and the non-cooperative approach is most important. International agreements must be based on voluntary (self-enforcing) actions taken by countries. This means essentially analysing games where cooperation can be a non-cooperative equilibrium. However, once cooperation is reached we immediately have an allocation problem, that is, a cooperative game where the benefits arising from cooperation should be shared in some fair way. It is clear that these two approaches are linked in many ways. Kronbak and Lindroos (2005) is one of the first steps in this direction. Many others are still needed.

There are already many practical applications of coalition games. More applications are, however, needed in order to gain further knowledge on what factors influence incentives and timing of cooperation or signing international fisheries agreements.

Finally, we should make a clear distinction between different interest groups in the fisheries. One way forward could be the approach adopted in Kronbak and Lindroos (2003) where countries negotiate in the first stage and then at the second stage we have fishermen competing or forming coalitions (producer organizations) with one another.

ACKNOWLEDGEMENTS

We gratefully thank the two anonymous referees for helpful comments.

5 REFERENCES

Arnason, R., G. Magnusson, and S. Agnarsson (2000), The Norwegian Spring-Spawning Herring Fishery: A Stylized Game Model, *Marine Resource Economics* 15, 293–319.

Brasão, A., C. C. Duarte, and M. A. Cunha-e-Sá (2000), Managing the Northern Atlantic Bluefin Tuna Fisheries: The stability of the UN Fish Stock Agreement Solution, *Marine Resource Economics* 15(4): 341–360.

Clark, C. (1980), Restricted access to common-property fishery resources: A game-theoretic analysis, in *Dynamic Optimisation and Mathematical Economics*, Liu, Pan-Tai (ed.), Plenum Press, New York, 117–132.

Fischer, R. and L. Mirman (1992), Strategic dynamic interaction, *Journal of Economic Dynamics and Control* 16, 267–287.

Fischer, R. and L. Mirman (1996), The complete fish wars: Biological and dynamic interactions, *Journal of Environmental Economics and Management* 30, 34–42.

Hannesson, R. (1997), 'Fishing as a Supergame', *Journal of Environmental Economics and Management* 32, 309–322.

Kaitala, V. (1993), Equilibria in a stochastic resource management game under imperfect information, *European Journal of Operations Research* 71,439–453.

Kaitala, V. and M. Lindroos (1998), Sharing the Benefits of Cooperation in High Seas Fishery Game: A Characteristic Function Game Approach, *Natural Resource Modeling* 11, 275–299.

Kaitala, V. and M. Lindroos (2004), When to sign an environmental agreement: The case of high sea fisheries, *International Game Theory Review* 6, 55–68.

Kaitala, V. and G. Munro (1995), The economic management of high sea fishery resources: Some game theoretic aspects, in *Control and Game-Theoretic Models of the Environment,* Carraro, Carlo and Filar, Jerzy A. (eds.), Annals of the International Society of Dynamic Games, Vol. 2, Birkhäuser, Boston, MA, 299–318.

Kaitala, V. and M. Pohjola (1988), Optimal recovery of a shared resource stock: A differential game model with efficient memory equilibria, *Natural Resource Modeling* 3, 91–119.

Kennedy, J. (2003), Scope for efficient exploitation of North-East Atlantic Mackerel, *Marine Resource Economics* 18, 55–80.

Kronbak, L. and M. Lindroos (2003), An enforcement-coalition model: Fishermen and authorities forming coalitions, *Environmental and Resource Economics* 35, 169–194.

Kronbak, L. and M. Lindroos (2005), Sharing rules and stability in coalition games with externalities: The case of the Baltic Sea cod fishery. Department of Economics and Management discussion paper no. 7. University of Helsinki, Finland.

Laukkanen, M. (2003), Cooperative and non-cooperative harvesting in a stochastic sequential fishery, *Journal of Environmental Economics and Management* 45, 454–473.

Levhari, D. and L. Mirman (1980), The Great Fish War: An example using a dynamic Cournot-Nash solution, *Bell Journal of Economics* 11, 322–334.

Lindroos, M. (2002), Coalitions in Fisheries, Helsinki School of Economics Working Papers W-321.

Lindroos, M. (2004), Restricted coalitions in the management of regional fisheries organizations. *Natural resource modeling* 17, 45–69.

Lindroos, M. and V. Kaitala (2000), Nash equilibria in a coalition game of the Norwegian spring-spawning herring fishery. *Marine Resource Economics* 15, 321–340.

Lindroos, M., L. Kronbak, L., and V. Kaitala (2007), Coalitions in fisheries games, in Advances in Fisheries Economics: Festschrift in Honour of Professor Gordon Munro, Blackwell.

McKelvey, R., L. Sandal, and S. Steinshamn (2002), Fish wars on the high seas: A straddling stock competition model, International Game Theory Review 4, 53–69.

Mesterton-Gibbons, M. (1993), Game-Theoretic Resource Modeling, *Natural Resource Modeling* 7, 93–146.
Munro, G. (1979), The optimal management of transboundary renewable resources, *Canadian Journal of Economics* 12, 355–376.
Pham Do, K. and H. Folmer (2003), International fisheries agreements: The feasibility and impacts of partial cooperation, CentER DP 2003–52, Tilburg University, the Netherlands.
Pintassilgo, P. (2003), A coalition approach to the management of high seas fisheries in the presence of externalities, *Natural Resource Modeling* 16, 175–197.
Pintassilgo, P. and C. C. Duarte (2000), The new-member problem in the cooperative management of high seas fisheries, *Marine Resource Economics* 15(4), 361–378.
Quinn, J. and G. Ruseski (2001), Effort subsidies and entry deterrence in transboundary fisheries. *Natural Resource Modeling* 14(3), 369–389.
Ruseski, G. (1998), International fish wars: The strategic roles for fleet licensing and effort subsidies, *Journal of Environmental Economics and Management* 36, 70–88.
Shapley, Lloyd (1953), 'A Value for n-person Games', in *The Shapley value - Essays in Honor of Lloyd S. Shapley*, A. Roth (ed.). Cambridge: Cambridge University Press, 1988.

Chapter 12

UNCERTAINTY IN BIOECONOMIC MODELLING

Linda Nøstbakken[1] and Jon M. Conrad[2]
[1]*Centre for Fisheries Economics, NHH/SNF Bergen, Norway*
[2]*Cornell University, Department of Applied Economics and Management, Ithaca, New York, USA*

Abstract This chapter reviews the large body of literature dealing with uncertainty in bioeconomic modelling of fisheries. The purpose is to provide an overview of some of the main developments in the field since its introduction in the early 1970s. We start by giving a detailed presentation of two fairly general stochastic bioeconomic models, one in continuous time and another in discrete time. These models serve as benchmarks for the other studies we discuss. This chapter then provides an overview of some of the achievements and issues that have been dealt with using stochastic bioeconomic models.

Keywords: Fisheries Economics, bioeconomic modelling, uncertainty

1 INTRODUCTION

Uncertainty was introduced in bioeconomic models in the early 1970s and an extensive literature has been generated since that time. The aim of this chapter is to review some of the main developments in stochastic bioeconomic modelling of fisheries. A complete survey of the literature is impossible in the space allocated for this chapter, and the aim is rather to present a sample of the literature to illustrate some of the main developments and to exemplify the range of topics analysed by use of stochastic bioeconomic models.[1]

Walters and Hilborn (1978) list the following three categories of uncertainty in fisheries management: (i) random effects, whose probability distribution can be determined from past experience, (ii) parameter uncertainty,

---

[1]For a more detailed survey of the earlier literature, the reader might refer to Andersen and Sutinen (1984).

and (iii) fundamental misunderstanding about variable choice and model form. The various forms of uncertainty along with methods used to analyse them are reviewed in Charles (1998). Most bioeconomic studies focus on the first two classes of uncertainty.

The standard bioeconomic model consists of a biological component, describing change in one or more resource stocks, and an economic part describing net revenue, net benefits or "social welfare". Uncertainty can be added to the model in several ways. The biological component can be made stochastic by allowing for random fluctuations in the stock-growth relationship. In addition, one might assume stock levels are observed with measurement error. Uncertainty can be introduced to the economic component by letting prices, costs, yield-effort relationships, and so on fluctuate according to some stochastic process.

The rest of this chapter is organised as follows. In Section 2 we present two studies of uncertainty in bioeconomic modelling; the first model is in discrete time whereas the second is modelled in continuous time. These models serve as benchmarks when reviewing other studies. Section 2 also gives a brief introduction to basic methods for solving stochastic dynamic optimisation problems. Section 3 gives an overview of applications of stochastic bioeconomic modelling. Section 4 concludes the chapter.

2 STOCHASTIC BIOECONOMIC MODELS

In the bioeconomic literature we find stochastic models both in discrete time and in continuous time. In this section, we present two models, a discrete-time model developed by Reed (1979) and a continuous-time model developed by Pindyck (1984). Conrad (2004) also provides a detailed description of both Reed's and Pindyck's models in his review of renewable resource management.

2.1 A Discrete-Time Model

Reed (1979) draws on the analyses in Jaquette (1972, 1974) and Reed (1974), and the model is used to derive an optimal harvest policy for a fishery. Reed (1979) uses a stochastic stock-recruitment function:

$$X_{t+1} = z_{t+1} G(S_t), \tag{1}$$

where X_t, $S_t = X_t - Y_t$, and Y_t are biomass, escapement, and harvest in period t, respectively. $\{z_{t+1}\}$ are independent and identically distributed (iid) random variables with mean one and constant variance, observed at the beginning of period $t + 1$. $G(S_t)$ is a growth function. Harvesting from the stock is explained by the Spence production function $Y_t = X_t\left(1 - e^{-qE_t}\right)$, where E_t represents effort and $q > 0$ is a catchability coefficient.[2] By assuming a constant cost per unit effort of c and a constant price p per unit harvest, net revenues are given by $\pi_t = pY_t - (c/q)\left[\ln(X_t) - \ln(S_t)\right]$. Using the fact that net revenues can be written as an additively separable function of X and S, we get the expression $\pi_t = N(X_t) - N(S_t)$, where $N(m) = pm - (c/q)\cdot\ln(m)$. The optimal policy is derived by maximising the expected present value of net revenues

$$\max_{\{S_t\}} E_0\left[\sum_{t=0}^{T} \rho^t \{N(X_t) - N(S_t)\}\right],$$

subject to (1), $0 \le Y_t \le X_t$, and X_0 given, where ρ is the discount factor. The maximisation problem is solved using stochastic dynamic programming. The optimal harvest policy is a constant-escapement policy where the optimal escapement level S^* must maximise the equation $W(S) = \rho E_z [N(zG(S))] - N(S)$. The optimal feedback policy can be expressed as

$$Y_t = \begin{cases} (X_t - S^*) & if\ X_t > S^* \\ 0 & if\ X_t \le S^* \end{cases}.$$

Reed is able to derive the optimal feedback policy analytically because net revenues in his model can be written as an additive separable function of the state and the control variables, that is he uses a linear control model. His choice of production function is crucial and a slightly different model specification would have made it impossible to derive a closed-form solution. In most stochastic bioeconomic models where net revenues are maximised, it is very difficult (or impossible) to derive closed-form solutions and numerical approximations must be used.

2.2 A Continuous-Time Model

One of the first bioeconomic studies in continuous time dealing with uncertainty is Ludwig (1979). Ludwig extends Clark's (1976) classic, deterministic fishery model by including stochastic change in the resource stock. Ludwig's model is similar to Reed (1979) with a linear control relationship and a fixed and exogenous resource price. Furthermore, the optimal feed-

[2] Spence (1974) used this production function in his study of blue whales.

back policy derived from Ludwig's model is similar to the constant-escapement policy derived from the Reed (1979) model. According to Ludwig, one should harvest either at the maximum or the minimum harvest rate depending on the current size of the stock (i.e., a bang-bang approach). Pindyck (1984) extends Ludwig's model by letting price be determined by a downward sloping demand curve.

In Pindyck (1984) the stock evolves according to

$$\mathrm{d}X = \left[F(X) - Y\right]\mathrm{d}t + \sigma(X)\mathrm{d}z\,, \tag{2}$$

where $\sigma'(X) > 0$, i.e., the variation in stock growth increases with the size of the stock, $\sigma(0) = 0$, and $\mathrm{d}z = \varepsilon(t)\sqrt{\mathrm{d}t}$ is the increment of a Wiener process. $Y(t)$ is the harvest rate. The stock-growth function $F(X)$ is assumed to be strictly concave with $F(0) = F(K) = 0$, where $K > 0$ is the carrying capacity of the resource in its natural environment.

Let net benefits at instant t be given by

$$U(X,Y) = \int_0^Y p(q)dq - c(X)Y, \tag{3}$$

where $p(Y)$ is the downward sloping demand curve, and $c(X)$ is the unit cost of harvesting from a stock of size X. $c(X)$ is assumed decreasing and strictly convex, and with $c(0) = \infty$.

Maximisation of discounted net benefits subject to (2) gives the following Hamilton-Jacobi-Bellman (HJB) equation:

$$\delta V(X) = \max_Y \left\{ \int_0^Y p(q)dq - c(X)Y + \left[F(X) - Y\right]V'(X) + \frac{1}{2}\sigma^2(X)V''(X) \right\}, \tag{4}$$

where δ is the discount rate. From the maximal condition $\partial\{\cdot\}/\partial Y = 0$, we have that $p(Y^*) - c(X) = V'(X)$.

Pindyck (1984) provides three examples where he specifies bioeconomic models and derives closed-form solutions for the optimal harvest policies. Linear feedback policies emerge in all three examples. The examples demonstrate, among other things, how an increase in σ_X can increase, decrease or leave harvest rates unchanged.

3 APPLICATIONS IN BIOECONOMICS

There has been an extensive development in the application of uncertainty in bioeconomic models. The topics and studies discussed in this section are not meant to give a comprehensive overview of the literature. The purpose is rather to illustrate the range of issues that have been analysed by use of stochastic bioeconomic modelling.

3.1 Optimal Harvest from a Fish Stock

In the deterministic setting the analysis of optimal harvesting typically involves finding the optimal steady-state harvest and biomass level along with the corresponding optimal approach path from the initial stock level (see e.g. Clark and Munro, 1975; Clark 1976).[3] In a stochastic fishery, there is no steady state. The system is randomly changing and as a result optimal harvest must be specified for every state that can possibly occur. Instead of deriving optimal steady-state harvest, the optimal harvest policy, that is harvest as a function of the stock or state, must be found.

In the studies by Reed (1979) and Pindyck (1984), presented in Section 3, stochastic dynamic programming was used to derive optimal harvest policies. A number of studies extend this research.

Lewis (1981) develops a discrete time, Markov model of a fishery. Whereas Reed (1979) introduced uncertainty to the stock-growth relationship, Lewis analyses the case of uncertain catchability. Lewis further assumes biomass can be described by a finite number of states represented by possible stock sizes. Population dynamics in Lewis (1981) are given by $X_{t+1} = X_t + G(X_t) - \eta_t a X_t K_t$, where $G(X_t)$ is a logistic growth function, and $\eta_t a X_t K_t$ is a production function giving catch in period t. Uncertainty is introduced by letting η_t be a uniformly distributed random variable with mean one. Markov transition probabilities are calculated and used to obtain the optimal solution through dynamic programming. The optimal strategy is seen to be a function of stock size (state), which is revealed to the fishery manager each period prior to decision making. Optimal strategies are derived for three different cost specifications. Lewis finds the optimal strategy to be pulse fishing if costs are increasing, and fishing at interiour harvest rates in two other cases. Lewis further analyses effects of attitude towards risk,

[3]It is common to make the assumptions that (i) stock growth is concave in the stock and (ii) the objective function is concave in harvest. If assumption (ii) is relaxed, continuous harvesting strategies may be outperformed by other harvesting policies (see e.g. Lewis and Schmalensee 1977).

increased uncertainty, and the difference between stochastic and deterministic results in terms of both the optimal policy and discounted value. While a deterministic analysis provides a good approximation to stochastic analysis in the case of increasing marginal costs, deterministic harvest rules are found to be poor substitutes for the optimal stochastic strategies when costs are decreasing in effort or zero.

Spulber (1982) extends Reed's (1979) model by letting the environmental disturbances follow a general Markov process, that is $z_{t+1} = \phi(\cdot|z_t)$, where $\phi(\cdot)$ is a probability distribution, and by assuming, like Reed (1974), that fishing firms face a fixed set-up cost of harvesting.

The optimal harvest rule is similar to that of Reed (1979) with some important distinctions. First, optimal escapement depends on the expected stock-recruitment as given by the value of the random variable z. Second, the net revenues from harvesting must cover the setup cost for harvesting to be optimal. The model reduces to the Reed model if $\{z\}$ are iid and the setup cost is zero.

Clark and Kirkwood (1986) model a fishery using a framework similar to Reed's but where the uncertainty is revealed after the harvest level has been determined. They thus assume that X_{t+1} in Eq. 1 is a random variable with a given probability distribution dependent upon the known escapement level S_t. Using this specification, Clark and Kirkwood show that the optimal harvest policy is not a constant-escapement policy as in the original Reed model. The optimal policy in Clark and Kirkwood's model can however only be approximated numerically.

Sandal and Steinshamn (1997) extend the Pindyck (1984) model by assuming non-linearity in the control variable Y. Instantaneous net revenues are then given by $\Pi(X,Y) = p(Y)Y - c(X,Y)$, where $p(Y)$ is the linear downward sloping demand curve and $c(X,Y)$ is a cost function increasing in Y. As in Pindyck, Sandal and Steinshamn seek to find the harvest rate that maximises the present value of net revenues subject to the dynamic constraint given by Eq. 2. By applying perturbation methods, they derive approximate expressions for optimal feedback policies, that is harvest rate as a function of stock size, for different cases.

Many other studies analyse optimal harvesting of a stock with stochastic stock growth. Lungu and Øksendal (1997) analyse what harvest policy maximises discounted harvest from a stock evolving according to the stochastic logistic equation $\mathrm{d}X = X(1 - X/K)(rdt + \sigma \mathrm{d}z) - Y$, which is slightly different from the stock dynamics Eq. 2 of Pindyck (1984). They show that optimal harvesting in this case is a constant-escapement policy. By maximising discounted harvest, they ignore harvesting costs.[4]

Sethi *et al.* (2005) develop a discrete model in which three sources of uncertainty are incorporated: growth, stock measurement, and harvest quota implementation. Stochastic stock growth follows Reed and is given by Eq. 1. Stock measurement and actual harvest are given by $X_t^m = z_t^m X_t$ and $Y_t = \min\left(X_t, z_t^i Y_t^q\right)$, respectively, where z_t^m and z_t^i are random variables, and Y_t^q is the harvest quota. The authors are able to numerically approximate the optimal policy of the problem of maximising expected present value of the fishery over an infinite horizon. They analyse how the optimal policy changes when one of the uncertainty sources is high whereas the others are low. If the growth or implementation uncertainties are high, the optimal policies are not qualitatively different from Reed's constant-escapement policy. With high measurement uncertainty however, Sethi *et al.* (2005) find that the optimal policy is not a constant-escapement policy and that, depending on the measurement of stock size, optimal escapement may be higher or lower than the constant optimal escapement in Reed's model.

Optimal harvesting has also been studied under price uncertainty. One example is Hanson and Ryan (1998) who study optimal harvesting from a fish stock subject to price and stock uncertainty. They find, not surprisingly, that price fluctuations have a big impact on the value of the fishery, but only a modest impact on the optimal harvest policy. Nøstbakken (2006) studies regime switching in a fishery subject to price and stock uncertainty, where increasing and decreasing the harvest rate incurs fixed adjustment costs.

Costello *et al.* (1998) and Costello *et al.* (2001) analyse optimal harvesting under environmental stock uncertainty and study the value of environmental prediction and how prediction changes optimal harvest. These studies find the effect on current harvest policy (and forecast value) of predictions beyond a 1-year forecast to be modest or non-existent.

3.2 Relative Efficiency of Management Instruments

The question of taxes versus quotas in fisheries management has been considered by several authors throughout the years. In a deterministic setting the two are equally good, but this might no longer be the case when uncertainty

---

[4]See e.g. Alvarez and Shepp (1998), Alvarez (2001), and Framstad (2003) for extensions of the analysis in Lungu and Øksendal (1997) and for alternative model specifications.

is introduced to the model. The efficiency of other management instruments has also been analysed and compared. In the following we will review some of the literature dealing with the relative effect of fisheries management instruments.

The classic studies on "prices vs. quantities" is Weitzman (1974). Koenig (1984a, b) follows along the lines of Weitzman (1974) and evaluates benefits and costs associated with different management instruments in a stochastic discrete-time model. He makes several simplifying assumptions to be able to solve the dynamic programming problem, including assuming a linear growth relationship. Both cost and benefit functions are quadratic and uncertainty is introduced by adding random disturbances to the linear terms. If there is no measurement error in the stock estimates, Koenig shows that taxes are at least as efficient as quotas and strictly better in the presence of demand or supply uncertainty. If stock size is observable only with error, harvest quotas can outperform landing taxes depending on the relative elasticities of market supply and demand (Koenig, 1984a). In a recent study, Jensen and Vestergaard (2003) discuss conditions for applying Weitzman's (1974) results to fisheries.

Androkovich and Stollery (1991) use a model very similar to Koenig's but with a slightly different treatment of risk. While Koenig assumes harvest decisions are made with full information whereas tax rates are set with incomplete information, Androkovich and Stollery (1991) assume decisions regarding tax rates and whether to harvest are taken before the realisation of the random variables. Under these assumptions, they show that a landing tax is always superior to harvest quotas.

Yet another analysis of taxes versus quotas is Anderson (1986). His approach differs from the studies presented earlier in that he combines discrete-time and continuous-time bioeconomic models. Regulatory decisions are made at discrete time steps, whereas fishing and stock dynamics are modelled in continuous time. Anderson (1986) finds that neither taxes nor quotas are generally superior; the optimal policy depends on the characteristics of the specific fishery.

Mirman and Spulber (1985) analyse fishery regulations under harvest uncertainty in a discrete-time model. As in the Reed model, the fishery regulator has perfect information on the size of the stock and makes regulatory decisions after observing last period's growth but before knowing next period's growth. Mirman and Spulber assume that the individual fishing firm does not necessarily know the current fish stock. The yield-effort relationship is therefore uncertain. With yield-effort uncertainty, both taxes on landings and vessel quotas might have unintended and unfortunate effects. Mirman and Spulber suggest applying taxes and quotas together and they show how this combination induces the fishing firms to choose both optimal effort and harvest levels.

In a recent study, Weitzman (2002) specifies a model similar to Clark and Kirkwood's (1986) by assuming that regulatory decisions are made before the recruitment level is known. He uses his model to compare two management instruments, a unit-landing fee and catch quotas, and he draws the conclusion that the landing fee is always superior to catch quotas. The conclusion is perhaps not surprising given that Weitzman's model includes environmental uncertainty but no economic uncertainty and therefore favours the landing tax. Weitzman's analysis, or perhaps rather his conclusion, has triggered renewed interest in studies of landings taxes versus harvest quotas (e.g. Hannesson and Kennedy, 2005).

We have seen several examples of stochastic bioeconomic models being used to evaluate the relative performance of landings taxes to catch quotas. These studies do however not give us an unambiguous answer. To prove analytically that one instrument is superior to the other, one has to make several rather restrictive assumptions. The work on the subject has therefore given us conditions for when an instrument is superior to the other rather than a general conclusion of superiority.

The relative efficiency of other management instruments has also been studied in the literature. Hannesson and Steinshamn (1991) use a one-period model to compare a constant harvest rule to a constant effort rule when faced with a stochastic varying stock. Hannesson and Steinshamn conclude that neither a constant effort rule nor a constant escapement rule is superior; high sensitivity of CPUE to changes in stock suggests that a constant effort rule is optimal. Quiggin (1992) extends Hannesson and Steinshamn's analysis by deriving conditions for superiority of constant effort rules to constant catch rules. As Hannesson and Steinshamn (1991), Quiggin (1992) builds his analysis on a one-period model.

Danielsson (2002) further extends the analysis of relative efficiency of catch quotas to effort quotas by including stock dynamics with uncertain stock growth, and stochastic variations in the CPUE. By including stock dynamics, the effect of the present period's harvest on next period's stock size is taken into account. Danielsson's model is to some extent related to the Reed (1979) model but with some exceptions. Instead of Eq. 1 stock dynamics are explained by $X_{t+1} = S_t + f(X_t, \varepsilon_t)$, where ε_t is a random variable representing uncertainty in stock growth. Danielsson uses a production function of the form $Y_t = H\left(K_t, X_t + f(X_t, \varepsilon_t), \eta_t\right)$, where η_t is a random variable reflecting variations in CPUE independent of stock size. In addition, Danielsson allows for measurement error in stock estimates by letting $X_t = m(X_t^m, \theta_t)$, where X_t^m is measured stock size, and θ_t is a random variable possibly correlated with η_t. Maximised expected present value of net benefits from the fishery, where benefits (utility or profits) are expressed as a function of stock size and fishing effort, are derived both for catch quotas and for effort quotas. Based on this, Danielsson derives sufficient conditions for when management with catch quotas is superior to management with effort quotas and *vice versa*.

Herrera (2005) analyses the relative efficiency of different management instruments focusing on bycatch and discarding. He develops a two-stock model of an input regulated fishery with stochastic bycatch. He evaluates the relative efficiency of four regimes: price instruments, trip-based value and quantity limits, and no output regulations. He concludes that price instruments (taxes or subsidies) are more efficient than the trip-based quotas. Comparing trip-based quotas, value limits are found to give better results than quantity limits, as they eliminate some of the incentives to discard.

Marine protected areas have recently received widespread attention as a management instrument that recognises the importance of spatial processes in the bioeconomic system. Marine reserves and spatial modelling of fish stocks will be discussed in Section 3.5.

3.3 Management of Shared Fish Stocks

Stochastic modelling can contribute to the understanding of game-theoretic aspects of the management of shared fish stocks. Information or beliefs on the sources and magnitude of variation may vary between the players. It may also be in the players' interest to conceal information from one another. Uncertainty might therefore, *inter alia*, destabilise otherwise satisfactory sharing agreements. An important part of the bioeconomics literature deals with international management of transboundary fish stocks. There are however few studies which incorporate uncertainty. One exception is Kaitala's (1993) application of stochastic game theory to the management of fisheries. Another exception is the recent study by Laukkanen (2003), who establishes a model of a sequential fishery based on the Reed (1979) model. Laukkanen models a fish stock which migrates between two areas: a feeding area and a breeding area. Two agents, one operating in each area, harvest the stock.

Agent 2 determines his harvest level based on the observed initial stock level in the breeding area. Before the stock migrates to the feeding area, the stock grows stochastically according to Eq. 1. Agent 1 observes the initial stock migrating to the feeding area and decides how much to harvest. What he does not know is the escapement from the breeding area, that is the stock left unharvested by agent 2 before recruitment. In contrast, agent 2 has full information on agent 1's escapement level. Laukkanen assumes risk neutral agents who seek to maximise profits. Harvest is explained by the Schaefer production function ($Y_t = qE_tX_t$). Both cooperative and non-cooperative harvest policies are analysed within this framework and Laukkanen is able to derive conditions under which cooperation is sustained as a self-enforcing equilibrium.

Considering that much attention has been focused on international management of shared fish stocks, it is somewhat surprising that only few have incorporated uncertainty in their models.

3.4 The Risk of Biomass Collapse

One strand of the literature deals with the risk of stock collapse. Similar analyses of the effects of catastrophic risks can be found in the forestry literature, for example Reed (1984) who considers the effects of the risk of fire on the optimal rotation period of a strand of trees. Returning to the bioeconomic literature, Clemhout and Wan (1985) study a renewable resource under the random threat of extinction. The model is in continuous time and the instantaneous probability of extinction is decreasing in stock size. Clemhout and Wan model individual fishing firms' harvesting from the stock in a game-theoretical framework and study both cooperative and non-cooperative stationary solutions. A stationary solution is defined as a situation with constant stock and harvest rates until the time of sudden resource extinction. Stationary solutions are derived analytically and show that the stationary cooperative stock is larger than the stationary non-cooperative stock. Consequently, cooperation increases the survival prospect of the resource.

Amundsen and Bjørndal (1999) develop a model where the biomass collapse is due to exogenous factors. This is similar to what is referred to as "environmental collapse" in an earlier study by Johnston and Sutinen (1996). The probability of collapse, provided that it has not already occurred, is assumed constant as time goes by and the size of the collapse is a function of the stock size prior to collapse. Amundsen and Bjørndal find that the optimal stock can be above, equal to, or below the no-collapse stock, depending on the size of the collapse and the failure rate. When harvest costs and the size of the collapse are independent of stock size, it is shown that the optimal pre-collapse stock is larger than the optimal no-collapse stock. If, instead, the collapse is a given percentage of the total stock, the optimal stock is always below the optimal no-collapse stock.

Bulte and van Kooten (2001) develop a bioeconomic model with stochastic stock growth and risk of downward shifts in stock caused by catastrophes, which are modelled as a Poisson jump process. They use their model to analyse the concept of minimum viable population size.

Several studies analyse sustainable harvesting where the risk of extintion typically is minimised given certain conditions, for example maximisation of discounted rents or annual yields. Ludwig (1995) models stock dynamics in a similar manner to Bulte and van Kooten (2001) and analyses the concept of sustainability. In Ludwig (1998) he continues the work on stocks under the threat of collapse, this time focusing on optimal management.[5]

3.5 Spatial Bioeconomic Models and Marine Reserves

Lately, spatial bioeconomic models have been given increased attention by fisheries economists and others, and the focus has in particular been on the study of marine reserves or marine protected areas. Deterministic models of marine reserves have shown that they, if anything, reduce the value of fisheries when harvest can be set optimally. Also stochastic bioeconomic models have been used to analyse the effects of marine reserves. One of the early rationales for marine reserves was the view by Lauck *et al.* (1998) that marine fisheries confront managers with "irreducible uncertainty"; that is uncertainty that cannot be further reduced with more information or predictive models, and that in the face of irreducible uncertainty, no fishing zones might be the best strategy. The common opinion in the fisheries economics literature is that protecting the source by establishing a marine reserve is effective in the case of sink-source systems (for a definition see e.g. Sanchirico and Wilen, 1999), where young individuals are found in one area (a source) before migrating to other areas (the sinks). In most other cases however, no unambiguous conclusions have been reached.

Several authors suggest marine reserves to secure the biomass at a sustainable level in the presence of harvest uncertainty (e.g. Mangel, 1998; Doyen and Béné, 2003).

---

[5]See also the work by Engen, Lande and Sæther on sustainable harvesting of stochastic stocks under the risk of resource collapse (see Lande *et al.*, 1997, and references therein).

Hannesson (2002) develops a continuous-time model of two patchy populations, neither being a source or a sink. The growth Eq. 2 is modified to describe growth in two interdependent sub-stocks. Hannesson explains harvest using the Schaefer production function. If the fishery is unregulated (open access), closing off one area is seen to reduce the variability of the catch and increase the total population. However, Hannesson finds no increase in expected rents from protecting one sub-population. While Hannesson considers the effects of marine reserves with open access elsewhere, Conrad (1999) analyses the effects of marine reserves under the assumption of a total allowable catch given by a linear policy in the open area (i.e., total allowable catch is a constant share of the stock size in the open area). As Hannesson (2002), Conrad finds that the variability in biomass is reduced when an area is closed off.

Grafton *et al.* (2004) develop a model of an uncertain fishery, where two sources of uncertainty are incorporated. Environmental variability is modelled as a Wiener process and the possibility of a negative shock is included as a Poisson process.[6] The model is used to analyse the value of a marine reserve when harvesting is optimal. Net economic return is maximised over harvest and reserve size. They find that marine reserves generate values that cannot be obtained through optimal choice of harvest and effort levels alone.[7]

Bulte and van Kooten (1999) analyse optimal harvesting of a stock consisting of two local subpopulations. Stock growth is stochastic in both subpopulations and the analysis is done in a continuous-time framework similar to Hannesson (2002). Instead of protecting one area, they consider the possibility of managing the two subpopulations independently. Using stochastic optimisation, they derive expressions for optimal harvest in each area and find that total harvest might increase or decrease compared to total harvest when treating the subpopulations as one stock. By managing the subpopulations independently, the fishery manager can take advantage of migration by choosing local harvest rates and thereby increase total harvest. Furthermore, if stock-growth in the two subpopulations is dependent, the manager can hedge against risk.

In a recent work by Costello and Polasky (2005), a spatial, discrete-time model of a fishery is developed, in which four sources of uncertainty are incorporated. The sources of uncertainty are biological: stochastic spatial dispersal and random environmental shocks to production of young, survival

---

[6]Sumaila (1998, 2002) analyses the optimal size and effects of a marine reserve in the Barents Sea cod fishery when a large shock is introduced to the system (irreducible uncertainty).

[7]See Grafton and Kompas (2005) for a presentation of this and other studies on marine reserves and uncertainty.

of adults, and survival of settlers. Using dynamic programming they manage to derive an interior solution to the fishery's rent maximisation problem. The existence of an interior solution implies that the harvest rate in each fishing area is positive and that no area should be closed. The problem is found to have an interior solution if the stock size in every patch is sufficiently large. The study also considers conditions for corner solutions, which mean that an area closure is optimal, and concludes that marine reserves can be optimal "under a number of different, and realistic, bioeconomic conditions".

Whereas most studies discussed thus far have been optimisation models, there is a significant literature on behavioural models of fisheries. Discrete choice models have been used to predict fisherman behaviour and an often-sited reference within the fisheries literature is Bockstael and Opaluch (1983) who analyse seasonal gear choice and target species. The key element of discrete choice models is that individual choice is driven by utility, where utility is assumed to consist of a deterministic part and a random component. The models further allow for heterogeneity among individuals. Discrete choice or random utility modelling can be used to describe spatial behaviour, for example choice of fishing ground, and is therefore very suitable for analysis of marine reserves as a management instrument or spatial management of fish stocks in general.

Smith and Wilen (2003) link a spatial behavioural model to a biological model of the northern California red sea urchin fishery and analyse how rent will be spatially dissipated by mobile divers in the fishery. The spatial behaviour of the divers is modelled and estimated in a repeated nested logit framework, where daily discrete participation and choice of fishing location are modelled jointly. The estimated model shows that fishermen adjust to spatial differences in expected returns. The biological model represents the sea urchin population as a metapopulation consisting of 11 fishing areas linked with a dispersal matrix. The implications of spatial closures are analysed by simulating the integrated model with and without a closure of one of the patches. The authors find that accounting for fishermen's spatial behaviour offsets the harvest gains from marine reserves in the sea urchin fishery and concludes that optimistic results obtained about reserves may be due to simplifying assumptions that ignore economic behaviour. In Smith and Wilen (2004), they extend the analysis by making the choice of fisher home port endogenous and thereby allowing for simulation of both short and long run diver behaviour. Although allowing for port switching gives some new insights, the main conclusion is the same, namely that traditional analysis of marine reserves as a management instrument might be biased in favour of reserves because of simplistic assumptions made about the behaviour of fishermen.

3.6 Other Issues

The literature deals with several issues beyond those covered in this chapter. A number of studies examine uncertainty in multi-cohort and multi-species models (e.g. Mendelssohn, 1978, 1980; Spulber, 1983; Reed, 1984; Kennedy, 1989). These models are similar to the single-cohort (biomass), single-species models discussed earlier although the inclusion of additional cohorts and/or species adds to the complexity of the models.

Extensive research has been done on the issue of investment in capacity in the fishing fleet. An often-sited reference on this is Charles (1983) who analyses optimal fleet investment in a stochastic framework. He models the change in biomass in a similar manner to Reed (1979).

The literature on other natural resources contains many studies related to bioeconomic modelling. There is, for instance, an extensive literature on real options and optimal stopping rules (see e.g. Clarke and Reed, 1990 for a review), a topic that has not been discussed here but, nevertheless, can be applied to bioeconomic models.

4 CONCLUDING REMARKS

In this chapter, we have tried to provide an overview of some of the development in stochastic bioeconomic modelling since its introduction in the early 1970s. We live in a stochastic world and have to deal with inaccurate data and unknown external disturbances in addition to the fundamental uncertainty of the future. To deal with this, uncertainty has been incorporated into bioeconomic models to do normative studies, to analyse industry behaviour, and to evaluate alternative management policies.

We have seen how incorporating uncertainty into bioeconomic models can make the models more realistic, provide additional insights, present new problems, and suggest solutions that would not arise in a deterministic model. However, the introduction of uncertainty into a bioeconomic model might not be worthwhile if the stochastic components do not significantly change the behaviour of the system. In such cases, one should consider whether it is possible to keep the analysis within a more straightforward deterministic setting, since the incorporation of uncertainty comes at the cost of increased complexity.

5 REFERENCES

Alvarez, L. H., 2001, Singular stochastic control, linear diffusion, and optimal stopping: a class of solvable problems, *SIAM Journal on Control and Optimization* **39**:1697–1710.

Alvarez, L. H. and L. A. Shepp, 1998, Optimal harvesting of stochastically fluctuating populations, *Journal of Mathematical Biology* **37**:155–177.

Amundsen, E. S. and T. Bjørndal, 1999, Optimal exploitation of a biomass confronted with the threat of collapse, *Land Economics* **75**:185–202.

Andersen, P. and J. G. Sutinen, 1984, Stochastic bioeconomics: a review of basic methods and results, *Marine Resource Economics* **1**:117–136.

Anderson, E. E., 1986, Taxes vs. quotas for regulating fisheries under uncertainty: a hybrid discrete-time continuous-time model, *Marine Resource Economics* **3**:183–207.

Androkovich, R. A. and K. R. Stollery, 1991, Tax versus quota regulation: a stochastic model of the fishery, *American Journal of Agricultural Economics* **73**:300–308.

Bockstael, N. E. and J. J. Opaluch, 1983, Discrete modelling of supply response under uncertainty: the case of the fishery, *Journal of Environmental Economics and Management* **10**:125–137.

Bulte, E. H. and B. C. van Kooten, 1999, Metapopulation dynamics and stochastic bioeconomic modeling, *Ecological Economics* **30**:293–299.

Bulte, E. H. and B. C. van Kooten, 2001, Harvesting and conserving a species when numbers are low: population viability and gambler's ruin in bioeconomic models, *Ecological Economics* **37**:87–100.

Charles, A. T., 1983, Optimal fleet investment under uncertainty, *Canadian Journal of Fisheries and Aquatic Sciences* **40**:2080–2091.

Charles, A. T., 1998, Living with uncertainty in fisheries: analytical methods, management priorities and the Canadian groundfishery experience, *Fisheries Research* **37**:37–50.

Clark, C. W., 1976, *Mathematical Bioeconomics*, New York: Wiley.

Clark, C. W. and G. P. Kirkwood, 1986, On uncertain renewable resource stocks: optimal harvest policies and the value of stock surveys, *Journal of Environmental Economics and Management* **13**:235–244.

Clark, C. W. and G. R. Munro, 1975, The economics of fishing and modern capital theory: a simplified approach, *Journal of Environmental Economics and Management* **2**:92–106.

Clarke, H. R. and W. J. Reed, 1990, Application of optimal stopping in resource economics, *The Economic Record* **66**:254–265.

Clemhout, S. and H. Wan, Jr., 1985, Common-property exploitations under risk of resource extinctions, in T. Basar (ed.), *Dynamic Games and Applications in Economics, Lecture Notes in Economics and Mathematical Systems*, pp. 267–288, Berlin: Springer-Verlag.

Conrad, J. M., 1999, The bioeconomics of marine sanctuaries, *Journal of Bioeconomics* **1**:205–217.

Conrad, J. M., 2004, Renewable resource management, *Encyclopedia of Life Support Systems*, Paris: United Nations, UNESCO.

Costello, C. J., R. M. Adams, and S. Polasky, 1998, The value of El Niño forecasts in the management of salmon: a stochastic dynamic assessment, *American Journal of Agricultural Economics* **80**:765–777.

Costello, C. J. and S. Polasky, 2005, Spatial Bioeconomics Under Uncertainty, Presented at AERE Workshop 2005: Natural Resources at Risk, Wyoming.

Costello, C. J., S. Polasky, and A. Solow, 2001, Renewable resource management with environmental protection, *Canadian Journal of Economics* **34**:196–211.

Danielsson, A., 2002, Efficiency of catch quotas in the presence of risk, *Journal of Environmental Economics and Management* **43**:20–33.

Doyen, L. and C. Béné, 2003, Sustainability of fisheries through marine reserves: a robust modelling analysis, *Journal of Environmental Management* **69**:1–13.

Framstad, N. C., 2003, Optimal harvesting of a jump diffusion population and the effect of jump uncertainty, *SIAM Journal of Control and Optimization* **42**:1451–1465.

Grafton, R., P. Ha, and T. Kompas, 2004, Saving the Seas: The Economic Justification of Marine Reserves. Working Paper EEN04-02, Economics and Environmental Network, Canberra: The Australian National University.

Grafton, R. and T. Kompas, 2005, Uncertainty and the active adaptive management of marine reserves, *Marine Policy* **29**:471–479.

Hannesson, R., 2002, The economics of marine reserves, *Natural Resource Modeling* **15**:273–290.

Hannesson, R. and J. O. S. Kennedy, 2005, Landing Fees versus fish Quotas, *Land Economics*, 81:518–529.

Hannesson, R. and S. I. Steinshamn, 1991, How to set catch quotas: constant effort or constant catch?, *Journal of Environmental Economics and Management* **20**:71–91.

Hanson, F. B. and D. Ryan, 1998, Optimal harvesting with both population and price dynamics, *Mathematical Biosciences* **148**:129–146.

Herrera, G. E., 2005, Stochastic bycatch, information asymmetry, and discarding, *Journal of Environmental Economics and Management* **49**:463–483.

Jaquette, D. L., 1972, A discrete time population control model, *Mathematical Biosciences* **15**:231–252.

Jaquette, D. L., 1974, A discrete-time population-control model with setup cost, *Operations Research* **20**:298–303.

Jensen, F. and N. Vestergaard, 2003, Prices versus quantities in fisheries models. *Land Economics* **79**:415–425.

Johnston, R. J. and J. G. Sutinen, 1996, Uncertain biomass shift and collapse: implications for harvest policy in the fishery, *Land Economics* **72**:500–518.

Kaitala, V., 1993, Equilibria in a stochastic resource management game under imperfect information, *European Journal of Operational Research* **71**:439–453.

Kennedy, J. O. S., 1989, Approximately optimal recovery of a multicohort fishery from depleted stocks and excess vessel capacity, *The Journal of the Operational Research Society* **40**:231–241.

Koenig, E. F., 1984a, Fisheries under uncertainty: a dynamic analysis, *Marine Resource Economics* **1**:193–208.

Koenig, E. F., 1984b, Controlling stock externalities in a common property fishery subject to uncertainty, *Journal of Environmental Economics and Management* **11**:124–138.

Lande, R., B. E. Sæther, and S. Engen, 1997, Threshold harvesting for sustainability of fluctuating resources, *Ecology* **78**:1341–1350.

Lauck, T., C. W. Clark, M. Mangel, and G. Munro, 1998, Implementing the precautionary principle in fisheries management through marine reserves, *Ecological Applications* **8**:S72–S78.

Laukkanen, M., 2003, Cooperative and non-cooperative harvesting in a stochastic sequential fishery, *Journal of Environmental Economics and Management* **45**:454–473.

Lewis, T. R., 1981, Exploitation of renewable resources under uncertainty, *Canadian Journal of Economics* **14**:422–439.

Lewis, T. R. and R. Schmalensee, 1977, Nonconvexity and optimal exhaustion of renewable resources, *International Economic Review* **18**:535–552.

Ludwig, D., 1979, Optimal harvesting of a random fluctuating resource. I: application of perturbation methods, *SIAM Journal on Applied Mathematics* **37**:166–184.

Ludwig, D., 1995, A theory of sustainable harvesting, *SIAM Journal of Applied Mathematics* **55**:564–474.

Ludwig, D., 1998, Management of stocks that may collapse, *Oikos* **83**:397–402.

Lungu, E. M. and B. Øksendal, 1997, Optimal harvesting from a population in a stochastic crowded environment, *Mathematical Biosciences* **145**:47–75.

Mangel, M., 1998, No-take areas for sustainability of harvested species and a conservation invariant for marine reserves, *Ecology Letters* **1**:87–90.

Mendelssohn, R., 1978, Optimal harvesting strategies for stochastic single-species, multiage class models, *Mathematical Biosciences* **41**:159–174.

Mendelssohn, R., 1980, Managing stochastic multispecies models, *Mathematical Biosciences* **49**:249–261.

Mirman, L. J. and D. F. Spulber, 1985, Fishery regulation with harvest uncertainty, *International Economic Review* **26**:731–746.

Nøstbakken, L, 2006, Regime switching in a fishery with stochastic stock and price, *Journal of Environmental Economics and Management* **51**:231–241.

Pindyck, R. S., 1984, Uncertainty in the theory of renewable resource markets, *Review of Economic Studies* **51**:289–303.

Quiggin, J., 1992, How to set catch quotas: a note on the superiority of constant effort rules, *Journal of Environmental Economics and Management* **22**:199–203.

Reed, W. J., 1974, A stochastic model for the economic management of a renewable animal resource, *Mathematical Biosciences* **22**:313–337.

Reed, W. J., 1979, Optimal escapement levels in stochastic and deterministic harvesting models, *Journal of Environmental Economics and Management* **6**:350–363.

Reed, W. J., 1983, Recruitment variability and age structure in harvested animal populations, *Mathematical Biosciences* **65**:239–268.

Sanchirico, J. N. and J. E. Wilen, 1999, Bioeconomics of spatial exploitation in a patchy environment, *Journal of Environmental Economics and Management* **37**:129–150.

Sandal, L. K. and S. I. Steinshamn, 1997, A stochastic feedback model for optimal management of renewable resources, *Natural Resource Modeling* **10**:31–52.

Sethi, G., C. Costello, A. Fisher, M. Hanemann, and L. Karp, 2005, Fishery management under multiple uncertainty, *Journal of Environmental Economics and Management*, 50: 300–318.

Smith, M. D. and J. E. Wilen, 2003, Economic impacts of marine reserves: the importance of spatial behavior, *Journal of Environmental Economics and Management* **46**:182–206.

Smith, M. D. and J. E. Wilen, 2004, Marine reserves with endogenous ports: empirical bioeconomics of the California sea urchin fishery, *Marine Resource Economics* **18**:85–112.

Spence, A. M., 1974, Blue whales and applied control theory, in H. Göttinger (ed.), *System Approaches and Environmental Problems*, Göttingen: Vanderhoeck & Ruprecht:.

Spulber, D. F., 1982, Adaptive harvesting of a renewable resource and stable equilibrium, in L. J. Mirman and D. F. Spulber (eds.), *Essays in the Economics of Renewable Resources*, 117–139, New York: North-Holland.

Spulber, D. F., 1983, Pulse-fishing and stochastic equilibrium in the multicohort fishery, *Journal of Economic Dynamics and Control* **6**:309–332.

Sumaila, U. R., 1998, Protected marine reserves as fisheries management tools: a bioeconomic analysis, *Fisheries Research* **37**:287–296.

Sumaila, U. R., 2002, Marine protected area performance in a model of the fishery, *Natural Resource Modeling* **15**:439–451.

Walters, C. J. and R. Hilborn, 1978, Ecological optimization and adaptive management, *Annual Review of Ecological Systems* **9**:157–188.

Weitzman, M. L., 1974, Prices vs. quantities, *The Review of Economic Studies* **41**:477–491.

Weitzman, M. L., 2002, Landing fees vs harvest quotas with uncertain fish stocks, *Journal of Environmental Economics and Management* **43**:325–338.

Chapter 13

PLANNING IN FISHERIES-RELATED SYSTEMS

Multicriteria models for decision support

Daniel E. Lane
School of Management, University of Ottawa, Ottawa, Canada

Abstract In this chapter, models for decision support in fisheries related systems are presented involving problems with many varied and often conflicting criteria. The fisheries systems problems and decision support opportunities include: (i) spatial-temporal decisions on the exploitation and use of marine zones defined by geographic information systems (GIS) to evaluate diverse activities such as traditional fisheries, aquaculture, and recreation; (ii) longer-term strategic evaluation of fisheries policy and planning exercises; (iii) operational assessment of stock status leading to commercial exploitation regulations, and (iv) deciding on the scientific activities for fisheries science research as a function of prespecified mandates. In each of these cases, the commercial fishery and coastal zone systems require the consideration of multiple criteria characterised by the conflicting issues of sustaining and conserving the marine ecosystem while carrying out exploitation on marine species for commercial and socioeconomic supporting enhancements. This chapter presents structured analysis of multiple criteria fisheries problems using pairwise comparative methodology for evaluating and ranking alternative policies in the fisheries system. Comparisons are developed through the integration of feedback from key decision-making participants in the fisheries system governance framework. The multicriteria decision support case results illustrate the application of the Analytic Hierarchy Process (AHP) methodology embedded in fisheries software models developed for each application arising from Canadian fisheries. The software models allow policy analysts to explore and compare alternative policies and their impacts while providing a logical, non-dominated and ranked decision evaluation procedure to support governance decision makers in their operational and strategic fisheries negotiations.

Keywords: Multicriteria analyses, decision-making in multicriteria problems, pairwise comparisons, analytic hierarchy process, group decision making, governance, risk, uncertainty, utility

1 INTRODUCTION

Decision support in fisheries-related systems involves many, varied, and conflicting criteria. Planning in these systems requires multicriteria models for decision support as part of an organisation's successful strategic and operational process (Wooster 1988). The objective of this chapter is to present fisheries system case studies in four important fisheries policy areas: (i) spatial-temporal decisions on the polyvalent use of marine zones including diverse activities such as traditional fisheries, aquaculture, and recreation; (ii) operational assessment of stock status for sustainable exploitation; (iii) strategic evaluation of fisheries policy; and (iv) identification of scientific research projects for achieving political mandates. These case studies from Canadian fisheries illustrate how success is achieved by logical integration of the multiple dimensions of fishery systems into a general planning context for problem formulation and opportunistic policy development (Keeney, 1992; Hammond *et al.*, 2002).

In preparation for describing successful fisheries-related systems requiring multicriteria models for decision support, it is understood throughout that there are necessary prerequisites for implementation. These prerequisites include: (i) selecting appropriate resources (e.g. expertise, involved stakeholders) for model building and decision support; (ii) defining clearly the problems, objectives and the alternative policy options; (iii) selecting models for policy evaluation and sensitivity analysis; and (iv) implementing and subsequent monitoring and tracking of the selected policy decision.

2 METHODOLOGIES IN MULTICRITERIA DECISION MAKING (MCDM)

Beginning with the problem-solving context, it is necessary to identify the set and form of the decisions to be made, to formulate the multiple objectives and identify the constraining factors. The development of multicriteria decision making (MCDM) methodologies and applied research falls within the domain of management science/operations research. MCDM is a relatively new and rapidly evolving field. The first conference devoted entirely to MCDM was held in 1973 (Cochrane and Zeleny, 1973). Since then, biannual international MCDM conferences have taken place and many special publications on the topic of MCDM have appeared (Starr and Zeleny, 1977; Hwang and Yoon, 1981; Spronk and Zionts, 1984; Steuer, 1986; Bodily, 1992; Kersten and Michalowski, 1992; Yu and Zhang, 1992). In 2005, the International Federation for Operations Research (IFORS, 2005) held their Triennial Conference on the full spectrum of topics in operational research. This conference included methodological and applied sessions on MCDM as well as a session specifically on fisheries and wildlife applications. MCDM research is supported through professional and academic societies such as the Institute for Operations Research and the Management Sciences (INFORMS), and the International Society for Multi-criteria Decision Making (MCDM).

MCDM methodologies (Triantophyllou, 2000) range from preference specification methods of multicriteria decision analysis (MCDA) using simple scoring, to constrained optimisation approaches of multi-objective programming (Keeny and Raiffa, 1976) and multiattribute utility theory (MAUT, Fishburn, 1970) to outranking methods (Roy and Vincke, 1981; Roy, 1985) that produce potentially many non-dominated "solutions". Interactive tradeoff approaches and preference specifications in MCDA have been popularised in hierarchical approaches (e.g. Saaty, 1980–AHP; and Ehtamo and Hamalainen, 1995), in the ideal point comparison methods of Zeleny (1982), fuzzy set analyses (Zadeh, 1965; Kosko, 1993; Sakawa, 1993), as well as through visual interactive programming methods (Belton and Vickers, 1989; Bell, 1991, Elder, 1992,).

In recent years, more and more "user-friendly" software has been made available as decision support for MCDM problems. These include Saaty's "Expert Choice", Istel's "SEE-WHY", and "VISIT", British Steel's "FORSIGHT", Insight International's "OPTIK" and "INORDA", Elder's "V.I.S.A", Hamalainen's "HIPRE 3+", Intelligent Decision Systems (IDS) Ltd.'s Evidential Reasoning (ER) approach, and Miettinen and Mäkelä's (2000) "NIMBUS" as well as other systems (including Internet freeware) that are commercially available and/or sold as modules of larger mathematical programming or decision support software systems.

Potential decision makers have a wide range of options when it comes to selecting a suitable choice of MCDM models both from the point of view of methodology and presentation of results. As in all problems in general, and in the spirit of decision support, different approaches preclude attempts to identify categorically a "best" methodological procedure. In fisheries management, challenges to applying MCDM methodologies are not due to a lack of appropriate methodology, but rather due to (i) a lack of awareness among fisheries decision makers that such decision support methods can assist policy development, and (ii) a lack of demonstrated formulation and operational success of applied fisheries problems as multicriteria decision problems.

In general, quantitative methods for problem solving can be classified into either single criterion methods or multiple criteria methods. Single criterion methods can be optimised directly with alternatives ranked relative to a single performance indicator. The history of policy making in fisheries management has tended to apply single objective methodology at the main advisory level, for example sustainability of estimated stock status. Single objective methods typically consider the conservation objective as top priority, and tacitly adjust decision options to consider other objectives including administrative, compliance, and socioeconomic considerations. The nature of integrated fisheries systems analysis requires a more complex methodology than the simple insight of the modified single criterion approach. Fisheries systems formulations require logical, multicriteria problem structures (Lane, 2001). To illustrate, three general multicritera methods for evaluating candidate policy alternatives for fisheries decision making are presented below. These include: (i) simple scoring – a form of subjective multiattribute valuation and ranking; (ii) multi-objective programming – a structured, functional-based form of multiattribute utility theory using mathematical programming, and (iii) pairwise comparisons and the analytic hierarchy process – a form of interactive tradeoff analysis and ranking (ISNAR, 2001).

The purpose of this chapter is to present a spectrum of fisheries case studies as MCDM problems and to illustrate their analyses and results. Throughout the cases presented below from Canadian fisheries experience, the focus is on the application of interactive tradeoffs using pairwise comparisons in MCDM and, in particular, the application of the Analytic Hierarchy Process AHP (Saaty, 1980) for fisheries policy support.

2.1 Scoring

Consider a fisheries problem with three identified objectives and their associated problem criteria: (i) economics of the fisheries; (ii) resource status; and (iii) sustainability, and three designated alternatives (Projects 1, 2 and 3) to be ranked. Scores for the relative importance of individual criteria are based on assigned subjective information. In scoring, each decision alternative is compared against the criteria using simple scores applied independently to measurable criteria or qualitative measures subjectively based on "expert knowledge". Criteria importance may be imputed by attaching a subjective weight to each criteria and rescaling as necessary. The alternatives are evaluated against the criteria by directly providing scaled criteria importance weights. Alternatives are also assigned scaled criteria scores based subjectively on the decision maker's perception or based on absolute values of the criteria for each alternative, where available. Examples of relative weighted scores are shown in Table 1to provide a ranking of the alternative projects. From Table 1, the final score for preferred Project 2 is calculated as 47.6. Sensitivity analysis of the initial rankings can be used subsequently to evaluate the impact of different weighting systems and assigned alternative scores.

Table 1. MCDM Scoring method for sample fisheries problem.

Criteria	Subcriteria	Units	Assigned weight	Project1 score	Project2 score	Project3 score	Scores
Economic	Production value	Million$	0.40	71.0	6.6	22.4	100.0
Resource Status	Biomass growth	1000 t	0.35	11.4	85.7	2.9	100.0
Sustainability	Commercial fisheries	Number	0.25	30.0	60.0	10.0	100.0
Projects' weighted final score (ranking)			1.00	39.9 (2)	47.6 (1)	12.5 (3)	100.0

In order to define scores and criteria weights, decision makers are required to structure their understanding of the problem. Scoring thus provides insight into problem component interactions, although the assignment of simple relative scores by criteria does not explicitly account for the tradeoffs among the criteria. Scoring provides a logical framework to improve the degree of consensus among different participants involved in decision making. However, scoring can be ineffective when the number of criteria is large, thereby making it harder to be consistent in considering explicitly how criteria are linked and scored relative to each other. Although the subjective nature of scoring permits a degree of flexibility in cases where data are not readily available, it may also undermine the reliability of scoring results due to issues of repeatability and consistency among independent experts' subjective opinions.

2.2 Multi-Objective Programming

Multi-objective programming methods require defined weighted scores for the criteria. Unlike scoring and AHP, they are not typically restricted to evaluating a discrete set of alternative policy options. Rather, multi-objective programming aims to optimise an explicit, prespecified and constrained objective function. A mathematical multi-objective programming model therefore includes constrained multiple objectives that quantify the nature of trade-offs among objectives that are explicitly defined.

The mathematical formulation for a constrained multiple criteria fisheries resource problem can be described as follows (Alston *et al.*, 1995):

Objective function:

$$\text{Maximise } Z(x) = G[z_1(x), z_2(x), \cdots z_k(x)] \tag{1}$$

subject to the constraints:

$$x \in X, \quad x \geq 0, \tag{2}$$

where

$Z(x)$ is the objective value function as a function of k criteria: $z_i(x),\ i=1,2,\ldots k$

G is the function defining the form of the overall objective value function to be maximized; and

x is the n-dimensional vector of nonnegative quantitative decision variables.

Suppose that three criteria (z_1, z_2, z_3) represent: economics, resource status and sustainability as in the previous example. Suppose also that the decision variable x is denoted by the three-dimensional policy variable with components: (i) the number of fishermen (x_1); (ii) the total allowable catch (TAC) (x_2); and (iii) age of first capture (x_3) (e.g. allowable net size). Suppose now that the value functions quantifying the three criteria are defined by simple linear functions in the x_i values, as follows:

Economics ($):

$$z_1(x_1, x_2, x_3) = p * x_2 + Dx_1 - Cx_3, \text{ for } p = 1.00, D = 0.50, C = 190 \tag{3}$$

Resource Status (t): $z_2(x_1, x_2, x_3) = a[1+(r-sx_1-x_2/B0)]B0$, for $a = 1.02$, $r = 0.05$, $s = .03$, $B0 = 10{,}000$ (4)

Sustainability (%) $z_3(x_1, x_2, x_3) = r+x_3/q-x_2/B0$, $q=.015$ (5)

Assuming that G is a linear function in the z_i's, then let the prespecified criteria weights be $(\alpha_1, \alpha_2, \alpha_3)$=(0.40, 0.35, 0.25). The objective function to be maximised can then be written as: max $Z(x) = 0.4z_1(x)+0.35z_2(x)+0.25z_3(x)$ or simplifying, we obtain the reduced linear form:

$$\max Z(x) = -0.871 * x_1 + 0.043 * x_2 - 59.33 * x_3 + 3748.5, \tag{6}$$

subject to the policy boundary constraints on the policy variables,

Constraint 1 (Numbers of fishermen): $900 \le x_1 \le 1200$ fishermen (7)

Constraint 2 (Total Allowable Catch): $5{,}000 \le x_2 \le 20{,}000$ t (8)

Constraint 3 (Age at first capture): $2.0 \le x_3 \le 5.0$ years (9)

Constraint 4 (Socioeconomic value): $1.00x_2 + 0.50x_1 - 190x_3 \ge \$20{,}100$ (10)

This simple linear programming problem has the unique solution: x_1= 960 fishermen, x_2= 20,000 t of annual catch, and x_3= 2 age of first capture, with maximum overall objective value of Z^*=3654.

Mathematical programming for multicriteria problems has a number of possible variations, such as goal programming in which the weighted deviation from stated desirable goals is minimised, or compromise programming in which the feasible solution closest to the optimal non-constrained solution is identified (Aouni, 2004). Multi-objective programming problems enlarge the decision space of feasible alternatives and then searches for an optimised feasible result. The solutions to these mathematical programming problems characterise the nature of optimal results for the problem, thus providing insight into how optimal can be described for the problem. The method depends on the decision maker's ability to define the appropriate functional forms to describe the multiattribute value function as well as the various forms of the constraints. In many cases, the specification of functional forms and the quantification of the multiple criteria are not known or are not easily definable, nor are the means of combining measures of different criteria. In poorly defined cases, symptomatic of many complex fisheries management problems, this renders the formulation of the mathematical problems difficult and the results rather unintuitive.

2.3 The Analytic Hierarchy Process

AHP (Saaty, 1980) is a more rigorous version of scoring that explicitly accounts for the consistency and the relative valuations between criteria. AHP provides a structural framework by decomposing the fisheries problem into a hierarchical structure. AHP also allows for the expertise of multiple participants in the multicriteria decision support problem by making use of their knowledge-base including their subjective judgements.

AHP for a single decision maker can be described by three basic procedures as follows:

1. Decompose the problem into a hierarchical structure of the important decision criteria. The goal of the decision appears at the top of the hierarchy; the next level(s) progressively breaks down the goal into selected relevant criteria and subcriteria; the set of policy alternatives to be evaluated appears at the bottom level of the hierarchy. Figure 1 illustrates the AHP hierarchy corresponding to the example of Table 1.

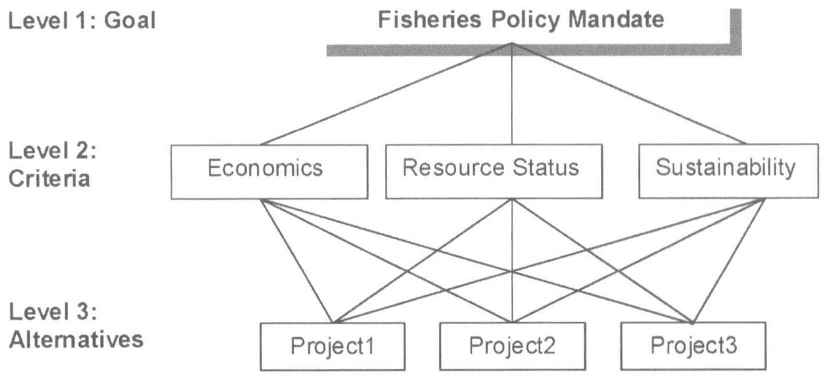

Figure 1. The hierarchical structure of the example MCDM fisheries system decision problem of Table 1.

2. Make comparative judgement of the elements in the hierarchy by pairwise comparison. Each of the criteria and the alternatives are compared in pairs with respect to each element of the next higher level. A common preference scale is used to express the paired comparisons in different terms (e.g. a descriptive verbal and/or numerical 9-point scale, or graphical continuous) that can then be translated into quantitative measures (Table 2).
3. Synthesise the pairwise comparisons to obtain the priorities, integrating subjective and objective judgements information. The overall priorities of the alternatives are calculated using the pairwise comparison matrix eigenvectors (assumed to be within measurable and acceptable levels of the calculated consistency) with respect to each criterion and the weights of each criterion with respect to the goal. The key of the AHP method is the determination of the relative weights used to rank the decision alternatives under the assumption that the comparisons are relatively consistent. For the example used above, the scaled local AHP global priorities (Table 3) calculated for the three projects are respectively: (0.416, 0.444, 0.140) with Project 2 preferred as before.

Table 2. Fundamental 9-point scale for comparing importance, likelihood or preference of pairwise comparisons in AHP.

Comparative Question: How important, likely or preferred is criteria/subcriteria/alternative A compared to criteria/subcriteria/alternative B relative to the specific Goal/criteria/subcriteria?	
Verbal descriptions	**Numerical values**
Equally important, likely or preferred	1
Moderately more important, likely or preferred	3
Strongly more important, likely or preferred	5
Very strongly more important, likely or preferred	7
Extremely more important, likely or preferred	9
Intermediate values to reflect interim measures	2, 4, 6, 8

Table 3. AHP Criteria weights and final priority rankings assigned to projects.

Criteria (weights)	Policy alternatives			Totals
	Project1	Project2	Project3	
Economics (0.403)	0.692	0.077	0.231	1.000
Resource status (0.345)	0.162	0.778	0.059	1.000
Sustainability (0.251)	0.300	0.600	0.100	1.000
Global AHP Priority (rank)	0.416 (2)	0.444 (1)	0.140 (3)	1.000

As a consequence of the need to specify consistent, multiple, pairwise comparisons, the AHP hierarchical formulation improves the decision maker's understanding of the factors influencing the overall valuation of the policy alternatives. It should be noted, however, that multicriteria problem hierarchies are not unique, and consequently, AHP results are dependent on the formulated hierarchical structure. For a given hierarchy, the method generally provides a broad agreement on the general ranking of policy alternatives, points out inherent inconsistencies among pairwise comparisons, and encourages multiple participants to pool their knowledge and expertise to develop consensual input. Sensitivity analysis in AHP can be carried out using computer software (Expert Choice, 2005) designed precisely for this analysis and is an effective and thorough means of further understanding of the MCDM problem formulation and the AHP results. Finally, in cases where more than one decision maker is considered explicitly in the decision problem, group decision making methods are required (Tavana, 2003).

3 MCDM CONSIDERATIONS FOR FISHERIES MANAGEMENT

Formulating complex fisheries problems as MCDM problems incorporate many different problem elements. Decision makers' understanding of problem complexity improves through the exercise of formally structuring different criteria in a hierarchy of the decision problem. Thus, the hierarchical approach associated with simple scoring and AHP are more attractive methods for fisheries problem solving than multi-objective programming, especially given the difficulties associated with defining appropriate functional forms.

As well, through interactive pairwise comparison methods, such as AHP, that define weights and ranks for criteria and alternatives, decision makers are able to quantify their perceptions of these problems. Pairwise comparison methods give fisheries decisions makers the ability to provide, with relative ease, objective quantitative scientific information as well as more subjective qualitative information as the basis for the comparison of criteria and alternatives. Given the mix of science-based and subjective-based perception of criteria and alternatives in fisheries problems, pairwise comparison methods such as AHP are an attractive approach for MCDM problem analysis.

In fisheries management problems involving multiple and diverse decision makers, structured MCDM methods capture the diverse position of each member of the group. Using interactive tradeoff approaches such as AHP, information on ranking of alternatives can be made readily available to all decision makers. This information can then be useful in explaining similarities and differences and promoting consensus positions among the participants towards operationalising MCDM methods decision aids.

Based on past successes and documented applications of MCDM approaches (DiNardo *et al.*, 1989; Kangas, 1995; Merritt and Criddle, 1993; Leung *et al.*, 1998; Tseng *et al.*, 2001; Soma, 2003; Mardle *et al.*, 2004), the fisheries-related policy case studies illustrated in this chapter adopt the MCDA pairwise comparison preference specification method using Saaty's AHP as an effective approach for fisheries problems formulation, solution exploration and decision support. The case studies in the following section are illustrated with examples from the Atlantic Canadian commercial fisheries. Note that these examples of the multicriteria fisheries problem can be readily applied to other fisheries cases in other countries.

4 CASE STUDIES

4.1 Case Study 1: Integrated Systems Analysis for Coastal Aquaculture

4.1.1 Case 1: Background

A multicriteria decision support system is presented for the evaluation of local marine sites under alternative use policies for Grand Manan Island, New Brunswick, in the Bay of Fundy region of Atlantic Canada. This case involves multiple participants in decision-making processes that include potentially conflicting use of coastal zone sites for aquaculture fish farms, for traditional commercial fisheries or for recreational activities (Zhao, 2004; Zhao, *et al.*, 2004).

There are four methodological components used in modelling this MCDM case. These are shown in Fig. 2. These components include: (i) spatial specification in a Geographic Information System (GIS) of alternative marine sites; (ii) valuation assignments to marine sites based on ecosystem inventory; (iii) interpretation of participants' perspectives on the importance of the ecosystem components; and (iv) ranking and sensitivity analyses of the marine sites in support of group decision making.

(1) Marine site GIS identification component – describes the geophysical state of the ecosystem using spatial and temporal indicators in a visual GIS; the ecosystem is described by four key dimensions: (i) Resources – the spatial distributions of natural resources for selected species; (ii) Habitat – the spatial distributions of important natural habitats; (iii) Effluents – spatial distributions of chemicals attributed to human activities or natural sources; and (iv) Activities – the spatial distribution of human commercial and recreational activities. Data are geocoded graphic objects with longitude and latitude coordinates and are compiled for Grand Manan Island to establish the effective ecosystem inventory status of the sites to be evaluated.

(2) Selected-site specific valuation component – assigns value to the spatial-temporal ecosystem inventories. Valuation of a selected marine site takes into account overlapping layers of resources, habitats, effluents and activities and their evaluated cumulative interactions based on "best available" scientific estimates. "Overlap rules" are determined for more complex valuation cases, that is overlapping area yield may increase or decrease compared with the original component yields, for example, when a Resources subcomponent overlaps another Resource or a Habitat

subcomponent, the effect is anticipated to be a positive increase on the overall yields for both the Resource and the Habitat subcomponents in the overlap area. When more than two overlapping layers occur, pairwise overlaps are computed, for example a triple overlapping of layers, ABC will be considered as an overlap of the paired areas: AB, AC and BC. Similarly, the yield determination is made based on the impacts of each layer by pairs. Table 4 summarises the overlap yields rules used in this case.

Table 4. Case 1: Directional yield valuation rules for overlapping layers in pairs (Source: Zhao *et al.*, 2004).

Yield affected components	Yield affecting components			
	Resources	**Habitat**	**Effluents**	**Activities**
Resources	+	+	–	–
Habitat	+	+	–	–
Effluents	0	0	+	0
Activities	+	+	–	–

*Note: Plus signs (+) denote that the overlapping effect on yields for the layer indicated at the left-hand side is positive; (–) denotes negative yields; 0 denotes no effect of the interaction. This is not a symmetric matrix. For example, the overlapping effect of resources and activities is negative for resources, but is positive for activities.

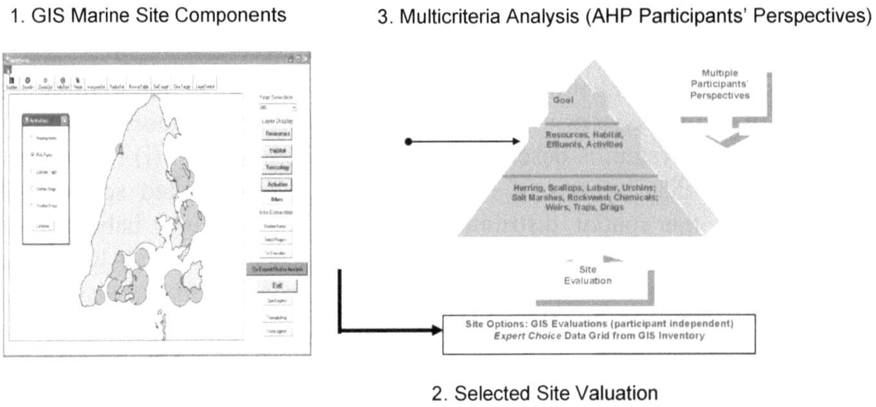

Figure 2. Case 1: Methodology components for the coastal site evaluation problem (Source: Zhao *et al.*, 2004).

(3) Participants' perspective – weights assigned to each level of the inventory hierarchy determined from pairwise comparisons provided by the multiple participants in the decision-making process. The hierarchy comprises of the four ecosystem dimensions and their respective components, for example species (including herring lobster), habitats (such as salt

marshes, rockweed), natural and industrial effluents, and human activities (recreation, fish farms, fisheries). Participant organisations are represented by different groups including: (i) communities; (ii) federal government marine scientists; (iii) commercial organisations; (iv) provincial government managers; and (v) environmental groups. The attributed importance weights with respect to the evaluated ecosystem components at each site are expected to differ among them, for example communities and non-governmental organizations (NGOs) attribute relatively more importance to effluents and habitat, and less importance to human commercial activities; scientists apply relatively more importance to resource abundance compared to recreational and exploitation activities, and commercial organisations attach more importance to resource exploitative activities (Table 5).

Table 5. Case 1: Attributed AHP weights to ecosystem components by the participants (Source: Zhao, 2004)

Components and subcomponents	Participants				
	Local communities	Federal scientists	Industrial organisations	Non-governmental organisations	Provincial governments
Resources	0.147	0.546	0.256	0.235	0.226
Habitat	0.302	0.217	0.124	0.235	0.075
Effluents	0.435	0.163	0.082	0.439	0.185
Activities	0.116	0.075	0.538	0.083	0.514

(4) Sensitivity analyses in support of group decision making – searches for group consensus among the site rankings arising from the different perspectives of the participants. The analysis provides support for strategic use of the marine sites by eliminating completely polarised positions, and searching for generally acceptable ranking positions among the participants for the specific sites evaluated.

4.1.2 Case 1: Application Results

Several different experiments are used to assist in ranking selected marine sites. These experiments include: (i) ranking sites compared to the "average" total study area, (ii) ranking sites compared to an "idealised" site, (iii) direct ranking of two sites, and (iv) ranking of several sites under differing conditions, for example with and without activities such as fishing or fish farm sites.

Consider for example the case of comparing 2 marine sites around Grand Manan Island identified as sites Area1 and Area2 in Fig. 3. Ecosystem weights are first obtained from querying the decision makers responsible for evaluating these two sites in the framework of the AHP hierarchy and based on the supporting data of each site's ecosystem status.

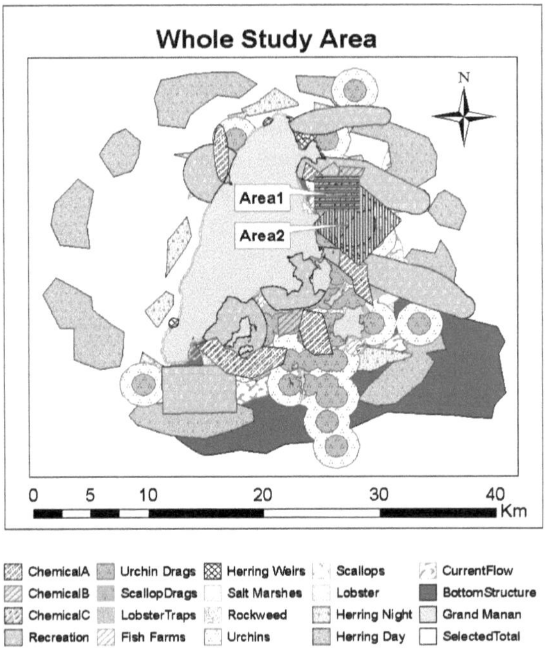

Figure 3. Case 1: Map of Grand Manan Island and selected marine sites Area1 and Area2 (Source: Zhao *et al.*, 2004).

Attributed weights for the importance of the ecosystem components are first obtained independent of the marine sites to be evaluated. Thus, the perspectives of the different participants are understood to be consistent across the marine ecosystem. For example, the data in Table 5 provide overall importance information of the ecosystem components for five major participants in this problem. In particular, it is noted that the scaled AHP weights differ among the participants, for example scientists provide higher importance weighting overall to the resources, communities and NGOs weight effluents relatively higher, whereas industrial organisations give most weight to human activities.

Applying the weights of the hierarchy for each participant to each of the selected sites, Area1 and Area2, yields a set of ranking results as shown in Fig. 4. This figure illustrates similarities and differences among the participants of the direct comparison of marine areas Area1 and Area2. This ranking information provides the starting point for further analysis of the sensitivity of ranking across the participants that would include negotiation, development and further analysis in the search for consensus opportunities for marine use, for example including analysis of dominant site rankings, group "average" (equally weighted) rankings over all participants, weighted participants' contribution rankings, and ranging of participants' criteria for determining dominance.

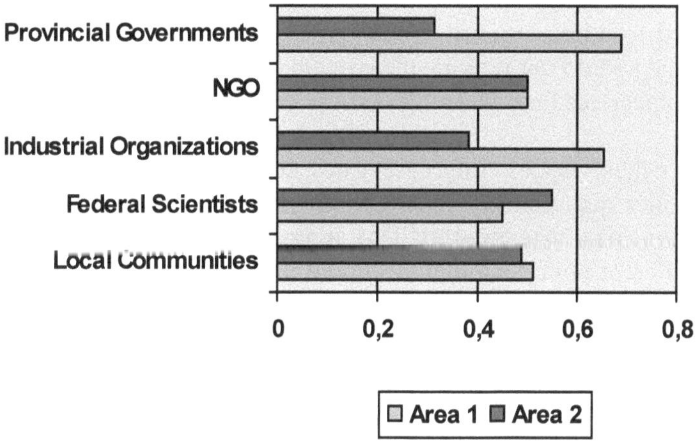

Figure 4. Case 1: Evaluation summary of participants in comparing marine sites Area1 and Area2 (Source: Zhao *et al.*, 2004).

4.1.3 Case 1: Conclusions

Pairwise comparison trade-offs of the multicriteria ecosystem components are extracted from each of the participant's independent of selected areas. A marine site valuation and cumulative effects analyses assign values to the discrete site alternatives. Ecosystem priority rankings result from combining participants' weightings from trade-offs and the yield valuation of sites to establish the participants' independent valuation of the discrete sites. Optional experimental valuation schemes allow for the examination of strategic decision components (e.g. comparing sites with and without fish farms) and provide ranking results as a key element of the decision support model in the search for consensus opportunities and decision support in marine sites use.

4.2 Case Study 2: Strategic Evaluation of Policies for Fisheries Systems

4.2.1 Case 2: Background

In complex resource management, decision makers often take opposing views on the relative importance of, for example, environmental issues (viewed as more important in the long-run), compared to social and financial considerations (viewed as more important politically in the short term). At the same time, fish stock uncertainties and market fluctuations complicate decision making with respect to these resources and sustainability criteria. Policy reviews designed to address conflicts are typically descriptive, propose no substantial change and do not provide clear operational direction from broad-based strategic objectives (e.g. DFO, 2004).

To deal with these issues, this case study describes: (i) an integrated model that provides quantitative valuations for multicriteria components of the fishery system; (ii) a structured MCDM approach for the exploration and evaluation of different policy alternatives under different scenarios that arise in the fisheries system. Planning in the commercial herring fishery along the Scotian Shelf and in the Bay of Fundy in Atlantic Canada is used to illustrate the application (Stephenson *et al.*, 1993; Xue 2003; Xue and Lane, 2004).

The integrated fishery model comprises of three linked methodological subsystems: (i) the database subsystem manages stock data queries and specifies policy inputs and data parameters; (ii) the spreadsheet subsystem uses the database to determine impacts on the biological, economic, social and administrative criteria based on the single decision maker's utility trade-offs relative to specified targets; and (iii) the MCDM subsystem uses AHP to compare the utilities of alternative policies in the hierarchy of the fishery system.

The database subsystem defines specific policies under uncertainty by selecting from a set of policy alternatives and scenarios for stochastic variables. These variables are defined, and transferred to the spreadsheet analysis subsystem. The spreadsheet subsystem then applies the policies and forecasts their expected fishery system impacts on the biological, economic, social and administrative dimensions of the fishery system. Finally, these impacts are transformed into policy performance measures through the decision maker's utility function. Utility function data are then used in the MCDM analysis to provide ranking of the policy alternatives.

Two utility transformations are applied here: (i) absolute indicator value transformations are used where comparisons are made to an "ideal" or "target" policy; and (ii) relative transformations (between alternatives) are used to compare the performances of different policy alternatives. Expected policy scenario results from the spreadsheet analysis are assigned a performance domain value in accordance with the degree to which they meet a specific target value (i.e., the domain of the utility function is the ratio of the calculated performance of the policy compared with its targeted value) for all the criteria measures of the system. Relative utility values for alternative scenario performances are similarly assigned a performance value by comparing the range of the policy's calculated performance to the worst-case performance over all policies for each criteria and relative to the range of the best versus worst-case performance values. These domain values are then mapped into an S-shaped utility curve (Lane and Stephenson, 1998) to yield scaled utility trade-off values as data used in the AHP formulation for each policy.

4.2.2 Case 2: Application Results

The MCDM problem is formulated using the hierarchy criteria: biological, economic, social and administrative characteristics for the herring fishery system. The hierarchy includes detailed subcriteria indicators for each criteria: (i) current herring stock assessment statistics and stock productivity for the biological perspective; (ii) cash after tax for each gear and the processing sector for the herring economic perspective; (iii) herring industry employment and labour earnings for the social perspective; and (iv) administrative employment and per capita cost of fisheries management (e.g. compliance and monitoring) for the administrative perspective. Figure 5 shows the AHP hierarchy for the illustrative strategic problem in the herring fishery in Atlantic Canada.

The integrated database and analysis system contains historical data for the herring fishery from 1989 to 2001, and projected data from 2002 to 2010. Policy objectives data are based on information for this fishery provided in OECD (2000). Validation and scenario analysis exploration are performed in the model using combinations of policy and stochastic variables (Table 6), as well as the decision maker's pairwise comparisons among hierarchy criteria and the subcriteria. For example, in an "equal" weighting scheme, each of the four system criteria (biological, economic, social and administrative) are assigned equal weights. Alternatively, in a "conservation" weighting scheme, biological criteria are weighted most heavily.

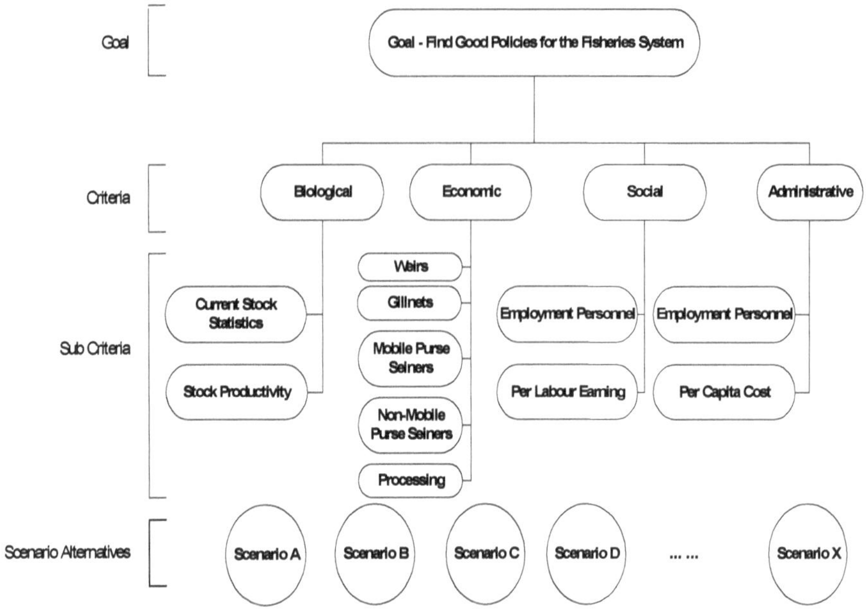

Figure 5. Case 2: AHP hierarchy for herring policy scenario planning (Source: Xue, 2003).

For a single year "conservation" analysis, the results rank lower fishing mortality (F) policies higher, whereas, in an "equal" weighting scheme, higher fishing mortality policies rank higher since the immediate benefits to economic and social perspectives from the greater harvest policies exceed the utility costs of exploiting fish in one particular year, *ceteris paribus*.

Table 6. Case 2: Variable definition for policy scenario definition and analysis (Source: Xue, 2003).

	Criteria category	**Variable name**	**Description**
Policy variables	Biological	(F) Fishing mortality	Total allowable catch (%)
	Economic	(G) Gear allocation	TAC quota allocation between Weirs, Gillnets, MPS and NMPS (%)
Stochastic variables	Biological	(M) Natural mortality	Natural fish stock mortality (%)
	Economic	(CR) Raw material cost factor	Factor for processing price by markets ($)
		(CR) Processing price factor	Factor for raw material cost for actual landed weights by markets ($)
		(CC) Cost inflation factor	Factor for processing cost by markets ($)

In a multiple year analysis, the "conservation" and "equal" weighting schemes are examined along with the combined effects of dynamically varying harvest policies and TAC allocation policies to herring gear. For both schemes, a conservation-oriented (lower) fishing mortality policy and an equal gear allocation policy consistently dominate policy rankings for the herring fishery (Fig. 6). When the environment is favourable (e.g. low natural mortality), preferred fishing mortality strategies are higher due to more attributed contribution profit; when the environment is hostile, the preferred fishing mortality policy element is lower and conservation-focused.

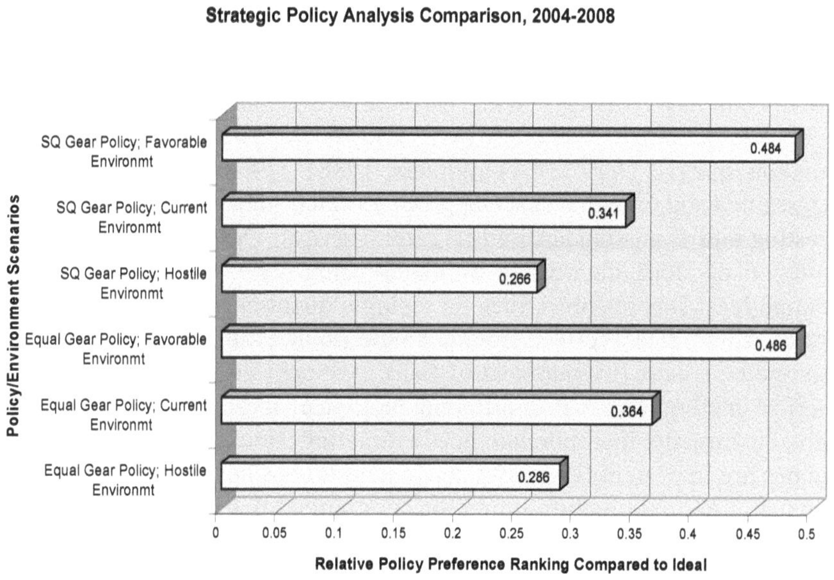

Figure 6. Case 2: Strategic Policy Analysis Comparison for Herring 2004–2008.

4.2.3 Case 2: Conclusions

The decision support integrated system for multicriteria policy evaluation is designed to provide a quantitative tool that facilitates exploration of multicriteria strategic policy evaluation. The herring case study illustrates the ability of the model to evaluate the expected impacts on the system for alternative policies under different environmental scenarios. The model also demonstrates that implications of multiple year policy analysis can be different from those of a short-term analysis due to the interaction of uncertainties over time, and the relative importance of operational considerations versus strategic impacts.

The integrated MCDM system supports decisions by: (i) presenting integrated system-wide results that improve the understanding of policy conflicts and trade-offs between short-term and long-term policy consequences; (ii) involving different participants as contributors to criteria weighting within the system hierarchy; (iii) making the policy setting process more structured and measurable; and (iv) allowing for environmental risk and utility analyses and encouraging investment in adaptive policies that contribute to ongoing information gathering and understanding about dynamic uncertainties.

4.3 Case 3: Operational Assessment of Stock Status

Annual estimation of fish stock abundance in highly variable marine environments is a most difficult task. The most widely and globally used fisheries stock assessment methodology is known as "virtual" or "sequential" population analysis (VPA/SPA) (Winters, 1988). The method relies on age-disaggregated and time-and space-aggregated catch statistics from commercial harvesting and from standardised research surveys. Deterministic numerical calculation of stock abundance estimates using VPA is carried out using minimum least squares objectives for a single quantitative measure based on historical catch data. Results provide single point estimates of the size of a given age or cohort (by numbers of fish). As such, under a single criterion numerical method like VPA, different biological indicators of stock status cannot be directly incorporated into estimating stock status. These stock estimates are important fisheries management tools since they are required to set annual TAC exploitation limits for commercial marine fisheries through the application of standardised biological and stock specific "reference points".

Recently, more restrictive budgetary issues on science activities, superceding issues (e.g. broader oceans management), high cost of maintaining stock surveys and dedicated research vessels, the gradual loss of highly trained scientists familiar with the numerical fish stock assessment process, as well as the decline in once lucrative commercial fish stocks, have caused governments to revisit the need for continued VPA stock assessments (DFO, 2003). For these reasons, traditional VPA-style fisheries stock assessments are considered to be in decline. With this background, this case presents an alternative, inclusive, and cost-effective framework for stock assessment using multicriteria decision analysis and applied to the haddock stock off the southwest of Nova Scotia along Canada's Atlantic coast.

Fisheries scientists examine a variety of stock indicators in their review of stock status and towards estimating stock abundance. For the purpose of defining the AHP hierarchy for stock assessment analysis, these indicators are grouped into general population status characteristics or criteria for: (i) abundance, (ii) production, (iii) fishing mortality, and (iv) management. Within each of these population characteristics or hierarchy criteria, scientific and stock specific measures are defined. For example, these criteria typically include measures attributed to:

- Research vessel survey results, for example numbers and weight per tow by age distribution;
- Commercial catch results, for example numbers and weight per catch by age distribution;
- Area occupied by predefined length or age groupings;
- Local area density by predefined length or age groupings;
- VPA estimates for stock biomass by age groups, and estimates of newly recruited young fish to the fishery;
- Fish condition measures (based on average size, shape, and weight of a fish and its measured fat content);
- Year-over-year growth rate of fish by age groupings;
- Estimates of the commercial exploitation rates by age groupings on the stock and estimates of total and relative mortality (fishing plus natural mortality) by age groupings.

These subcriteria measures are used to define the stock assessment problem hierarchy for this case as illustrated in Fig. 7, the AHP hierarchy for the case of haddock.

Figure 7 associates specific indicator measures to the stock status criteria. In the pairwise comparison analysis of the AHP, expert feedback from fisheries scientists is provided to establish the importance weights for the criteria. Level 4 of Fig. 7, the indicators of stock status alternatives, are used to represent different comparative scenarios for the absolute indicator measures. These scenarios include: (i) year-over-year comparison of the indicator values; (ii) comparative stock status hypotheses for the indicator values; (iii) comparison of ideal indicator values versus actual values.

(i) Year-over-year ranking. Historical review of past annual stock status and past indicator values will lead to a comparative ranking of each year's stock status compared to all others. In this way, the historical stock status can be traced based on the historical records incorporating

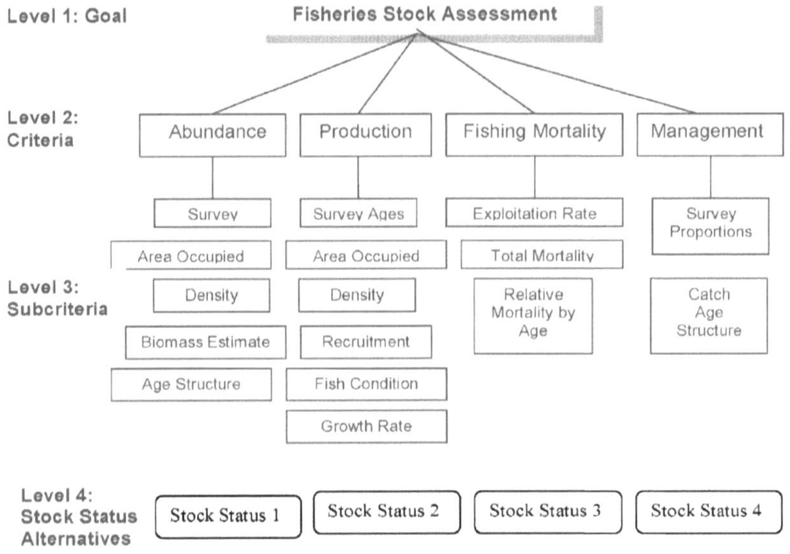

Figure 7. Case 3: The hierarchical structure of the haddock stock assessment problem.

all stock indicator measures. The analysis assists in verifying and validating estimates of past stock status.

(ii) Comparative hypotheses for indicator values. This analysis illustrates the sensitivity of stock status for changes in the indicator values. Different hypotheses are posited having different indicator values (e.g. in cases of uncertainty about indicator values, different hypotheses about "actual" values can be used). Comparison of ranked results shows the proximity of stock status estimates for changes in the indicator values. Should small changes in indicator values result in large relative changes in hypotheses rankings, then further investigation of the sensitivity of the indicators is warranted.

(iii) Comparison of ideal versus actual stock status. In this analysis, a set of indicator values that reflect the stock targets, that is designated as the "ideal" alternative, is used to rank and compare actual indicator values to the ideal. The ranked results provide an indication of the degree to which the actual stock status approaches the targeted ideal.

4.3.1 Case 3: Application Results

The stock of haddock in NAFO Division 4X5Y in Atlantic Canadian waters is one of the few success stories in the litany of catastrophic declines that have occurred to Atlantic groundfish stocks since the early 1990s. The 4X5Y haddock stock abundance has, on the contrary, grown substantially since this period such that the traditional VPA estimate of the stock has more than doubled between 1995 and 2001 (DFO, 2002).

To illustrate the stock assessment multicriteria model, consider the year-over-year stock indicators design for annual periods 1997 – 2001. Data for the annual stock indicator values used in the multicriteria problem are provided in Table 7 along with (i) indicator threshold values; and (ii) indicator stock characteristics for abundance, production, fishing mortality and management.

Table 7. Case 3: Haddock stock characteristics weights for different policy weighting schemes.

Criteria	Stock characteristics	Fisheries management-policy weighting schemes		
		Conservation policy	Traffic light policy	Consolidated policy
1	Abundance	0.584	0.250	0.417
2	Production	0.233	0.250	0.221
3	Fishing mortality	0.130	0.250	0.264
4	Management	0.053	0.250	0.098
	Weight totals	1.000	1.000	1.000

The data set of Table 7 provides data to determine the AHP ranking of the alternatives (years). The rankings of the years depend therefore on the pairwise comparison of the criteria (stock characteristics) and subcriteria (stock indicator values) and the resulting weights for the hierarchy (Fig. 7). The software procedure developed specifically for the stock assessment model establishes a set of different weighting schemes for the hierarchy criteria to be applied to determine the stock characteristics and indicator values weights. These weighting schemes recognise different general fisheries management policies that are attributed to the weights and include: (i) conservation weighting scheme; (ii) "traffic light" weighting scheme posited by the Department of Fisheries and Oceans (DFO), Canada (Halliday *et al.*, 2001); and (iii) consolidated weighting scheme. A summary of the AHP stock characteristic weights for each weighting scheme is presented in Table 8.

The combination of a selected management-policy weighting scheme, and the application of the policy alternative stock indicators values data set provides the overall ranking of the alternatives with respect to the selected policy interpreted as the attributed weighting scheme. For example, Fig. 8 provides the rankings by year for the 4X5Y haddock stock under the conservation weighting scheme. The results show that year-over-year changes in

Table 8. Case 3: Haddock stock annual indicators dataset for 1997–2001 (Source: DFO, 2002).

Stock:	4X5Y Haddock	Indicator	Indicator	Annual indicator data for 1997-2001				
Stock Indicators	Indicator units	Ideal	Character	1997	1998	1999	2000	2001
Summer RV	Kg/tow age4+	50	Abundance	17.73	22.93	23.58	15.63	22.88
ITQ survey	Kg/tow age4+	35	Abundance	25.78	26.88	22.32	25.34	25.46
Summer RV Area occupied	43cm+ fish, sq km	0.6	Abundance	0.38	0.41	0.41	0.47	0.48
Summer RV Local density	43cm+ fish, sq km	3	Abundance	2.19	2.12	2.29	2.28	2.44
VPA biomass	Ages 4+ in kg	66,000	Abundance	34,656	40,349	36,279	40,261	45,061
Summer RV Age structure	No. above avg yc age 3–10	8	Abundance	2.00	3.00	5.00	2.00	6.00
VPA Pop age structure	No. above avg yc age 3–10	8	Abundance	3.00	4.00	3.00	4.00	7.00
Summer RV Area occupied	26-43cm fish, sq km	0.5	Productivity	0.41	0.34	0.36	0.44	0.46
Summer RV Local density	26-43cm fish, sq km	3.5	Productivity	3.33	2.93	3.44	3.25	3.28
Summer RV	qc age 2+3	120000	Productivity	101,149	22,262	35,760	59,214	178,963
ITQ	qc age2+3	120000	Productivity	71,035	17,055	44,885	47,300	1,25,146
VPA Recruitment	Age 2 In 000s of fish	55000	Productivity	13,609	18,966	28,868	93,705	66,887
Summer RV Condition	Fulton's K	1.14	Productivity	1.01	1.00	1.03	0.99	1.01
Summer RV Growth	Length in cm at age7	66	Productivity	50.38	49.18	44.91	47.19	46.06
Summer RV Growth	Length in cm Age 2–4	25	Productivity	22.00	18.00	13.00	20.00	19.00
Summer RVZ5-7	Z5-7 (3yr ave)	0.3	Fishing mort	0.49	0.53	0.44	0.36	0.36
Exploitation (%)	% for ages 5–7	15	Fishing mort	14.98	20.13	11.07	9.46	7.81
Rel F5-7	(3yrsmooth)	0.2	Fishing mort	0.21	0.21	0.22	0.21	0.21
Summer RV proport	% 42cm+	1	Management	0.7	0.5	0.45	0.39	0.6

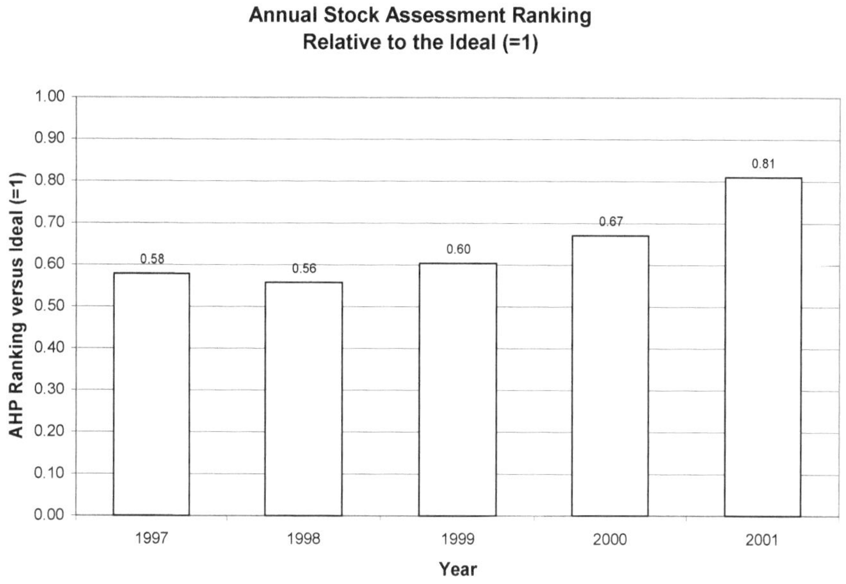

Figure 8. Case 3: 4X5Y Haddock annual stock assessment comparison 1997–2001.

the stock status improves during the planning period from 1998–2001. This result is consistent with traditional stock assessment indications for this same period.

4.3.2 Case 3: Conclusions

The multicriteria stock assessment model presented here provides an opportunity to use independent multiple sources of stock status indicators as integrated information on stock status. The approach illustrated here provides a structured mechanism that takes into account all the various data sources and their relative importance. The results of the multicriteria model can be used to validate the traditional historical aggregated estimates of stock status, and to adjust these accordingly to take account of the inherent uncertainty associated with stock estimation. The proposed approach also permits the inclusion of "experts" in the stock assessment problem. These experts include fisheries scientists who develop and present different indicators of on-going stock status, as well as fishermen and other stakeholders who inevitably bring their own important and regular observations as indicators of stock status to the problems of estimating stock abundance.

Finally, this approach represents the opportunity to explore a comprehensive stock assessment methodology that reflects fisheries management policy while incurring cost savings (compared to the traditional deterministic numerical stock assessment methods) and providing a logical basis to develop and defend a consensus position.

4.4 Case 4: Fisheries Science Prioritisation

4.4.1 Case 4: Background

Science prioritisation interprets the scientific mandate of the organisation by identifying specific programmes that are expected to best fulfill the mandate. Science prioritisation is carried out subject to a limited set of resources for budgets, human capital, and ultimately, science programmes alternatives. In this case, the prioritisation exercise of the Science Branch of the Canadian government's Department of Fisheries and Oceans (DFO) is presented as a decision-making problem in which science programmes are selected among a set of alternatives designed to fulfill the multicriteria science mandate (Chen, 2004).

Since the collapse of the Canadian North Atlantic groundfish stocks in the 1990s, international tensions, growing recognition of aboriginal and treaty rights and unprecedented expansion of broader oceans and ecosystem concerns over the past decade, DFO has been required to operate in a more constrained and more demanding science delivery system than ever before. These additional pressures have required that DFO review its scientific strategic plan, re-examine its science branch organisational structure, revise its fundamental mandate and undergo a process of science re-prioritisation.

The scientific mandate at DFO is broadly stated as "ensuring safe, healthy, productive waters and aquatic ecosystems for the benefit of present and future generations" (http://www.dfo-mpo.gc.ca/science/strategic-strategique/strategy_e.htm) and is constructed around three strategic objectives: (i) sustainable fisheries and aquaculture; (ii) healthy and productive aquatic ecosystems; and (iii) safe and accessible waterways.

Furthermore, to achieve these objectives within the science prioritisation exercise, DFO applies a systematic approach to assessing how these objectives are characterised and how they are delivered in an environment of uncertainty. The characterisations of the objectives that are explicitly identified by DFO are: (i) scientific excellence and credibility; (ii) long-term efficiency; and (iii) sound and relevant decision-making.

These characterisations are the subcriteria in the AHP hierarchy formulated for the problem (Fig. 9).

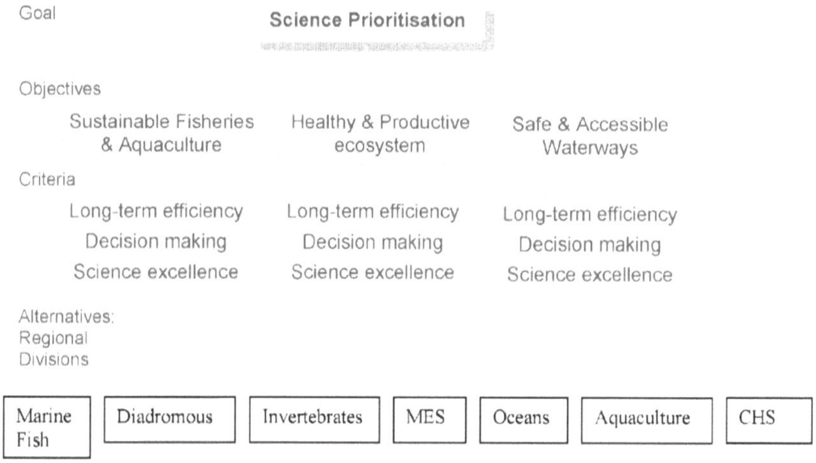

Figure 9. Case 4: Regional prioritisation hierarchy reflecting national mandates applied to divisions; MES – Marine Environmental Science; CHS – Canadian Hydrographic Survey (Source: Chen, 2004).

Science prioritisation at the different levels of the organisational structure proceeds from the strategic goals of the science organisation. The ultimate goal of the science prioritisation exercise is to develop a procedure for attaching "importance" to the science functions vis-à-vis the mandate at each level of the organisational hierarchy. Consequently, this importance weighting system establishes resource allocation down to the level of projects throughout the organisation. DFO Science delivers its mandate and services from an organisation consisting of national headquarters and six regional offices. A chain of responsibility links the regions to their own autonomous divisions. Within each region, the divisions deliver and are in turn responsible for the delivery of their own science programmes that are composed of specific science projects. The hierarchical structure of DFO science (Fig. 10) can be described as a "top-down" structure from national to regional to divisional to programmes and finally to project levels carried out by individual scientists.

The pie chart is Fig. 11 illustrates the link between regions and divisions within one of the regions (Maritimes region) for this case.

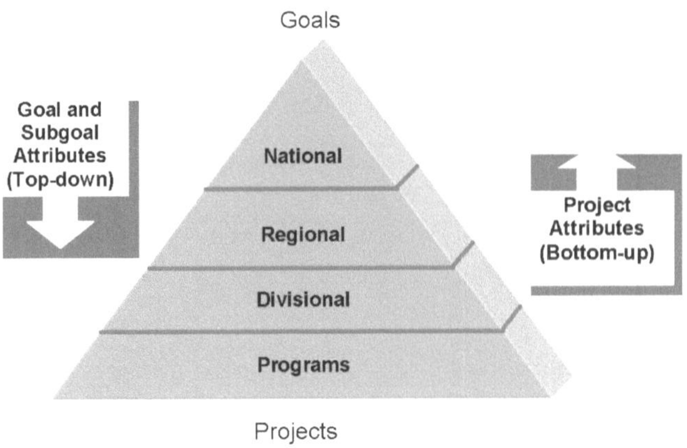

Figure 10. Case 4: Government fisheries science organisational structure.

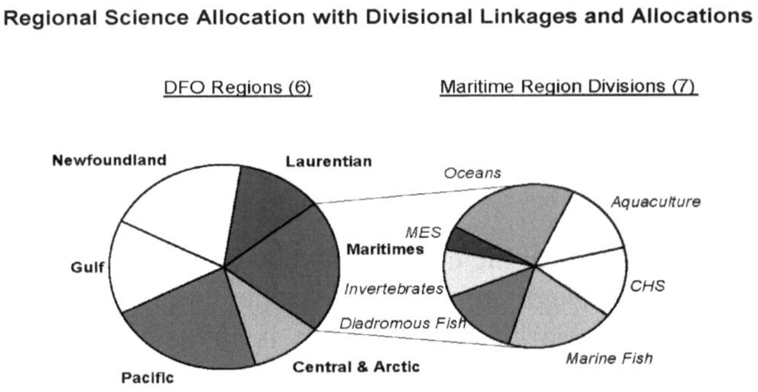

Figure 11. Case 4: DFO Maritimes Region and Divisional linkages; MES – Marine Environmental Science; CHS – Canadian Hydrographic Survey (Source: Chen, 2004).

The goal of the science prioritisation exercise is to describe a procedure for attaching "importance" to the science functions at each level of the hierarchy. This importance metric will enable the relative allocation of resources including budgets, down to the level of projects as the operational means of delivering the national overarching science mandate.

4.4.2 Case 4: Application Results

At each level of the organisational structure (National, Regional, Divisional, and Program), a multicriteria problem is defined for the allocation of resources to be made at each level according to the hierarchy defined by the science mandate. This first problem is to allocate total annual DFO science resources to each of the regions from the national level. The end result is a weighted ranking of the regions that evaluates the capability of each Region to contribute to the overall National Goal.

Similarly, the second-level problem is carried out within each region. Given a budget allocation from the national problem, each region independently evaluates and allocates its budget within its own divisions. The division-level problem then directs allocation from this stage down to programmes. As for the regional-divisional problem, the allocation problem of divisional resources to programmes is independent. Although each level may specify it own goals, objectives, and criteria, for the science prioritisation problem, it is natural to make the linkage back to these national goals that can be made explicit at each level problem. Finally, the last level problem allocates resources of the divisional programmes to specific projects. For illustration, Fig. 12 shows the allocation of budget from the AHP ranking for the set of seven DFO regional divisons (as for Fig. 10) denoted as: (1) Oceans Science, (2) Marine Environmental Science (MES), (3) Aquaculture, (4) Invertebrates, (5) Canadian Hydrographic Service (CHS), (6) Marine Fisheries, and (7) Diadromous Fish.

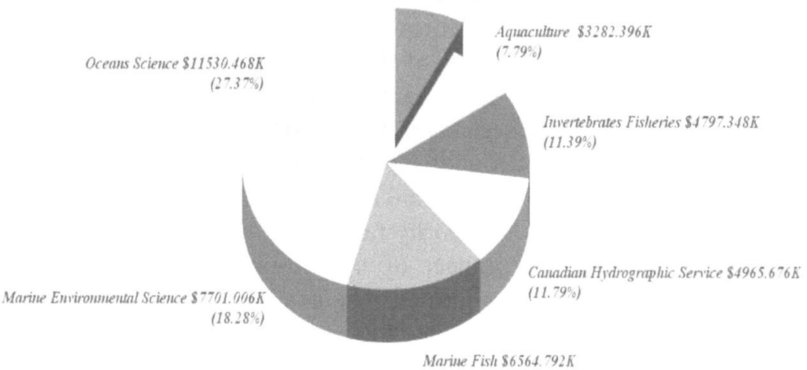

Figure 12. Case 4: Divisional allocations based on AHP rankings and the assigned regional budget (Source: Chen, 2004).

The analysis of risk in the science prioritisation problem has become an increasingly important one for DFO (DFO, 2003). The software procedures developed for this case incorporate the integration of risk analysis through sensitivity analysis of the AHP model rankings. In this problem, risk is associated with the probability of achieving the expected outcomes of the projects, programmes, divisions, and regions toward meeting the overall mandate or goal. Risk analysis is interpreted in two components: risk assessment and risk management (Lane and Stephenson 1998). For this case, risk analyses are carried out through sensitivity analysis on the AHP results via 3 analyses: (i) risk assessment of the top-down hierarchy objectives analysis, (ii) risk assessment of the bottom-up alternatives' performance analysis, and (iii) risk management of the overall mandate through utility analysis.

(i) Risk assessment of objectives. This procedure explores the impact of changes to the alternative rankings of the AHP model subject to changes in the weights of the top level criteria. For example, an increase in the original priority weighting for "Safe and Accessible Waterways" favours programmes (i.e. provides higher overall rankings) with high evaluation with respect to this criterion, for example monitoring programmes for Oceans Science and Hydrographic Monitoring, while reducing the rankings of other programmes that contribute less to "Safe and Accessible Waterways", for example Groundfish Monitoring. This analysis of changes to the top level criteria shows the sensitivities of the programme priorities to these changes. Given the potential shifts that may occur in top-down policy, this analysis provides the direct impact of these policy changes.

(ii) Risk assessment of the alternatives. A second sensitivity procedure developed for this case explores science prioritisation sensitivities due to changes in the evaluation of the programme alternatives data. The procedure examines a percentage shift as input to the original data values for a selected alternative. An increase for example in the performance value of the Groundfish Monitoring programme has a direct effect on its marginal priority ranking, for example doubling its original data values, improves its priority ranking by 60% *ceteris paribus*. This analysis provides the impact of bottom-up data changes and their impacts on programme priorities. The analysis is useful for improving values to satisfy desirable outcomes, or for exploring the uncertainties inherent in the data gathering exercise.

(iii) Risk management through utility analysis. Budgetary analysis requires specifying low, medium, high and maximum budget allocation thresholds for each alternative along with associated utility using an S-shaped utility curve. This analysis assigns utilities based on programme rankings and programme budget allocation. Overall utility is computed as the sum of the utilities for all programmes. It is noted that this simple sum is used to compare overall utilities for different possible thresholds that decision makers may assign, or for different total budgets. Current budget allocations from the prioritisation process are assumed to achieve an overall "medium threshold" acceptable utility level. Consequently, changes in the thresholds or changes in the total budget cause changes in the utility value. If the budget decreases, then programmes' performance, as measured by the utility, may fall to an unacceptable level; similarly, budget increases may show a gain in programmes' overall utility performance, and successful mandate delivery becomes less uncertain. This analysis is used to defend budget requirements to meet acceptable degrees of achieving the science mandate.

4.4.3 Case 4: Conclusions

For this case, the AHP model provides (i) an effective means of prioritising objectives and projects, (ii) a structured means of presenting and defending budgetary requirements, and (iii) the means of evaluating and scoring projects as part of science mandate delivery. The results of the DFO risk-based science prioritisation case provide alternative and effective solutions for scientific resources allocation designed to achieve the strategic goals of the organisation. As well, the risk-based science prioritisation analysis anticipates the overall ability of the prioritisation strategy to deliver the science mandate.

5 DISCUSSION

This chapter presented four multicriteria problem formulations for fisheries management decision-making. The structured multicriteria methodology of the AHP is shown to be a rich and generally applicable framework for problem formulation and solution exploration from strategic to operational fisheries problems. Moreover, in all instances, the multicriteria formulation exercise invariably leads to improved insights about the complex problem and a better understanding by decision makers of important trade-offs in the fishery system under scrutiny. This is enhanced by the ability of the methods to consider different perspectives involving multiple participants in the decision-making exercise.

Among stumbling blocks to the application of effective MCDM methods to fisheries problems, the need for interdisciplinary data and analysis persists as an issue. This is so since fisheries organisations tend to be highly disciplinary and do not easily integrate and share data and decision making (Lane and Stephenson, 2000). Without more opportunity for interdisciplinary organisation, the necessary dialogue and discussions among biologists, economists, sociologists, ecologists of the fishery system towards a more concerted multiple criteria problem-solving setting, cannot take place.

Analysis of interdisciplinary management problems require the definition of appropriate roles and responsibilities for all participants in problem solving. In particular, it is not necessarily government fisheries agencies who will be ultimately "responsible" for defining particular objectives and constraints of the MCDM problems. Government agencies should act as decision support experts charged with providing the stakeholders who operate the fishery system (and produce and enjoy value from it) with a range of interpretations and strategic opportunities arising from both strategic and operational decision problems (Hammond *et al.*, 2002). Appropriately presented MCDM exploratory analyses, such as presented here, will enable stakeholder-decision makers to make effective decisions in a multicriteria governance consensus-setting environment.

Further work on applying MCDM support involves several key areas. These include:

1. Development of integrated databases that provide measures and indicators on multiple criteria in fisheries;
2. Organisation of fisheries institutions designed to deal with integrated analysis and multidisciplinary systems including fisheries scientists working side-by-side with fisheries economists and sociologists, policy makers, and fisheries managers;
3. Organisation of inclusive governance structures involving government experts in a decision support role and all stakeholders with responsibility and accountability for decision making and marine stewardship;
4. Production of computer software dedicated to specific fisheries problem formulations and designed for in-house use in the regular governance framework for policy exploration and analysis;
5. Enhanced explanation of post-ranking results analysis related to problem sensitivity to the input data, risk-based approaches to policy performance; and
6. Improved presentation and explanation of multicriteria analyses for consolidating multiple participants in group decision-making towards achieving a consensus position especially among conflicting participants.

6 REFERENCES

Alston, J.M., Norton, G.W. and Pardey, P.G. 1995. Science under scarcity. Cornell University Press.

Aouni, B. 2004. Goal programming model: Theory and applications (Part I). Special Issue of INFOR. 42(3), August.

Bell, P.C. 1991. Visual interactive modelling: the past, the present, and the prospects. Mini-Euro Conference on Visual Interactive Modelling, Warwick, 1988. *European Journal of Operational Research* 54, 307–317.

Belton, V. and Vickers, S.P. 1989. V.I.S.A. – VIM for MCDA. In Lockett, A.G. and Islei, G. (eds.) *Improving Decision Making in Organisations*. Springer-Verlag, Berlin, 287–304.

Bodily, S.E. (ed.) 1992. Decision and risk analysis. Special Issue of Interfaces. 22(6), 186.

Chen, R. 2004. *Multicriteria Approach for Risk-based Science Prioritisation*. M.Sc. Thesis. Master's Program in Systems Science, Faculty of Graduate and Postdoctoral Studies, University of Ottawa.

Cochrane, J.L. and Zeleny, M. (eds.) 1973. *Multiple Criteria Decision Making*. Columbia, University of South Carolina Press.

DFO. 2002. Haddock on the Southern Scotian Shelf and Bay of Fundy (Div. 4X/5Y). DFO Sci. Stock Status Report. A3–07(2002).

DFO. 2003. Financial stability, program renewal and modern management. The DAAP Final Report. Department of Fisheries and Oceans Canada. Ottawa,

DFO. 2004. A Policy Framework for the Management of Fisheries on Canada's Atlantic Coast. Department of Fisheries and Oceans Canada. Ottawa, March 46p.

DiNardo, G., D. Levy, and B. Golden. 1989. Using decision analysis to manage Maryland's river Herring fishery: An application of AHP. *Journal of Environmental Management* 29: 193–213.

Ehtamo, H. and Hamalainen, R. 1995. Credibility of linear equilibrium strategies in a discrete time fishery management game. *Group Decisions and Negotiation* 4, 27–35.

Elder, M. 1992. *Visual Interactive Modelling: Some Guidelines for its Implementation and Some Aspects of its Potential Impact on Operational Research*. Doctoral Dissertation, University of Strathclyde, Scotland.

Expert Choice. 2005. Expert Choice 2000. http://www.expertchoice.com/software/.

Fishburn, P.C. 1970. *Utility Theory for Decision Making*. Wiley, New York.

Halliday, R.G., Fanning P.F. and Mohn, R.K. 2001. Use of the Traffic Light Method in Fishery Management Planning. DFO Canadian Science Advisory Secretariat Research Document 2001/108.

Hammond, J.S., Keeny, R.L. and Raiffa, H. 2002. Smart Choices: A Practical Guide to Making Better Life Decisions. Broadway Books, New York.

Hwang, C.L. and Yoon, K. 1981. Multiple Objective Decision Making – Methods and Applications: A State-of-the-Art Survey. Lecture notes in Economics and Mathematical Systems. Springer-Verlag, Berlin.

IFORS. 2005. International Federation of Operational Research Triennial 2005 Conference. http://www.ifors.org/

ISNAR. 2001. Information and Discussion Forum on Priority Setting in Agricultural Research. http://www.isnar.cgiar.org/fora/index2.htm

Kangas, J. 1995. Supporting the choice of the sports fishing site. *Journal of Environmental Management* 43: 219–31.

Keeny, R.L. 1992. *Value-Focused Thinking: A Path to Creative Decisionmaking*. Harvard University Press, Cambridge, MA.

Keeny, R.L. and Raiffa, H. 1976. *Decisions with Multiple Objectives: Preferences and Value Tradeoffs*. Wiley, New York.

Kersten, G. and Michalowski, W. (eds.) 1992. Special Issue on Multiple-Criteria Decision Making. *INFOR* 30(2), 65–191.

Kosko, B. 1993. *Fuzzy Thinking: The New Science of Fuzzy Logic*. Hyperion, New York.

Lane, D.E. 2001. Fisheries Management Science: Toward the Development of Decision Support for Multicriteria Systems in Conflicts. School of Management, University of Ottawa. PRISM Working Paper 01–47.

Lane, D.E. 2003. Multicriteria Procedures for Science Program Prioritization at Fisheries and Oceans, Canada. School of Management, University of Ottawa Working paper No.04–11.

Lane, D.E. and Stephenson, R.L. 1998. Toward a framework for risk analysis in fisheries decision making. *ICES Journal of Marine Science* 55(1), 1–13.

Lane, D.E. and Stephenson, R.L. 2000. Institutional Arrangements for Fisheries: Alternate Structures and Impediments to Change. *Marine Policy* 24, 385–393.

Leung, P., J. Muraoka, S.T. Nakamoto, and S. Pooley. 1998. Evaluating fisheries management options in Hawaii using analytic hierarchy process (AHP). *Fisheries Research* 36(2–3): 171–183.

Mardle, S., Pascoe, S. and Herrero, I. 2004. Management objective importance in fisheries: An evaluation using the analytic hierarchy process (AHP). *Environmental Management* 33(1): 1–11.

Merritt, M.F. and K.R. Criddle. 1993. Evaluation of the analytic hierarchy process for aiding management decisions in recreational fisheries: a case study of the Chinook salmon fishery in the Kenai River, Alaska. In G. Kruse, D.M. Eggers, R.J. Marasco, C. Pautzke and T.J. Quinn II (eds.), *Proceedings of the International Symposium on Management Strategies for Exploited Fish Populations, 93–02*, University of Alaska Fairbanks, 683–703.

Miettinen K. and Mäkelä M.M. 2000. Interactive multiobjective optimization system WWW-NIMBUS on the Internet. *Computers and Operations Research*, 27, 709–723.

OECD 2000. Transition to Responsible Fisheries: Economic and Policy Implications, Paris.

Roy, B. 1985. Methodologie Multicritere d'Aide a la Decision. Economica. Paris.

Saaty, T.L. 1980. *The Analytic Hierarchy Process*. McGraw-Hill, New York.

Sakawa, M. 1993. *Fuzzy Sets and Interactive Multiobjective Optimization*. Plenum Press, NY

Soma, K. 2003. How to involve stakeholders in fisheries management – a country case study in Trinidad and Tobago. *Marine Policy* 27(1): 47–58.

Starr, M.K. and Zeleny, M. (eds.) 1977. Multiple criteria decision making. TIMS Studies in the Management Sciences, 6. North Holland, Amsterdam.

Stephenson, R.L., Lane, D.E., Aldous, D.G., and Nowak, R. 1993. Management of the 4WX Atlantic Herring (Clupea harengus) Fishery: An Evaluation of Recent Events. *Canadian Journal of Fisheries and Aquatic Sciences* 50:2742–2757.

Steuer, R.E. 1986. *Multiple Criteria Optimization: Theory, Computation, and Application*. Wiley, New York.

Spronk, J. and Zionts, S. 1984. Special Issue on Multiple Criteria Decision Making. Management Science. 30 (November), 1265–1387. University of South Carolina Press, Columbia.

Tavana, M. 2003. CROSS: A multicriteria group-decision-making model for evaluating and prioritizing advanced-technology projects at NASA. *Interfaces* 33(3): 40–56.

Tseng, C.T., Chen, S.C., Huang, C.S., and Liu, C.C. 2001. GIS-assisted site selection for artificial reefs. *Fisheries Science* 67(6): 1015–1022.

Triantophyllou, E. 2000. *Multi-Criteria Decision Making Methodologies: A Comparative Study. Applied Optimization, 44*. Kluwer Academic Publishers, Dordrecht.

Winters, G.(1988). The development and utility of sequential population analysis in stock assessments in Collected Papers on Stock Assessment Methods. In D. Rivard (ed.), *Canadian Atlantic Fisheries Scientific Advisory Committee Res Doc. 88/61*, 349–368.

Wooster, W.S. (ed.) 1988. Fishery Science and Management: Objectives and Limitations. Lecture Notes on Coastal and Estuarine Studies, Vol 28. Springer-Verlag, New York.

Xue, L. 2003. *Decision Support Integrated System for Multicriteria Policy Evaluation for Fisheries Management.* M.Sc. Thesis. Master's Program in Systems Science, Faculty of Graduate and Postdoctoral Studies, University of Ottawa, 84.

Xue, L. and Lane, D.E. 2004. Evaluation of Strategic Policies for Fisheries Systems. Paper submitted to the Proceedings of the 4th World Fisheries Congress, Vancouver, B.C., June. 25p.

Yu, P.L. and Zhang, D. 1992. Multicriteria optimization. In Salvendy, G. (ed.) *Handbook of Industrial Engineering*, 2nd ed., Wiley, New York, 2672–2692.

Zadeh, L.A. 1965. Fuzzy sets. *Information and Control* 8, 338–353.

Zeleny, M. 1982. *Multiple Criteria Decision Making*. McGraw-Hill, New York.

Zhao, Y. 2004. Integrated Systems Analysis for Coastal Aquaculture. M.Sc. Thesis. Master's Program in Systems Science, Faculty of Graduate and Postdoctoral Studies, University of Ottawa. 84 pp.

Zhao, Y., Lane, D.E., Michalowski, W., Stephenson, R.L. and Page, F. 2004. Integrated Systems Analysis for Coastal Aquaculture. Paper submitted to the Proceedings of the 4th World Fisheries Congress, Vancouver, B.C., June. 26p.

Chapter 14

CAPACITY AND TECHNICAL EFFICIENCY ESTIMATION IN FISHERIES: PARAMETRIC AND NON-PARAMETRIC TECHNIQUES

Sean Pascoe and Diana Tingley

Centre for the Economics and Management of Aquatic Resources (CEMARE), University of Portsmouth, Portsmouth, UK

Abstract The measurement and analysis of efficiency and capacity utilisation in fisheries has seen increased attention over the last decade. Two key approaches are available to estimate these measures – stochastic production frontiers and data envelopment analysis (DEA). In this chapter, the methods are outlined, and examples of their use in fisheries are presented. Issues relating to the application of both techniques in fisheries are outlined.

Keywords: Capacity, technical efficiency, stochastic production frontier, data envelopment analysis, fisheries

1 INTRODUCTION

Input controls are a common feature of most fisheries management programmes throughout the world. An implicit assumption in the use of input controls is that vessels are relatively homogeneous in terms of efficiency and capacity utilisation. If such was the case, then a given reduction in fleet numbers would result in a proportional reduction in fishing effort and, consequently, fishing mortality. However, heterogeneity in efficiency and capacity utilisation has been demonstrated to be a feature of many fisheries (see Pascoe *et al.*, 2001a; Kirkley *et al.*, 2003; Tingley *et al.*, 2003; Vestergaard *et al.*, 2003 for recent examples of capacity utilisation; and Pascoe *et al.*, 2001b; Herrero and Pascoe, 2003; Kompas *et al.*, 2004 for recent examples of efficiency). Given this heterogeneity, the choice of which vessel to remove has an impact on the efficacy of the management measure. Further, understanding the factors that affect the level of efficiency and capacity utilisation is important, as changes in these factors may reduce the benefit of any input control programme.

Capacity utilisation and efficiency are similar in concept as each represents the degree to which vessels are performing relative to other vessels using similar levels of inputs. The capacity of a vessel can be defined as the maximum level of output that it could be expected to produce under normal working conditions (FAO, 2000). Capacity output therefore takes into account periods of maintenance, poor weather, seasonal factors and other normal breaks in activity. *Capacity utilisation* is the degree to which the vessel is achieving its potential (capacity) output given its physical characteristics (i.e. fixed inputs such as size, engine power, etc.). Capacity underutilisation may be a result of using fewer variable inputs (e.g. days fished and crew) than it otherwise could.

In contrast, *technical efficiency* is related to the difference between the actual and potential output given both fixed and variable input use. A vessel may be operating at below its capacity level due to underutilisation of the fixed inputs, or the inefficient use of these inputs, or some combination of the two. Differences in efficiency may be related to differences in the skill of the skipper and crew, age of the vessel, differences in search and navigational aids, etc.

The two concepts are illustrated in Figure 1, in which a vessel of a given size is observed to be producing O_0 level of output as a result of using V_0 levels of inputs. If all inputs were fully utilised (i.e. using V_c rather than V_0

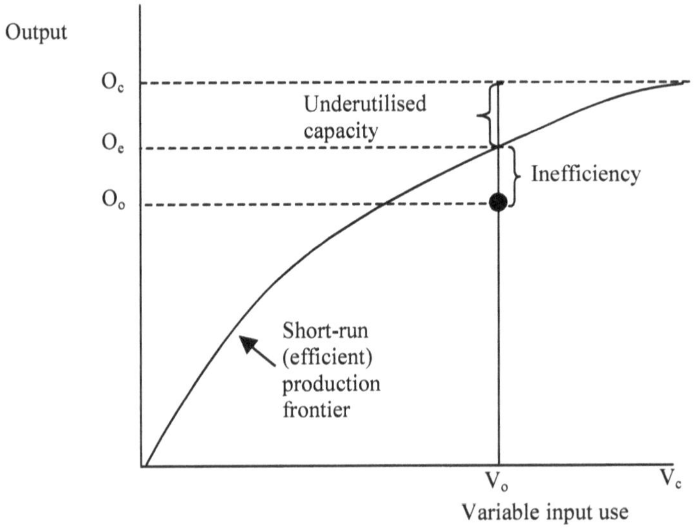

Figure 1. Capacity underutilisation and inefficiency.

potential (capacity) output would be O_c. Even at the lower level of input usage, if the vessel was operating efficiently it would be expected to produce O_e level of output. Hence, the difference O_c–O_e is due to capacity under-utilisation, and the difference O_e–O_o is due to inefficiency.

The distinction between the two concepts, while subtle, is important in terms of its consequence for fisheries management. A fleet that is inefficient but fully utilised would respond to management changes differently than one that is efficient but underutilised even though initial output levels may be similar.

Both capacity and technical efficiency are relative measures. That is, the efficiency of one vessel, for example, is assessed against the other vessels in the fleet, the most efficient of which will be taken as perfectly efficient. It is conceivable that all vessels could be inefficient or underutilised relative to some idealised vessel, but if such a vessel does not appear in the data, then the level of inefficiency or underutilised capacity will be underestimated.

Empirical measurement of capacity utilisation and technical efficiency falls into two principal areas: stochastic production frontiers (SPFs) and data envelopment analysis (DEA). (SPFs) are estimated using econometric techniques that allow for random error as well as inefficiency and capacity underutilisation. Data DEA is non-parametric, frontier-based and in most applications non-stochastic. The DEA frontier is estimated using linear programming.

In this chapter, the basic methodology underlying SPFs and DEA will be presented. Issues relating to application of these approaches in fisheries, and the relative advantages and disadvantages, are discussed. Finally, an example of both techniques is presented to highlight the potential influence of the methods on estimates of the potential output in a fishery. More details on the theory and empirical estimation of both methods can be found in Pascoe *et al.* (2003).

2 ESTIMATION OF EFFICIENCY AND CAPACITY UTILISATION

Farrell (1957) developed a conceptual model involving the contraction of inputs to an efficient frontier that laid the foundations for the development of both the parametric (SPF) and non-parametric (DEA) estimation of efficiency. The efficient frontier can be considered in terms of either the maximum output for a given set of inputs (an output orientation), or the minimum set of inputs required to produce a given set of output (an input orientation). These both relate to the assumption of profit maximisation, the former approach representing the maximisation of revenue for a given level of costs, whereas the latter approach minimises the costs for a given level of revenue.

The study of capacity utilisation and technical efficiency in fisheries, as in many other industries, has largely adopted an output oriented (or primal) approach, on the assumption that fishers aim to maximise their revenue each trip. Given such an assumption, the level of efficiency of a particular firm is characterised by the relationship between observed production and some ideal or potential level of production. The measurement of firm-specific technical efficiency is based upon deviations of observed output from the best production or efficient production frontier. If a firm's actual production point lies on the frontier, it is perfectly efficient. If it lies below the frontier then it is technically inefficient, with the ratio of actual to potential production defining the level of efficiency of the individual firm.

2.1 Stochastic Production Frontier

The general specification of the SPF model is identical for both the estimation of capacity utilisation and technical efficiency. The models differ only in terms of the inputs used in the analysis. A general SPF model can be given by

$$\ln \boldsymbol{y} = f(\ln \boldsymbol{x}) + \boldsymbol{v} - \boldsymbol{u}, \tag{1}$$

where $\boldsymbol{y}$ is the output produced by firm j, $\boldsymbol{x}$ is a vector of factor inputs, $\boldsymbol{v}$ is the stochastic error term and $\boldsymbol{u}$ is a one-sided error term that represents either technical inefficiency *or* capacity underutilisation of firm j, depending on which inputs are included in the production function. Estimation of technical efficiency requires both variable and fixed inputs to be included in the production function, whereas estimation of capacity utilisation requires only fixed inputs. In the former case (i.e. with both variable and fixed inputs), $\boldsymbol{u}$ represents technical inefficiency (where technical efficiency is given by $\mathrm{TE}_j = e^{-u_j}$), whereas in the latter case (i.e. only fixed inputs), $\boldsymbol{u}$, represents capacity underutilisation (with capacity utilisation being given by $\mathrm{CU}_j = e^{-u_j}$). The stochastic error term, $\boldsymbol{v}$, is assumed to be normally distributed ($N[0,\sigma_v]$), whereas the inefficiency term has a distribution truncated at zero.

In order to separate the stochastic and inefficiency effects in the model (i.e. $\boldsymbol{v}$ and $\boldsymbol{u}$ from the combined error term, v_j-u_j), a distributional assumption has to be made for u_j. Several different distributional assumptions have been proposed, the most common being a normal distribution truncated at zero, for example, $\boldsymbol{u} \approx |N(\boldsymbol{\mu}, \sigma_u^2)|$ (Aigner *et al.*, 1977) and a half-normal distribution truncated at zero i.e., $\boldsymbol{u} \approx |N(0, \sigma_u^2)|$ (Jondrow *et al.*, 1982). A further approach is to define the inefficiency as a function of the firm specific factors such that:

$$\boldsymbol{u} = \boldsymbol{z}\boldsymbol{\delta} + \boldsymbol{w}, \tag{2}$$

where z is the vector of firm-specific variables which may influence the firm's efficiency, $\boldsymbol{\delta}$ is the associated matrix of coefficients and w is a matrix of random error terms ($N[0,\sigma_w]$). The parameters of the inefficiency model are estimated in a one-step procedure (Battese and Coelli, 1992) along with the parameters of the production frontier.

There are several potential functional forms of the production frontier, the most common being the translog production function, given by:

$$\ln y_j = \beta_0 + \sum_i \beta_i \ln x_{j,i} + \frac{1}{2}\sum_i \sum_k \beta_{i,k} \ln x_{j,i} \ln x_{j,k} - u_j + v_j, \tag{3}$$

The Cobb-Douglas production function is a special case of the translog production function, where all $\beta_{i,k} = 0$.

Further details on the theory underlying SPF models can be found in Greene (1993), Coelli *et al.* (1998) and Kumbhakar and Lovell (2000).

2.2 Data Envelopment Analysis

DEA is a non-parametric, linear-programming approach to the estimation of technical efficiency and capacity utilisation. The technique does not require any pre-described structural relationship between the inputs and resultant outputs, as is the case with the SPF analysis, so allowing greater flexibility in the frontier estimation. It can also accommodate multiple outputs into the analysis. A disadvantage of the technique, however, is that it does not account for random variation in the output, and so attributes any apparent shortfall in output to technical inefficiency (i.e. the estimated efficiency score is effectively $v_j - u_j$ following the same terminology as for the SPF model, where the true values of both v_j and u_j are unknown). Random error can, and typically will, push out the frontier as well as lead to misinterpreting error for inefficiency at the individual boat level.[1]

---

[1]For example, if $v_j > u_j$, the frontier is effectively pushed out by $v_j - u_j$, resulting in a downward bias in the estimated efficiency scores of other vessels. Thus an inefficient vessel might be

The DEA model is formulated as a linear programming (LP) model, where the value of θ for each vessel can be estimated from the set of available input and output data. Following Färe *et al.* (1989, 1994), the DEA model of technically efficient output requires both variable and fixed inputs to be considered, whereas capacity output considers the fixed inputs only. The DEA model for both measures is given as

$$\begin{aligned} &\text{Max } \theta \\ &\text{subject to} \\ &\theta y_{0,m} \le \sum_j z_j y_{j,m} \quad \forall m \\ &\sum_j z_j x_{j,n} \le x_{0,n} \quad \forall n, \\ &\sum_j z_j = 1 \\ &z_j \ge 0 \end{aligned} \tag{4}$$

where θ ($\geq$1) is a scalar denoting the multiplier that describes by how much the output of the target boat (i.e. j=0) can be expanded using inputs in a technically efficient configuration. Further, $y_{j,m}$ is the output m produced by boat j, $x_{j,n}$ is the amount of input n used by boat j, and z_j are weighting factors such that technically efficient output is the weighted sum of the output of other vessels in the data set, including itself. The value of θ is estimated for each vessel separately, with the target vessel's outputs and inputs being denoted by $y_{0,m}$ and $x_{0,n}$, respectively. Inputs include both fixed and variable factors, which are constrained to their current levels.

The shape of the frontier will differ depending on the scale assumptions that underlie the model. The restriction Σz_j=1 imposes variable returns to scale. In contrast, excluding this constraint implicitly imposes constant returns to scale, while $\Sigma z_j > 1$, $\Sigma z_j < 1$ and $\Sigma z_j \leq 1$ imposes increasing, decreasing and non-increasing returns to scale respectively. (Färe *et al.*, 1989).[2]

---

determined to be fully efficient (on the frontier) due to random error, while an efficient vessel may be determined to be inefficient (e.g. if v_j<0, u_j=0). This error is further compounded by the frontier being pushed out by 'lucky' but potentially inefficient vessels. Other vessels' efficiency might be either under or overestimated both because of the shift in the frontier and where it is placed relative to the frontier. Random error in its own production i.e., v_j<0 results in the efficiency being underestimated, while $v_j<u_j$ results in the efficiency being overestimated.

[2]Further details on the theory underlying DEA, including the return to scale assumptions, can be found in Cooper *et al.* (2000)

When both fixed and variable inputs are included in the analysis, the technically efficient level of output is defined as θ multiplied by observed output (i.e., $y_{0,m}^{\mathrm{TE}} = \theta y_{0,m}$). The level of TE is estimated as $\mathrm{TE} = 1/\theta$, which has a value $0 \le \mathrm{TE} \le 1$. Similarly, if only fixed inputs are included in the model, θ relates to capacity rather than efficiency. Capacity output can be given by $y_{0,m}^{C} = \theta y_{0,m}$, while the level of capacity utilisation is given as $\mathrm{CU} = 1/\theta$.

2.2.1 Economic Capacity

The traditional DEA model of capacity implicitly assumes that the additional income from increasing output exceeds the additional cost of the additional use of variable inputs necessary to produce it. This is not necessarily the case, particularly if decreasing returns to fishing effort exist. An alternative specification of the DEA model has been developed to take into account the additional costs and benefits of increasing output, thereby defining an "economic" rather than technical measure of output. The estimation of economic capacity is based on a DEA model that determines the level of output and variable inputs that maximise the firm's profits. For a firm producing a set of m outputs from n inputs (where $n \in \alpha$ is the subset of fixed inputs and $n \in \hat{\alpha}$ is the subset of variable inputs), this is given by:

$$\max_{\theta_1, \lambda_{0,n}, z_j} \left[\sum_m \theta p_m y_{o,m} - \sum_{n \in \hat{\alpha}} v_{0,n} \lambda_{0,n} x_n \right]$$

subject to

$$\begin{aligned} &\theta y_{0,m} \le \sum_j z_j y_{j,m} \quad \forall m \\ &\sum_j z_j x_{j,n} \le x_{0,n} \quad n \in \alpha \\ &\sum_j z_j x_{j,n} = \lambda_{0,n} x_{0,n} \quad n \in \hat{\alpha}, \\ &\sum_j z_j = 1 \\ &z_j \ge 0, \lambda_{o,n} \ge 0 \end{aligned} \tag{5}$$

where θ is again a scalar denoting how much each output, $y_{j,m}$, of the target firm (i.e. j=0) must be (radially) increased, and $\lambda_{0,n}$ is the factor by which each variable input, $x_{j,n\in\hat{\alpha}}$, must be increased (or decreased), to achieve the profit maximising level of output given the level of fixed inputs ($x_{j,n\in\alpha}$).[3] Output and variable input prices – p_m and v_{0n} – are included in the objective function to estimate revenues and variable costs. Fixed costs do not influence the level of capacity utilisation as the level of fixed inputs is given. z_j are the weights that relate the target firm to its set of peers (i.e. the firms against which it is compared, including itself).

As the additional output is produced through increasing the use of variable inputs, these are included in the model. However, these are not constrained as in the case of technical efficiency.

The economic capacity output of each firm is determined by $y'_{j,m} = \theta y_{j,m}$, where $y_{j,m}$ is the current level of each output m produced by firm j and $y'_{j,m}$ is the potential full capacity level of output m by firm j. The ray measure of capacity assumes that each individual vessel's outputs are produced in fixed proportions – a fairly realistic assumption in most fisheries but not necessarily in all industries.

Economic capacity utilisation is given by $1/\theta$. As noted previously, the measures of physical CU and TE range from 0 to 1. However, the economic capacity utilisation measure may take a value greater than 1 if a contraction in both outputs and inputs would increase profitability of the firm. As noted by Nelson (1989), the physical measure of capacity utilisation will be less than or equal to the economic measure. Both measures of capacity utilisation will be less than or equal to the measure of technical efficiency.

2.3 Unbiased Capacity Utilisation

The measure of capacity utilisation estimated using either SPF or DEA, by default, includes the measure of inefficiency. For example, in Figure 1, capacity utilisation was defined as O_c–O_e. However, the empirical estimate measures the distance from the actual (observed) to the potential catch, i.e., O_c–O_o.

Färe *et al.* (1989) suggest that capacity and capacity utilisation should be adjusted to remove the effects of inefficiency, as the firms would not achieve the full potential output if their inputs were fully utilized due to their ineffi-

[3] The model does not necessarily guarantee that profits of an individual firm are maximised as the condition that marginal revenue equals marginal costs cannot be imposed. Further, firms on the frontier may not necessarily be at their profit maximising level of output. Consequently, the model is more appropriately described as profit increasing (rather than maximising) to at least the maximum comparable level observed in the fishery.

ciency. Further, in the case of DEA estimates of capacity utilisation, random variation in output would manifest itself in terms of both lower average capacity utilisation and lower average efficiency. The ratio of these measures would therefore also cancel out the effects of this random variation (Holland and Lee, 2002). Consequently, Färe *et al.* (1989, 1994) propose an "unbiased" measure of capacity utilisation (CU*) given by

$$CU^* = \frac{CU}{TE}, \tag{6}$$

where CU may be either the economic or physical measure of capacity utilisation. As $TE \leq 1$, $CU \leq CU^* \leq 1$. That is, this "unbiased" measure of capacity utilisation is greater than the original measure (which includes efficiency effects), but less than 1. The difference between the measures reflects the degree to which technical inefficiency, and in the case of DEA estimates of CU, random variation, affects the output levels of the different firms.

Given this, capacity underutilisation can be considered to consist of two components: a "pure" capacity underutilisation (i.e. the "unbiased measure") that arises as a result of the underutilisation of variable inputs, and technical inefficiency. However, there is some disagreement as to the treatment of inefficiency in relation to a firm's capacity. Coelli *et al.* (2002) argues against the use of such "unbiased" measures as they consider that the estimation of different levels of capacity for firms with the same level of fixed inputs is unintuitive. They therefore consider that technical inefficiency is a component of unused capacity and need not be separated out.

In the case of fisheries, differences in apparent technical efficiency may reflect managerial skill or spatial differences in stock abundance, as well as random events such as adverse weather conditions or mechanical failure (Felthoven and Morrison Paul, 2004). The latter two events may be categorised as "bad luck", and it could be argued along the lines proposed by Coelli *et al.* (2002) that their output would be substantially higher if they had not been unlucky. The first two factors, however, are more likely to be capacity-limiting factors, at least in the short term. A poor skipper would not catch as much as good skipper even if operating identical vessels in the same conditions. Similarly, the increased revenue achieved from moving from poor fishing grounds to better fishing grounds may not offset the increased costs, restricting activity to the poorer grounds. These differences in environmental conditions and skipper skill are effectively unobservable fixed inputs in the production process that are apparent only as inefficiency (Pascoe and Coglan, 2002). Consequently, the apparent inefficiencies are unlikely to be eliminated, and estimating capacity on the assumption that they could would therefore overestimate output (Felthoven and Morrison Paul, 2004). Hence, the potential production would be different for identical vessels operating under differing unobservable environmental and managerial conditions. As a consequence, the approach proposed by Färe *et al.* (1994) (i.e. the measure of "unbiased" capacity utilisation) is considered more appropriate in the case of fisheries.

3 EXPERIENCES OF EFFICIENCY AND CAPACITY ANALYSIS IN FISHERIES

Both approaches described here have advantages and disadvantages. Stochastic estimations incorporate a measure of random error. However, they impose an explicit functional form on the relationship between inputs and outputs and require assumption about the error term to separate out the random error from inefficiency. In contrast, linear programming techniques, in particular DEA, do not impose a specific functional form of the relationship between inputs and outputs. Hence, DEA is less prone to errors arising from mis-specification of the functional form. However, as DEA is a non-parametric approach, it does not take into account random error. As a consequence, the efficiency estimate also includes random error as well as inefficiency. As a result, the efficiency estimates may be biased if the production process is thought to be largely characterised by stochastic elements.

DEA has an additional advantage in that it can incorporate the possibility of multiple outputs. In contrast, the SPF approach commonly only incorporates a single output. As most fisheries are multi-species, most fishing fleets are multi-output. Assuming a single output, as is commonly the case for SPF, may result in bias in the efficiency estimation. More recently, multi-output SPF models have been developed, but have had only limited application.

Both techniques have been applied in fisheries, although the choice of technique has tended to depend on the main objective of the analysis. Studies of technical efficiency have tended to use the SPF approach (e.g. Kirkley *et al.*, 1995, 1998; Campbell and Hand, 1998; Sharma and Leung, 1999, Grafton *et al.*, 2000; Pascoe *et al.*, 2001b; Pascoe and Coglan, 2002; Fox *et al.*, 2003; Squires *et al.*, 2003; Kompas *et al.*, 2004) as the effect of random error of the efficiency measure is removed in the estimation process. In contrast, studies of capacity and capacity utilisation have tended to favour DEA (e.g. Pascoe *et al.*, 2001a; Dupont *et al.*, 2002; Felthoven, 2002; Kirkley *et al.*, 2003; Tingley *et al.*, 2003; Vestergaard *et al.*, 2003; Walden *et al.*, 2003, Tingley and Pascoe, 2005a, b). In a limited number of cases, both

DEA and SPFs have been applied (e.g. Tingley *et al.*, 2003, 2005; Kirkley *et al.*, 2004).

3.1 Outputs

The output measure presents certain challenges when modelling efficiency and capacity in fisheries. In most studies of efficiency (and when estimation production functions), output is a physical measure of volume. For most industries, the production process for individual outputs can generally be identified. However, most fisheries are characterised by joint production, that is, a combination of different types of output (i.e. different species) is produced for a given set of inputs. This is particularly a problem for SPFs, which have been largely estimated assuming a single output.

The approach that has been adopted in previous studies of efficiency and production frontier estimation in fisheries has been to use landed weight when a single species fishery is examined and value of total catch when multiple species are harvested. For example, Kirkley *et al.* (1995, 1998) used the weight of meat landed per trip when modelling the Atlantic Scallop fishery; Sharma and Leung (1998) used value of catch per trip when modelling the mixed Hawaiian long-line fishery; and Pascoe and Coglan (2002) used value of catch per month when modelling the English channel demersal trawl fishery. Herrero and Pascoe (2003) compared the use of both aggregate weight and value as an output measure and found consistent results for most vessels. Where differences in efficiency scores were apparent, these were believed to be linked to different risk preferences.

A perceived advantage of using value as a measure of aggregate output is that it takes into account both the quantity of the catch and the importance of the catch to the overall output (i.e. its contribution to revenue). However, the use of aggregate value of the multi-product firm as the output measure has implications for the analysis. Firstly, value is a factor of prices as well as quantity, so that price changes may affect the measurement of technical efficiency. Further, assuming fishers are profit maximisers, then a change in relative prices may result in a change in their fishing strategy. As a result, the function is not truly a production function and the efficiency scores may represent a combination of allocative as well as technical efficiency.

The potential biases introduced into the analysis from using value as the output measure are not likely to be large. Squires (1987) and Sharma and Leung (1998) note that fishers base their fishing strategies on expected prices, the level of technology and resource abundance. However, price expectations are not always accurate, fishing gear is not species selective (so the species mix is function of seasonal abundance) and changing gear types from otter trawl to beam (or vice versa) is time consuming and needs to be done on shore rather than at sea (as only one gear type is taken to sea at any one time). Hence, the ability of fishers to respond to changes in relative prices is limited.

An alternative to using value as the output measure directly is to derive an aggregate quantity measure weighted by revenue share. This can be derived directly from quantity and price data. A common approach is to weight the output by the relative revenue shares. This is given by

$$\text{Composite Output} = Q_t^C = \sum_{i=1}^{n} q_{it} w_{it} , \qquad (7)$$

where

$$w_{it} = p_{it} q_{it} \Big/ \sum_{i=1}^{n} p_{it} q_{it} . \qquad (8)$$

An advantage of this approach is that the catch is weighted by its relative importance, but changes in aggregate output are not directly affected by changes in prices.[4]

Aggregation of outputs is less of an issue for DEA, as it is capable of handling multiple outputs directly. However, some level of aggregation is still often required. In many multi-species fisheries, most of the value of the catch is obtained by only a few species, although the vessels may catch a much larger number of species. Aggregating the less important species into an "other" category is common practice, and is usually undertaken using the revenue-share approach.

3.2 Inputs

3.2.1 Vessel Characteristics

The inputs used to explain production generally include a measure representing the level of capital employed in the fishery and a measure of capital

[4] Changes in relative prices will change the relative weightings of each species in the composite measure. However, this measure still reflects the relative importance of each species in the catch.

utilisation.[5] Other technically relevant inputs are also employed. The exact choice of variables, however, differs between studies. Sharma and Leung (1998) used crew size and trip length, as well as a variable representing the cost of other variable inputs (e.g. fuel, bait, ice, etc. on a dollar/trip basis). As the vessels used static gear (i.e. long lines), the size of the vessels was not considered important in the production process. In contrast, both Kirkley *et al.* (1995, 1998) and Pascoe and Coglan (2002) modelled fleets using mobile gear, in which case size of the boat can be assumed to be a major factor in the production process. Kirkley *et al.* (1995, 1998) chose a set of boats with similar physical characteristics to overcome the need to develop a measure of capital employed, and estimated the frontier using variable factors only (representing capital utilisation). These included crew size and days at sea. Dredge size was also used as a technology-based input. In contrast, Pascoe and Coglan (2002) included measures of the size of the boat (represented by the product of the length and breadth), the engine power of the boat (in kW) and the number of trips each month.

3.2.2 Stock Size

Output from fishing is not just a function of the inputs employed by the fisher, but also a function of the available resource. As a result, a measure of the relative stock abundance is generally required. A number of different approaches have been applied in the literature for incorporating stock into the production frontier. While some studies have been able to derive stock-abundance indices directly from estimates of stock, these have been limited to fisheries where the catch consists of only one (e.g. Kirkley *et al.*, 1995, 1998) or two (e.g. Pascoe *et al.*, 2001b) species. Further, a time series of stock abundance information in these examples corresponding to the period of the analyses was available. In many fisheries, particularly multi-species fisheries, information on stock abundance of all species (or in some cases any of the species) may not be available. As a result, other means of estimating the effect of changes in stock abundance on production need to be employed.

These differences can be incorporated into the analysis through the use of dummy variables. Coglan *et al.* (1999) used dummy variables representing the

---

[5] Capital utilisation differs from capacity utilisation. The former is a measure of the intensity of use of the fixed inputs (e.g. days fished, number of trips etc), and may include variable inputs (e.g. fuel use and crew-days at sea). The measures are usually considered as variable inputs for the purposes of the efficiency and capacity estimation. The latter is a measure of the degree to which the vessel is achieving the maximum output given normal working conditions.

different months, years and métiers. However, incorporating sufficient dummy variables to allow also for interactions between months, years and métiers would result in a substantial loss of degrees of freedom. For example, 4 years, 12 months and 5 metiérs would require 240 separate dummy variables to cover the combinations. Kompas *et al.* (2004) used annual dummy variables to represent changes in stock conditions, developing a model using annual catch and effort data.

An alternative approach is to derive an index of stock abundance based on relative catch rates. Kirkley *et al.* (1995, 1998) developed such an index based on the catch rate of survey vessels undertaking routine stock monitoring. Pascoe and Coglan (2002) developed an index based on the average value per hour fished of the boats that operated in the same month in the same métier. Hence, it takes into account the differences in the composition of the catches taken by the different gear types at each point in time and in each area, as well as the different set of prices in each time period. Were price changes not accounted for in the model, then changes in the set of prices may have affected the estimates of efficiency (as the output measure may change without any change in the physical inputs). The index was calculated as a geometric mean of the observed values in each period/métier to limit the effects of extreme observations on the mean.

Sharma and Leung (1998) argue against the use of catch per unit effort (CPUE) as a measure of stock abundance on the basis that average CPUE is affected by the characteristics of the boats in the area at the time. A change in CPUE from one period to the next may reflect the different composition of the boats from which the CPUE was derived as well as changes in the stock abundance.

Similarly, Álverez (2003) demonstrated that the use of CPUE required implicit assumptions about the stock elasticity that may not be valid for the fishery concerned. Further, failure to impose these restrictions may result in biased results.

Pascoe and Herrero (2004) developed a method for deriving a "stock effect" that can be used to modify the output measure. The stock effect is first estimated using DEA, then applied to the output measure for the purposes of estimating on SPF. The key advantage of the measure is that it allows for variations in stock density to be captured (both spatial and temporal) without the problems associated with both dummy variables and CPUE indexes.

Stock size is less of an issue for DEA. Usually, efficiency and capacity utilisation is estimated only within a common time period and area. This, however, creates difficulties for inter-period comparisons.

3.3 Single or Multiple Outputs?

As noted earlier, most applications of SPFs in fisheries have been estimated using a single composite output measure. Consistent aggregate measures of output, however, require input–output separability. This has been found not to be appropriate in several fisheries studies using alternative specifications (e.g. cost or profit functions, see Jensen, 2002). An alternative, therefore, is to estimate multi-output production frontiers. Attempts at estimating multi-output production functions in fisheries have been relatively limited (e.g. Fousekis, 2002, Weninger and Strand, 2003; Orea *et al.*, 2005). Difficulties include treatment of the individual stocks within the multi-output function, and problems of multi-collinearity as the outputs are often correlated. Further, as noted previously, the large number of species caught in many fisheries requires some degree of aggregation of the less-valuable species to develop manageable models. These problems not withstanding, the studies have concluded that a multi-output specification of the production function is more appropriate and should be undertaken where possible.

With DEA, the more common approach has been to incorporate multiple outputs into the analysis directly, although some aggregated categories are often incorporated to ensure sufficient degrees of freedom in the analysis. Tingley *et al.* (2003) compared the results of using aggregated single outputs versus multiple outputs in a DEA analysis of efficiency and capacity utilisation in several different fleets operating in the English Channel. They found that the estimates of both technical efficiency and capacity utilisation were higher when incorporating multiple outputs into the model, although the "unbiased" measures of capacity utilisation (i.e. CU/TE) were fairly consistent using both approaches.

3.4 Data Limitations

Repeated observations for the same boat are required in order to separate out the effects of random fluctuations in output from systematic differences due to inefficiency or capacity underutilisation using the SPF approach. This requires a time series of information for a cross section of boats in the population. This is generally referred to as panel data. Panel data may be balanced or unbalanced. Balanced panel data exists where there is an equal number of observations for all boats in the sample and every boat operates in every time period of the data. Unbalanced panel data occurs when there are not an equal number of observations for each boat, and/or the boats do not operate in every time period of the data.

A difficulty with unbalanced panel data is that different sets of boats may be compared in different time periods, and there may be instances where some boats are not directly compared. As efficiency is a relative (rather than absolute) measure, this may be problematic if there are only a few boats in the sample for given time periods, such that the boats are only compared to a small number of other boats in the same period. Ideally, the data set should be broad enough for this not to occur, and ideally every boat should operate in the same period with every other boat (not all at the same time necessarily) at least once (and preferably more times). Time periods where only a few boats are operating should be excluded from the data set. Similarly, boats that have only a few observations should be excluded from the sample, as their efficiency score will be measured relative to only a few other boats in few time periods, with the result that the efficiency and capacity utilisation scores are artificially inflated (see Tingley *et al.*, 2003). This requires a subjective assessment as to how many to exclude. For example, Pascoe and Coglan (2002) included boats that had observations for at least 4 months a year in at least 3 of the 4 years of the data. This resulted in only 63 boats out of a possible 457 being included in the analysis. In contrast, Kirkley *et al.* (1995, 1998) limited their analysis to only ten boats for which a long and consistent time series were available.

Where cross-sectional data only are available (i.e. only one observation per boat), a strict assumption about the distribution of the inefficiency term is required. The resultant estimates of efficiency will conform to the imposed distribution, and it is not possible to statistically distinguish between the nested distributions (i.e. half-normal and truncated normal). Similarly, if an inefficiency model is imposed, the inefficiency measures will conform to the model. Statistical interpretation of the parameters in the inefficiency model are not possible. Consequently, there is little benefit in imposing such a distribution onto the data, and it is preferable to use the standard distributions (i.e. half- or truncated normal). For example, Sharma and Leung (1998) developed their model using cross-sectional data only and imposed an inefficiency model onto the data. As would be expected, most of the parameters were non-significant, with only one variable defining the inefficiency distribution at 5% level of significance.

For DEA, each time period is considered separately, so a panel data set is less necessary. However, within each time period the estimated inefficiency measure includes both inefficiency and random error, so is potentially less reliable than the SPF measure when considering a single year.

Data quality has a major impact on the estimated efficiency or capacity utilisation measures. Discarded, misreported or under-reported catch by a vessel will manifest itself as either inefficiency or underutilisation (or both), provided other boats do report all of their catch. This is particularly a problem in fisheries managed through quota controls, where some vessels may be forced to discard over-quota catch (or risk landing it illegally and not report it). Similarly, when not all species in the catch are required to be recorded, then vessels of fishers who chose not to record their catch may appear inefficient or underutilised compared with vessels of fishers who record the catch of all species. This is again a problem in quota-managed species where non-quota species are also caught but reporting the catch of non-quota species is not mandatory.

4 EXAMPLE: EFFICIENCY AND CAPACITY UTILISATION ESTIMATION

A comparison of the different approaches to estimating efficiency and capacity utilisation can be illustrated through application of the different techniques to a common data set. For the purposes of illustration, both DEA and SPFs are used to estimate technical efficiency and capacity utilisation of the Scottish fleet operating in 2001. Full details of the Scottish analyses are presented in Tingley and Pascoe (2003) and Pascoe and Tingley (2006).

4.1 Efficiency and Capacity Utilisation of the Scottish Fleet

The analysis was undertaken for 1823 Scottish registered vessels that operated in 2001, although only the results for the larger vessels (over 10 m in length) are presented here, representing 943 individual vessels. The annual quantities and values of landings by each vessel of 36 individual species were derived from logbook data. These species accounted for around 96% of the total recorded value landed of all species in 2001. Not all species were caught by all vessels in the different fleet segments. The technical inputs used in the analysis were boat size (defined as the product of length and breadth), engine power and days at sea.

The capacity and efficiency measures were estimated using both DEA and SPFs. For the SPF analysis, the aggregate value of landings was used as a single output measure. For the DEA analysis, all 36 species were included as outputs. As the data set was cross section, a truncated normal distribution was assumed for the SPF analysis. All vessels were compared together to overcome problems associated with limited degrees of freedom in some fleet segments.

The key results averaged over the different fleet segments are summarised in Table 1. From Table 1, the DEA estimates of technical efficiency and capacity utilisation were higher in some instances than those estimated using SPFs, and lower in other instances. Further, there is little correlation between the SPF and DEA results, with the highest positive correlation between the results being r=0.06 (between SPF and DEA "unbiased" measures of economic capacity utilisation).

In this example, the poor correlation between the results is largely an artefact of the approach used when estimating SPF. Incorporating all vessels into a single model implicitly assumes that each has a similar production technology. That is, the impact of an additional unit of engine power, for example, is the same for beam trawlers (using mobile trawl gear) as it is for gill netters (using static gear). This is a criticism often applied to the SPF approach, although it is accentuated in this example. Even within a relatively homogeneous fleet segment (e.g. beam trawlers), a common production technology may not be a valid assumption. The use of a translog production frontier reduces this problem by allowing greater flexibility, although it still imposes a common production function on all vessels. In contrast, DEA imposes neither efficiency distributional assumptions nor production technology assumptions.

Table 1. Average efficiency and capacity utilisation scores.

	No. boats	Technical efficiency		Capacity utilisation			"Unbiased" CU		
		SPF	DEA	SPF	DEA		SPF	DEA	
					Physical	Econ.		Physical	Econ.
Beam trawl	15	0.59	0.98	0.53	0.97	0.97	0.89	0.99	0.99
Pelagic nets	39	0.73	0.86	0.50	0.84	0.84	0.68	0.98	0.98
Pelagic lines	5	0.73	0.96	0.72	0.95	0.95	0.99	0.99	0.99
Nephrops trawl	225	0.66	0.46	0.54	0.41	0.41	0.83	0.90	0.91
Demersal seine/trawl	443	0.65	0.78	0.57	0.71	0.72	0.88	0.92	0.92
Dredge	89	0.70	0.59	0.65	0.53	0.53	0.93	0.93	0.94
Pots	117	0.72	0.54	0.67	0.49	0.49	0.94	0.94	0.95
Gill nets	10	0.65	0.87	0.60	0.84	0.85	0.92	0.97	0.98

The relatively high levels of efficiency and capacity utilisation by some fleet segments in the DEA analysis may also be an artefact of the large number of outputs used. For example, the pelagic line vessels were found to be highly efficient and have high capacity utilisation. These vessels catch a limited number of species that are also caught as bycatch by other fleet segments. By comparison, then, their output levels would appear substantially higher than other vessels. The small number of vessels in the fleet segment would mean that most of the comparison was against other fleet types that caught lower quantities of the pelagic species. Tingley *et al.* (2003) looked at the impact of degrees of freedom on estimates of capacity utilisation in fisheries and found, as expected, that fewer vessels for comparison results in higher capacity utilisation scores.

The example in Table 1 illustrates that considerable care needs to be given when estimating efficiency and capacity utilisation using either approach.

5 CONCLUSIONS

The predominance of the use of input controls in the management of fisheries requires fisheries managers to have a detailed understanding of the production processes that occur in the fisheries. A lack of understanding of these processes in the past is a major contributing factor to the overexploitation of many fisheries throughout the world. Heterogeneity in efficiency and capacity utilisation has substantial implications for measures to reduce overexploitation. For example, removing inefficient vessels from a fleet through a buy-back programme will have a less than proportional impact on catch rates. Further, if the remaining vessels were previously underutilised, then increases in the capacity utilisation of these vessels may offset the reduction in vessels, resulting in no real reduction in fishing mortality.

Two approaches have been illustrated that allow estimates of capacity utilisation and efficiency. From the examples presented, these approaches may result in differing results. Both approaches are based on similar theoretical principles, but employ different approaches in the estimation of efficiency and capacity. Both approaches have advantages and disadvantages.

In many respects, DEA has some theoretical advantages over SPFs. These include the ability to incorporate multiple outputs more readily, and the avoidance of the need to impose a common production technology on all vessels. However, random variation is also captured as inefficiency in DEA. Fisheries are often thought to be subject to high levels of stochasticity; fishers are harvesting an unseen, fugitive resource, and "luck" may play a major role in the final output level.

Given this, SPFs may be more appropriate for estimating technical efficiency in fisheries, and DEA more appropriate for estimating capacity and capacity utilisation (as the "unbiased" measure reduces the effects of random variation on the measures).

6 ACKNOWLEDGMENT

This work was undertaken as part of the European funded project CAFE (Capacity, fishing mortality and effort), Contract No. 022644.

7 REFERENCES

Aigner D., Lovell, C.A.K. and Schmidt, P., 1977. Formulation and estimation of stochastic frontier production function models. *Journal of Econometrics* 6: 21–37.

Álvarez, A., 2003. *Econometric Estimation of Fishing Production Functions when Stock is Unknown: A Monte Carlo Analysis*, Efficiency Series Paper 09/2003, University of Oviedo, Spain.

Battese, G.E. and Coelli, T.J., 1992. Frontier Production Function, technical efficiency and panel data: with application to Paddy Farmers in India. *Journal of Productivity Analysis* 3: 153–169.

Campbell, H.F. and Hand, A.J., 1998. Joint ventures and technology transfer: the Solomon Islands pole-and-line fishery. *Journal of Development Economics* 57: 421–442.

Coelli, T., Grifell-Tatje, E. and Perelman, S., 2002. Capacity utilisation and profitability: a decomposition of short-run profit efficiency. *International Journal of Production Economics* 79: 261–278.

Coelli, T.J., Rao, D.S.P. and Battese, G.E., 1998. *An Introduction to Efficiency and Productivity Analysis*, Kluwer Academic Publishers, USA.

Coglan, L., Pascoe S. and Harris, R.I.D., 1999. Measuring efficiency in demersal trawlers using a frontier production function approach, in *Proceedings of the Xth Annual Conference of the European Association of Fisheries Economists, The Hague, Netherlands, 1–4 April 1998*, Pavel Salz, ed. Agricultural Economics Research Institute (LEI), The Netherlands, 236–257.

Dupont, D.P., Grafton, R.Q., Kirkley, J. and Squires, D., 2002. Capacity utilisation measures and excess capacity in multi-product privatized fisheries. *Resource and Energy Economics* 24(3): 193–210.

FAO, 2000. *Report of the Technical Consultation on the Measurement of Fishing Capacity*. FAO Fisheries Report No. 615 (FIPP/R615(En)). FAO, Rome.

Färe, R., Grosskopf, S. and Kokkelenberg, E.C., 1989. Measuring plant capacity, utilisation and technical change: a non-parametric approach. *International Economic Review* 30(3): 655–666.

Färe, R., Grosskopf, S. and Lovell, C.A.K., 1994. *Production Frontiers*. Cambridge University Press, Cambridge.

Farrel, M.J., 1957. The measurement of productive efficiency. *Journal of the Royal Statistical Society Series A, General* 125: 252–267.

Felthoven, R.G., 2002. Effects of the American Fisheries Act on capacity, utilisation and technical efficiency. *Marine Resource Economics* 17: 181–206.

Felthoven, R.G. and Morrison Paul, C.J., 2004. Multi-output, non-frontier primal measures of capacity and capacity utilisation. *American Journal of Agricultural Economics* 86(3): 619–633.

Fousekis, P., 2002. Distance vs. ray functions: an application to the inshore fishery of Greece. *Marine Resource Economics* 17: 251–268.

Fox, K.J., Grafton, R.Q., Kirkley, J. and Squires, D., 2003. Property rights in a fishery: regulatory change and firm performance. *Journal of Environmental Economics and Management* 46(1): 156–177.

Grafton, Q., Squires, D. and Fox, K.J., 2000. Private property and economic efficiency: a study of a common pool resource. *The Journal of Law and Economics* 43: 671–714.

Greene, W.H., 1993. *Frontier Production Functions*, EC-93-20. Stern School of Business, New York University.

Herrero, I. and Pascoe, S., 2003. Value versus volume in the catch of the Spanish South-Atlantic Trawl Fishery. *Journal of Agricultural Economics* 54(2): 325–341.

Holland, D.S. and Lee, S.T., 2002. Impacts of random noise and specification on estimates of capacity derived from data envelopment analysis. *European Journal of Operational Research* 137(1): 10–21

Jensen, C.L., 2002. Applications of dual theory in fisheries: a survey. *Marine Resource Economics* 17(4): 309–334.

Jondrow, J., Lovell, C.A.K., Materov, I.S. and Schmidt, P., 1982. On the estimation of technical inefficiency in the stochastic frontier production function model. *Journal of Econometrics* 19: 223–238.

Kirkley, J., Morrison-Paul, C.J. and Squires, D., 2004. Deterministic and stochastic capacity estimation for fishery capacity reduction. *Marine Resource Economics* 19: 271–294.

Kirkley, J., Squires, D. and Strand, I.E., 1995. Assessing technical efficiency in commercial fisheries: The Mid-Atlantic Sea Scallop Fishery. *American Journal of Agricultural Economics* 77(4): 686–697.

Kirkley, J.E., Squires, D. and Strand, I.E., 1998. Characterizing managerial skill and technical efficiency in a fishery. *Journal of Productivity Analysis* 9: 145–160.

Kirkley, J.E., Squires, D., Alam, M.F. and Ishak, H.O., 2003. Excess capacity and asymmetric information in developing country fisheries: the Malaysian purse seine fishery. *American Journal of Agricultural Economics* 85(3): 647–662.

Kompas, T., Che, T.N. and Grafton, R.Q., 2004. Technical efficiency effects of input controls: evidence from Australia's Banana Prawn Fishery. *Applied Economics* 36: 1631–1641.

Kumbhakar, S.C. and Lovell, C.A.K., 2000. *Stochastic Frontier Analysis.* Cambridge University Press, Cambridge.

Nelson, R.A., 1989. On the measurement of capacity utilisation. *The Journal of Industrial Economics* 37: 273–286.

Orea, L., Alvarez, A. and Morrison Paul, C.J., 2005. Modeling and measuring production processes for a multi-species fishery: alternative technical efficiency estimates for the Northern Spain Hake Fishery. *Natural Resource Modeling* 18(2): 183–213.

Pascoe, S. and Coglan, L., 2002. Contribution of unmeasurable factors to the efficiency of fishing vessels: an analysis of technical efficiency of fishing vessels in the English Channel. *American Journal of Agricultural Economics* 84(3), 45–57.

Pascoe, S., Coglan, L. and Mardle, S., 2001a. Physical versus harvest based measures of capacity: the case of the UK vessel capacity unit system. *ICES Journal of Marine Science* 58(6), 1243–1252.

Pascoe, S., Andersen, J.L. and de Wilde, J., 2001b. The impact of management regulation on the technical efficiency of vessels in the Dutch Beam Trawl Fishery. *European Review of Agricultural Economics* 28(June): 187–206.

Pascoe, S. and Herrero, I., 2004. Estimation of a composite fish stock index using data envelopment analysis. *Fisheries Research* 69: 91–105.

Pascoe, S., Kirkley, J., Greboval, D. and Morrison-Paul, C., 2003. *Measuring and Assessing Capacity in Fisheries. 2: Issues and Methods.* FAO Fisheries Technical Paper 433/2, FAO, Rome, 129 pp.

Pascoe, S. and Tingley, D., 2006. Economic capacity estimation in fisheries: A non-parametric ray approach, *Resource and Energy Economics* 28(2): 124–138.

Sharma, K.R. and Leung, P., 1999. Technical efficiency of the longline fishery in Hawaii: an application of a stochastic production frontier. *Marine Resource Economics* 13: 259–274.

Squires, D., 1987. Public regulation and the structure of production in multiproduct industries and application to the New England otter trawl industry. *Rand Journal of Economics* 18(2): 232–247.

Squires, D., Grafton, R.Q., Alam, M.F. and Omar, I.H., 2003. Technical efficiency in the Malaysian Gill Net Artisanal Fishery. *Environment and Development Economics* 8(3): 481–504.

Tingley, D. and Pascoe, S., 2003. *Estimating the level of excess capacity in the Scottish Fishing Fleet.* Report prepared for the Scottish Executive Environment and Rural Affairs Department, CEMARE Research Report 66, University of Portsmouth, UK.

Tingley, D. and Pascoe, S., 2005a. Factors affecting capacity utilisation in fisheries, *Journal of Agricultural Economics* 56(2): 287–307.

Tingley, D. and Pascoe, S., 2005b, Eliminating excess capacity: implications for the Scottish fishing industry. *Marine Resource Economics* 20(4): 407–424.

Tingley, D., Pascoe, S. and Coglan, L., 2005. Factors affecting technical efficiency in fisheries: stochastic production frontier versus data envelopment analysis approaches. *Fisheries Research* 73(3): 363–376.

Tingley, D., Pascoe, S. and Mardle, S., 2003. Estimating capacity utilisation in multi-purpose, multi-metier fisheries. *Fisheries Research* 63(1): 121–134.

Vestergaard, N., Squires, D. and Kirkley, J.E., 2003. Measuring capacity and capacity utilisation in fisheries: the case of the Danish gillnet fleet. *Fisheries Research* 60(2–3): 357–368.

Walden, J.B., Kirkley, J.E. and Kitts, A.W., 2003. A limited economic assessment of the Northeast groundfish fishery buyout programme. *Land Economics* 79(3): 426–439.

Weninger, Q. and Strand, I.E., 2003. An empirical analysis of production distortions in the mid-Atlantic Surf Clam and Ocean Quahog Fishery, *Applied Economics* 35(10): 1191–1197.

Chapter 15

STUDIES IN THE DEMAND STRUCTURE FOR FISH AND SEAFOOD PRODUCTS

Frank Asche[1] , Trond Bjørndal[2], and Daniel V. Gordon[3]
[1]*University of Stavanger and Centre for Fisheries Economics, SNF, Bergen, Norway;* [2]*Centre for Fisheries Economics, SNF, Bergen, Norway and CEMARE, University of Portsmouth, Portsmouth, UK;* [3]*Department of Economics, University of Calgary, Canada and Centre for Fisheries Economics, SNF, Bergen, Norway*

Abstract The purpose of this chapter is to provide a review of demand and market integration studies with respect to fish and seafood products, focusing on methods used, on information that is obtained and on how this information varies with the approach used. Within the general demand structure for fish, we review the Rotterdam system and the almost ideal demand system as well as single equation specifications. Market integration studies have become an increasingly common approach to obtain information about the demand structure when the data availability is limited. We also provide a general assessment of the demand elasticities for fish and seafood products.

Keywords: Rotterdam system, demand analysis, almost ideal demand system, market integration, demand elasticities

1 INTRODUCTION

Until the mid-1980s, the structure of demand for seafood received little academic attention. During the last decades, there has been a virtual explosion in the number of studies of the demand structure for seafood markets. This is due to several factors including the expansion of the exclusive economic zone to 200 miles and increased trade with seafood due to improved logistics and the expansion of aquaculture. The most common research approach is demand analysis, where demand equations are estimated either individually or in a system of demand equations. These studies of the demand structure focus on the price sensitivity of demand, on the degree of substitution between potentially competing products and on income/expenditure effects. However, as price information is often more easily available than quantity, there have been a number of market integration studies that primarily focus on the competition between different products.

The different studies are empirical and are conducted on a specific data set. This gives, strictly speaking, information about the demand structure for some specific products or species in a specific market for the time period covered by the data. The purpose of this chapter is to give a review of demand and market integration studies with respect to fish, focusing on methods used, on information that is obtained, and on how this information varies with the approach used. Are there patterns that become apparent when one looks at the results obtained in a number of demand studies? What can we say about the demand for fish in general or about the demand for specific groups of species or markets?

To present results from many different studies creates a number of problems that one should be aware of when comparing the results. In addition to the different markets and species studied, a number of different modelling methods have also been used. Since the methods used affect the interpretation of the results, it is also important to be aware of the potential differences. Moreover, measuring data at different market levels, for example import or retail, has important implications for interpretation of the results.

The different methods used for data measurement at different market levels make the empirical results difficult to compare, in a strict sense, but some comparisons are possible. In particular, one might observe whether the price responsiveness for fish is in a specific range, or whether this varies systematically with species, markets or measurement level for the data.

Some implications of economic theory for the magnitudes of the elasticities are worthwhile to note. An own-price elasticity of –1 is a focal point. A good with constant budget share and no substitutes will have an elasticity of –1, so that a 1% increase in the price will lead to a 1% reduction in the quantity demanded and vice versa. In particular for aggregated goods, the budget shares are relatively constant with few substitutes. This indicates that one should expect many demand elasticities to be close to –1. It is also of interest to note that the value of a market is at its highest when the demand elasticity is –1. If the quantity supplied increases above the level that gives a demand elasticity of –1, the value of the market will fall. Finally, the more elastic the demand for the good, the greater substitution possibilities there will be and therefore the keener the competition.

This chapter will be limited to the markets that have been studied and this will unfortunately leave out some important market areas. In particular, few studies[1] have been carried out on the demand for fish in developing countries. Moreover, we cannot hope to cover the substantial number of reports and working papers on the demand for seafood. In Section 2, we provide a brief description of the approaches used in estimation. In Section 3, we discuss market integration studies and show to what extent demand analysis and market integration provide complementary information. In Section 4, we provide a review of a number of demand studies. We try to emphasise main trends, as we do not attempt to give an exhaustive review of the literature. While we do not give much attention to cross-price effects, these are also important when considering demand structure, and the degree of competition is commented on briefly. In Section 5, we discuss the results form market integration studies before concluding in Section 6.

2 DEMAND ANALYSIS

In this section, the most common functional forms for demand system specifications are presented. We start with single equation specifications and then we review flexible functional forms; the Rotterdam system and the almost ideal demand system (AIDS).

The first empirical demand studies were primarily concerned with estimating elasticities and paid little attention to consumer theory. The researchers specified single equation demand functions linear in the parameters and quantity-dependent, of which the double log was most common. Letting q_{it} be the quantity consumed of good i at time t, p_{jt} the price of good j at time t and X_t the expenditure at time t, the equation to be estimated is:

$$\ln q_{it} = \alpha_i + \sum_j e_{ij} \ln p_{jt} + e_i \ln X_t . \tag{1}$$

The advantage with this specification is that the estimated parameters can be interpreted as elasticities as $e_{ij} = \partial \ln q_{it} / \partial \ln p_{jt}$ (the own and cross-price elasticities) and $e_i = \partial \ln q_{it} / \partial \ln X_t$ (the expenditure elasticity).

The range of j varies, and typically includes commodities that are assumed to be closely associated with good i. The measure of expenditure X_t is typically a measure of the consumer's income, often highly aggregated.

---

[1]See e.g. Ali (2005).

Economists had early on discovered that dynamic adjustment might be important in consumer behaviour. The first explicit attempt to specify demand functions that distinguished between short- and long-run behaviour, to the authors' knowledge, was Houthakker and Taylor's (1966) habit formation model. This model is based on the double log:

$$\ln q_{it} = \alpha_i + c_i \ln q_{it-1} + \sum_j e_{ij} \ln p_{jt} + e_i \ln X_t . \tag{2}$$

The dynamics are introduced in the lagged consumption variable, q_{it-1}, which makes current consumption dependent on the previous period's consumption. The short-run elasticities are e_{ij} and e_i, and the long-run elasticities are found by setting $\ln q_i$ equal at all t, as implied by long-run equilibrium. The long run elasticities may then be computed from Eq. 2 as $\eta_i = e_i(1-c_i)^{-1}$. To be consistent with utility maximisation, the parameter c_i must be between zero and one, and this is empirically observed.

During the 1970s, dynamic models, motivated primarily by problems with persistent autocorrelation and poor forecasting abilities, appeared in the macro-economic literature; particularly in reference to the consumption function. The work of Davidson *et al.* (1978) left a major impact, not only on macro-economic research, but on time series empirical economic research in general. The basic formulation is an autoregressive distributed lag model with a functional form linear in the logarithms of the variables:

$$\ln q_{it} = \alpha_i + \sum_{k=1}^{r} c_{ik} \ln q_{it-k} + \sum_j \sum_{l=0}^{s} e_{ijl} \ln p_{jt-l} + \sum_{l=0}^{s} e_{il} \ln X_{t-l} . \tag{3}$$

Lag lengths, r and s, are an empirical question and chosen to ensure that all dynamics are accounted for and the residuals are white noise.

There are both statistical and economic arguments for including lags in a model such as Eq. 3. The statistical arguments are founded on the observation that often in time series data there exists dependencies in the data over time. To capture these dependencies, dynamic specifications are necessary. Economic arguments focus on the lagged or dynamic adjustment to changes in economic variables. As instantaneous adjustment implies a static model, thus, the arguments against instantaneous adjustment are also arguments against a static model.

Habit formation is one argument for a dynamic model, but restrictions on the adjustment process, such as contractual obligations and imperfect information that induce adjustment costs, can also invalidate the hypothesis of instantaneous adjustment. These restrictions require more general dynamic specifications than the habit formation model. To model demand when these features are present, a general dynamic model is necessary. The advantage with Eq. 3 is that all linear dynamic structures are included as special cases.

Note that the habit formation model in Eq. 2 is a special case of the lag model in Eq. 3 with $r = 1$ and $s = 0$. Each parameter in (3) gives the elasticity of one variable at a particular lag with respect to current consumption. The long-run elasticities are found by summing over all lags. Hence, the long-run elasticities from Eq. 3 are $\eta_{ij} = \sum_l e_{ijl}(1-\sum_k c_{ik})^{-1}$ and $\eta_i = \sum_l e_{il}(1-\sum_k c_{ik})^{-1}$ An inconvenience with this model is that the long-run elasticities (often the elasticities of interest) must be computed after estimation. The long-run parameters (elasticities) can be estimated directly by transforming Eq. 3 to an error correction model (ECM) or;

$$\begin{aligned} \Delta \ln q_{it} = \alpha_i + \sum_{k=1}^{r-1} C_{ik} \Delta \ln q_{it-k} + \sum_j \sum_{l=0}^{s-1} E_{ijl} \ln p_{jt-l} + \sum_{l=0}^{s-1} E_{il} \ln X_{t-l} \\ - \omega(\ln q_{t-r} - \sum_j \eta_{ij} \ln p_{jt-s} - \eta_i \ln X_{t-s}) \end{aligned} \quad . \tag{4}$$

The parameter ω is also of interest as it may be interpreted as the adjustment speed towards equilibrium. Equation 4 is nonlinear in parameters and requires more computationally difficult nonlinear estimation techniques.

Other single equation specifications have appeared in the literature. These models specify variables in level form or with a Box–Cox transformation. The advantage with the Box–Cox transform is that the double-log and linear in levels forms are nested as limit cases. An empirical example may be found in Bjørndal *et al.* (1992).

Although most demand function estimation with single equation specifications have used quantity-dependent models, there are examples where price is the dependent variable. These inverse demand curves are common in agricultural and fishery studies where quantity is restricted by quota or other regulations. It is worth pointing out that the endogeneity problem for price and quantity variables in demand (and supply) studies has usually been studied with single equation demand (and supply) functions. This endogeneity problem has generally been ignored in demand specifications based on an assumption that either price or quantity is exogenous.

There exist two major problems with single equation models. First, in general, they are not theoretically consistent. The most common of these specifications, the double log, is theoretically consistent only when demand is independent of expenditure, that is the consumer's preferences are homothetic. This also violates Engel's law, which claims that the propensity to consume a particular group of goods varies with total expenditure. It is sometimes argued that in the analysis of a single commodity, where the functional form of the other goods in the system remains unspecified, the double-log specification may give a satisfactory local approximation, particularly if there is not too much variation in total expenditure. For specifications linear in the variables and using the Box–Cox transformation, it is not possible to be theoretically consistent (possibly with the exception of an approximation point). This can be seen by noting that the demand equation cannot be homogenous of degree zero when using these specifications.

Second, the single equation models specify uncompensated demand equations. The prices of the goods omitted from the specification may then cause problems because any change in either of them causes changes in demand for the commodity in question through changes in expenditure. This problem may be reduced if one specifies a compensated demand function (Stone, 1954). In empirical work, this problem may not be too serious, as the effect is small if the particular good represents a small portion of the budget.

In order to estimate demand functions that are consistent with utility maximisation, the concept of weak separability is used to separate a group of goods from the rest of the consumer's bundle. The demand functions for the goods inside the group are then specified in a system of demand functions where the restrictions associated with consumer theory can be tested or imposed (i.e. adding up, homogeneity and symmetry). These conditions, together with the trivial assumptions of positive prices and consumption, ensure that the demand system is consistent with consumer theory.[2] Most, but not all systems are derived from an explicitly formulated utility, indirect utility or cost function. However, this is not a necessary condition for theoretical consistency. In addition, only demand systems are used in empirical work as it is not possible to measure or compare utility. For a discussion of the connection between the functional form of a utility, indirect utility or cost function and each of the demand systems where this can be explicitly formulated, see Pollak and Wales (1992).

---

[2]Positive consumption is not absolutely necessary, and in some studies using cross section data at a micro level, zero consumption is allowed, see e.g. and Salvanes and DeVoretz (1997).

2.1 The Rotterdam System

In the Rotterdam system of Theil (1965) and Barten (1968), the demand equations are in budget share form and satisfy the adding-up condition. The symmetry and homogeneity restrictions implied by consumer theory may be expressed as linear functions of the estimated parameters. Consequently, one may either test if the data are in accordance with consumer theory or impose these restrictions on the estimated parameters to ensure theoretical consistency. Note that this, and most other empirical specifications, is an approximation to the underlying demand equations.[3] The results are dependent on the functional form. In particular, a rejection of the hypothesis of symmetry and homogeneity does not necessarily imply that the consumer theory is false. It could be caused by model specification problems, including functional form.

Another improvement with the Rotterdam system compared with the linear expenditure system is that it allows for estimation of price effects, including complements and inferior goods, without losing theoretical consistency. The Rotterdam system may be written as

$$w_i d \ln q_i = b_i d \ln \bar{x} + \sum_j c_{ij} d \ln p_j , \tag{5}$$

where

$$w_{it} = \frac{p_{it} q_i}{x},$$

$$d \ln \bar{x} = d \ln x - \sum_j w_j d \ln p_j = \sum_j w_j d \ln q_j ,$$

$$b_i = w e_i = p_i \frac{\partial q_i}{\partial x} \text{ and}$$

$$c_{ij} = w_i e_{ij}^* = \frac{p_i p_j s_{ij}}{x}.$$

Remember that e_i is the expenditure elasticity for good *i*. In addition, e_{ij}^* is the compensated price elasticity, which is related to the uncompens

---

[3]It is of course possible to postulate that the consumers' preferences actually correspond to the demand equations from a particular functional form.

ated and expenditure elasticities by Slutsky's equation on elasticity form $e_{ij} = e_{ij}^{*} - e_i w_j$. The continuous difference operators *d*, in applied work, are replaced by their discrete approximation Δ.

The adding-up restrictions imply that

$$\sum_i b_i = 1, \ \sum_i c_{ij} = 0. \tag{6}$$

These restrictions are satisfied when the budget shares in the data set add to unity. However, this restriction makes the covariance matrix singular. One must therefore delete one equation from the demand system before estimation. With correct estimation technique and an *iid* $(0, I \otimes \Sigma)$ error term, the system is invariant to which equation is deleted (Barten, 1968), and the adding-up restrictions from Eq. 6 are used to retrieve the parameters in the deleted equation. This is also a feature the Rotterdam system has in common with all the other systems of demand equations formulated in their budget share equations. The symmetry and homogeneity restrictions may be expressed as functions of the parameters:

$$\text{Symmetry:} \quad c_{ij} = c_{ji}$$
$$\text{Homogeneity:} \ \sum_j c_{ij} = 0. \tag{7}$$

The Rotterdam system is common in the literature, and this work has been extended to an inverse demand approach (Barten and Bettendorf, 1989). The Rotterdam system differs from most other functional forms in that the underlying utility or cost functions have never been explicitly formulated, and that differential demand functions are used instead of functions formulated in the levels of the variables.

2.2 The Almost Ideal Demand System

The most common functional form in demand system specification since the early 1980s has been AIDS of Deaton and Muellbauer (1980). As with the Rotterdam and translog systems, it is formulated in terms of the budget shares, and each demand equation can be written as

$$w_i = \alpha_i + \sum_j \gamma_{ij} \ln p_j + \beta_i \ln\left(\frac{X}{P}\right), \tag{8}$$

where

$$\ln P = \alpha_0 + \sum_i \alpha_i \ln p_i + \frac{1}{2}\sum_i \sum_j \gamma_{ij} \ln p_i \ln p_j \,.$$

AIDS is linear except for the translog price index $\ln P_t$. This problem has traditionally been circumvented in most applied work as suggested by Deaton and Muellbauer, by using a Stone price index that is $\ln P_t^* = \sum_i w_{it} \ln p_{it}$, which makes the system linear. Recently the use of the Stone price index has been shown to be inappropriate as it causes the estimated parameters to be inconsistent (Moschini, 1995). Moschini attributes this problem to the fact that the Stone price index does not satisfy the commensurability property, and suggests that the problem may be solved by using a price index that satisfies this property.[4] Moschini suggests several other price indices that satisfy this property and may be used to keep a linear specification of AIDS. He also shows that these indices perform as well as the translog index in a Monte Carlo experiment.

The restrictions to ensure theoretical consistency for AIDS are

Adding up: $$\sum_i \alpha_i = 1, \quad \sum_i \gamma_{ij} = 0. \tag{9a}$$

Symmetry: $$\gamma_{ij} = \gamma_{ji}. \tag{9b}$$

Homogeneity: $$\sum_j \gamma_{ij} = 0. \tag{9c}$$

AIDS is parallel to the Rotterdam and translog systems in that the adding-up restrictions are automatically imposed, and one equation must be deleted before estimation to avoid a singular covariance matrix. The symmetry and homogeneity restrictions may be tested or imposed. There exist no clear criteria for choosing among AIDS and the other two systems, and which functional form will perform best depends on the true structure in the underlying data. AIDS has the advantage that it is linear and formulated in levels. It may accordingly be encountered as more intuitive and easier to use than the Rotterdam systems. In common with the Rotterdam system, AIDS has an inverse demand representation.

[4]The commensurability property means that a price index should be invariant to the unit of measurement for the prices.

3 MARKET INTEGRATION

While measuring, the degree of substitution is the preferred way of determining to what extent commodities compete; the development or changes in prices over time provides valuable information on the relationship among commodities. The importance of prices in defining markets was recognised early on by Cournot, who in 1938 defined a market:

It is evident that an article capable of transportation must flow from the market where its value is less to the market where its value is greater, until difference in value, from one market to the other, represents no more than the cost of transportation (Cournot, 1971).

Similar definitions have been provided by, for example Stigler (1969) and others. Stigler maintains the spirit of Cournot in defining a market as "the area within which the price of a commodity tends to uniformity, allowance being made for transportation costs". The concept also applies to product space, where quality differences take the place of transportation costs (Stigler and Sherwin, 1985).

To motivate the Law of One Price (LOP) and price-founded definitions of a market, Figure 1 sketches the equilibrium for two markets. Prices in both markets are initially normalised at P. Assume then that there is a supply shock in Market 1 that shifts the supply schedule to $S1'$, giving p' and $q1'$ as new price and quantity. This causes the price to decrease while the quantity increases. What happens in Market 2 depends on the degree of substitution between the two commodities.[5] If there is no substitution possibility between the two markets/commodities, there will be no change in price and quantity in Market 2. If the goods are perfect substitutes, the demand schedule in Market 2 is shifted down to $D2'$ as consumers substitute commodity 1 for commodity 2, and the fall in price is just enough to equilibrate prices in both markets at P'. (This is LOP). If the goods are imperfect substitutes, the demand schedule is shifted down somewhat, say to $D2''$, but not enough to equate prices in the two markets.

As mentioned, the strength of the influence of the shock in Market 1 on Market 2 is normally measured by the cross-price elasticities.[6] However, one

[5]For completeness one should also mention that if the demand schedule in Market 2 shifts upwards, the two goods are complements

[6]The same story can be told based on a demand shock, but here it is the producers that potentially adjust their supply.

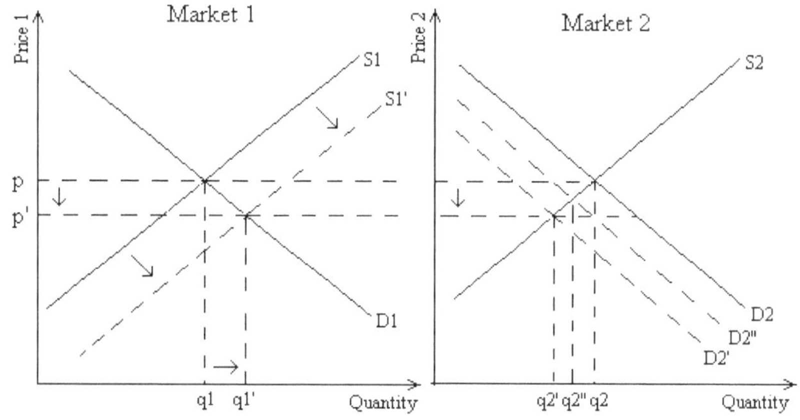

Figure 1. Potential Market Interactions between Two Markets.

can also look at the effect of the supply shock only from the price space. The price change in Market 1 can impact price in the other market in a number of ways. If there is no substitution effect, the demand schedule does not shift and there is no movement in price in Market 2. If there is a substitution effect the demand schedule in Market 2 shifts down, and the price in this market shifts in the same direction as the price in Market 1. At most the price in Market 2 can shift by the same percentage as the price in Market 1, that is LOP holds, and relative prices are constant. Hence, with respect to structural information about a market, analysis of relationships between prices can provide us with information on whether the two markets (goods) do not compete, whether they are imperfect substitutes or whether they are perfect substitutes so that the relative price is constant. This is the basis for the hypotheses we want to test when investigating price relationships.[7]

Several studies have pointed out that the adjustment towards a new equilibrium can be delayed by adjustment costs (Goodwin, *et al.*, 1990). A dynamic adjustment model is used to investigate this relationship and can also be used to determine whether the adjustment process is bi- or unidirectional. If causality goes only in one direction, this can be interpreted as price leadership for the price that does not adjust. This is possible if there is one central market that affects the price in smaller regional markets.[8]

[7]A negative relationship between the prices implies complements.

[8]In product space, the quality of one commodity is the reference quality.

It is common in studies of market integration to perform the analysis on the logarithms of prices, and we will proceed using this transformation. Given time series on two prices, say, p_t^1 and p_t^2, the simplest specification to test for market integration is

$$p_t^1 = a + bp_t^2 + e_t . \qquad (10)$$

A null hypothesis that $b = 0$ is a test that no substitution possibilities exist. A null hypothesis that $b = 1$ is a test for constant relative prices and LOP. The constant term a is the logarithm of a proportionality coefficient, and is zero if the prices are identical with exception of the arbitrary deviations caused by the error term. A non-zero constant term is in most cases interpreted as transportation costs or quality differences, which then are assumed to be constant.[9] Economic theory gives little guidance as to the choice of dependent variable, and the test is therefore often repeated by interchanging price variables in Eq. 10.[10]

In the early 1980s, several authors argued that adjustment could be costly and therefore take time. To account for this, models were introduced with variable specifications that could distinguish between short- and long-run effects. This test is performed by first running the regression:[11]

$$p_t^1 = a + \sum_{j=1}^{m} b_j p_{t-j}^1 + \sum_{i=0}^{n} c_i p_{t-i}^2 + e_t . \qquad (11)$$

The lag structure on prices is chosen so that e_t is white noise. The data support a hypothesis that there is a relationship, or in statistical terms that p_t^2 causes p_t^1, if a joint test that all *ci* parameters are zero is rejected.[12] Interchanging price variables in Eq. 11 allows a test of the null hypothesis that p_t^1 causes p_t^2. In this dynamic specification, test results based on different dependent variables have an economic interpretation. If one price causes the other while the opposite causality does not hold, this is evidence of price leadership. If causality is not observed in any of the equations, this is

---

[9]Some authors argue that the assumption of constant transportation cost is too restrictive, and can at times cause tests to show less market integration than there actually is. For instance, Goodwin, Grennes and Wohlgenant (1990) show closer market integration when transportation costs are explicitly modelled.

[10]This also gives rise to a simultaneity problem that often is acknowledged, but otherwise ignored. A good discussion can be found in Goodwin, Grennes and Wohlgenant (1990).

[11]In some cases, exogenous variables that represent common trends for the prices are also included.

[12]This is in econometric terms a test for Granger causality (Granger, 1969).

evidence that the goods are not in the same market. A test for a long-run LOP relationship corresponds to a test that the restriction $\sum b_j + \sum c_i = 1$ holds. What is more, if the restrictions $c_o = 1$, $c_i = 0$ and $b_j = 0$, for all $\forall ij > 0$ cannot be rejected, this is evidence that LOP holds in a static sense, and hence Eq. 11 nests Eq. 10.

In the 1980s, economists became increasingly aware that most economic time series are non-stationary. This means that normal statistical inference is not valid for linear regressions on nonstationary data and casts doubt on the reliability of early results obtained using the approach described earlier. In general, for non-stationary data there will be no linear long-run relationship. However, if the data series in question have common stochastic trends, the linear combination of two non-stationary data series can be stationary and the data series are said to be cointegrated (Engle and Granger, 1987).

There are two common approaches to testing for cointegration: the Engle and Granger (1987) test and the Johansen (1988) test. The Engle and Granger test is a straightforward regression procedure. However, there are two problems with this test. First, it is subject to the same normalisation problem in setting the dependent variable as with stationary data. Second, and more serious, is that normal statistical inference and tests for LOP are not valid, although cointegration tests for a (substitution) relationship between two commodities are possible. These problems are avoided when the Johansen approach is used.

4 DEMAND ELASTICITIES FOR FISH

Modelling and estimating demand elasticities for fish have a long history starting with the classic study by Bell (1968). Over time, and particularly from the 1980s, there has been a multitude of demand studies covering a wide rage of markets and species.[13] Wessells and Anderson (1992) provide a good review of this literature through the early 1990s. There are also a number of studies that investigate the relationship between seafood and other food commodities (e.g. Salvanes and DeVoretz, 1997; Johnson *et al.*, 1998). Depending on the model specification used, elasticities or flexibilities are

[13]See e.g. Asche, Salvanes and Steen (1997), Barten and Bettendorf (1989), Bjørndal, Gordon and Salvanes (1994), Bjørndal, Salvanes and Andreassen (1992), Eales and Wessells (1999), Hermann, Mittelhammer and Lin (1992), Jaffry, Pascoe and Robinson (1999), Kinnucan and Myrland (2002a), Zidack, Kinnucan and Hatch (1992).

reported. In some studies, flexibilities are estimated in the model and used to calculate and report elasticities.[14]

A review of the literature shows substantial variation in reported demand elasticities over species, model specifications, market levels investigated and level of aggregation over species (Schrank and Roy, 1991). In general and for most species, product groups and product report both elasticities and flexibilities from a common data set. The reported summary statistics are to some extent comparable, although for this data set the inverse demand system is supported by statistical testing. In any case, they report that the magnitude of the inverted flexibilities is substantially higher than the unitary elasticities measured. This suggests that elasticities measured directly from the model may be substantially lower than indirect elasticities calculated as inverse flexibilities. Moreover, it is likely that the difference is larger the more elastic the inverted flexibilities.

Estimates of elasticities at different market levels appear to show less elastic response at the retail level compared with the ex-vessel level. This generalisation must be interpreted carefully as different model specifications are commonly used at different points in the marketing chain. In addition, more recent studies report less elastic results, which may be caused by an identification problem between shifts in the demand curve (the result of increased and more competitive advertising) and movements along a demand curve.

In studies[15] where retail level data are used in quantity-dependent demand systems, reported elasticities vary closely around a value of –1.0. There are certainly deviations and more valuable fish have more elastic demand. However, the aggregation level for the data used in these studies is relatively high, and this would tend to make demand less elastic. This is because substitution possibilities are larger the less aggregated the data.

Whitefish and related species, particularly plaice and sole, were the group of species that obtained most attention early on. This can be explained by the large importance of these species when measured by value. Research attention was first directed at the fisheries of the Georgia and Grand Banks off the Atlantic coast of the USA and Canada. The seminal study of Bell

[14]The inverse of a flexibility will be a consistent estimate of the elasticity only if the good in question has no substitutes. Otherwise, the inverted flexibility will provide a lower bound for the elasticity (Houck, 1965).

[15]See e.g. Wessells and Wilen (1994), Eales, Durham and Wessells (1997), Johnson, Durham and Wessells, (1998).

(1968) indicates elastic demand for all the species using price-dependent models. However, with the exception of ocean perch, which seems to be very odd, the magnitude of the elasticities is not too high. When one takes into account that a price dependent specification is used, the true elasticities are not likely to be very elastic and most likely not smaller than –2. Tsoa *et al.* (1982) contradict Bell's results in indicating that the demand elasticity for cod fillet is highly inelastic (–0.46), and also find demand for redfish fillets to be inelastic. However, it should also be noted that the results of Tsoa, Schrank and Roy have been disputed (see Crutchfield, 1986). This dispute is interesting as it describes some of the difficulties confronted when estimating elasticities for seafood. Other studies of whitefish and flounders vary in their estimates, but in general the elasticities are either about –1.0 or more elastic.

Demand for salmon has received serious attention by researchers. This is not surprising given that Pacific salmon has always been among the world's most valuable fisheries, and salmon is one of the most successful species in intensive aquaculture. The first studies were carried out in Canada and the USA, with focus on wild Pacific salmon and the potential competition from salmon aquaculture (see, e.g. DeVoretz, 1982; Kinnucan and Myrland, 2002b). Most of these studies report a demand elasticity for salmon that is highly elastic. However, DeVoretz found that the demand for canned salmon is substantially less elastic than the demand for fresh/frozen salmon.

Asche (1996) reports a general trend in the demand for salmon being less elastic. This is possible due to the total supply of salmon (both wild and farmed) increasing threefold from the early 1980s. However, Bjørndal *et al.* (1992) argue that generic marketing has led to an outward shift in demand. Based on these studies, it seems reasonable to assume that the demand elasticity for salmon is near –1.0. However, the elasticity does vary by product form and species, and demand for frozen Pacific salmon seems to be inelastic.

Catfish is another species where aquaculture production has increased substantially. Catfish is a low-value species and despite successful generic advertising, Kinnucan and Miao (1999) argue that the elasticity has become less elastic with increased supply, indicating a movement along the demand schedule.

It is somewhat surprising that we do not observe the same tendency for whitefish. One of the main features of the whitefish market since the mid-1980s has been the increased internationalisation and the introduction of

Alaska pollock and Pacific hake to this market (Myrland and Vassdal, 1998). Further investigations are needed to account for the changes in the market.

Tuna is a species of major importance, yet demand for tuna has received little attention. Wessells and Wilen (1994) and Johnson *et al.* (1998) indicate that retail demand for tuna in Japan is close to –1, but inelastic. Wallström and Wessells (1995) indicate that demand for canned tuna in the USA is highly inelastic.

Several other species like crawfish, scallops, shrimp, shellfish, halibut, lobster, cuttlefish, crabs, crustaceans have received some attention. However, as estimates exist from only one or a few studies it is not possible to generalise. The only obvious trend is that high-valued species tends to have more elastic demand.

In several studies, particular for whitefish and salmon, different product forms are also studied. It seems hard to generalise the results, with the exception that demand for canned products is more inelastic than demand for other product forms. It also seems like the fresh product form tends be the most elastic.

Johnson *et al.* (1998) address the issue of competition between meat products and seafood products. Estimating systems which contain both types of product is important if the two types of products are not separable for the consumers. While the results are somewhat mixed, one can conclude that the substitutability between seafood and meat products is rather limited.

So far we have focused only on own-price effects. However, in most cases one also needs information about substitution effects as measured by cross-price elasticities or flexibilities. Although it is difficult to generalise, it is clear that most seafood products have substitutes. As expected, similar species and product forms tend to be the closest substitutes. For instance, different species and product forms of salmon tend to be closer substitutes than any given salmon category and other seafood species/products.

5 EMPIRICAL MARKET INTEGRATION STUDIES

There are a number of variations in the econometric approach used for market integration, but the common feature is a test for cointegration between at least two different prices. Some studies provide additional testing for LOP, leading prices, central markets and speed of adjustment after a price shock.

The first study to appear with respect to seafood that we are aware of is Squires *et al.* (1989) who studied the relationship between sablefish prices in Japan and the USA. They find that the Japanese and Alaska markets are integrated, whereas the US west coast is a separate market. This study is also notable as being the only one that treats prices as stationary.

Gordon *et al.* (1993) is the first in a string of studies that investigate the relationship between salmon, cod and other species, and are also the first to find that salmon is a separate market from wild fish (see also Asche *et al.*, 2002).

As in demand studies, salmon is the most studied species using cointegration analysis. Asche (2001) and others provide evidence that there is a global market for salmon including farmed as well as wild salmon. However, Clay and Gordon (1999) show that in the USA, the different regional markets are segmented. Asche and Guttormsen (2001) look further into the micro-structure showing that although there are seasonal variations in the price for different weight classes of salmon, prices are highly correlated. In total, these studies indicate that there is an integrated market for salmon both globally and for different product forms, and as such, all forms of salmon are competing in the same market. Each product form or species need not be directly substitutable with any other, but there are so many species and product forms that are substitutable, that there is a link in the price formation process.

The whitefish market has also received substantial attention. These studies indicate that all product forms of cod compete, although fresh cod is somewhat weaker related to the other product forms such as frozen, frozen fillets, wet salted and dried salted cod. Cod is also a part of a larger whitefish market that includes haddock, saithe, hake and pollock. The keenest competition seems to be at the international trade/wholesale level, as the competition appears less intense at the ex-vessel level. Still, Asche *et al.* (2002) find a high degree of price transmission between the different levels in the supply chain for cod (Gordon and Hannesson, 1996).

While salmon and whitefish are the most studied species, there are also studies investigating market integration either spatially or in product space for several other species and product forms, for example the study by Bose and McIlgrom (1996) of tuna in Japan.

6 CONCLUDING REMARKS

In this chapter, a review of demand and market integration studies for fish and seafood products is presented. With a few notable exceptions, the demand for fish and seafood received little attention until the mid-1980s. However, henceforth a number of studies, using a number of different methodological approaches, have been carried out. This research has vastly increased our knowledge about fish and seafood markets, in particular for salmon and whitefish.

Estimating demand elasticities and testing for market integration is an empirical exercise with each study focusing on a specific market in a given period of time. This is a problem as it provides information only about a given market for a given time period, and there is no reason why the demand for salmon in Japan should show similar characteristics for example to the demand for cod in the UK. Moreover, a number of different model specifications have been used, making it even more difficult to compare across studies.

Nevertheless, the demand in most markets appears to be price-elastic. This is good news for the seafood industry in general as it implies that total revenue will increase as production continues to increase. However, it also implies that the market will give little help for conservation measures, as fishermen's income will fall if landings have to be reduced.

For species with a rapidly increasing production, like new aquaculture species such as salmon and catfish, the demand gets less elastic with increases in supply. Hence, even though there is substantial evidence of successful generic marketing campaigns, it seems like lower prices facilitated by productivity improvements are more important in terms of increasing the quantity sold of these species.

Notwithstanding the many demand studies that have been carried out, there are gaps in our knowledge about the demand for seafood products. Perhaps the most serious is the lack of studies on the demand for seafood products from developing countries. Developing countries are of increasing importance in seafood production, international trade in fish and in own demand.

7 REFERENCES

Ali, Z. (2005) Disaggregated demand for fish in Bangladesh: An analysis using the almost ideal demand system, *The Bangladesh Development Studies*, 28(172), 1–45.

Asche, F. (1996) A system approach to the demand for salmon in the European Union, *Applied Economics*, 28, 97–101.

Asche, F. (2001) Testing the effect of an anti-dumping duty: The US salmon market, *Empirical Economics*, 26, 343–355.

Asche, F., D. V. Gordon, and R. Hannesson (2002) Searching for price parity in the European Whitefish Market, *Applied Economics*, 34, 1017–1024.

Asche, F. and A. G. Guttormsen (2001) Patterns in the relative price for different sizes of farmed fish, *Marine Resource Economics*, 16, 235–247.

Asche, F., K. G. Salvanes, and F. Steen (1997) Market delineation and demand structure, *American Journal of Agricultural Economics*, 79, 139–150.

Barten, A. P. (1968) Estimating demand equations, *Econometrica*, 36, 213–251.

Barten, A. P. and L. J. Bettendorf (1989) Price formation of fish, *European Economic Review*, 33, 1509–1525.

Bell, F. (1968) The pope and the price of fish, *American Economic Review*, 58, 1346–1350.

Bjørndal, T., D. V. Gordon, and K. G. Salvanes (1994) Elasticity estimates of farmed salmon demand in Spain and Italy, *Empirical Economics*, 4, 419–428.

Bjørndal, T., K. G. Salvanes, and J. H. Andreassen (1992) The demand for salmon in France: The effects of marketing and structural change, *Applied Economics*, 24, 1027–1034.

Bose, S. and A. McIlgrom (1996) Substitutability among species in the Japanese Tuna Market: A cointegration analysis, *Marine Resource Economics*, 11, 143–156.

Clay, P. and D. V. Gordon (1999) From Atlantic to Pacific: Price links in the wild and farmed salmon market, *Aquaculture Economics and Management*, 3(2), 93–105.

Cournot, A. A. (1971) *Researches into the Mathematical Principles of the Theory of Wealth*, New York: A. M. Kelly.

Crutchfield, S. (1986) U.S. demand for selected groundfish products 1967–80: Comment, *American Journal of Agricultural Economics*, 68, 1018–1020.

Davidson, J. E. H., D. F. Hendry, F. Srba, and S. Yeo (1978) Econometric modelling of the aggregate time–series relationship between consumers' expenditure and income in the United Kingdom, *Economic Journal*, 88, 661–692.

Deaton, A. S. and J. Muellbauer (1980) An almost ideal demand system, *American Economic Review*, 70, 312–326.

DeVoretz, D. (1982) An econometric demand model for Canadian salmon, *Canadian Journal of Agricultural Economics*, 30, 49–60.

Eales, J., C. Durham, and C. R. Wessells (1997) Generalized models of Japanese demand for fish, *American Journal of Agricultural Economics*, 79, 1153–1163.

Eales, J. and C. R. Wessells (1999) Testing separability of Japanese demand for meat and fish within differential demand systems, *Journal of Agricultural and Resource Economics*, 24, 114–126.

Engle, R. F. and C. W. J. Granger (1987) Co-integration and error correction: Representation, estimation and testing, *Econometrica*, 55, 251–276.

Goodwin, B. K., T. J. Grennes, and M. K. Wohlgenant (1990) Testing the Law of One Price when trade takes time, *Journal of International Money and Finance*, 5, 682–693.

Gordon, D. V. and R. Hannesson (1996) On prices of fresh and frozen cod, *Marine Resource Economics*, 11, 223–238.

Gordon, D. V., K. G. Salvanes, and F. Atkins (1993) A fish is a fish is a fish: Testing for market linkage on the Paris fish market, *Marine Resource Economics*, 8, 331–343.

Granger, C. W. J. (1969) Investigating causal relations by econometric models and cross-spectral models, *Econometrica*, 37, 424–438.

Herrmann, M., R. C. Mittelhammer, and B. H. Lin (1992) Applying almon-type polynomials in modelling seasonality of the Japanese demand for salmon, *Marine Resource Economics*, 7, 3–13.
Houck, J. P. (1965) The relationship of direct price flexibilities to direct price elasticities, *Journal of Farm Economics*, 47, 789–792.
Houthakker, H. S. and L. D. Taylor (1966) *Consumer Demand in the United States: Analysis and Projections*, Cambridge, MA: Harvard University Press.
Johansen, S. (1988) Statistical analysis of cointegration vectors, *Journal of Economic Dynamics and Control*, 12, 231–254.
Jaffry, S. A., S. Pascoe, and C. Robinson (1999) Long run price flexibilities for high valued UK fish species: A cointegration systems approach. *Applied Economics*, 31, 473–481.
Johnson, A., C. A. Durham, and C. R. Wessells (1998) Seasonality in Japanese household demand for meat and seafood, *Agribusiness*, 14, 337–351.
Kinnucan, H. W. and Y. Miao (1999) Media-specific returns to generic advertising: The case of catfish, *Agribusiness*, 15, 81–99.
Kinnucan, H. and Ø. Myrland (2002a) Seasonal allocation of an advertising budget. *Marine Resource Economics*, 17, 103–120.
Kinnucan, H. W. and Ø. Myrland (2002b) The relative impact of the Norway–EU salmon agreement: A mid-term assessment. *Journal of Agricultural Economics*, 53, 195–220.
Moschini, G. (1995) Units of measurement and the Stone price index in demand system estimation, *American Journal of Agricultural Economics*, 77, 63–68.
Myrland, Ø. and T. Vassdal (1998) Import demand for shrimp and whitefish fillets in the UK. In *Proceedings of the 9th International Conference of the International Institute of Fisheries Economics and Trade* (A. Eide and T. Vassdal, eds.), Tromsø, Norway.
Pollak, R. A. and T. J. Wales (1992) *Demand System Specification and Estimation*, Oxford: Oxford University Press.
Salvanes, K. G. and D. J. DeVoretz (1997) Household demand for fish and meat products: Separability and demographic effects, *Marine Resource Economics*, 12, 37–55.
Schrank, W. E. and N. Roy (1991) *Econometric Modelling of the World Trade in Groundfish*, Dordrecht: Kluwer.
Stone, J. R. N. (1954) Linear expenditure systems and demand analysis: An application to the pattern of British demand, *Economic Journal*, 64, 511–527.
Squires, D., S. F. Herrick, Jr., and J. Hastie (1989) Integration of Japanese and United States sablefish markets, *Fishery Bulletin*, 87, 341–351.
Stigler, G. J. (1969) *The Theory of Price*, London: Macmillan.
Stigler, G. J. and R. A. Sherwin (1985) The extent of a market, *Journal of Law and Economics*, 28, 555–585.
Theil, H. (1965) The information approach to demand analysis, *Econometrica*, 33, 67–87.
Tsoa, E., W. E. Schrank, and N. Roy (1982) U.S. demand for selected groundfish products 1967–80, *American Journal of Agricultural Economics*, 64, 483–489.
Wallström, P. and C. R. Wessells (1995) Analysis of U.S. consumer demand for canned tuna: Impact of dolphin-safe controversy, in *International Cooperation for Fisheries and Aquaculture Development: Proceedings of the 7th Biennial Conference of the International Institute of Fisheries Economics and Trade*, Vol. 3 (D. S. Liao, ed.).
Wessells, C. R. and J. L. Anderson (1992) Innovations and progress in seafood demand and market analysis, *Marine Resource Economics*, 7, 209–288.
Wessells, C. R. and J. E. Wilen (1994) Seasonal patterns and regional preferences in Japanese household demand for seafood, *Canadian Journal of Agricultural Economics*, 42, 87–103.
Zidack, W., H. W. Kinnucan, and U. Hatch (1992) Wholesale- and farm-level impacts of generic advertising: The case of catfish, *Applied Economics*, 24, 959–968.

PART III

FORESTRY

The articles presented in forestry aim to cover main aspects of research and application of OR at different levels of decision, as well as user concerns, such as environment preservation, wildlife, fire containment, and so on.

The first article, "Models for strategic forest management" by Eldon Gunn deals with high-level, long-term decision processes. Main issues discussed include sustainability, looking at forests as ecosystems, and forests as economic and managerial systems. Most used OR tools at strategic level are linear programmes and simulation.

Following the hierarchical structure, Richard Church presents "Tactical-level forest management", which bridges strategic and operational decisions. The chapter presents several approaches and models developed at that intermediate level to consider decisions on harvesting, road building and environment. Spatial representation becomes particularly important at this level

At the operational level we have three articles: "Harvest operational models in forestry" by Epstein, Karlsson, Ronnqvist and Weintraub, where the authors present successful implementations of OR tools to support short-term decisions related to harvesting and bucking, location of harvesting machinery and road accesses and integration of harvesting and transportation, leading to analysis of the forest supply chain. In a similar operational setting, Hamish Marshall in "Log merchandizing models used in mechanical harvesting" considers how to cut trees into logs at destinations such as plants, in order to obtain best value, given prices and demands for specific logs. Mechanized processes use OR models to optimize this stage, with significant improvements in value obtained from each stem.

Finally, in "Forest transportation", by Epstein, Ronnqvist and Weintraub, the issue of using models for short- and medium-range transportation decisions is presented. Successful implementations of heuristics mostly for daily dispatching of up to several hundred trucks have led to important savings and show promise for future use in real time.

Environmental issues play a very significant and increasing role in forestry. In "Optimization of forest wildlife objectives" John Hof and Robert Haight discuss the importance of considering spatial optimization to integrate wildlife preservation to habitat possibilities. The dimension and form of areas with mature forests as well as recently harvested areas play important roles to preserve different animal species. They show how exact and heuristic algorithms have been developed in this area.

In a related chapter, Alan Murray in "Spatial environmental concerns" discusses specific spatial models to enhance environmental quality. In particular restricting the maximum contiguous area that can be harvested is a common practice, which leads to difficulties in solving combinatorial problems.

Heuristic as well as exact approaches have been proposed to solve these problems. In "Heuristics in forest planning" by Sessions, Bettinger and Murphy show how different OR heuristic techniques, including some based on LPs, simulation and metaheuristics have been applied to solve forest planning problems at different decision levels. The difficulties in data capturing and uncertainties are discussed, and several real planning exercises are shown.

Rainukar and Buongiorno in "Forest economics: historical background and current issues" present forest analysis from an economic theory perspective. Issues such as optimal rotation, first analysed in the mid-nineteenth century, externalities, market equilibrium, taxes and incentives are analysed, as well as non-market valuation, in particular when dealing with tourist amenities and environment.

Clearly forest planning involves multiple objectives, ranging from harvest production to environmental protection and biodiversity to visitor amenities. In "Multiple criteria decision making in forest planning: recent results and current challenges", Diaz-Baltiero and Romero discuss the different approaches and techniques that have been proposed to deal in an explicit form with them.

The next chapters deal with the menace to forests caused by fires and insects and how OR can support decisions to deal with these problems.

In "Forest fire management: current practices and new challenges for operational researchers" David Martell presents contributions made by OR at different levels of fire protection and prevention, going from planning firebreaks and fire fighting crews and airplanes, to actual operations to contain fires in real time.

In "A model for the space-time spread of Pine Shoot Moth" Cominetti and San Martin analyse the particular problem of an insect species which causes severe damage to Chilean pine plantations and develop a non-linear model to predict successfully its diffusion in time and space.

Finally, Lohmander in "Adaptive optimization of forest management in a stochastic world" presents different techniques to deal with uncertainties typically present in forest planning, such as tree growth and in particular future prizes.

Chapter 16

MODELS FOR STRATEGIC FOREST MANAGEMENT

Informed by Strategic Perspectives

Eldon A. Gunn

Department of Industrial Engineering, Dalhousie University, Halifax, N.S., Canada

Abstract Strategic forest-management models are models that assist strategic decision makers in examining forest strategy. There is history of long-term linear programming and related models being seen as strategic. However, strategy is broader than just the forest-management process. As a result, a large number of ecological and economic models may also inform the forest-management process. As we change our perspective on what strategic issues are important, this may require us to think about both the formulation and use of the strategic models.

Keywords: Management, linear programming, simulation, strategy

1 INTRODUCTION

In forestry, "strategic", in reference to planning and modelling, is often used as a synonym for long term, even though not all strategies are long term, nor is all long-term planning strategic. The term strategic forest-management model usually refers to the linear programming and simulation models, used by almost every corporate body and government agency that manages forest land holdings. These are among the most widely used models in operations research. Much has been written on the use of certain models in the forest-management planning process in government agencies. On the other hand, although most forestry companies use these models, relatively little is written about this use by private industry.

In spite of this history, some negativity can to be found in the literature. Some may be unavoidable in a milieu of government policy formation, but some is due to a failure to distinguish between strategy in forest-management and strategic models. Strategy is something management does; it is not the result of running a computer model, although models can be useful to analyse consequences of a strategy. Models used in strategic planning are typically models of management control of a strategy.

Strategic forest-management models focus on the interaction between forest-management decisions, such as harvest and silviculture scheduling, and issues such as sustainability and economic returns from the forest. The long-term supply of timber products has often been a primary focus. One strategic theme that permeates forestry today is sustainable forest-management (SFM) with an increased emphasis on ecosystem management and a decreased emphasis on timber.

Interestingly, the strategic models have tended to treat both the economic environment and the productive capability of the land base as givens that are unaffected by the forest management decisions. The use of simulation models aimed at understanding forest productivity has been growing in both analytical and advocacy roles. These models have had limited involvement from operations researchers and even foresters. The emphasis has been on forest growth dynamics and ecology. In forest economics, the emphasis has been less on forest management than on land values and the markets for forest products. However, important models, based on partial equilibrium analysis, attempt to understanding how scarce resources eventually translate into market prices for both products and forest land. Some of this resource scarcity is, of course, due to changing views on ecological sustainability. Sections 3 and 4 provide a short discussion of ecological and economic models, before we examine the linear programming and simulation models in Section 5.

Much of the focus of the strategic models has been on a formulation that describes what types of treatments, in which stand types, and when to carry them out. However, the strategic context has been that of land availability, definitions of sustainability, required cover conditions, and allowable treatments. What seems obvious is that the ecological context will require more attention to where treatments occur. This can also be true as economic issues come to the fore. In Section 5, our discussion of the strategic model points out some of the relations between formulation and strategy

There is a very large literature on the use of models for analysis of forest management strategy. Davis *et al.* (2001) textbook is required reading. Greber and Johnson's (1991) theme of examining the perspective behind the analysis is important for strategic forest decision makers, although the emphasis on "overcutting" is a little dated. Weintraub and Bare (1996), and Martell *et al.* (1998) and are two overview papers with an emphasis on strategic forest-management models that can help give additional perspective.

2 STRATEGY, SUSTAINABILITY, AND MODELS

2.1 What is Strategy?

Strategic planning is often depicted as a formal rigorous process, designed to position the enterprise in some advantageous way. However, Mintzberg (1989) depicts strategy as a craft in which past experience and present creativity combine to develop strategy for the future. Strategy formation is seldom if ever programmed (see Mintzberg, 1994). Realized strategy is a combination of intended strategy and emergent strategy, a pattern of behaviour that emerges from the organizational culture. Anthony's (1965) juxtaposes strategic planning with "management control", the "process by which managers assure that resources are obtained and used effectively and efficiently in the accomplishments of the organizations objectives". Anthony, like Mintzberg, is clear that there can be no normative theory of strategy development. Strategy development is a creative, irregular, opportunistic activity.

An issue that inevitably comes up in discussions of forest-management strategy is "whose strategy?". If strategy is not a computation, then there are strategic decision makers who develop and adopt the strategy, and the analysis needs to be put in terms of their responsibilities. Typically, people who deal with long-term and large-scale strategy are not the same people who deal with short- and medium-term operations and tactics.

2.2 Sustainability

The public goods aspect of forests provides a second perspective of "whose strategy". Forests provide multiple benefits to society. Many argue that forests need to be managed taking into account the interests of stakeholder groups. The Working Group on Criteria and Indicators for the Conservation and Sustainable Management of Temperate and Boreal Forests (the "Montreal Process") was formed in Geneva in June 1994. The criteria and indicators from this international agreement (see Montreal Process Working Group, 2004) now lie behind the strategic efforts of many organizations. The Montreal Process Criteria include i) Conservation of biological diversity, ii) Maintenance of productive capacity of forest ecosystems, iii) Maintenance of forest ecosystem health and vitality, iv) Conservation and maintenance of soil and water resources, v) Maintenance of forest contribution to global carbon cycles, vi) Maintenance and enhancement of long-term multiple socio-economic benefits to meet the needs of societies and vii) Legal, institutional, and economic framework for forest conservation and sustainable management. Only two criteria (vi and vii) mention economic issues. The public goods aspect of these indicators implies that there are likely to be many players, each with their own strategies. Thus strategic planning may include not only those who make strategic decisions but also stakeholder groups whose interests are affected by these decisions.

Some have argued that adaptive forest management involves feedback processes that, using similar indicators to the Montreal process, lead to a form of automatic control (see Gunn, 2004). However, leading economists (see Arrow *et al.*, 2000) have argued that tradeoffs exist between economic and ecologic values and markets do not resolve these tradeoffs in an automatic fashion. Gunn (2004) indicates that it is unlikely feedback systems can be controllable or stable. Thus, SFM needs management to be achieved. As Mintzberg points out, the essence of management is making strategic choice.

2.3 Models in the Strategy Process

What then is the role of models in the strategy process? Strategy needs management control in its implementation. Implementation sheds light on the strengths and weaknesses, costs, and benefits of a strategy and suggests to management the need for modifications to the strategy, closing the feedback loop. Models of the management control process, similarly, serve to shed light on the strategy. Models cannot set our values and wants; the framing of these is strategy. However, models can reveal some of the biological and economic consequences of proposed actions.

There has been discussion of decision support systems (DSS) for strategic planning of forest ecological management (e.g. Rauscher, 1999). This suggests a view that strategic decision-making can be systematized. Most successful implementations of DSS technology are in routine operational decision-making (see Holsapple and Whinston, 2000), not strategy. Quantitative analysis of the consequences of strategic decision is consistent with ideas in Anthony and Mintzberg; systematized strategic decision-making is not.

An important question is the level of detail required in strategic analysis. Sometimes, emphasizing the details of a process loses the perspective of simple concepts and large-scale trends (Nelson, 2003). Anthony (1965) points out the issue of confusion by detail. Modellers, in an effort at "accuracy" often attempt to include detail at spatial and temporal resolutions that are not meaningful to the strategic decision makers. This occurred in the manufacturing literature more than 40 years ago (Silver *et al.*, 1998, Chapter 13). Hierarchical planning (Gunn, 1991) has been a way of ensuring that the model focus and modelling detail are consistent with the needs of the strategic decision maker.

3 MODELS OF THE ECOSYSTEM

The first four criteria of the Montreal process require that SFM maintains the ecosystem in a healthy condition. This suggests that models of ecosystem condition will be an important part of developing a forest-management strategy. Historically, forest-management strategy has taken the ecosystem as exogenous and mostly worried about managing trees. A number of models have been developed for projecting how the ecosystem will evolve over time. Deutschman *et al.*, 1997 gives a nice introduction to the field. The primary aim of these models is to understand forest succession and productivity, starting at the stand level and extending to the landscape level. The literature is very large and a serious review is outside the scope of this paper. Forest operations researchers have also tended to view these models as outside the scope of strategic models.

3.1 Gap Models

The best known of such models are the "gap" models, also referred to as the JABOWA models (see Deutschman *et al.*, 1997). One of the later models in this series is Zelig (Urban, 1990). Gap models are aimed at a stochastic simulation of detailed ecosystem processes. A forest stand is simulated as a series of cells, where one cell is about the size that would be occupied by a single mature tree (approximately 10 m^2 in many models). Typically, individual tree characteristics, as well as the characteristics of the canopy, are simulated. Initial distributions of tree species, sizes, and heights, as well as detailed soil characteristics, are the starting point. As trees grow and leaf litter accumulates, soil characteristics change. Light access for the trees is a key issue. The simulation takes stand slope and latitude as exogenous. As trees die, they create gaps with increased access to light. Each species is described as having different requirements for light and nutrients. A single cell is not sufficient to model light interference from tall trees and seed dispersal. For temperate forests, the zone of interaction may extend over 5–6 cells (Urban *et al.*, 2000). Simulations typically involve grids ranging from

100 to 2,500 cells and may take a long time to arrive at a steady-state distribution of key measures, more than 2000 1-year periods in some cases.

Forest ecosystems are modelled as a coupled set of stand models (e.g. FACET, Urban *et al.*, 2000). Although the gap models are highly dependent on spatial relationships, the ecosystem models developed from them are often not. The SORTIE model (see Deutschman *et al.*,) is an example of a model that is spatial at both the individual tree and stand levels.

3.2 Landscape Models

Because some ecosystem processes, such as fire, wind, and insects, are themselves highly spatial, some landscape ecology models emphasize spatial issues. LANDIS (Mladenoff, 2004) is a cellular simulation model but the cells are typically larger and cover the entire landscape. The within-cell succession is generally not as detailed as the gap models, although He *et al.* (1999) report using a gap model to estimate the within-cell processes in a LANDIS model of landscape response to climate warming. Ecosystem/ landscape models typically have been used to study natural disturbance, but this is changing. Gustafson *et al.* (2000) report using harvest rules within LANDIS to model joint forest succession and timber harvesting. Simulations of 50 ten-year time periods for 25,143 stands on a map representing 262,080 ha (60-m cell, 728,000 cells) with 23 tree species, 4 size classes (seedling, sapling, pole, saw log) and 6 prescriptions to 6 management areas are reported to take about 6 h using a 450-MHz Pentium processor. The model was used to study the landscape response to wind and fire regimes under no harvest, clearcutting, and uneven age management options.

There has been ongoing debate on accuracy and appropriateness of landscape process models (see Larocque, 1999) and the papers in that volume). Mladenoff and Baker (1999) is a comprehensive survey of ecosystem models, tracing the development from gap tree level and stand level models through to complete landscape level models. Urban *et al.* (1999) explains the process of putting together a landscape level simulation from a gap model. The gap models are not the only way of getting at forest development. The Forecyte/Forecast series of models (Kimmins *et al.*, 1999) emphasize the ability to model other processes, such as soil and moss development, in addition to tree processes.

To date, these simulations have been developed primarily from a perspective of ecological modellers, with a limited orientation to management control. The methodology is not the discrete event simulation familiar to most operations researchers but rather more a simulation of a set of stochastic differential equations. However, forest operations researchers need to be aware of this work. The linear programming models and simulation models discussed later assume that site capability is constant and unaffected by management. They also implicitly assume that characterization of ecological health can be expressed as aggregate constraints on cover characteristics. The models discussed here are based on a view that site capability and ecological health are more complicated than this.

Traditionally the forest-management community has used simulation models to investigate various strategies of forest regulation (see Davis *et al.*, 2001) using concepts such as binary search on harvest levels to develop a notion of sustainability. The ecological models discussed here are much more detailed in their simulation of ecological processes, although their simulation of forest growth and the consideration of a host of management scenarios is still less sophisticated. Nonetheless, there does appear to be a trend developing of using these types of models as part of the strategy process (e.g. see Gustafson and Rasmussen, 2002). Whether the level of detail in these models is appropriate for strategic planning remains to be seen.

4 MODELS OF THE FOREST ECONOMIC SYSTEM

If current forest strategy is focusing on the ecosystem, much of the past strategic forest management has been based on the view of the forest as a resource base for industry. Economic issues start with the economic rent that can be earned from forested land. If the forest rent is not higher that that from other land uses, it is unlikely to remain forested. Land is not the only issue. The relative location of forest production and the nature and capacity of markets are also important strategic issues.

4.1 Stand Level Economics

Economic perspectives have had a dual viewpoint of both determining the optimal sustained economic (timber) yields of the forest and also determining the economic rent that land is capable of earning. This work has a very long history (e.g. Hellig and Linddal, 1997) Samuelson (1976) is a modern source, while Buongiorno (2001) and Zhang (2001) continue the analysis, emphasizing the role of uncertainty.

The basic problem is the "rotation age" of a forested stand. Under some prescribed method of forest management, there is a function $\mathbf{V}(t) = (V_1(t), V_2(t), \ldots, V_K(t))$ giving the volume per hectare for up to K classes of timber for a stand that has been allowed to grow for t periods. A function $p(\mathbf{V}(t))$ gives the end of period discounted future value of all cash flows in the interval $[0,t]$ if the stand is managed during this interval and harvested in period t. For a given forest type and site capability, an infinite series of rotations, starting from bare land, which will then have a net present value of

$$\mathrm{NPV}(T,r) = p(\mathbf{V}(T))(e^{-rT} + e^{-r2T} + \ldots) = p(\mathbf{V}(T))\left(\frac{1}{(1-e^{-rT})} - 1\right) = p(\mathbf{V}(T))\left(\frac{e^{-rT}}{(1-e^{-rT})}\right)$$

The maximizing value T* is the optimal rotation and NPV(T*,r) is the maximum net present value per hectare that this stand is capable of producing. The average annual yields V(T*)/T* give a concept of sustainable yields. These models also produce policy in the form: "if T ≥ T*, then end the rotation through a regeneration harvest and start again from age 0".

The T* and the NPV(T*,r) depend on many factors such as (i) r, the time value of money, (ii) the silvicultural techniques applied and the wood volume response, (iii) the set of products produced, (iv) the harvesting, sorting and manufacturing choices available, and (v) the price of forest products. Different strategies will produce different values of T*. By modifying the model slightly, using a Hartman-like analysis such as Koskela and Ollikainen (2001), amenity value of standing timber influence both T* and rent. These amenity values range from recreation, scenic beauty, and ecological and watershed effects.

Note that each stand type and site capability will have its own optimal policy and yields. Applying these across all the stands in the forests gives state dependent, forest level analysis. Since the rotation period and yield depend on a plan of stand management, a variety of management plans/ silviculture strategies can be investigated. The analysis can be extended to stochastic settings (Buongiorno, 2001).

Since these models are based on infinite horizon analysis, the policies are inherently sustainable at the stand level, so long as the rotations chosen do not lead to deterioration of site capability. The response to price, to discount rates or to constraints on rotation periods is clear and unambiguous and provides an upper bound on the rent the actual system can achieve. These models probably are not used as much as they should be for strategy analysis. The reason they aren't is mainly because of irregularity of wood supply, not forest sustainability. With no inherent coordination mechanism, the irregular wood supply may not be compatible with the needs of a forest products industry. Habitat requirements for certain species (e.g. Rempel and Kushneriuk, 2003) and the requirements for forest cover to provide appropriate water quality and quantity (e.g. Alila and Beckers, 2001) may also require patterns of forest cover across various landscapes that do not naturally arise from stand level analysis.

Strategies that require a steady flow of wood products, or that consider habitats and watershed conditions, require different modelling capabilities. This has lead to the linear programming and simulation models. However as Pareds and Brodic (1989) Print out, there are strong connections between these stand level models and the forest level models.

4.2 Models of Forest Products Markets

There is another side of forest economics that the standard strategic models do not address. Production of forest products assumes a forest product market. Supply and demand economics imply that production levels affect market prices. Substitution is also an issue. Timber of a given type and size can may be restricted to certain markets; some markets can use a variety of types and sizes.

There is a long history of the use of partial equilibrium models that integrate forest-management strategies and economic development strategies. Perhaps the best known is the Timber Assessment Market Model (TAMM) (Adams and Haynes, 1980). This model uses a partial equilibrium modelling calculation to calculate a set of prices, timber supplies, and timber demands that are in economic equilibrium in a given period. TAMM deals with solid wood product uses in the economy. There are also models of the pulp and paper industry North American Pulp and Paper (NAPAP) that use a similar equilibrium concept but a different mathematical approach based on Gilless and Buongiorno (1987). These models have been in long-term use in the US Resources Planning Act timber Assessments (see Adams, 2002). In this work, they are assembled into a large, complex suite of models which includes not only NAPAP and TAMM, but also AREACHANGE, a model designed to forecast the potential changes in forested area and ATLAS, a model designed to provide an estimate of timber availability in each region by period and to project the change in forest timber availability after the harvests and following growth.

Both TAMM and NAPAP are spatial equilibrium models (Takayama and Judge, 1971). The basic model structure can be represented as the following mathematical programming problem, although in the case of NAPAP, there is also a manufacturing component in addition to the transportation component:

$$\text{Maximize} \quad \sum_i \int_0^{Q_i^D} D_i(q)\,\mathrm{d}q - \sum_j \int_0^{Q_i^S} S_j(q)\,\mathrm{d}q + \sum_{i,j} c_{ij} Q_{ij}$$

$$\text{Subject to:} \quad Q_i^D \le \sum_j Q_{ij} \qquad Q_j^S \ge \sum_i Q_{ij},$$

Where $D_i(.)$ and $S_j(.)$ are demand and supply curves respectively, Q_{ij} is the amount transported to demand region i from supply region j, Q_i^D is the quantity consumed in region i, Q_j^S is the quantity produced in region j, and c_{ij} is the cost of transportation from region i to region j. The rational behind the objective is a maximization of consumer surplus plus producer surplus minus transportation costs as discussed in Samuelson (1952). The supply sector is actually not just a simple curve. It involves capacity, inventory, and stumpage price interactions.

The computation of equilibrium can be challenging in such models. The TAMM model is based on reactive programming in which prices are adjusted in response to an imbalance in supply and demand until equilibrium is obtained. These methods have been used in TAMM over many years. However, they are difficult to extend to more complex markets. The NAPAP model is based on the "price endogenous linear programming" (PELPS) (Lebow *et al.*, 2003) concept originally developed by Buongiorno and Gilless (1984). Here, the demand curves and the supply curves are replaced by step functions:

$$\int_0^{Q_i^D} D_i(q)\,\mathrm{d}q \approx \sum_{k=1,K_i} p_{ik} q_{ik}, \quad \sum_{k=1,K_i} q_{ik} = Q_i^D, \quad 0 \le q_{ik} \le q_{ik}^{\text{Max}}$$

with $p_{i1} > p_{i2} > \ldots$. Since the linear programming model will choose the high-value consumption q_{i1} before it chooses the q_{i2}, the solution will correspond to the demand curve. Thus, linear programming represents a robust solution algorithm for these equilibrium problems, subject only to the errors introduced from the step function approximation.

These equilibrium models have found wide application in a variety of settings. Sohngen and Sedjo (1998) review four large-scale models. Buongiorno *et al.* (2003) and Kallio *et al.* (2004) are similar examples in a European and global context. Although these large scale econometric models go beyond what many foresters think of as forest-management strategic issues, clearly modelling capabilities do exist that can take into account the market price effects and transportation costs due to modifying forest outputs. These models take as given the available forest production in a single period and calculate the market response. This is in contrast to the models discussed below that assume markets for forest products are not an issue.

5 MODELS OF FOREST LAND AND ECOSYSTEM MANAGEMENT

What most foresters think of as the basic tools of strategic forest planning are models that give a long-term response to forest-management inputs. Two classes of such models are linear programming and simulation models, although the distinction between the two blurs at times. We often see linear programming used as a way of simulating the consequences of restricting harvest techniques or constraining the management in some way. We also see simulation models used in a binary search method to optimize harvest levels.

Referring back to our discussion of ecological and economic models, it interesting to observe that these forest-management models have simplified views of both ends of the spectrum. The essential assumption is that forest-management changes neither the capability of the soil to produce fibre nor the capability of the economic system to absorb the forest products produced.

5.1 The Linear Programming Models

In few fields has the use of linear programming received more use than forestry. Its use for the analysis of strategic decisions has been one of the prime areas of application. The USDA Forest Service has been particularly active in supporting the development of analysis tools based on linear programming. Kent *et al.* (1991) give an outline of the development of the FORPLAN system for forest-management modelling. FORPLAN began with Multiple Use Sustained Yield Calculation (MUSYC) (Johnson and Jones, 1979), evolved to FORPLAN as considerations other than timber became important and has since continued in its evolution to SPECTRUM (Greer and Meneghin, 2000). The Timber RAM package (Navon, 1971) has had a long history of government use and the MaxMillion package (Ware and Clutter, 1971) has influenced several generations of industrial application. These applications have spread internationally including the New Zealand FOLPI system (Garcia, 1984) and the JLP (Lappi, 1992) system in Finland. JLP is in turn part of the larger Finnish forest-management package called MELA (Siitonen *et al.*, 2001). One outcome of all this development has been a commercial software industry for forest-management modelling. The Woodstock/Stanley package from REMSOFT is one example.

In spite of, or perhaps because of, its extensive use, FORPLAN has received much criticism as a strategic tool. Both Kent *et al.* (1991) and Rauscher (1999) have many references. Some of these criticisms are directed at the USDA Forest Service processes. Others are directed at the concept of a "normative, rational, optimization" approach (Rauscher, 1999). This appears to be a case of confusing the modelling tools with the process of strategy formation.

The linear programming models have three distinct parts. The first models the process of forest growth and management. The second models the sustainability of forest products. The third models the requirement to provide certain types of forest cover, usually associated explicitly or implicitly with some type of habitat consideration. As discussed earlier, a strategy is often expressed as constraints. The strategic modelling provides a framework for a strategic decision maker to examine the tradeoffs as these constraints are imposed. Linear programming is particularly useful for this analysis (i) because of its unambiguous calculation of feasibility or infeasibility and (ii) because of the availability of shadow costs that tell how much a constraint is costing at the margin (Dantzig, 1963).

There are three separate modelling approaches to forest growth and management (Fig. 1). Two are the well-known Model I and Model II (see Davis *et al.*, 2001). What some people call Model III is less common (see Garcia, 1990) but still the basis of widely used packages such as FOLPI. Briefly, in Model III, all stands of the same age class are aggregated. In each period, the land in an age class is either harvested, reverting to the regeneration age class, or not harvested becoming one age class older. The process of growth and harvesting of the forest can be represented as the flow through a network. Model II involves a similar aggregation of all stands. However the network is less detailed. The arcs (i,j) of the network correspond to a stand originating in a certain period i and being regeneration harvested in a subsequent period j. An arc of the Model II network corresponds to a path between two regeneration nodes in the Model III, or a path from one of the initial nodes to a regeneration node. Model I can be thought of as either aggregated or as individual stands. If aggregated, then all stands of a given age class are aggregated to a single node. Each arc of Model I correspond to a path from the equivalent node in the Model II or Model III network. From this point of view, the Model I, Model II and Model III networks thus are equivalent. However, in practice the arcs of Model II do not correspond to every possible path of Model III and the arcs of Model I are usually only a small fraction of all possible paths.

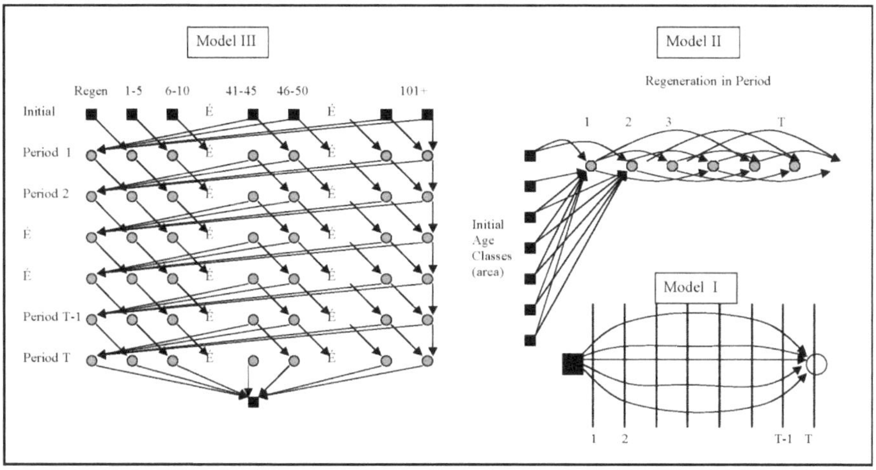

Figure 1. Models III, II, and I.

At first glance, Models II and III appear to be the more efficient modelling frameworks. However, because Models II and III merge stands at harvest, validly representing growth requires a separate network for every different site capability (site index) and cover type (see Davis *et al.*, 2001). To account for different management regions, government jurisdictions, biophysical zones such as riparian zones and steep slopes, or ecological districts, again requires separate Model II or Model III networks for each unique combination of attributes. This can result in very large LP models with substantial network constraints. Such models are known to be relatively difficult to solve.

All three models are described above as if there is only one silvicultural regime, namely harvesting at a certain point. It is easy to model a variety of silvicultural regimes in Model I. Any particular path of regeneration, pre-commercial thinning, commercial thinning, and regeneration harvest and subsequent treatments, is just one arc, commonly referred to as one prescription. Obviously many such arcs are possible. In most practice, only a few are examined, although codes such as JPL (Lappi, 1992) use column generation techniques to generate as many as needed as the algorithm proceeds. In Model II a similar comment applies in that there can be several alternate paths from one regeneration node to the next. In Model III, it is more complicated. Separate networks need to be created for each type of silviculture, and a transition from one network to another is created for each treatment.

5.1.1 Overall Model Structure

Using a Model I formulation for forest growth and management, forest-management models typically have the following structure

$$\begin{aligned} &\text{Maximize (minimize)} \quad \sum_{i=1}^{I}\sum_{k=1}^{P_i} C_{ik}x_{ik} \\ &\text{Subject to:} \quad \sum_{k=1}^{P_i} x_{ik} = A_i, \quad i=1,I, \end{aligned} \tag{1}$$

where the principal decision variables are x_{ik}, the hectares allocated to prescription k from analysis area I and the data are c_{ik} , the net present value of all future returns if prescription k is used on analysis area i, and A_{i} the total area of analysis area i.

This Model I formulation, with its very simple constraints structure (1), can be replaced by the Model II or Model III structure, usually with a substantial reduction in the number of variables but at the cost of many more constraints and more complicated constraints. There are strategic issues in the choice what prescriptions to include in the various P_i. For example, some environmental strategies might eliminate clearcutting as a permitted treatment or establish a minimum rotation period?

Historically the first issue has been *flow constraints* to ensure some sort of regularity of harvest flow. Thus the following equations describe the amount of forest products produced:

$$\sum_{i=1}^{I}\sum_{k=1}^{P_i} h_{iklrt}x_{ik} = H_{lrt} \qquad l=1,L, \quad r=1,R_l, \quad t=1,T \tag{2}$$

where h_{iklrt} is the volume of timber type l produced in region r in period t if prescription k is used for analysis area i, L is the number of forest product types, R_{l} is the number of regions for which product flows are of interest for forest product of type l, and T is the number of time periods of interest in the model.

Then we have various possible flow constraints, examples of which are

$$H_{lrt} = H_{lr(t-1)}, \qquad t = 2,T, \quad l = 1,L, \quad r = 1,R_l \tag{3a}$$

$$H_{lrt} \geq H_{lr(t-1)}, \qquad t = 2,T, \quad l = 1,L, \quad r = 1,R_l \tag{3b}$$

$$\alpha H_{lr(t-1)} \geq H_{lrt} \geq \beta H_{lr(t-1)}, \quad t = 2,T, \quad l = 1,L, \quad r = 1,R_l. \tag{3c}$$

The constraints of type (3a), (3b) are referred to as *level flow* and *non-declining yield*, respectively. The constraints (3c) generalized the previous two with $\alpha > 1$ and $\beta < 1$. The choice of how many timber regions and how many timber types are modelled is obviously part of the decision maker's strategic outlook.

The other types of constraints will be called *habitat constraints.* Others might call these *forest cover constraints* or *landscape constraints.* These constraints are of the form

$$\sum_{i=1}^{I}\sum_{k=1}^{P_i} w_{iksdt} x_{ik} \geq W_{sdt}, \qquad s = 1,S, d = 1,D_s \qquad t = 1,T, \tag{4}$$

where w_{iksdt} is a habitat condition measure for wildlife type s in district d in period t if prescription k is used for analysis area i. W_{sdt} is a lower limit on the total amount of the condition measure, S is the number of wildlife habitats of strategic interest, and D_s is the number of wildlife districts (landscapes) of strategic interest for habitat of type s.

There is obviously considerable flexibility in defining what habitats, districts, and wildlife type mean. Some of the habitat districts can be landscapes corresponding to watersheds and the constraints can correspond to a constraint on forest cover on the watershed. There is no reason that districts have to be disjoint even for a given habitat type. Neither is there any reason that districts need to align themselves for different habitat types nor is there any necessity for alignment with the forest product regions.

It is worth commenting on the objective coefficients C_{ik}. If the objective is maximization of net present value, then these are of the form

$$\sum_{t=1}^{T} \delta^t \sum_{l=1}^{L} c_{iklt} x_{ik} = C_{ik} x_{ik},$$

where δ is a discount factor and c_{iklt} is the net revenue produced if prescription k applied to analysis unit i produces forest product type l in period t. If a Model I form is used, the C_{ik} can reflect transportation costs but this will be problematic with Model II or Model III forms.

Since these terms are linear, this implies that markets are perfectly elastic and no substitution of product between alternative industrial consumption. Surprisingly, in many applications, these issues do not seem to matter because the model is only directed to maximizing volume. That is C_{ik} is just V_{ik}, the total volume of all forest products produced if prescription k is used on analysis area i. Barros and Weintraub (1982) and Gunn and Rai (1987) have both used other approaches where an additional industrial sector is added to the model with the option of allocating one type of forest product to several alternative uses with differing prices. The demand sector representation ideas in PELPS (Lebow *et al.*, 2003) could also be applied here.

5.1.2 Sustainability, Uncertainty, Spatial representation, and Habitat

These LP models have had a very large influence on how government foresters think about sustainability (Kent *et al.*, 1991). Moreover, most forestry students have been taught to use such a model and the flow constraints (3a and/or 3b) are generally portrayed as the "sustainability" constraints. This raises interesting questions when the levels of harvest from the model, are lower than current. In this case, non-declining (or level flow) yield has required an immediate decline in harvest! Practitioners have had the experience of replacing the (3a) constraints with (3c) and observing a solution that starts at current levels, experiences a modest early decline and then returns to harvest levels that are both higher than initial levels and maintain higher standing volumes. Daugherty (1991) has also reported on the "declining non-declining yield". Here the model solution is assumed to be implemented exactly in the first period and the model run again as if at the beginning of the second period. The harvests in the first period of the new model will usually be lower than in the second period of the original model because the initial model solution scheduled expensive harvests late in the planning horizon, at low net present value cost because of discounting, in order to have a high initial harvest. Once the initial harvest is taken, the high-cost later harvests are no longer worthwhile. These are two sides of the same coin. The strategic decision maker needs to decide how to deal with intergenerational issues (Church and Daugherty, 1999) of historical harvests and future harvests.

Traditionally models have used relatively simple breakdowns, both in terms of forest products, regions, habitat, and divisions. Many models use only one forest products region and a very small number of product classes, making it impossible to represent transportation. As foresters explore strategic issues of sustainability, it may be important to be able to constrain harvest levels or forest cover conditions within riparian zones, watersheds, and/or various types of ecozones, as well as on a political district and management district basis. This has considerable impact on the utility of the three model frameworks. A typical situation of a 500,000 ha land base with about 100,000 stands could easily produce as many as a thousand zone combinations. Managing on the basis of ecologic and economic zones implies not aggregating across zone combinations, requiring a separate Model II (or Model III) network for each. Because the Model I framework preserves stand location, it is more adaptable as strategic decision makers feel a need to cope with increased spatial specificity. Users of Model II report very large models even with far less spatial representation than the example above.

The three model frameworks also have different implications in treating uncertainty. Gunn (1991) discusses hierarchical planning using a rolling planning horizon as a way of dealing with uncertainty such as fire, or insect infestations. The essence of rolling planning is to replace the uncertain fire/insect consumption by a mean value and being prepared to re-plan on a regular basis. Reed and Errico (1986) and Boychuk and Martell (1996) have shown that a Model III framework works very well as a way of representing the mean fire/insect consumption. However, there seems to be no nice way of representing mean fire/insect consumption in a Model I or Model II framework. This has left planners with the options of either ignoring these effects or of just treating them as an extra harvest (equivalent to reducing the implemented harvest from the computed H_{lrt} by mean fire/insect amounts). However, since harvests are optimized, this latter action amounts to optimizing the placement of the fire and insect outbreaks.

The linear constraints (4) on landscape cover are unlikely to model in any exact way the requirements to protect the landscape or provide habitat. They neither approximate wildlife nor hydrological processes; they only attempt to constrain forest cover. As Hof and Bevers (2002) have discussed, it is often possible to build linear programming models that represent wildlife and watershed processes. However, such models are more complicated and are not models of the management control of strategy. Implementing a strategy that requires special treatment of a watershed or ecodistrict usually means setting some type of aggregate target constraint. Although the constraint does not necessarily reflect the landscape effects, the shadow cost on the constraint reflects the consequences on the objective function.

5.2 Simulation Models

There is quite a long history of simulation used to model forest-management strategy. Much of the earlier work was used to model different strategies to achieve forest regulation using area or volume regulation (see Chapelle as cited in Davis *et al.*, 2001).

In most simulations, stands are given an initial description in terms of site capability and forest cover characteristics. Some sort of goal state is specified for each period. This is typically total harvest. Using some harvest rules, stands are designated for harvest until the goal is attained or until no eligible stands are left. The designated stands are harvested. Then all stands are "grown" for a period and the process repeats itself.

This apparent simplicity hides the fact there is considerable flexibility in how a simulation is designed. Early simulations were non-spatial with land sharing certain cover, site, and other attributes combined together into a single "macro-stand". At every period, stands of the same type, but different ages that are harvested during the period were merged at harvest. This is exactly the same as the Model II philosophy of the linear programming models. However, models such as LAMPS (Bettinger and Lenette, 2004), where individual stands are spatially represented, are now more common.

Simulation can offer a lot more flexibility than LP. Among other things, it is possible to use detailed growth models and to implicitly offer more prescription choices. In Model I or Model II LP models, harvest/silviculture prescriptions have to be defined in advance. With simulation, the prescriptions can be defined in terms of eligibility (age, diameter, etc.). In each period, all eligible stands can be considered for silvicultural treatment or harvest. Thus, it is possible for a simulation model to sometimes achieve larger values of an objective than the optimal solution of a similar linear programming model, because, over the course of the simulation, each stand has more "prescriptions" available to it.

There are other areas of flexibility as well. In some cases, assuming that management choices are made "optimally" is unrealistic. For example, non-industrial private forest (NIPF) owners are unlikely to target their harvest to age or site classes in a way that mimics optimizing behaviour. The LAMPS model (Bettinger and Lennette, 2004) includes some mechanisms for modelling NIPF. More generally, an important aspect of a simulation model is the ability to specify rules and priorities that control which stands should be treated and harvested in a given period. These rules and priorities can extend the users ability to control not just the level of harvest but also spatial issues associated with the harvest. Put simply, if the modeller can think of a harvest policy, he/she can simulate it! A key area of strengths in simulation models is the ability to model stochastic phenomena. These include ongoing growth stochastics, delays in regeneration, and cover type changes at regeneration, and the larger stochastic events such as fire, insect, and disease outbreaks. As we observed earlier, it is difficult to model even mean effects in the Model I and Model II LP models.

In spite of the obvious differences, in some ways the use of simulation models and optimization models in forest management has been surprisingly similar. This is partly due to the tendency of foresters to choose an objective of volume maximization and level flow constraints in their LP models. As Davis *et al.* (2001) discuss, similar analyses have often carried out using binary search on harvest levels within a simulation model. As a modern example, LAMPS has a variety of simulation search options available. Adding to the confusion, the Hoganson and Rose (1984) simulation model is really a Lagrangian dual decomposition to an implicitly defined large LP in which the flow constraints are removed using Lagrange multipliers. Each stand will have its own optimal harvest strategy in response to the prices. Calculating these strategies gives a forest "simulation".

Where optimization models have strong advantages over simulation models is in their ability to deal with sophisticated objective functions, while at the same time finding feasible solutions to complex constraints over and above the basic harvest flow constraints.

5.3 Spatial Models

There is an ongoing concern with spatial issues, often phrased as an ability to put the strategic forest-management plan "on the ground" in terms of road building and stand access. Other concerns include the assignment of inappropriate prescriptions and inappropriate harvest levels to sensitive regions and landscapes. Nelson (2003) discusses the increasing trend to using spatially specific models for strategic analysis and comments on the excessive detail that is being placed on the strategic decision maker.

Models that focus on the detailed placement of management treatments to account for opening size restrictions and for road access are properly in the realm of tactical and operational planning. On the other hand, many of the current forest-management challenges, requiring spatial analysis in terms of levels of protection and special status for watersheds, wildlife habitat and in terms of location of large facilities, do require spatial placement of the treatments. As discussed above, this is relatively easy using Model I LP formulations or simulation models like LAMPS. Many authors interpret spatial as discrete, necessitating solution using integer programming and heuristics. At a strategic level, there seems to be little reason why this should be necessary. Hof and Bevers (2002) discuss a variety of models where linear programming gives useful spatial analysis. Simulation models such as LAMPS and the hybrid optimization simulation methods such as Hoganson and Rose (1984) and Lappi (1992) naturally produce discrete spatial (Model I) solutions but require that constraints be treated with some flexibility.

It has become practice to define the strategic models aspatially and then try to replicate the strategic solution in a spatial fashion using a tactical model of a portion of the land base. Church, Murray and Barber (2000) give some of the issues. One point about which there has been little discussion is the meaning of the solutions from the strategic models. If we use a stochastic simulation model to produce harvests and other silvicultural treatments that meet strategic objectives, simulation modellers would recognize that the model solution is just one realization that demonstrates that the strategic constraints are achievable. There is no particular reason to believe that this solution will, or should, be implemented. The same actually holds true of linear programming. Linear programming models typically have alternate optimal solutions and all convex combinations of these solutions are also optimal. Moreover, in the simplex algorithm, there will be a large number of iterations with objective function values very close to optimal. Since the data underlying the objective function and harvest flow constraint coefficients are very imprecise, any of the basic feasible solutions produced in the later stages of the algorithm could have been declared optimal, given minor perturbations of the data. This implies that achievable objective levels, feasibility of constraints and shadow costs on constrains are meaningful assessments of the strategy, but precise allocation of harvest and other silviculture to stand types in each period have little meaning, particularly in a hierarchical planning context.

6 SUMMARY

Foresters are faced with an increasingly complex strategic environment. Although they have long used both simulation and linear programming models to investigate harvest strategies, and strategies for land availability and timber sustainability, foresters may well need extensions to these models to deal with this complexity. At the same time, modellers will need to resist the tendency to add such complexity to the models that they are no longer useful for strategy. This will be an ongoing challenge.

The tradition in most forest-management strategic models has been to assume that forest productivity does not change with forest management. Ecologists have developed a variety of simulation models of ecological processes and these are increasingly being used to look at the effects of various types of management. Insights gained here may lead to new constraints for the linear programming models. It also appears that the simulation models normally used for forest-management modelling are evolving in the direction of the ecosystem models.

Similarly, forest-management models have tended to ignore issues of prices and markets for forest products. Often the models look at the production of forest products in terms of volume instead of looking at commodities that can be consumed in a market, a market often characterized by spatial location and capacity limitations. PELPS and related methods show that linear programming models can be extended to deal with market issues. For the solutions to make sense in such models, the LP models may require more spatial representation than has been traditional.

In thinking about strategic models, it is important to stress the role of the model. Models that are useful for strategy are unlikely to be at the level of detail of implementation. Given the time horizon, scale, complexity and uncertainty surrounding the strategic decision process, it is unrealistic to expect these models to predict how the chosen strategy will be implemented. They simply cannot. What the strategic models can do is assess the effect of the constraints imposed by strategy. They tell if the constraints are feasible and, if feasible, the costs and benefits of tightening and relaxing these constraints. Used in this way, strategic models can be important sources of insight to the strategic decision maker.

7 REFERENCES

Adams, D. M. 2002. RPA5: Harvest, inventory and stumpage prices. *Journal of Forestry* **100**(2):26–31.

Adams, D. M. and Haynes, R.W. 1980. The 1980 softwood timber assessment market model; structure, projections, and policy simulations. Monograph 22. *Supplement to Forest Science* **26**(3):64.

Alila, Y. and Beckers, J. 2001. Using numerical modelling to address hydrologic forest management issues in British Columbia. *Hydrological Processes* **15**:3371–3387.

Anthony, R.N. 1965. *Planning and Control Systems: A Framework for Analysis*. Harvard University Graduate School of Business Administration. Boston. MA.

Arrow, K., Daily, G.C., Dasgupta, P., Levin, S., Maler, K-G, Maskin, E., Starrett, D., Sterner, T., and Tietenberg, T. 2000. Managing ecosystem resources. *Environmental Science and Technology* **34**:1401–1406.

Barros O. and Weintraub A. 1982. Planning for vertically integrated forest industries. *Operations Research* **30**(5):1168–1182.

Bettinger, P. and Lennette, M. 2004. *Landscape Management Policy Simulator (LAMPS), Version 1.1. USER GUIDE.* Research Contribution 43, Forest Research Laboratory. Oregon State University. Corvallis. OR.

Boychuk, D. and Martell, D. L. 1996. A multistage stochastic programming model for sustainable forest-level timber supply under risk of fire. *Forest Science* **42**:10–26.

Buongiorno, J. 2001. Generalization of Faustmann's formula for stochastic forest growth and prices with Markov decision process models. *Forest Science* **47**:466–474.

Buongiorno, J. and Gilless, J. K. 1984. A model of international trade of forest products, with an application to newsprint. *Journal of World Forest Resource Management* **1**:65–80.

Buongiorno, J., Zhu,S., Zhang, D., and Tomberlin, D. 2003. *The Global Forest Products Model.* Academic Press. UK. p. 301.

Church, R. L., Murray, A. T., and Barber, K. H. 2000. Forest planning at the tactical level. *Annals of Operations Research* **95**:3–18.

Church, R. and Daugherty, P. J. 1999. Considering intergenerational equity in linear programming-based forest planning models with MAXMIN objective functions. *Forest Science* **45**:366–373.

Dantzig, G. B. 1963. *Linear Programming and Extensions.* Princeton University Press. Princeton. NJ. p. 627.

Davis, L. S., Johnson, K. N., Bettinger, P. S., and Howard, T. E. 2001. *Forest Management. Fourth edition.* McGraw-Hill. New York. p. 804.

Daugherty, P. J. 1991. *Credibility of Long Term Forest Planning: Dynamic Inconsistency in Linear Programming Based Forest Planning Models.* Ph.D. Thesis. Dept of Env. Sci. Policy and Management. University of California. Berkeley. p. 158.

Deutschman, D. H., Levin, S. A., Devine, C., and Buttel, L. A. 1997. Scaling from trees to forests: analysis of a complex simulation model. Science Online. Last viewed Mar. 2005 at URL: http://www.sciencemag.org/feature/data/deutschman/index.htm.

Garcia, O. 1984. FOLPI, a forestry-oriented linear programming interpreter. In: H. Nagumo *et al.* (eds.) *Proceedings IUFRO Symposium on Forest Management Planning and Managerial Economics*. University of Tokyo. Japan. pp. 293–305.

Garcia, O. 1990. Linear programming and related approaches in forest planning. *New Zealand Journal of Forestry Science* **20**:307–331. (a more detailed version is found at http://web.unbc.ca/~garcia/unpub/unpub.htm, viewed Nov. 24, 2004).

Gilless, J. K. and Buongiorno, J. 1987. *PAPYRUS: A Model of the North American Pulp and Paper Industry.* Forest Science Monograph 28. p. 37.

Greber, B. J. and Johnson, K. N. 1991. What's all this debate about overcutting? *Journal of Forestry* **89**(11):25–30.

Greer, K. and Meneghin, B. 2000. SPECTRUM: an analytical tool for building natural resource management models, In: J. Fried, M. Vasiewvich, and L. Leefers (eds.) *Proceedings of the Seventh Symposium on Systems Analysis in Forest Resources, May 28–31,*

1997. Shanty Creek Resort, Bellaire, Michigan USDA North Central Research Station General Tech. Rep. NC-205, pp. 376–384.

Gunn, E. A. 1991. Some aspects of hierarchical planning in forest management. In: *Proceedings of the 1991 Symposium on Systems Analysis in Forest Resources*. Compiled by M. A. Buford. USDA For. Serv. Gen. Tech. Rep. SE-74. pp. 54–62.

Gunn, E. A. 2004. Sustainable forest management: control, adaptive management, hierarchical planning. In: M. Bevers, T. M. Barrett (eds.) tech. comps. *Systems Analysis In Forest Resources: Proceedings of the 2003 Symposium*. Gen. Tech. Rep. PNW-GTR-656. USDA, Forest Service. Pacific Northwest Research Station. pp. 7–14. Portland, OR.

Gunn, E. A. and Rai, A. K. 1987. Modelling and decomposition for planning long-term forest harvesting in an integrated industry structure. *Canadian Journal of Forest Research* **17**:1507–1518.

Gustafson, E. J. and Rasmussen, L. V. 2002. *HARVEST for Windows, Timber Harvest Simulation Model Version 6.0 User's Guide*. North Central Research Station. Rhinelander. WI 54501.

Gustafson, E. J., Shifley, S. R., Mladenoff, D. J., Nimerfro, K. K., and He, H. S. 2000. Spatial simulation of forest succession and timber harvesting using LANDIS. *Canadian Journal of Forest Research* **30**:32–43.

He, H. S., Mladenoff, D. J., and Crow, T. R. 1999. Linking an ecosystem model and a landscape model to study forest species response to climate warming. *Ecological Modelling* **114**:213–233.

Hellig, F. and Linddal, M. 1997. Early Danish contributions to forest economics. *Journal of Forest Economics* **3**(1):87–103.

Hof, J. G. and Bevers, M. 2002. *Spatial Optimization in Ecological Applications*. Columbia University Press. New York.

Hoganson, H. M. and Rose, D. W. 1984. A simulation approach for optimal timber management scheduling. *Forest Science* **30**(1):220–238.

Holsapple, C. W. and Whinston, A. B. 2000. *Decision Support Systems: A Knowledge-based Approach*. West Publishing Company. Minneapolis/St Paul. p. 713.

Johnson, K. N. and Jones, D. B. 1979. *A User's Guide to Multiple Use—Sustained Yield Resource Scheduling Calculation (MUSYC)*. Land Management Planning System Section. USDA Forest Service. Fort Collins. CO.

Kallio, A. M. I., Moiseyev, A., and Solberg, B. 2004. *The Global Forest Sector Model EFI-GTM— the Model Structure*. European Forest Institute Internal Report No. 15.

Kent, B., Bare, B. B., Field, R., and Bradley, G. 1991. Natural resource land management planning using large-scale linear programs: the USDA forest service experience with FORPLAN. *Operations Research* **39**(1):13–27.

Kimmins, J. P., Mailly, D., and Seely, B. 1999. Modelling forest ecosystem net primary production: the hybrid simulation approach used in FORECAST. *Ecological modelling* **122**:195–224.

Koskela, E. and Ollikainen, M. 2001. Optimal private and public harvesting under spatial and temporal interdependence. *Forest Science* **47**(4):484–496.

Lappi, J. 1992. *JLP: A Linear Programming System for Management Planning*. The Finnish Forest Research Institute. Research Papers 414. p. 134.

Larocque, G. R. 1999. Editorial. *Ecological Modelling* **122**:135–137.

Lebow, P. K., Spelter, H., and Ince, P. J. 2003. *FPL-PELPS, A Price Endogenous Linear Programing System for Economic Modelling; Supplement to PELPS III, V 1.1*. USDA Forest Service, Forest Products Laboratory Research Paper FPL-RP-614. Madison. WI. p. 32.

Martell, D., Gunn, E., and Weintraub, A. 1998. Forest management challenges for operational researchers. *European Journal of Operations Research* **104**(1):1–17.

Mladenoff, D. J. 2004. LANDIS and forest landscape models. *Ecological Modelling* **180**: 7–19.

Mladenoff, D. J., Baker, W. L. (eds.) 1999. *Spatial Modeling of Forest Landscape Change: Approaches and Applications*. Cambridge University Press. Cambridge. UK. p. 364.

Mintzberg, H. 1989. *Mintzberg on Management: Inside our Strange World of Organizations*. Free Press. New York.

Mintzberg, H. 1994. *The Rise and Fall of Strategic Planning: Reconceiving Roles for Planning, Plans, Planners*. Free Press. New York.

Montreal Process Working Group. 2004. The Montréal Process site. http://www.mpci.org/home_e.html last viewed Nov. 22, 2004.

Navon, D. I. 1971. *Timber RAM - a long-range planning method for commercial timber lands under multiple-use management*. USDA Forest Service. Research Paper PNW-70. Albany, CA: Pacific Southeast Research Station, Forest Service, U.S. Department of Agriculture. p. 30.

Nelson, J. 2003. Forest-level models and challenges for their successful application. *Canadian Journal of Forest Research* **33**:422–429.

Paredes, G. L. and Brodie, J. D. 1989. Land values and the linkage between stand and forest analysis. *Land Economics* 65(2):158–166.

Rauscher, H. M. 1999. Ecosystem management decision support for federal forests in the United States: a review. *Forest Ecology and Management* **114**:173–197.

Reed, W. J. and Errico, D. 1986. Optimal harvest scheduling at the forest level in the presence of the risk of fire. *Canadian Journal of Forest Research* **16**:266–278.

Rempel, R. S. and Kushneriuk, R. S. 2003. The influence of sampling scheme and interpolation method on the power to detect spatial effects of forest birds in Ontario (Canada). *Landscape Ecology* **18**:741–757.

Samuelson, P. 1952. Spatial price equilibrium and linear programming. *The American economic review* **42**:283–303.

Samuelson, P. 1976. The economics of forestry in an evolving society. *Economic Enquiry* **14**:466–492. (also *Journal of Forest Economics* **1**(1):115–149).

Siitonen, M., Anola-Pukkila, A., Haara, A., Harkonen, K., Redsven, V., Salminen, O., and Suokas, A. (eds.) 2001. *MELA Handbook 2000 Edition*. The Finnish Forest Research Institute. Helsinki. Finland. p. 498.

Silver, E. A., Pyke, D. F., and Peterson, R. 1998. *Inventory Management and Production Planning and Scheduling*. Wiley. New York. p. 754.

Sohngen, B. and Sedjo, R. 1998. A comparison of timber market models: static simulation and optimal control approaches. *Forest Science* **44**:24–36.

Takayama, T. and Judge, G. G. 1971. *Spatial and Temporal Price and Allocation Models*. North Holland. Amsterdam.

Urban, D. L. 1990. *A Versatile Model to Simulate Forest Pattern: A User's Guide to ZELIG Version 1.0*. University of Virginia. Environmental Sciences Department. Charlottesville, VA.

Urban, D. L., Miller, C., Halpin, P. N., and Stephenson, N. L. 2000. Forest gradient response in Sierran landscapes: the physical template. *Landscape Ecology* **15**:603–620.

Urban, D. L., Acevedo, M. F., and Garman, S.L. 1999. Scaling fine-scale processes to large-scale patterns using models derived from models: meta-models. In: D. J. Mladenoff and W. L. Baker (eds.). *Spatial Modeling of Forest Landscape Change: Approaches and Applications*. pp. 79–98.

Ware, G. O. and Clutter, J. L. 1971. A mathematical programming system for the management of industrial forests. *Forest Science* **17**:428–445.

Weintraub, A. and Bare, B. 1996. New issues in forest management from an operations research perspectives. *Interfaces* **26**(5):9–25.

Zhang, D. 2001. Faustmann in an uncertain policy environment. *Forest Policy and Economics* **2**:203–210.

Chapter 17

TACTICAL-LEVEL FOREST MANAGEMENT MODELS

Bridging between strategic and operational problems

Richard L. Church
Department of Geography, University of California at Santa Barbara, Santa Barbara, CA 93106-4060

Abstract Tactical analysis represents a bridge between strategic modeling and operations modeling. Whereas strategic models represent space with large tracks and production areas, operational models represent a space as a set of stands, riparian zones, feasible positions for road segments, and the terrain surface so that logging activities can be laid out. Strategic models allow a broad-scale analysis over a long horizon to optimize forest-level outputs and costs. Operational-level models make location-specific decisions over a much smaller time frame. The task of a tactical-level model is to bridge the disconnect that exists between the strategic domain and the operational domain.

Keywords: Forest management, optimization, hierarchical models, strategic, tactical, and operational models, linear programming, integer programming

1 INTRODUCTION

Managing a forest involves a variety of objectives, including the production of forest products, the provision of recreational opportunities, the protection of biological elements (e.g., endangered or threatened species), the maintenance of watershed values, the conservation of soil, the control of pests and disease, and the reduction of losses due to wildfire. Activities that support such objectives involve a wide ranging set of possible activities that typically involve prescriptions of forest cutting and thinning, road building for access, designating areas for wilderness, controlled burning of excess fuel conditions, and the placement of sediment traps to protect riparian zones. This list of activities is by no means exhaustive, as forest operations tend to be complex. Thus, capturing the essence of all elements of forest management in a single model over a large tract of land is virtually impossible. This is due, not to the lack of ingenuity on the part of systems analysts, but in large measure to shear size and complexity of elements that could be modeled. Further, uncertainty and stochastic elements, which have often been ignored, can add a significant degree of complexity to modeling forest systems. Modeling is by its very nature abstract and simplified when compared to the real problem. The approach taken to model a forested region is a function of scale and the areas represented, the number of time periods and the modeled components (e.g., habitat protection, fire losses, road building, and harvesting operations), and the number of spatial interaction terms and constraints. The logical approach to modeling such a complex landscape is to break the problem into a hierarchy of problems, each supporting specific aspects of forest management (Hof and Pickens, 1987; Weintraub and Cholaky, 1991; Church *et al.*, 1994). At the highest level of the hierarchy are "strategic" models that are designed to analyze broad-scale planning decisions over a large landscape and over a long period of time. The FORPLAN and Spectrum models of the U.S. Forest Service are good examples of strategic models (Johnson *et al.*, 1986; Bare and Field, 1987; Johnson and Stuart, 1987; Kent *et al.*, 1991). The objective of such models is to devise a trajectory of decisions that leads to long-term production levels, measured in terms of metrics such as timber volume, return on investment, and acres of habitat protected. For example, in the strategic level, decisions over 10 decades or more are optimized for large planning units, represented by strata aggregated into age-classes. The outputs from strategic models are poorly specified in terms of the spatial dimension. For example, a solution from a strategic model may call for 500 acres of 100-year-old Douglas fir trees to be harvested from a 10,000 acre planning area, but it will not specify specific stands to be harvested. This is because all stands of similar type and age have been aggregated into one planning unit, about which specific prescription/activities are considered. Aggregating stands across large planning units allows a wide variety of activities to be considered (e.g., thinning in year 10 followed by harvesting in year 20). Simply put, at the strategic level there is a greater emphasis on the types of activities and outputs than representing details of spatial layout (Boungiorno and Svanqvist, 1982; Weintraub *et al.*, 1986). Strategic plans cannot be implemented without further analysis, but they help support the decision-making process for long-term planning by estimating possible levels of performance over a number of different metrics.

At the other end of the spectrum are operational models. Operational models are oriented towards supporting decision making that deals with forest operations over a week, a season or two, up to perhaps a decade.

Whereas, strategic models represent forest stands in an aggregated form, operational models are spatially explicit. For example, Weintraub *et al.* (1996) developed a model to schedule trucks to haul timber from specific harvest stands to specific destinations such as pulp mills, saw mills, sorting yards, and ports. Epstein *et al.* (1995) developed a model integrated with a Geographical Information System to optimize the layout of harvest areas with the placement of machinery and road location. Another important problem addresses road building/scheduling and harvesting/scheduling as the cost and maintenance of roads to harvest units can exceed the value of specific units (see as a recent example, Andalaft *et al.*, 2003). Another example of an operational-level problem is the cutting block problem. The cutting block problem schedules harvesting of individual stands of a forested area. The adjacent stands harvested at the same time represent a harvest block. The major constraint is that the size of any given harvest block is limited to a predefined area limitation (see, e.g., McDill *et al.*, 2002; Murray *et al.*, 2004). Such spatially explicit details are important in making operational decisions, but are not necessary in making long-term forest management decisions at the strategic level.

Defining decision-making models for strategic and operational levels makes a great deal of sense. If no spatial constraints exist to constrain the positioning of activities or the amount of activity that takes place in a set of adjacent stands, then it is relatively straight forward to determine where strategic activities should be allocated at the operational level. This is because of the fact that if no spatial condition restricts the level of activities in a given region of the forest, then harvest activities of the strategic level can be concentrated in the areas of the forest that can be extracted at the least cost. Unfortunately, there are three conditions which tend to restrict activities spatially across a forest area. First, forest stands of a given age and condition are not evenly distributed across a region. Second, most forest operations have limits on the size of a given operation, so that a harvest block is restricted to be no larger than a given size and so that two adjacent harvest blocks are not scheduled to be harvested in the same decade. Finally, there exist constraints that may limit the amount of activity in a given area (e.g., watershed) so that specific environmental conditions may be maintained (e.g., limiting the amount of sediment that may reach riparian zones). Over time, a number of environmental guidelines have been added to forest management protocols in the U.S. Forest Service. Such standards and guidelines are often translated into constraints which limit the amount of activity that can take place in each planning unit. Such constraints are often tracked at a level of detail and spatial resolution that does not exit in the strategic-level model. Consequently, it is possible that a solution generated in a strategic-level model may not be feasible on the ground, as the level of activities generated in the strategic level for specific strata-age classes concentrate too much of the total activity in specific planning units and simply violate standards applied to a level of spatial detail not present in the strategic model (Church and Barber, 1992). Thus, the two levels of modeling, strategic and operational, cannot be modeled independently of each other. That is, the decisions made at the strategic level need to be translated to appropriate feasible targets at the operational level (Nelson *et al.*, 1991). This is the task of a tactical model. A second type of tactical model has also emerged over the last decade. This model is oriented towards identifying the impacts on forest lands of maintaining specific levels of biodiversity protection. For example, how much of the landscape would be needed to protect 40% of the habitat that supports a given species, in such a way that such protected areas are as connected as possible (see, e.g., Nalle *et al.*, 2002; Fisher and Church, 2003). On a strategic level, we might estimate how much of a habitat is necessary to support a population of a given size. Tactically, we need to determine possible arrangements across the landscape that provide this level of protection. Both types of tactical-level models will be described in greater detail in subsequent sections of this chapter.

2 STRATEGIC MODELING: A CLASSIC EXAMPLE

There have been a number of strategic models that have been developed (see, e.g., Buongiorno and Svanqvist, 1982; Johnson and Stuart, 1986; Weintraub *et al.*, 1986; Sleavin and Camenson, 1994). As described earlier, these models tend to model forest operations over a long period of time. The emphasis tends to be on the different types of activities and management alternatives, where planning units tend to be large, represented by composite totals of stand types and age classes. For example, Buongiorno and Svanqvist (1982) model the Indonesian forest industry, where each forest region is represented as a point source, with maximum levels of harvest productivity per year. Their model focuses on the flows of logs to sawmills and ports, to minimize the costs of transport of logs and products and minimize the costs of port and plant operations while meeting desired levels of demand for logs and other products. Weintraub *et al.* (1986) define a strategic model for forest industries where stands in a forest region are represented as "macro stands." Each macro stand is an aggregation of like stands in terms of age and class owned by a forest company in the same region. Each macro stand is represented by a number of management alternatives, where each alternative yields certain outputs per acre per time period. The overall objective is to maximize net present revenue, based on decisions of stand management, mill investment and operations, acquisition of additional forest stands, product sales, and debt loads. Just as in the Buongiorno and Svanqvist model, the forest area is described as a relatively small number of large area sources. The reason for this is that the characterization of the stands in detail is not necessary when attempting to optimize investment within the context of mills, sales, overall production levels, etc. Perhaps the most widely known strategic model in forestry is FORPLAN. The FORPLAN model represents an extension of the earlier model of Navon (1971) called Timber-RAM. FORPLAN represented one of the first models to characterize forest management in terms of multiple use, where timber production represented one of the outputs. FORPLAN (along with extended

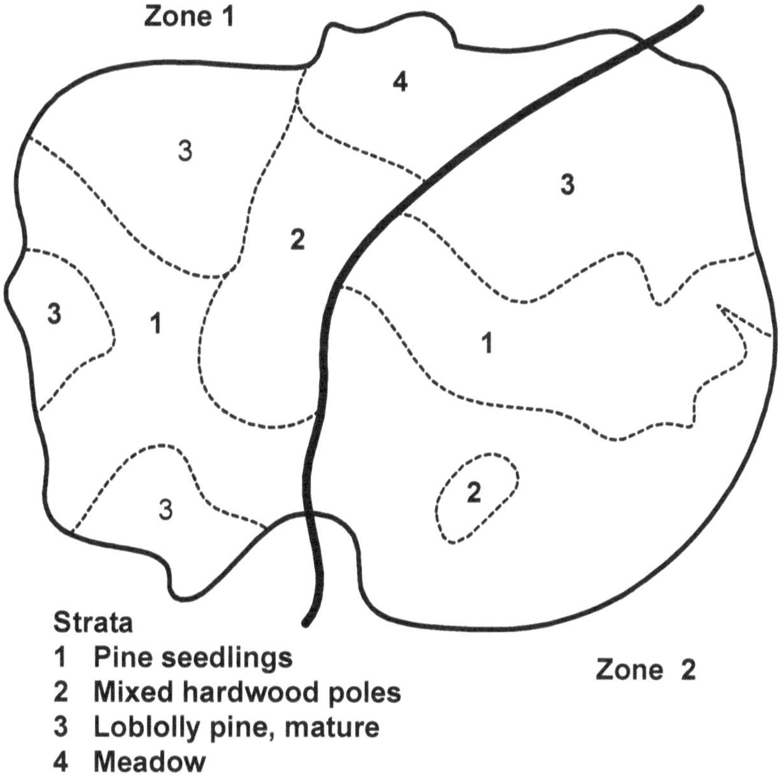

Figure 1. A forest region of two zones.

versions like SPECTRUM) became one of the most widely used models in the USA for forest management by government agencies and private companies (Kent *et al.*, 1991). We will use the FORPLAN model to demonstrate the problems that one must over come in applying the results of a strategic model. To describe FORPLAN, consider the forest region depicted in Fig. 1. This region is divided into two zones and there are four different strata-age classes distributed across the two zones. In each zone, the stands of the same type are aggregated into macro stands. For example, zone 2 contains three macro stands, 1, 2, and 3. Zones represent large divisions of a forest and are used to restrict operations so that management alternatives selected for a given zone are all compatible. For example, if zone 1 is allocated to wilderness, then macro stands within zone 1 are restricted to management alternatives that are compatible to a wilderness zone designation. Thus, macro stand management alternatives are coordinated to selected zone management strategies. We can use this spatial arrangement to define a strategic model for forest management using the following notation:

s is the index of analysis areas (or strata) represented as macro stands
z is the index of management zones in the forest
m is the index of coordinated allocation of choices (CAC)
n is the index of timing choices for zones
p is the index of prescription sets
t is the index of time periods
k is the index of timing choice associated with a prescription on an area
p_i is the prescription i of the prescription set indexed as p
$\hat{X}_{sp_ik}$ is the acres allocated to timing choice k of prescription i in the set p defined for analysis area s
Y_{zmn} is the proportion of zone z allocated to timing choice n of the CAC m
c_{zmn} is the contribution to the objective function of timing choice n of the CAC m for zone z
$\hat{c}_{sp_ik}$ is the per acre contribution to the objective function of timing choice k of prescriptions i in set p defined for analysis area s
U_p is the set of prescriptions in set p
a_{zmnspt} is the acres made available in time period t to prescription set p for analysis area s if timing choice n of CAC m is chosen in zone z
W_{p_it} is the set of timing choices k of the prescription i in set p that have their first management action in period t
$T_{sp(t-1)t}$ is the number of acres made available for but not allocated to analysis area s under prescription set p in period $(t\text{–}1)$.

2.1 Coordinated Allocation of Choices (CAC)

$$\text{Maximize } Z = \sum_z \sum_m \sum_n c_{zmn} Y_{zmn} + \sum_s \sum_t \sum_{i \in U_p} \sum_{k \in W_{p_i t}} \hat{c}_{sp_i k} \hat{X}_{sp_i k}, \quad (1)$$

subject to

$$\sum_m \sum_n Y_{zmn} = 1 \quad \forall z, \quad (2)$$

$$-\sum_z \sum_m \sum_n a_{zmnspt} Y_{zmn} + \sum_{i \in U_p} \sum_{k \in W_{p_i t}} \hat{X}_{sp_i k} - T_{sp(t-1)t} + T_{spt(t+1)} = 0$$

$$\forall\ s, p, t, \quad (3)$$

$$0 \leq Y_{zmn} \leq 1 \quad \forall z, m, n, \quad (4)$$

$$0 \leq \hat{X}_{sp_i k} \quad \forall s, k, p_i,$$

$$0 \leq T_{sp(t-1)t} \quad \forall s, p, 1 < t < T,$$

The model given here features two types of management decision variables, X and Y, where X represents decisions associated with the prescriptions and timing choices on given stands and Y represents zone decisions and timing choices. Such a model can be used to coordinate harvest activities within a zone as well as to eliminate harvest activities within a zone, depending upon the zone designation. The objective involves maximizing the return or value associated with the zoning and stand-based decisions. Constraint (2) restricts each zone to be allocated to a management strategy and timing choice. Constraint (3) tracks a stand allocation within a zone, and restricts activity to be compatible to the zone management strategy. Acres of a stand that are not allocated to a given prescription in time period t are carried forward to time period t+1 for subsequent allocation as long as it falls within the compatible set of prescriptions p, given the zonal designation. Altogether constraint (3) ensures that the sum of the allocations to a stand cannot exceed the area of that stand. Constraints (4) specify the appropriate bounds on the decision variables. Additional constraints are often added to the earlier model; the most common of which is a constraint that specifies that the yield of timber over time does not decline. A second type of constraint is often added to reflect environmental restrictions. For example, harvesting activities disturbs the soil, and some of that ends up as sediment in the streams. Sediment yields from harvested stands within a given zone or across the entire region can be limited by a constraint for each time period, so that the sediment/erosion is controlled over time stream degradation is limited. Such restrictions are often added to represent each standard and guideline restriction of the U.S. Forest Service.

For all intents and purposes, the CAC model became the model of choice for a number of forest applications in the USA. The major problem with the CAC model in application involved the notion that zonal designations should be discrete, that is, 0 or 1. To use the model in this fashion required a mixed integer—linear solver, which was not generally available to forest analysts. Thus, the full capabilities of this type of model were somewhat limited in application.

3 TACTICAL MODELING

If a solution to the CAC model specifies that 500 acres of mature Loblolly pine are to be harvested from zone 1 in time period 1, it does not specify which 500 acres, because all mature Loblolly pine is represented as a composite total or macro stand within the zone. To show that this creates a problem in knowing exactly where to apply the treatment, consider the detail given in Fig. 2. Figure 2 depicts the same region as that given in Fig. 1 except that each zone is further divided into two subunits. These subunits may represent planning areas like small watersheds. The watersheds are defined so that constraints such as the erosion constraints can be applied at the appropriate scale. That is, even though the CAC model may have a sediment constraint, it covers a larger area than the actual watershed unit that the constraint is applied when developing an operation plan. One approach to fix this dilemma is to construct greater spatial detail in the strategic model. Unfortunately, this cannot be done without pushing the size of the strategic model beyond what can be easily solved. Thus, the real crux of the problem is that the erosion constraint cannot be constructed at the appropriate level of spatial detail in the strategic model so that the condition can be guaranteed to hold when developing an operational plan. The real problem is just where should the 500 acres of treatment be placed so that the erosion condition is met at each of the appropriate watershed units. What is needed is a method of translating activities generated in the strategic model to appropriate spatial units so that spatial constraints such as the erosion constraint can be maintained when working at the operation level.

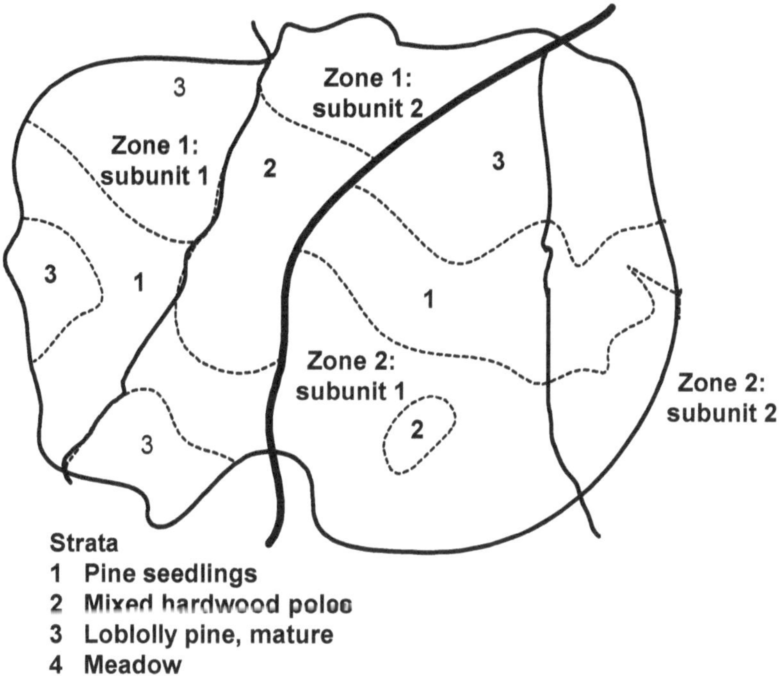

Figure 2. Forest region depicted at the zone and subunit levels.

This is the principal goal of a tactical model. Tactical models, in general, bridge strategic and operational models by:

1. Translating decisions made at the strategic level from large areas and macro stands to operational planning areas, such as small watersheds and forest tracts, appropriate for tracking and maintaining standard and guideline conditions
2. Attempting to meet targets and goals generated at the strategic level of analysis

The tactical model represents the details of a zone with greater spatial definition. The tactical model, however, does not represent all of the issues that an operation-level model might involve. Even though the treatments are assigned to smaller spatial units, like small watersheds in the tactical-level model, decision making does not focus on individual stands as in operational models. Thus, conditions such as block size and adjacency are not modeled explicitly at the tactical level. The task of the tactical-level model is simple, attempt to translate decisions to smaller spatial units, so that a feasible operation plan will likely result, or so that the strategic model solution can be tested for "feasibility" when taking into account a greater level of spatial detail. In fact, the tactical model presented later was first developed, because forest analysts started to think that their simple schemes for prorating decisions to smaller spatial units led to problems at the operation level (Church and Barber, 1992). It makes sense to use a simple proportioning rule as a tactical model, as long as it is feasible. To show how the proportioning process can be accomplished, we first assume that each zone z of the strategic problem have been further subdivided into subunits u, called planning units or analysis areas. From forest data, we can determine the amount of a given strata-age class, s, in each zone and subunit. These values can be represented as:

a_{zs} is the number of acres of strata s in zone z
a^u_{zs} is the number of acres of strata s in zone z that are in subunit u.

From the strategic model, we know:

θ_{zsp_ik} is the number of acres of strata s in zone z that are to be treated using prescription i in set p with timing choice k.

At the tactical level, we have to determine:

$\theta^u_{zsp_ik}$ is the number of acres of strata s in subunit u of zone z that are to be treated using prescription i in set p with timing choice k

We can simply prorate activities in a zone to the smaller subunits u. This can be represented in a general form as

$$\theta^u_{zsp_ik} = \theta_{zsp_ik} \frac{a^u_{zs}}{a_{zs}}, \tag{5}$$

As an example, if 35% of strata s of zone z falls within subunit u, then 35% of each treatment activity for strata s in the zone is assigned to subunit u. Each prescription has a corresponding set of activities and scheduled times. This means that from the definition of the prescription, we can then compute just in which period each activity occurs (Church and Barber, 1992):

$$H^u_{zsp_it} = \alpha^t_{zsp_ik} \theta^u_{zsp_ik}, \tag{6}$$

where

$H^u_{zsp_it}$ is the acres treated in strata s in subunit u of zone z with prescription i in set p in period t *and*

$$\alpha^t_{zsp_ik} = \begin{cases} 1, & \text{if prescription } i \text{ with timing choice } k \text{ begins in period } t \\ 0, & \text{otherwise.} \end{cases}$$

Suppose, for example, that the prescription represents a thinning in period t, then $H^u_{zsp_it}$ represents the number of acres of strata s in sub-unit u in zone z that are thinned in period t. We can use this proportioning process to translate all activities for each strata to the subunits level. Since we can determine the level and type of activity, the time at which each activity takes place, and the strata at which it is applied at the subunit level, we can model the impacts at the subunit level. For example, assume that the impact of concern is sediment generation in time period 1. We can estimate the level of sediment generation in given subunits for period 1 as:

$$\sum_s \sum_p \sum_{i \in U_p} \beta_{sp} H^u_{zsp_i1}, \tag{7}$$

where

β_{sp} is the amount of sediment generated under prescriptive action p per acre of strata s.

We can test to see if this level of sediment generation is within acceptable limits for subunit u, by comparing the generation level to the maximum acceptable level of sediment generation:

$$\sum_s \sum_p \sum_{i \in U_p} \beta_{sp} H^u_{zsp_i1} \leq \Gamma_u, \tag{8}$$

where

Γ_u is the upper limit on sediment produced within subunit u.

If the condition holds then sediment generation is within acceptable limits. In general, we can structure constraints for any standard and guideline, v, in the following manner:

$$\sum_s \sum_p \sum_{i \in U_p} \beta_{spv} H^u_{zsp_i1} \leq \Gamma_{uv}, \tag{9}$$

where

Γ_{uv} is the upper limit for impact on subunit u for standard or guideline v and β_{spv} is the amount of impact type v produced by the prescriptive action p on an acre of area s.

We can further expand the earlier conditions for any time period. If a simple proportioning scheme meets all of the additional threshold constraints, then it represents a feasible tactical plan. However, if threshold conditions are not met, then a simple proportion plan does not yield a feasible tactical plan. The U.S. Forest Service used simple proportioning schemes in a number of forests, assuming that such conditions were met. In 1991, this assumption was questioned and this model was developed (Church and Barber, 1992). A software program called visualization of the implementation process (VIP) was developed to map each planning unit, where each unit was colored based upon the level of feasibility for the most constraining condition in each planning unit. For example, assume that for unit u, that the most constraining condition was the sediment limitation. The unit is then colored based upon the slack of that constraint. If a considerable amount of slack was present, then the unit was colored green. If the constraint was barely met, then the unit was colored a pale yellow. If the constraint was violated, but only by a small amount, then the unit was colored a pale red. If the sediment threshold constraint was violated by a large amount, then the unit was colored a bright red. Intermediate shades of green and red were used for intermediate levels of constraint slack or violation. Thus, the VIP program could show the results of simple proportioning across the region. In most applications, the VIP program demonstrated that past proportioning schemes led to standards violations at the operational level. The color "red" appeared quite often in application. If simple proportioning schemes did not yield feasible solutions at the tactical level, then how can treatments at the strategic level be translated to the tactical planning unit and still meet threshold constraints? This question led to the development of a set of tactical level models, of which we will describe one.

Understanding that simple proportioning rules often produced activity levels that violated threshold conditions was an important step, but equally important was the development of an approach to identify feasible tactical-level solutions. To show how this might be accomplished, we can begin with the relationships described earlier:

$$H^{u}_{zsp_it} = \alpha^{t}_{zsp_ik}\theta^{u}_{zsp_ik} = \alpha^{t}_{zsp_ik}\theta_{zsp_ik}\frac{a^{u}_{zs}}{a_{zs}}, \qquad (10)$$

Taking the threshold constraint and substituting for the value of $H^u_{zsp_it}$ yields:

$$\sum_s \sum_p \sum_{i\in U_p} \sum_k \beta_{spv} \alpha^1_{zsp_ik} \theta_{zsp_ik} \frac{a^u_{zs}}{a_{zs}} \leq \Gamma_{uv}. \tag{11}$$

This equation fixes the proportioning in terms of the ration of the areas, a^u_{zs} and a_{zs}. If we allow for flexibility, then we can define this ratio as a variable rather than a fixed quantity in the following manner:

$X^u_{zsp_ik}$ is the fraction of analysis area *s* in sub-basin *u* of zone *z* that is assigned prescription *i* in set *p* with timing choice *k*.

Using this variable, we can restructure the threshold constraint as:

$$\sum_s \sum_p \sum_{i\in U_p} \sum_k \beta_{spv} \alpha^1_{zsp_ik} \theta_{zsp_ik} X^u_{zsp_ik} \leq \Gamma_{uv}. \tag{12}$$

This new condition allows for the proportioning scheme to vary from the fixed ratio. We can build a tactical-level model using this structure as follows (Church *et al.*, 2000):

3.1 Bridging Analysis Model A (BAM-A)

$$\text{Minimize } Z = \sum_z \sum_s \sum_i \sum_k \sum_u \left(S^u_{zsp_ik} + R^u_{zsp_ik} \right), \tag{13}$$

Subject to

$$\sum_s \sum_p \sum_{i\in U_p} \sum_k \beta_{spv} \alpha^1_{zsp_ik} \theta_{zsp_ik} X^u_{zsp_ik} \leq \Gamma_{uv} \qquad \forall u, v, \tag{14}$$

$$\sum \theta^u_{zsp_ik} X^u_{zsp_ik} \leq \theta_{zsp_ik} \qquad \forall z, s, p_i, k, \tag{15}$$

$$\sum_p \sum_{i\in U_p} \sum_u \sum_k X^u_{zsp_ik} \leq 1 \qquad \forall z, s, \tag{16}$$

$$X^u_{zsp_ik} - R^u_{zsp_ik} + S^u_{zsp_ik} = \frac{a^u_{zs}}{a_{zs}} \qquad \forall z, u, s, k, p_i, \tag{17}$$

$$0 \leq X^u_{zsp_ik} \leq 1 \qquad \forall z, u, s, k, p_i, \tag{18}$$

$$0 \leq R^{u}_{zsp_ik},$$

$$0 \leq S^{u}_{zsp_ik},$$

where

$R^{u}_{zsp_ik}$ is the amount that assigned activity exceeds area proportion

and

$S^{u}_{zsp_ik}$ is the amount that assigned activity is under area proportion.

The objective of BAM-A is to allocate strategic level prescriptions so that threshold conditions are met at the tactical level where the allocation is as close as possible to the simple proportioning scheme. Constraints (14) ensure that all activities meet the standards and guideline threshold conditions applied at the subunit level. Constraints (15) ensure that the strata assigned to prescription p and timing choice k in a zone is the upper limit on what can be assigned to that strata (using same prescription and timing choice) among the subunits of that zone. This means that BAM-A will not substitute acreage use on one strata with another type of strata, nor will it substitute treatment of acreage of the same strata over different zones (Church *et al.*, 2000). Constraint (16) restricts the sum of treatments on a given strata in a zone to be less than or equal to the area of that strata in the zone. Constraint (17) defines the amount of the level of deviation between the solution and a simple proportioning ratio. Constraint (18) imposes appropriate conditions on the decision variables.

BAM-A attempts to translate all strategic decisions to the tactical level and maintain all threshold conditions among smaller spatial units. Unfortunately, BAM-A is very restrictive and it allows little flexibility in substituting activities to meet strategic outputs or goals. It is designed so that all activities on a given strata remain on that strata, rather than substituting activities on various strata to efficiently meet the strategic model output targets (Church, 2001). We can structure a similar model to BAM-A which allows greater flexibility in determining the type and location of the activities as long as the sum of the outputs reaches the level of the outputs given in the strategic model. This model has been dubbed BAM-B. Over, time a number of bridging models have been developed. Alternate forms of the BAM models have been applied where the objective was to maximize the smallest level of slack across all planning units for all threshold constraints (Church *et al.*, 2000). This form of BAM model attempts to create a solution that does not have binding environmental constraints.

The bridging models described earlier represent one style of tactical-level modeling. Weintraub *et al.* (1986) describe an alternate form for tactical-level analysis. In fact, Weintraub *et al.* were the first to succinctly describe the need for a so-called tactical model. They described how an aggregation of stands into "macro" stands in strategic modeling can lead to results that are not feasible when tracking constraints at the operational level. The importance of tactical models is founded on the fact that environmental constraints may prevent strategic solutions from being easily translated to smaller planning areas in preparation for the application of operational models. In fact, one of the basic issues in strategic modeling is the extent to which aggregation is required to make a large-scale planning model operational. Cea and Jofre (2000) have also proposed a strategic-tactical model linkage, whereby the tactical model can be used to support and test strategically derived outputs. They also discuss the problem introduced by "macro" stands, when harvesting actually takes place at the stand level. Cea and Jofre's (2000) tactical model characterizes the forest at the stand level and optimizes road building, silvicultural activities, and transport. Their "tactical-level" model appears to be similar to many operational models that have been developed. The main purpose of a tactical-level model is to test the potential feasibility of a strategic model solution, within a finer mesh of constraints written at the appropriate level for application. That is, the greatest concern at the strategic level is the question of feasibility. Carroll *et al.* (1995) have also presented an approach which attempts to make a FORPLAN solution feasible associated with adjacency conditions. This tactical model is based upon a greedy/random stand selection method which attempts to create a feasible set of stands that meet harvesting goals. All of these examples really address a simple question: namely can a specific strategic solution be applied at the operational level and produce the appropriate amounts outputs and services, or will spatial-based constraints written at the operational-level limit exclude elements of that solution. It should be somewhat obvious that the importance of the task of translating activities from the strategic level to the operational-level cannot be overstated.

4 TACTICAL MODELING FOR BIODIVERSITY PROTECTION AND ENHANCEMENT

Another type of tactical-level model has begun to emerge when attempting to integrate biological preservation with traditional forest management goals. In public forests in North America, management objectives have changed over the decades, moving from managing for multiple uses, to managing for preservation and protection. For example, in California the U.S. Forest Service has been directed by the courts to reduce fuels, lower the risk of catastrophic fires, and protect natural habitat. Because of this, harvesting activities on forest service lands has been reduced to very low levels. The main objective is to identify those areas that should be thinned so that the severity of possible fires can be reduced and habitats containing endangered or threatened species can be protected as much as possible. Thus, the typical activities focused on harvesting and road building have been replaced by improving forest health and species protection. This has caused tension between timber companies who have relied on harvests from public lands, public agencies, and special interest groups. This has also led to the development of a different type of tactical model. Whereas, the traditional tactical model links strategic-based harvesting on macro stands to operational activities such as harvesting and road building, the new class of tactical models support planning for forest species preservation, including birds and animals (see, e.g., Hof and Bevers, 2000).

The underlying premise of much of the change in forest management over the past decade has been based upon what would it take to protect threatened species (such as the California spotted owl)? What changes in land management would be necessary to keep species from being considered at risk? If a forested region is managed primarily for its natural resources, then it might be possible to protect species by concentrating protection activities in specialized management areas and rely on the rest of the natural landscape for additional habitat values and connectivity. This premise formed the basis for the Biodiversity Management Area Selection (BMAS) model developed as a part of the Sierra Nevada Ecosystem Project (SNEP) (Davis *et al.*, 1996).

We can define biodiversity management areas as (BMAs) specially designated areas with an active ecosystem management plan in operation whose purpose is to contribute to regional maintenance of native genetic, species, and community levels of biodiversity. Economic activities are not necessarily precluded, but they are subordinate to the goal of biodiversity protection. The major underlying problem is to identify which areas should serve as BMAs in order to protect biodiversity in the region (Davis *et al.*, 1996). Before such a problem can be solved, it is necessary to define exactly what the basic unit of land is for biodiversity selection. For the SNEP study, it was decided to utilize small watersheds as the basic planning unit, although the model described later can easily handle planning units of other types. The choice was predicated on the fact that a small watershed represents a contiguous unit of land interrelated with a drainage and riparian area.

The major question is which watershed units should be chosen to help protect those elements which were found to be at risk. At the strategic level, an estimate is made in terms of the amount of habitat necessary to protect threatened elements. For example, this may be 15% of the existing range of a species. Such targets might be defined through the use of population viability models, whereby an estimate of what it takes to support a desired population is estimated, or it may be a politically defined goal. Since habitat or threatened elements are distributed unevenly across a landscape, the amount of land and the specific units that are needed to efficiently protect elements at risk may differ depending upon the aforementioned goals. It is important to make efficient choices so as to keep the costs as low as possible and keep the land under biodiversity management at a manageable and politically feasible level. This is the goal of the BMAS model. We can define this model using the following notation:

k , K is the index and set of elements at risk

j , J is the index and set of planning units

MIN_k is the minimum area containing element k that is needed within BMA

HD_j is the human density of planning unit j

PPI_j is the the density of public-private land interface in unit j

PLO_j is the the percentage of private ownership for unit j

A_j is the area of unit j

R_j is the percentage of area in unit j that is impacted by roads

a_{jk} is the area in unit j which contains element k

w_l is the weight attached to term l in the objective

$$x_j = \begin{cases} 1, & \text{if watershed unit } j \text{ is selected for a BMA.} \\ 0, & \text{if not} \end{cases}$$

We can now formulate the BMAS model in the following manner:

BMAS:

$$\text{Minimize} \quad Z = \sum_{j \in J} \left(w_1 A_j + w_2 HD_j + w_3 R_j + w_4 PPI_j + w_5 PLO_j \right) x_j, \tag{19}$$

subject to

Ensure that enough land containing element *k* is protected in a BMA:

$$\sum_{j \in J} a_{jk} x_j \geq MIN_k \qquad \text{for each } k \in K. \tag{20}$$

Integer requirements on decision variables:

$$x_j = 0,1 \qquad \text{for each } j \in J \tag{21}$$

This formulation is very simple in both structure and size and is equivalent to a multidimensional knapsack problem. There is one decision variable for each planning unit and one constraint for each element at risk. The model is structured as a binary integer programming problem. The objective minimizes a combination of a number of different factors, each weighted by their importance. The constraint set ensures that at least a minimum amount of area containing a given element at risk can be found among the planning units chosen for BMA status. Planning units are either chosen as a whole unit for BMA status or they are not chosen at all.

The cost of selecting a watershed for BMA status is a function of a number of different terms, like population density. Each term was selected as a surrogate measure for the ease at which it might be converted into and managed as a BMA. For example, selecting areas which are owned privately are discouraged over selecting suitable watersheds which are publicly owned. Selecting a planning unit with a low population density is encouraged over selecting a planning unit with a higher population density. Other factors of concern deal with the length of road segments per unit area (road density) and the fragmentation of public and private land within a watershed (public-private interface). The greater road density and land ownership fragmentation is, the greater the perceived difficulty in managing for biodiversity. Finally, the area of a unit is important as the larger the watershed, the higher the costs of management. Each of these terms can be weighted to reflect their importance in selecting watershed units for BMA designation.

BMAS can be solved optimally by off the shelf optimization software like CPLEX. The BMAS model has also been extended to encourage clustering of selected management areas (Fischer and Church, 2003). The BMAS model is also related to the SITES model which was developed for the Nature Conservancy (Andelman *et al.*, 1999). Nalle *et al.* (2002) present cell-selection model to zone land for reserves in support of strategic goals.

These models are representative of tactical-level models for biodiversity protection. They do not, in themselves, generate an operations plan, but help to delineate which areas are zoned for strategic goals.

5 CONCLUSIONS

Forest lands comprise one of the largest natural landscapes on earth. Many of the forested areas have been managed as a natural resource by industry and governments. In the USA alone, the U.S. Forest Service manages more than 190 million acres of forest and grassland. To manage such large tracts of land, large-scale optimization has been all but required. Strategic-level decisions involve devising forest plans and goals for a long-planning horizon (e.g., 150 years or more) and at a high level of data aggregation. Unfortunately, strategic plans may not be feasible when attempting to identify just where such activity should be allocated. Tactical-level modeling helps to bridge solutions reached at a strategic level with the some of details and scale of operational planning. This chapter has described tactical level models for both harvesting-oriented operations and operations that are oriented towards ecosystem protection. Since forested areas and their riparian zones support a wide range of biota, they play a significant role in world biodiversity protection. The tactical level of analysis represents an ideal level in which to mesh both species protection and forest management, including harvesting, thinning, and fuels treatments. It is indeed difficult to represent species protection levels without significant spatial detail. It is the tactical level in which the spatial detail supports planning for conservation goals.

The defined boundaries of strategic, tactical, and operations models overlap, as researchers have attempted to add greater levels of spatial detail in all management models. This trend is supported by larger and faster computers, common enterprise-level data bases involving Geographical Information Systems, new innovative solution strategies (which include tabu search, simulated annealing, column generation schemes), and better off the shelf optimization software. As forest modelers write applications integrated with large data bases, strategic models can be executed on aggregated data whereas tactical-level models can be applied to that same data involving smaller and more detailed spatial data. This has allowed greater linkages between levels of analysis, allowing an analyst to solve almost simultaneously each level and use information from each level to refine goals and targets between levels.

REFERENCES

Anadlaft, N., Andalaft, P., Guignard, M., Magendzo, A., Wainer, A., and Weintraub, A., 2003, A problem of forest harvesting and road building solved through model strengthening and lagrangean relaxation, *Operations Research* 51: 613–628.

Andleman, S.J., Ball, I., Davis, F.W., and Stoms, D.M., 1999, SITES V: an analytical toolbox for designing ecoregional conservation portfolios, Manual prepared for the *Nature Conservancy*: 1–43.

Bare, B.B. and Field, R., 1987, An evaluation of FORPLAN from an operations research perspective, *USDA Forest Service General Technical Report RM-140*, Fort Collins, CO.

Buongiorno, J. and Svanqvist, N., 1982, A separable goal-programming model of the Indonesian forestry sector, *Forest Ecology and Management* 4: 67–78.

Carroll, B., Landrum, V., and Pious, L., 1995, Timber harvest scheduling with adjacency constraints: using ARC/INFO to make Forplan realistic, in *1995 ESRI User Conference Proceedings*, Redlands, California, USA (http://www.esri.com/base/common/userconf/proc95/prochome.html).

Cea, C. and Jofre, A., 2000, Linking strategic and tactical forestry planning decisions, *Annals of Operations Research* 95: 131–158.

Church, R. and Barber, K., 1992, Disaggregating forest management plans to treatment areas, *Stand Inventory Technologies* 92: 287–295.

Church, R., Murray, A., and Barber, K., 1994, Designing a hierarchical planning model for USDA Forest Service planning, in *Proceedings of the 1994 Symposium on Systems Analysis and Forest Resources*: 401–409.

Church, R., Murray, A., and Barber, K., 2000, Forest planning at the tactical level, *Annals of Operations Research* 95: 3–18.

Church, R.L., 2001, GIS and large-scale linear programming: evolution of a spatial decision support system for land use management, in Regional Science and Business. Clarke G. and Madden M. (eds.) Springer-Verlag, Berlin.

Davis, F.W., Stoms, D.M., Church, R.L., Okin, W.J., and Johnson, K.N., 1996, Selecting Biodiversity Management Areas, in *Sierra Nevada Ecosystem Project: Final Report to Congress Vol. 2*, University of California at Davis.

Epstein, R., Weintraub, A., Sapunar, P., Nieto, E., Sessions, J., and Sessions, B., 1995, PLANEX: a system for optimal assignment of harvesting, in *Proceedings Harvest Planning Conference*. LIRO, New Zealand.

Fischer, D. and Church, R.L., 2003, Clustering and compactness in reserve site selection: an extension of the Biodiversity Management Area Selection Model, *Forest Science* 49: 1–11.

Hof, J. and Bevers, M., 2000, Direct spatial optimization in natural resource management: four linear programming examples, *Annals of Operations Research* 95: 67–81.

Hof, J.G. and Pickens, J.B., 1987, A pragmatic multilevel approach to large-scale renewable resource optimization: a test case, *Natural Resource Modeling* 1: 245–264.

Johnson, K.N. and Stuart, T., 1987, *FORPLAN Version 2: Mathematical Programmer's Guide*, Land Management Planning Systems Section, U.S. Forest Service, Washington, D.C.

Johnson, K.N., Stuart, T., and Crimm, S.A., 1986, *FORPLAN Version 2: An Overview*, Land Management Planning Systems Section, U.S. Forest Service, Washington, D.C.

Kent, B., Bare, B.B., Field, R., and Bradley, G., 1991, Natural resource land management planning using large-scale linear programs: the USDA Forest Service experience with FORPLAN, *Operations Research* 39: 13–27.

McDill, M.E., Rabin, S.A., and Braze, J. 2002, Harvest scheduling with area-based adjacency constraints, *Forest Science* 48: 631–642.

Murray, A.T., Goycoolea, M., and Weintraub, A., 2004, Incorporating average and maximum area restrictions in harvest scheduling models, *Canadian Journal of Forestry* 34: 456–464.

Nalle, D.J., Arthur, J.L., and Sessions, J., 2002, Designing compact and contiguous reserve networks with a hybrid heuristic algorithm, *Forest Science* 48: 59–68.

Navon, D., 1971, Timber AM...A Long Range Planning Method for Commercial Timber Lands under Multiple Use Management. Research paper PSW-70. U.S. Department of Agriculture, Forest Service, Pacific Southwest Forest and Range Experiment Station, Berkeley, California.

Nelson, J., Brodie, J.D., and Sessions, J., 1991, Integrating short-term, area-based logging plans with long-term harvest schedules, *Forest Science* 37: 101–122.

Sleavin, K.E. and Camenson, D., 1994, *Spectrum, Proceedings of the 1994 Symposium on Systems Analysis in Forest Resources*. Pacific Grove, Ca, September.

Weintraub, A. and Bare, B.B., 1996, New issues in forest land management from an operations research perspective, *Interfaces* 26: 9–25.

Weintraub, A. and Cholaky, A., 1991, A hierarchical approach to forest planning, *Forest Science* 37: 439–460.

Weintraub, A., Guitart, S., and Kohn, V., 1986, Strategic planning in forest industries, *European Journal of Operational Research* 24: 152–162.

Weintraub, A., Epstein, R., Morales, R., Seron, J., and Traverso, P., 1996, A truck scheduling system improves efficiency in the forest industries, *Interfaces* 26: 1–2.

Chapter 18

HARVEST OPERATIONAL MODELS IN FORESTRY

Rafael Epstein[1], Jenny Karlsson[2], Mikael Rönnqvist[3], and Andres Weintraub[1]
[1]*Department of Industrial Engineering, University of Chile, Santiago, Chile;* [2]*Division of Optimization, Linköping University, Sweden;* [3]*The Norwegian School of Economics and Business Administration, Bergen, Norway*

Abstract Harvest operations have provided many important operations research (OR) applications. The harvesting process incorporates decisions on areas to harvest, how to buck (or cross cut) trees to obtain demanded products by length and diameter, how to locate harvesting machinery, how to build and maintain roads to haul away logs, and how to distribute and stock logs. We discuss how OR has impacted successfully on these decision processes.

Keywords: Forest supply chain , harvesting, short-term planning

1 INTRODUCTION

This chapter deals with operational decisions that are related to the actual operations that need to be carried out in the harvesting process. Strategic and tactical decisions discussed in other chapters of this Handbook deal with long and medium range planning processes. As in other areas most of the strategic and tactical decisions do not lead directly to actual activities being carried out. This happens at operational level, basically related to harvesting activities. Whereas in other areas, operational decisions are typically very short term, forestry operational decisions can have planning horizons of up to several months.

We can distinguish several aspects at operational level

1. Harvesting decisions relating to which areas will be harvested in the planning horizon.

2. Bucking (or bucking control) decisions carried out at the stands (or smaller cutting units) or at destinations where logs are bucked.
3. The location and use of harvesting machinery.
4. Transportation of logs to primary destinations.
5. Handling of logs at primary destinations.
6. Road building or upgrading decisions to secure log supply throughout the planning horizon.
7. Downstream operations, which relate to the supply chain.

In this chapter we deal with most of these aspects. Transportation and handling of logs at primary destinations are taken care of in other chapters of this Handbook.

The basic operational activities we consider are related to harvesting. A typical situation will be where demand is known for the next 6–16 weeks, in terms of logs of well-defined quality, length, diameter and a delivery date. The scheduler also has knowledge of standing timber that can be harvested in this time frame and that is mature for harvesting and accessible through existing primary roads. Inventory models allow reasonably good estimates of yields of specific logs given bucking or cutting patterns used. Harvesting is usually carried out using harvesters and skidders for flat areas and towers or cable logging for steeper slopes. Transportation is usually done with trucks, although rail and even water barges are also used at times. Costs and productivity of harvesting machinery and transportation means are known.

The problem then can be stated as: which cutting units should be harvested in each time period (e.g. week), which machinery should be used, which secondary access roads need to be built, how should transportation be scheduled, and how should trees be bucked in order to satisfy the demand efficiently. This problem is particularly relevant to private companies handling plantations or regular forests. Similar decisions are also involved when native forests are harvested.

Environmental issues are also present and constrain the management options. Care needs to be taken in preserving soil and water quality in particular. These decisions need to be coherent with tactical plans which define areas to be harvested along a longer time horizon. This means care in not using heavy harvesting machinery on fragile soils, how to build roads to avoid erosion and other similar measures.

2 HARVESTING AND BUCKING

The units to harvest need not be identical to stands (homogeneous areas in terms of tree species and age, site quality) which are the basic units defined at strategic and tactical level. Cutting units are defined based on logistic considerations and environmental norms, for example, adjacency constraints for harvesting as detailed in another chapter. LP models have been proposed to solve this problem (Burger and Jamnick, 1991). An LP model has advantages in matching supply of standing timber with demand for specific products as it is efficiently solvable and accurately represent the actual problem. In most cases, the handling of tree bucking has been incorporated into the models. One important aspect is how to incorporate bucking patterns. A bucking pattern is defined by the length of each log and a minimum diameter at the smaller end. A typical pattern may consist of two larger pieces 12- and 8-m long, 30- and 27-cm minimum diameter, respectively, two medium pieces 4-m long and minimum diameter of 24 and 20 cm and a last thin piece to use for pulp. The bucking process tries to obtain as many high-value logs as possible in descending order, as the market value of larger, higher diameter logs is significantly higher.

There is an exponential number of bucking patterns, so it is not possible to include all patterns explicitly in an LP model. Mc Guigan (1984) proposed to include a given set of patterns generated externally. Eng *et al.* (1986) proposed a Dantzig Wolfe decomposition approach where a sub-problem generated the patterns. Column generation was proposed in Mendoza and Bare (1986), Dynamic Programming in Briggs (1989) and heuristics in Sessions *et al.* (1989).

In terms of applications, an LP model has been in use in New Zealand (Garcia, 1990). An LP model, using a branch and bound approach to generate bucking patterns, has been used by Chilean forest companies (Epstein *et al.*, 1999). It was shown that the column generation allowed to improve solutions by up to 5% over solutions based on a given set of externally generated patterns.

Bucking is often carried out at the forest, where operators measure trees to define how to buck each tree based on a specific pattern defined for all trees in the unit or with mechanized harvesters with online bucking computers where each tree is bucked independently. Bucking also occurs at processing plants, where each tree is scanned and analysed and where a best bucking pattern is found based on Dynamic Programming models. The latter requires transportation of full stems to the processing plant, for example, sawmills.

Algorithms have also been defined specifically, for bucking decisions, based on DP or Metaheuristics (Laroze and Greber 1997). In many harvesters there are built-in bucking-optimization routines that given price lists for combinations of species, length and diameter, maximize the value of each tree. One issue is to dynamically adjust this price list to produce a given demand of products.

Bucking and harvesting decisions are at times taken jointly, when the best possible cutting units are selected simultaneously with cutting patterns to match a specific demand (Epstein *et al.*, 1999). On the other hand, it has been shown that good bucking and handling of logs at destinations can add significant value. This aspect is discussed in a separate chapter of this Handbook.

3 MACHINE LOCATION

Given an area to be harvested, an important short-term decision process is related to detailed use of harvesting machinery. First, trees are felled and de-branched, then, they must be brought to areas adjacent to a road to be loaded on trucks. For steep slopes, towers or cable-logging are used to bring trees up the slope to the loading area. On flat areas, skidders or forwarders are used.

Short secondary roads are built to join loading areas to main roads. Loading areas are defined by the location of towers, and for skidders, the economic need not to haul logs for distances longer than 200 m.

Traditionally, forest engineers planned machine locations manually, based on topographical maps. This process is time consuming and basically does not allow to explore alternatives. The advent of GIS systems has significantly improved this planning process. PLANS (Twito *et al.*, 1987) was introduced by the US Forest Service and a similar system PLANZ (Cossens, 1992) in New Zealand. Based on GIS information, these systems act as simulators. The decision maker indicates the desired location of harvesting machinery and the system, in a visual interactive form determines areas that can be harvested from those machines, access roads to be built and timber volumes harvested. Logger PC (Jarmer and Session, 1992) provides a physical feasibility analysis for harvesting using log cable. PLANEX (Epstein *et al.*, 2006) determines a near optimal location of harvesting machinery, the corresponding access roads, areas and volumes to be harvested combining a visual interactive approach with a heuristic algorithm. In the GIS, cells of 10×10 m are defined with information on timber availability and topography. The system, being used by firms in Chile and Colombia is a significant improvement over manual planning. This problem is a combinatorial one, combining a plant location problem (where machines act as plants and cells with timber act as customers) and a fixed charge network flow problems for the roads. So far, only moderate size problems have been solved in exact formulations (Vera *et al.*, 2003) using Lagrangean relaxation and strengthening of the LP formulation. As alternative, Tabu search has provided good solutions for large size problems in small amounts of CPU time (Diaz *et al.*, 2005).

4 HARVESTING AND TRANSPORTATION

Transportation is an essential element in operations and is also being dealt in detail in a separate chapter of this Handbook.

Harvesting and transportation planning are often combined. A typical problem case, relevant in, for example, Sweden is a harvesting planning problem where the time horizon is 6-12 months. It starts from a list or a pool of cutting units that correspond to 6-18 months of harvesting. The main decisions deal with which units to harvest and which crews to assign. The selection of cutting units strongly affects the production level of different assortments (logs defined by species, type, dimension and quality), as each unit has a particular assortment mix and related harvest volumes. It also affects the choice of crews as each crew has a machinery capable to harvest a specified average tree size and a given capacity and efficiency depending on this size. The planning also includes transportation planning and control of storage of logs in the forest and at terminals. The decisions are which volumes and assortments should be transported from cutting areas to plants, each time period. Road maintenance can also be included. In a central planning process, the demand at paper-, pulp- and saw-mills are distributed typically on a monthly level. Weather conditions may vary significantly during the year and this fact needs to be considered. Some roads are closed and some areas cannot be harvested during thawing in Spring and heavy rainy periods in Autumn. The output from this plan is a sequencing of harvest areas distributed over the year, assuring accessibility and that the monthly harvested amount corresponds to demand of the wood-processing facilities. This problem, which leads to a mixed integer linear programming (MILP) model, is studied in Karlsson *et al.* (2004).

In many cases it is important to decide the sequence in which cutting units are to be harvested. This in order to account for moving costs of equipment and teams. For such considerations there is a need to use shorter time periods, say 1 week in the planning, and a total planning period of 4-8 weeks. For this time of planning period it can also be important to consider the time logs are stored as the properties and values change. The main decisions deal with the same aspects as the annual planning problem, but each crew is given a schedule, which means a defined sequence of areas to be harvested during the planning period. Here, costs for moving equipment are considered as the actual sequencing of areas is explicit. As for the storage, more variable and constraints are needed. For example, if four discretized storage times are used, four times as many storage variables and constraints are needed. Note that this short-term problem can be considered in the context of the harvesting and bucking problems of Section 2. What is added in this case is the explicit consideration of handling of crews, transportation and storage. A model dealing with this problem is described in Karlsson *et al.* (2003).

There are many papers dealing with harvesting planning on different levels. A medium-range tactical harvest schedule model called OPTIMED is described in Andalaft *et al.* (2003). This model supports decisions concerning which stands to harvest and how much timber is needed to satisfy projected demands, and which roads are needed to gain access to the harvest areas for a total planning horizon of 2–5 years. Annual harvesting planning is found in Newham (1991), where a version of the system LOGPLAN II is described. This is an LP-based model that can be used to schedule timber harvesting and regeneration activities given available equipment, wood resources, planting stock and mill demands to minimize cost. A related annual planning problem is described in Gunnarsson *et al.* (2003). Here the problem is to decide where and when forest residues (after harvesting) are to be converted into forest fuel, and how the residues are to be transported and stored in order to satisfy demand at heating plants.

There is often a large number of cutting units which leads to a large number of binary and continuous variables. Spatial environmental constraints related to harvesting may also be present. The presence of integer variables and the size of real-world problems often lead to heuristic methods to achieve practical solution times. The issues, which have led to the combinatorial nature of some forest management problems and solution algorithms proposed for these problems, are discussed in another chapter of this Handbook.

5 ROAD UPGRADING

The forest industry is highly dependent on an efficient road network since almost all logs are transported some distance by trucks. Weather conditions vary during the year and during some periods, accessibility becomes difficult due to rains and thawing. In order to secure a continuous supply to the mills, forest companies have safety stocks of raw material. During periods of stock building an increased harvest and transportation capacity is needed. The storage of raw material outdoors involves considerable costs, due to quality deterioration. Road blocking lead to longer hauling distances and an increased transportation cost. For example, in Sweden it is estimated that about 6% of total cost to deliver timber and pulp logs originates from increased storage and transportation costs due to insufficient road accessibility. There is a trade-off between restoration cost to improve standard and accessibility and losses due to road blocking.

At present, much of the road upgrading planning is carried out manually and the planning is strongly dependent on the experience of a senior planner. It is difficult to obtain an overview of the road network, the accessible volumes or necessary restorations. The planning periods are often several years but there are operational decisions on which roads to upgrade next. New GIS systems together with national road data-bases provide opportunities for using decision support and OR in forest applications.

The first models integrating land management and road building are presented in Weintraub and Navon (1976), Kirby *et al.* (1986). Murray and Church (1995) present a model including road building and other environmental constraints. Gunn and Richards (1997) discuss the trade-off between productivity losses and road building costs for a planning horizon of 10–30 years. Consideration of road building (and other spatial aspects) in forest planning models leads to MILP models, typically difficult to solve due to the large size of realistic problems. Weintraub *et al.* (1994) presents an LP based heuristic solution procedure for a problem as presented in Kirby *et al.* (1986). Gunn and Richards (2003) present a Tabu search algorithm for their problem. Guignard *et al.* (1998) show how solvability is improved for the integrated harvest/transportation model, by use of different valid inequalities and careful selection of B&B branching priorities.

Andalaft *et al.* (2003) present a model including short-term decisions concerning the areas to harvest, the amount of timber to produce, the roads to build or upgrade for access and storage of timber. The planning horizon is a couple of years, which imply consideration of different seasons (winter/

summer). There are two different road standards, defining which season the roads are open. Harvested timber can be stocked from summer to winter. Solution strategies involve model strengthening and applying Lagrangian relaxation. Cea and Jofre (2000) present a two-level model and optimization algorithm to assist simultaneously strategic and tactical planning. The tactical planning problem (time horizon is a couple of years) includes decisions about harvesting, transportation, road construction and upgrading considering two different road standards. A two-stage solution algorithm is presented. First the road network is designed and second a linear programming model corresponding to the harvest planning is solved.

Olsson (2004) presents an MILP model, which starts from the tactical overall harvest planning and includes decisions concerning restorations of existing forest roads and transportation. The problem is to provide available harvest areas during the part of the year when only roads of the highest standard are accessible (typically 6-10 weeks year) considering transportation and restoration costs. Henningsson *et al.* (2007) develop a MILP model to support forest road restorations. This chapter includes decisions on road upgrade with transportation between harvest areas and plants. Models using both link and path flows are tested in order to get detailed description of the upgrading decisions. The models are used as a basis for a decision support system RoadOpt (Frisk *et al.*, 2006) developed by the Forestry Research Institute of Sweden.

6 THE GLOBAL HARVESTING OPERATION

We note that the described decisions in previous sections are not independent. Machine locations depend on the areas where there is a decision to harvest in the next 6-12 months. Short-term harvesting to supply demand for the next 6-16 weeks will consider where harvesting machinery is presently located and costs and times of moving these machines. Harvesting and transportation also interact; given the high costs and the capacity constraints involved in transportation.

Time frame is also a factor. We have defined short term harvesting operation planning, with a horizon of 6-16 weeks, for which successful models have been developed. However, an important aspect is decisions in daily operations. No models have been proposed so far for harvesting decisions on a daily basis. These decisions are taken anyway by field managers, and consider basic aspects such as which crews will work with which machines today, on given areas, with given bucking patterns. What needs to be done on an hourly basis?

In transportation there is an equivalent. Plan on use of transport can be made for several weeks in advance, in terms of requirements of vehicles and when timber will be moved. The models implemented deal mostly with daily vehicle routing. As described in the transportation chapter of this Handbook, this scheduling can be done on a daily basis, with corrections as needed along the day, or dispatching that can be done in real time, where tasks are assigned dynamically. There has been little work integrating in detail harvesting and transportation; both at longer horizons of weeks or on a daily, real time basis. This is an approach which needs to be worked on to improve harvest management.

7 THE OPERATIONAL FOREST SUPPLY CHAIN

There is a whole wood chain or supply chain of forest production starting from standing trees which are bucked into logs and sent to primary destinations such as sawmills, pulp plants, or plants developing other wood products. The supply chain is complex. Sawdust and other by-products in plants are used for energy. Chips from sawmills are used in other plants. Best quality boards from sawmills are used in remanufacturing for building or furniture pieces. Panels are also used in manufacturing process further down the line. And transportation plays an important role along the supply chain. In addition some tasks in the chain are managed by different organizations. Figure 1 shows a systematic forest supply chain.

The supply chain can be very extensive geographically, where final manufacturing factories may be in different continents than the original forests. It can be readily seen that the forest supply chain: forests, plants, factories, transportation should be designed in coherent way at strategic level, and planned in an integrated way at tactical level.

At the operational level few efforts have been developed to synchronize the supply chain. Again, we can consider two operational levels: one with horizon of up to 4 months and another on a daily basis. Models have been proposed to coordinate at least parts of the supply chain (Carlsson and

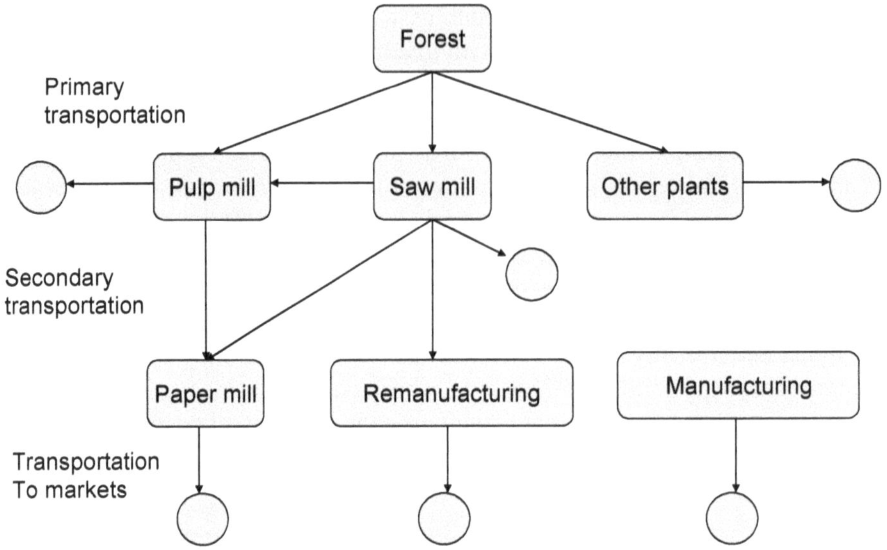

Figure 1. Illustration of the forest supply chain.

Rönnqvist, 2005), in particular harvesting and transportation. But, as noted in Weintraub and Epstein (2002) evaluating Chilean forest firms, there are areas that need stronger coordination. Sawmills and other plants have production schedules very weakly coordinated with forest supply to make sure sawmills receive the type of logs needed given their production demand. An integration of decisions, where logs supplied to each plant are coordinated with production plans and sale commitments of plants would clearly improve overall decisions. An attempt is made in Bredström *et al.* (2003). Models which consider all these decisions jointly can be built, but at present there is no strong coordination at management level.

One interesting fact in the analysis of the forest supply chain is the minimal effect of the well-known bull whip effect, where minor variations in final demand in conventional industries get significantly amplified upstream the supply chain. The reason for this low impact in forestry is due to the fact the production is based on standing timber, which provides high flexibility to produce different products, and carries basically no inventory cost. Moreover, process times in this industry are measured in weeks, so there is normally characteristic, which would deteriorate quite fast anyway. So, inventories of primary products, such as logs are usually small. Only at the level of manufactured products are stocks carried in conventional ways.enough time to produce an order from standing timber. Since variations in market demands for primary forest products can be quite readily fulfilled from standing trees, there is no need to have stocks of logs of specific

8 CONCLUSIONS

OR models have been successfully implemented by industry to support main decisions for harvesting, bucking, handling machinery, road building and upgrading and transportation space. There is still ample scope for improvement.

As technology becomes more sophisticated and cheaper, there will be a possibility of following each log from forest to destination, for example via code bars and coordinating in real-time harvesting and transportation. Developing models, information technology and communication in real-time to optimize operations along the chain may prove advantageous in the future.

Operational decisions relate to time frames of several months to 1 day. They include which units to harvest each period, how to buck trees to satisfy demand, location and use of harvesting machinery, building or upgrading secondary roads, use of stocks and transportation.

Use of OR has been successful in supporting these decisions. Application of such models in the Chilean timber industry was awarded the INFORMS Edelman Prize for best OR application in 1998. New challenges include making these applications more extensively used, integrating the whole forest supply chain, and incorporating new technology to support decision in real time.

ACKNOWLEDGEMENTS

The work of Rafael Epstein and Andres Weintraub was supported by several grants from Fondecyt and Fondef. The work by Mikael Ronnqvist was partially supported by Södra Skogsägarna and Holmen Skog.

REFERENCES

Andalaft, N., Andalaft, P., Guignard, M., Magendzo, A., Wainer, A. and Weintraub, A. 2003. A problem of forest harvesting and road building solved through model strengthening and Lagrangean relaxation. *Operations Research 51(4)*, 613–628.

Briggs, D.G., 1989. Tree value system: Description and assumptions, USDA Forest Service, General Technical Reprot PNW-239, Portland, Oregon.

Bredström, D., Carlsson, D., Lundgren, J.T., Mason, A. and Rönnqvist, M. 2004. Supply chain optimization in the pulp mill industry – IP models, column generation and novel constraint branches, *European Journal of Operational Research 156*, 2–22.

Burger, D. and Jamnick, M.S. 1991. Analysis of wood procurement strategies: Supplying multiple mills from multiple sources. Proceedings of 1991 Symposium on System Analysis in Forest Resources, Charleston, South Carolina, 3–6 March 1991, USDA Forest Service, General Technical Report PNW-239, Portland, Oregon.

Cossens, P. 1992. Planning and control of short-term log allocation in New Zealand, Integrated decision-making in planning and control of forest operations. Proceedings of IUFRO Conference, 27–31 January 1992. Christuchurch, New Zealand, New Zealand School oof Forestry University of Canterbury, Christchur, New Zealand, 46–56.

Carlsson, D. and Rönnqvist, M. 2005, Supply chain management in forestry – case studies at Södra Cell AB, *European Journal of Operational Research 163*, 589–616.

Cea, C. and Jofre, A. 2000. Linking strategic and tactical forestry planning decisions. *Annals of Operations Research 95(1–4)*, 131–158.

Diaz, A., Ferland, J., Ribeiro, C., Vera, J. and Weintraub, A. 2005. A tabu search approach for solving a difficult forest harvesting machine location problem. *European Journal of Operational Research*, Available online 20 December 2005.

Eng, G., Daellenbranch, H.G. and Whyte, A.G.D. 1986. Bucking tree-length stems optimally. *Canadian Journal of Forest Research 16*, 1030–1035.

Epstein, R., Nieto, E., Weintraub, A., Chevalier, P. and Gabarró, J. 1999, A System for Short Term Harvesting. *European Journal of Operations Research 119*, 427–439.

Epstein, R., Sapunar, P., Nieto, J., Sessions, B., Sessions, J., Bustamante, F., Musante, H. and Weintraub, A., 2006. A Combinatorial Heuristic Approach for Solving Real Size Machinery Location and Road Design Problems in Forestry Planning. *Operations Research.* 54(6): 1017–1027.

Frisk, M., Karlsson, J. and Rönnqvist, M. 2006, RoadOpt – a decision support system for road upgrading in forestry. To appear in *Scandinavian Journal of Forest Research.* 21(7): 5–15.

Garcia, O. 1990. Linear programming and related approaches in forest planning. *New Zealand Journal of Forest Science 20(31)*, 307–331.

Guignard, M., Ryu, C. and Spielberg, K. 1998. Model tightening for integrated timber harvest and transportion planning. *European Journal of Operations Research 111*, 448–460.

Gunn, E.A. and Richards, E. 1997. Optimizing stand level forest harvesting using a tradeoff objective function. In: Freid, J., Vasievich, J.M., Leefers, L. (eds.). *Proceeding of the Seventh Symposium On Systems Analysis in Forest Resources*, 28–31 May 1997, Bellaire, Michigan.

Gunnarsson, H., Lundgren, J.T. and Rönnqvist, M. 2004, Supply chain modelling of forest fuel, *European Journal of Operational Research 158*, 101–123.

Henningsson, M., Karlsson, J. and Rönnqvist, M. 2007. Optimization for forest road upgrade planning, To appear in *Journal of Mathematical Models and Algorithms.* 6(1): 3–23.

Jarmer, C. and Session, J. 1992. Logger PC-for improved logging planning. In: Schiess, P. Sessions, J. (eds). *Proceeding of the Planning and Implementing Future Forest Operations*, International Mountain Logging and 8th Pacific Northwest Skyline Symposium, 14–16 December 1992. Bellevue, Washington, USDA, Published by University of Washington, Forestry Continuing Education Department, 241–247.

Karlsson, J., Rönnqvist, M. and Bergström, J. 2004. An optimization model for annual harvest planning, *Canadian Journal of Forest Research 34(8)*, 1747–1754.

Karlsson, J., Rönnqvist, M. and Bergström, J. 2003. Short-term harvest planning including scheduling of harvest crews, *International Transactions of Operations Research 10*, 413–431.

Kirby, M.W., Hager, W.A., Wong, P. 1986. Simultaneous planning of wildland management and transportation alterations. *TIMS Studies in the Management Sciences 21*, 371–387, Elsevier Science Publisher B. V. (North Holland).

Laroze, A. and Greber, B.J. 1997. Using Tabu search to generate stand-level, rule-based bucking patterns. *Forest Science 43(2)*, 157–169.

Mc Guigan, B.N. 1984. A log resource allocation model to assit the forest industry manganer in process selection and location, wood allocation and transportation, and production planning. *APPITTA 37(4)*, 289–296.

Mendoza, G.A. and Bare, B.B. 1986. A two-stage decision model for bucking and allocation. *Forest Products Journal 36(10)*, 70–74.

Murray, A.T. and Church, R.L. 1995. Measuring the efficacy of adjacency constraint structure in forest planning models. *Canadian Journal of Forest Research 25*, 1416–1424.

Newham, R.M. 1991. LOGPLAN II: A model for planning logging and regeneration activities. Petawawa National Forestry Institute, Information Report PI-X-102, Forestry Canada, Chalk River, Ontario, 1–39.

Olsson, L. 2004. Optimization of forest road investments and the roundwood supply chain. Ph.D. Thesis, The Swedish University of Agricultural Sciences, Sweden.

Richards, E.W., Gunn, E.A., 2003. Tabu Search Design for Difficult Forest Management Optimization Problems. *Can. J. For. Res.* 33: 1126–1133.

Sessions, J., Olsen, E. and Garland, J. 1989. Tree bucking for optimal stand value with log allocation constraints. *Forest Science 35*, 271–276.

Twito, R., Reutebuch, S., McGaurghey, R. and Mann, C. 1987. Preliminary Logging Analysis Systems (PLANS), Overview, USDA. Forest Service. General Techincal Report PNW-199, Portland OR.

Vera, J., Weintraub, A., Koening, M., Bravo, G., Guignard, M. and Barahona, F. 2003. A lagrangian relaxation approach for a machinery location problem in forest harvesting. *Pesquisa Operacional 23(1)*, 111–128.

Weintraub, A. and Epstein, R. 2002. The supply chain in the forest industry: Models and linkages. In: Joseph, G., Pardalos, P.M. and Romeijn, H.E. (eds.) *Applied Optimization, Supply Chain Management Models, Applications, and Research Directions*, Kluwer Academic Publishers, 343–362.

Weintraub, A. and Navon, D. 1976. A forest management planning model integrating silvicultural and transportaion activities. *Management Science, 22(12)*, 1299–1309.

Weintraub, A., Jones, G., Magendzo, A. Meacham, M. and Kirby, M. 1994. Heuristic system to solve mixed integer forest planning models. *Operations Research 42(6)*, 1010–1024.

Chapter 19

LOG MERCHANDIZING MODEL USED IN MECHANICAL HARVESTING

Hamish Marshall
New Zealand Forest Research Institute Ltd.

Abstract Harvesting is a key component of the industrial forestry supply chain. One of the key decisions made during harvesting is how to cut the trees into logs. A number of mathematical models have been developed to optimally solve this problem. Increasingly around the world, harvesting of timber is becoming mechanized. This mechanization provides a platform for the use of state-of-the-art measurement and monitoring technologies and the application of increasingly powerful on-board computers. These technologies are now allowing these log merchandizing models to increasingly be implemented in the forest during harvesting.

Keywords: Mechanical harvesting, log merchandizing, buck-to-value, buck-to-order, forestry, dynamic programming, linear programming

1 INTRODUCTION

The harvesting process is a key component in the forestry supply chain. It is the time when a forest owner realizes a return from decades of investment. During the harvesting process many key decisions need to be made. As the entire volume of a tree is seldom merchantable (Brown, 1950), one of the key decisions is cutting the tree into sections, or segments, commonly referred to as logs. This process is often referred to as log making, log bucking or log merchandizing.

Different wood markets require different log characteristics and are willing to pay significantly different amounts (Conway, 1979). Such price differentials for different log types, means that considerable value can be lost through incorrect log-making. In New Zealand, initial value recovery studies in the early 1980s indicated that 40% of the total stem's value could be lost during harvesting operations (Murphy and Twaddle, 1986). The largest single source of loss was from log making where up to 26% of the potential value could be lost.

Mechanization of forest harvesting operations is increasing worldwide, particularly as tree size decreases. In Scandinavia, almost 100% of logging is carried out using mechanical harvesting systems (Nordlund, 1996). In other areas of the world such as eastern Canada, New Zealand, Australia and the Pacific Northwest of North America, the majority of harvesting is now carried out using mechanized harvesting systems. The drivers for this trend generally are productivity goals or labour-related issues, for example enhancing worker safety or overcoming labour shortages, and in some cases reducing environmental impacts.

In modern mechanical harvesting operations log merchandizing is normally carried out by either a harvester or a processor. A harvester is a machine that has the ability to fell, delimb, buck and sort stems. Processors can carry out the same tasks as harvesters with the exception of felling. Both these machines place their operators in completely enclosed cabs. These cabs not only provide a safe environment for the operator but they also allow powerful computers and communication systems to operate protected from the harsh harvesting environment. (Sondell *et al.*, 2002)

The modern on-board computers are capable of running a number of complex applications to control the harvester head, timber measurement and base machine functions. It controls, adjusts and monitors the power transmission and hydraulic systems. These computers can also run applications such as geographical information systems as well as optimization models designed to help in the log-merchandizing decision-making process.

2 OPTIMAL LOG MERCHANDIZING

The objective of optimal merchandizing is to section each stem into logs to obtain maximum monetary value from the resource. A stem can be cut into logs in numerous ways; with each set of logs yielding a different financial return. However, there is, in many cases one unique bucking pattern that yields the maximum value. There are two main optimal log-bucking problems; the individual tree optimization problem (buck-to-value) and the multiple trees with demand constraints problem (bucking-to-order). The objective of buck-to-value optimal bucking is to obtain from each individual stem the maximum monetary value. In buck-to-order the objective is to optimally buck the trees to maximize the monetary value at harvest unit or forest level while meeting market and operational constraints.

The value and logs cut using the optimal bucking pattern depends on the species, diameter, taper rate and quality of the stem plus the properties and relative market values of log grades being cut. When strictly bucking-to-value, value loss from log making occurs either when logs do not meet specification or when the combination of logs cut from a stem is sub-optimal. In buck-to-order, value can be lost when a product is either produced in excess or less than the market demand for that product.

Although optimizing the value produced when trees are cut into logs was thought about as early as 1913 (Bryant, 1913), the mathematical techniques to solve the problem were not researched until the early 1960s. Over the last 45 years a number of mathematical optimization techniques have been developed and used to solve the optimal log-bucking problems.

2.1 OR Techniques used in Log Merchandizing Models

2.1.1 Buck-to-Value

The most common optimization techniques that have been used to solve the individual tree-bucking problems are dynamic programming (DP), network programming, simulation and, to a lesser extent, branch and bound integer programming and linear programming. Clemmons (1966) was the first to propose the use of dynamic programming techniques to solve the buck-to-value problem. However, Pnevmaticos and Mann's 1972 paper is credited as the first detailed, published description of using DP to solve the buck-to-value problem. Pnevmaticos and Mann's approach is now considered simplistic as they defined their stage length as the minimum log length. They also calculated the stem volume using a truncated cone which, for many tree species, is an oversimplification. Log quality was dealt with in a probabilistic way.

Briggs (1980) redeveloped the Pnevmaticos and Mann formulation removing the restriction that a stage length must be integer multiples of the minimum log length. In Brigg's formulation every log length must be an integer multiple of the stage length. He also introduced the concept of a "dummy" log, which has no value; however, he did not consider any quality information.

Geerts and Twaddle (1984) developed a dynamic programming algorithm, which was implemented in a software product called AVIS; 'Assessment of Value by Individual Stems'. This formulation used the Briggs definition of a stage. The Geerts and Twaddle formulation considered stem quality in a deterministic way. Each potential log was checked to make sure that the stem qualities at that state did not violate the required log type quality. Stem shape such as sweep was dealt with by including cut zones within which a cut must be made.

The optimal bucking problem can also be solved using network programming algorithms that efficiently find either the shortest or longest path through a network. In the mid-1980s, network analysis was introduced as a potential way to solve the buck-to-value problem (Nasberg, 1985; Sessions, 1988). Sessions (1988) formulated the problem as a network where all the possible bucking points along a stem represented the nodes on a network and the arcs connecting the nodes represent the value of the log that might be cut between them. The objective of this formulation is to find the path of arcs and nodes that gives the maximum value. Sessions implemented an algorithm similar to Dijkstra's node-labelling algorithm (Sessions, 1988). Nasberg (1985) also used a longest path algorithm to solve the same network discussed above.

Many of these formulations have been implemented into software, mostly for educational and training purpose. Examples of these programs are: OSU BUCK© (Sessions, 1988), AVIS (Geerts and Twaddle, 1984), VISION (Lembersky and Chi, 1984), HW Buck (Pickens *et al.*, 1992) and others. In 1986, Weyerhaeuser reported that the VISION log bucking training and decision simulator had produced operational benefits that exceeded $ 100 million in increased profits since its implementation in 1977 (Lembersky and Chi, 1986). In 1985, VISION won the Edelman Prize from the Operations Research and Management Science Society (now know as INFORMS) of the USA. In the mid-1990's the idea of placing an optimal log-bucking algorithm onto a set of digital calipers for optimizing individual stems on a landing was commercialized by a New Zealand company (Boston, 2001).

2.1.2 Buck-to-Order

In the literature there are generally two approaches that have been taken to solve the buck-to-order problem:

- Selecting cutting instructions, either before or during the bucking process, for each tree that will produce the required volume for each product

- Finding the correct price list (in some cases the correct specification) that will be applied to the stand of trees to produce the required, volume for each product

The first published optimal bucking formulation (Smith and Harrell, 1961) was actually solving the buck-to-order problem, using linear programming. However, as Pnevmaticos and Mann (1972) stated, the Smith and Harrell linear programming formulation was limited by the requirement that all relationships be linear and by the limited number of cutting instructions available for each diameter class.

The limited number of cutting patterns, problem was solved by Nasberg (1985), Mendoza and Bare (1986), Eng *et al.* (1986) and Laroze and Greber (1997) by using a two-stage iterative formulation of the problem. The first stage, or master problem, uses linear programming to solve the constrained market problem and the second stage, or sub-problem, uses a dynamic programming or network algorithm to solve the individual tree problem. The shadow prices from the master problem were used in the second stage to generate new cutting patterns. These were then used to form new columns in the master problem using column generation techniques. This general approach is theoretically correct and computationally efficient (Laroze, 1993), but as many authors (Sessions *et al.*, 1989; Laroze, 1993, etc.) have pointed out, the solutions produced are not particularly practical as they produce large numbers of cutting instructions. Sessions *et al.* (1989) also noted that the requirement of these techniques to subdivide the stand into stem classes makes these solutions hard to implement.

The second approach does not suffer from these same problems; however, it can not theoretically guarantee that maximum revenue is gained from the bucking of the stand. Duffner (1980) has first published work on adjusting the price list in a bucking algorithm to meet market demands. There was, however, very little detail in the Duffners (1980) paper on exactly how he adjusted the prices. Sessions *et al.* (1989) developed a system to adjust prices iteratively using a shortest path algorithm to solve the sub-problem and a binary search to find the price multipliers to obtain the correct ratio of long logs to short logs. A number of other approaches have been tried, such as using an LP solution at the upper level to adjust the prices in the DP lower level, or using a heuristic to find the correct prices so the demand constraints are met in the master problem (Laroze and Greber, 1993; Murphy *et al.*, 2004). Kivenen and Unsitalo, 2002 developed a fuzzy logic controller to adjust the prices specifically for a mechanical harvester.

2.2 Models used on Mechanical Harvesters/ Processors

The potential of using optimal log-making algorithms in mechanical log-making systems was realized early in their development. Pnevmaticos and Mann noted in their 1972 (p. 26) paper that this procedure [a DP optimal bucking algorithm] can be applied either with the use of tables or a system of computerized control of the slasher.

In 1973, Ösa, a Finnish company, developed a processor called the Ösa 710. This processor head incorporated a relay-controlled system designed to calculate the optimum use for each tree. However, these systems were cumbersome and prone to breakdowns caused by vibration. Five years later, in 1978, the first microprocessor-based measuring systems appeared on Volvo's new 900 harvesters. This development caused a rapid evolution of automated measuring and bucking technology (Drushka and Konttinen, 1997).

A Swedish study showed that a change to computerized bucking could increase the value of the wood by about 4%. It concluded that there are few areas in forestry that have the potential for such large high return on investment as that of computerized bucking equipment (Berglund and Sondell, 1985). The use of optimal bucking computer algorithms on Swedish harvesters was first tried in 1986. The algorithm was a simple heuristic model, which optimized over 9-m lengths. In 1987, a range of different log-bucking techniques were used on different mechanical harvesters/processors. Many machines had simple operator selection or predetermined selection of log lengths. Some harvesters/processors used automatic bucking to length, automatic bucking to taper and automatic bucking to value (Sondell, 1987).

In 1989, Olsen *et al.* tried using the OSU-BUCK software developed by Oregon State University on a Hahn Harvester. The trials showed that a 7.5% increase in value could be obtained through the use of an optimal bucking algorithm. A study in Australia showed that by using a harvester with an optimizer enabled increased the value cut from a stand by at least 11% (Murphy *et al.* 2006).

Most of the literature on the implementation of optimal bucking on mechanical processing systems is based on Scandinavian harvesters/ processors. Modern single-grip harvesters employ either buck-to-value or buck-to-order approach (Sondell *et al.*, 2002 and Uusitalo *et al.*, 2003). There is little in the literature on what mathematical techniques the different mechanical harvest manufacturers use. It is known that Ponsse use the Nasberg's network formulation in their log-merchandizing computers (Kivinen and Uusitalo, 2003).

There are currently two main approaches to buck-to-order optimization (also known as apportionment bucking) that have been developed by the harvester manufacturers. In the 'adaptive price list' approach, the value of each log grade is changed in accordance with how well the demand for each product is being met as harvesting progresses through the stand. In the 'near optimum' approach a cutting solution is selected from the top 5 %, based on value, of the buck-to-value solutions that best fulfils the demand requirements (Uusitalo and Kivinen, 2001). The Dasa4, Timbermatic 300, Valmet and Motomit computers all use the near optimum approach; Ponsses' computer uses the 'adaptive price list' approach (Sondell *et al.*, 2002).

2.3 Implementing the Models onto Mechanical Harvesters/Processors

The on-board computers that run the optimal bucking models are connected to diameter and length measuring systems on the processing head. Modern mechanical harvesters/processors use mechanical sensors, and in some cases photocells, to measure diameter and length. The length measurements are commonly done using a measuring wheel (90 % of the time) or the feed rollers that are connected to an encoder (Andersson and Dyson, 2002; Gellerstedt, 2002). The encoder generates a fixed number of pulses each time the wheel is turned. The wheel is kept in contact with the stem either by using a spring or a hydraulic cylinder (Makkonen, 2001). The wheel is either reset using the action of the cut-off saw or in some cases using photocells located near the cut-off saw. The diameter of the log is measured using one or two potentiometers or encoders connected to the feed rollers or delimber arms (Makkonen, 2001; Andersson and Dyson, 2002). The measuring systems are connected to the on-board computers and the measuring sensors provide them with a continual stream of length and diameter measurements.

The majority of harvesters use adaptive functions for stem form prediction. These functions are able to 'learn' the mean taper in the stand as they work their way through it (Sondell *et al.*, 2002). This allows the stem to be bucked based on only partial information about stem shape. Uusitalo *et al.* (2003) state that harvester operators can apply optimal bucking in three ways:

1. Automatic bucking – if no significant changes in quality exist within the stem, it can be bucked automatically using the cross-cutting decisions from the optimization system.
2. Automatic quality bucking – changes in quality are entered into the optimization system, the optimization system then takes the quality changes into account when calculating the optimal cross-cutting decisions. The decisions are then automatically carried out by the harvester.
3. Quality bucking – In this case, bucking is carried out manually. Preselected species and log lengths or diameters (Coyner, 2004) are entered into the computer and can be assigned to 'hot keys' on the operator's controls.

When harvesting Norway spruce in Scandinavia 'automatic bucking' is commonly used as there is little variation in quality, and the value differences between lumber-quality grades are quite small. In Sweden 'automatic quality bucking' is quite popular (Uusitalo *et al.*, 2003). In other species where there is large variation in quality between trees, and the value of log grades depends heavily on quality, automatic optimization is considered inefficient and has led to economic losses (Uusitalo *et al.*, 2003; Murphy, 2003; Marshall and Murphy, 2004).

Many of the new harvesters have wireless connection to the internet, which provides a platform for better supply chain management in the forestry industry. Wireless communication between the forest machine and the company offices mean that current production can be communicated in real time to the logging company, transportation company and customer. Möller *et al.* (2003, p. 66) gave the following examples of data exchanges:

> bucking instructions (e.g. customer orders, price lists, and restrictions for different assortments), reporting of production results (volume of logs per assortments and dimensions).

In Scandinavia, Skogforsk administers StandForD, which is a common Nordic and North European Standard for Forest Data and communication (Anon., 2003). All modern harvesters in Nordic countries now operate with the same standard of open data exchange which allows production and demand data to be easily exchanged among all members of the supply chain (Möller *et al.*, 2002).

3 CONCLUSION

In the past the logging industry has concentrated its efforts on reducing costs in an attempt to maximize profits. In designing mechanical harvesting and processing machines, increasing productivity has been a key motivation for new developments. Today's machines have faster processing speed and are more reliable than in the past. Logging, however, can be one of the major areas where the potential value of a forest can be lost.

In Scandinavia, where many of the new mechanical harvesters are being developed, the forestry and forest equipment industries have invested and still are investing, considerable amounts of money and time in new hardware and software that can help obtain the optimal distribution of logs both in terms of their value and order fulfillment. However, unlike the productivity improvement technology that seems to have been successfully applied to other areas of the world, the value-maximizing technology has not been so successful. This is largely due to differences in tree species, forest types and markets.

As the level of competition in forestry increases, value maximization at the time of harvesting trees is vitally important for improving or maintaining international competitiveness. The Scandinavians have shown that mechanical harvesters/processors provide a great platform from which to better manage the forest supply chain. In many parts of the world, however, the way the technology is applied will need to be customized for different forest types and markets.

REFERENCES

Andersson, B. and Dyson, P. 2002. Evaluating the measuring accuracy of harvesters and processors. *Forest Engineering Research Institute of Canada, Advantage* 3(4), 19.

Anon. 2003. The Standard of Forest Data (StanForD). Running updates from Skogforsk at www.skogforsk.se.

Berglund, H. and Sondell, J. 1985. Computerized bucking – one way to increase value of the wood in mechanized logging systems. *Skogsarbeten Report. NR6*. 51 p.

Briggs, D.G. 1980. A dynamic programming approach to optimizing stem conversion. Ph.D. Thesis, University of Washington, 409 p.

Brown, N.C. 1950. *Logging: The Principles and Methods of Harvesting Timber in the United States and Canada*. John Wiley & Sons, New York, 418 p.

Bryant, B.C. 1913. *Logging*. Stanhope Press, Boston, MA, 590 p.

Clemmons, J.W. 1966. New techniques in individual tree description for the basis of economic decision making. *Proceedings Society of American Foresters, Annual Meeting. Seattle, Washington*: 197–200.

Conway, S. 1979. *Logging Practices: Principles of Timber Harvesting Systems*. Miller Freeman Publications, USA, 416 p.

Coyner, B. 2004. Moving from cut-to-length to cut-to-diameter. *Journal of Logging & Sawmills: Timber West*. www.forestnet.com/timberwest/archives/May_Jun_04. [Accessed on 10/04/2004]

Drushka, K. and Konttinen, H. 1997. *Tracks in the Forest*. Timberjack Group Oy, Helsinki, Finland, 254 p.

Duffner, W.W. 1980. Decision making from market to stump. *Proceedings of Weyerhaeuser Science Symposium. Tacoma, Washington, USA,* 81–95.

Eng, G.; Daellenbach, H. and Whyte, A.G.D. 1986. Bucking tree-length stems optimally. *Canadian Journal of Forest Research* 16:1030–1035.

Geerts, J.M.P. and Twaddle, A.A. 1984. A method to assess log value loss caused by cross-cutting practice on the skidsite. *New Zealand Journal of Forestry Science* 29(2):173–184.

Gellerstedt, S. 2002. Operation of the single-grip harvester: motor-sensory and cognitive work. *International Journal of Forest Engineering* 13(2):35–47.

Kivinen, V. and Uusitalo, J. 2002. Applying fuzzy logic to tree bucking control. *Forest Science* 48(4):673–684.

Laroze, A.J. 1993. Development and Comparison of Stand-Level Bucking Optimization Methods. PhD Thesis. Oregon State University, 95 p.

Laroze, A.J. and Greber, B. 1997. Using tabu search to generate stand-level, rule-based bucking patterns. *Forest Science* 43(2):157–169.

Lembersky, M.R. and Chi, U.H. 1986. Weyerhaeuser decision simulator improves timber profits. *Interface* 16(1):6–15.

Makkonen, I. 2001. Factors affecting measurement accuracy on processing heads. *Forest Engineering Research Institute of Canada Advantage* 2(24):10.

Marshall, H.D. and Murphy, G.E. 2004. Economic evaluation of implementing improved stem scanning systems on mechanical harvesters/processors. *New Zealand Journal of Forestry Science* 34(2):158–174.

Mendoza, G. and Bare, B. 1986. A two-stage decision model for bucking and allocation. *Forest Products Journal* 36(10):70–74.

Möller, J.J., Wilhelmsson, L., Arlinger, J., Moberg, L. and Sondell, J. 2003. Automatic characterization of wood properties by harvesters to improve customer orientated bucking and processing. *Skogforsk. Arbetsrapport. Nr 537 12 May 2003*: 64–76.

Murphy, G.E. 2003. Procedures for scanning radiata pine stems dimensions and quality on mechanized processors. *International Journal of Forest Engineering* 14(2):11–21.

Murphy, G.E. and Twaddle, A.A. 1986. Techniques for the assessment and control of log value recovery in the New Zealand forest harvesting industry. *Improving Productivity through Forest Engineering, Proceedings of the Council on Forest Engineering, September, Mobile, AL., USA,* 43–47.

Murphy, G.E., Marshall, H.D. and Bolding, M.C. 2004. Adaptive control of bucking on harvesters to meet order book constraints. *Forest Products Journal* 54(12):114–121.

Murphy, G.E., Wilson, I., Barr, K., 2006. Developing methods for pre-harvest inventories which use the harvester as the sampling tool. *Australian Journal of Forestry*. 69(1):9–15.

Nasberg, M. 1985. Mathematical Programming Model for Optimal Log Bucking. Dissertation No. 132. Linköping University, Sweden, 200 p.

Nordlund S. 1996. Logging technology and methods: trends in large-scale forestry. Results No. 2. P. 4 Skogforsk Uppsala, Sweden.

Pickens, J.B., Lee, A. and Lyon, G.W. 1992. Optimal bucking of hardwoods. *Northern Journal of Applied Forestry* 9(4):149–152.

Pnevmaticos, S.M. and Mann, S.H. 1972. Dynamic programming in tree bucking. *Forest Products Journal* 22(2):26–30.

Sessions, J. 1988. Making better tree-bucking decisions in the woods. *Journal of Forestry* 86(10):43–45.

Sessions, J., Olsen, E.D. and Garland, J.J. 1989. Tree bucking for optimal stand value with log allocation constraints. *Forest Science* 35(1):271–276.

Smith G.W. and Harrell, C. 1961. Linear programming in log production. *Forest Products Journal* 11(1):8–11.

Sondell, J. 1987: Computerised bucking: techniques and systems. *Skogsarbeten Results, Forskningsstiftelsen No. 3.* 4 p.

Sondell, J., Möller, J.J. and Arlinder, J. 2002. Third-generation merchandising computers. *Skogforsk Results No 2.* 6 p.

Uusitalo, J. and Kivinen, V.P. 2001. Factors affecting the output distribution of logs with modern tree bucking optimizing. In: K. Sjöström and L. Rask (eds.) *Proceedings from 2nd World Symposium on Logistics in Forest Sector. Timber Logistics 2001 "Supply Chain Management for Paper and Timber Industries"*. 133–141.

Uusitalo, J., Kokko, S. and Vivinen, V.P. 2003. Comparison of various tree-bucking principles for Scots pine. *Proceedings of the Wood for Africa Forest Engineering Conference*. July 2002 Pietermanitzbury, South Africa. Forest Engineering Department, Corvallis, Oregon. 141–148.

Chapter 20

FOREST TRANSPORTATION

Rafael Epstein[1], Mikael Rönnqvist[2], and Andres Weintraub[1]
[1]*Department of Industrial Engineering, University of Chile, Santiago, Chile;* [2]*The Norwegian School of Economics and Business Administration, N-5045, Bergen, Norway, The Forestry Research Institute of Sweden, Sweden*

Abstract Transportation decisions in forestry appear in many planning situations and are often integrated with harvest planning as this step links supply with demand. At the same time, transportation corresponds to a large proportion of total operational costs and thus it is important to manage it efficiently. In this chapter we discuss how OR models and methods have been proposed and used to support forest transportation decisions in planning and scheduling.

Keywords: Routing, forest transportation, multimodal transportation, backhauling, dispatching

1 INTRODUCTION

In forestry several modes of transportation are used; truck, train, ship and water. All logs are carried by trucks for parts of the transportation; either directly to customers or indirectly to storage locations, train terminals or harbours. The raw material can be divided into three main categories: pulp logs to pulp and paper mills, sawlogs to saw mills and energy wood to heating plants. In addition, wood chips are carried from saw mills to pulp, paper mills and heating plants. Moreover, sawn timber and other final products are distributed to domestic and international markets. Different types of trucks are used depending on the type of raw material, operational conditions and each firm's policy. In general, there are separate truck fleets for logs, wood chips and final products. Fuel wood has the potential to be chipped or packed together in the form of large bundles directly to production areas. In this case it is possible to integrate it with log transportation as the same trucks can be used. In some operations, full stems are transported to saw mills. In this way, a better utilization is possible as scanning and optimized breakdown processes can be used directly. These trucks are generally larger and are restricted to private road networks. The latter is also true for trucks with very large capacity; there are trucks with a capacity of up to 200 tons.

The Industrial mills are dependent on continuous deliveries throughout the year. On a time horizon of several years, the transportation planning is integrated with forest management, road building/upgrading/maintenance and harvesting. Inadequate quality standard of the road network reduces the possibility to use roads for heavy traffic during certain parts of the years. This is the case in for example Sweden during spring, when frozen soils are thawing or during winter periods of heavy rains, as is the case for example in Chile. These cases correspond to tactical and strategical planning described in other chapters of this handbook.

How the transportation planning is done, by whom and to what level of detail vary significantly among companies. One case is when a forest company organizes and makes central planning for one fleet of trucks. These trucks may belong to one or several independent transporters. A second case is when one large transport company organizes the transportation for several forest companies or organizations. A third and more decentralized case is when several transportation managers, each responsible for a region, provide target quotas for the number of loads that should be carried from specified roadsides to different mills. The transporters may then organize the work independently from other transporters and just keep in contact with harvesting personnel to ensure that there is enough wood at the landing ready for haulage. The trend is, however, for planning to become more centralized and trucks to increase their working area. This trend is supported by the development of geographical information systems (GIS) road databases and algorithms.

Environmental issues are increasingly important aspects of transportation planning. To minimize transportation costs is often equivalent to minimizing the actual transportation work. This in turn often decreases emissions, thus reducing the negative impact on the environment. Usage of other transportation modes, such as trains and ships, often has advantages, in terms of costs and environmental impacts, over trucks for long distances and high volumes. Other negative impacts due to the use of trucks that should be considered correspond to, for example, traffic congestion and risks, disturbance in population centers, noise and dust on unpaved roads.

In this chapter, we focus on planning problems arising in annual, monthly and shorter time periods and we study the following planning problems/ issues.

1. Destination of logs
2. Backhauling
3. Daily/weekly scheduling (and dispatching)
4. Train and ship integration
5. Organizational aspects

The first three problems are described from a planning horizon perspective. Destination of logs is a network flow problem, which basically balances demand against supply. Backhauling is also a flow problem but where routes are included to explicitly consider efficient routing. In scheduling and dispatching, detailed routes for individual trucks are used. Here binary variables are needed as decision variables. Train and ship integration consider planning with several modes, and organizational aspects consider integrated planning between several organizations or companies.

2 DESTINATIONS OF LOGS

The destination of logs is based on a network flow problem with the structure given in Fig. 1. In this problem we use decision variables x_{ij} representing the flow from supply point (or harvest area) i to destination point (or industry) j. At each supply point i there is a supply s_i and at demand point j a demand d_j.

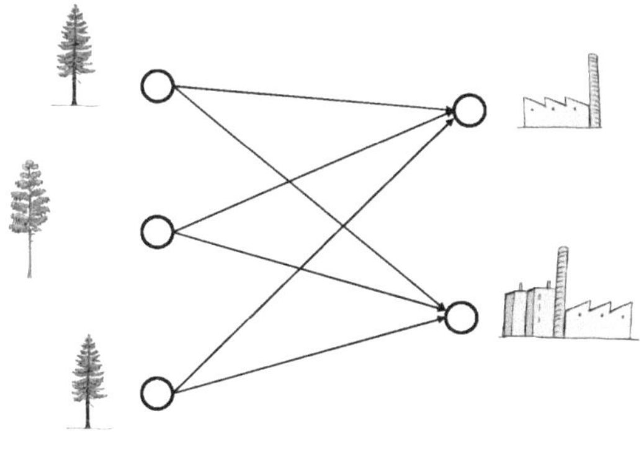

Figure 1. Illustration of the network structure for the destination problem.

The objective is to minimize the total cost, where c_{ij} denotes the unit flow cost between supply point i and demand point j. The index sets I and J denote the sets of supply and demand points, respectively. The model can be stated as

$$[\text{P1}] \min \sum_{i \in I} \sum_{j \in J} c_{ij} x_{ij} ,$$

$$\text{subject to} \sum_{j \in J} x_{ij} \leq s_i, \quad \forall\ i \in I, \tag{1}$$

$$\sum_{i \in I} x_{ij} = d_j, \quad \forall\ j \in J, \tag{2}$$

$$x_{ij} \geq 0, \ \forall\ i \in I, j \in J, \tag{3}$$

Constraint set (1) represents the limited supply, set (2) the demand in industries and (3) is the non-negativity restrictions on variables. The formulation corresponds to a linear programming (LP) model and is efficiently solved using any commercial LP package. The model is a basis for many other more complicated and integrated models. In practical situations we often have several assortments and the model is modified into a multi-commodity flow problem. Moreover, demand can often be satisfied by several (similar) assortments and this linkage must be included. In addition, often there are special restrictions on which supply points can be used to satisfy a particular demand point. One example is when special bucking patterns are used at the harvest points to satisfy special demands. Moreover, often there are several time periods. In this case we also need to include storage balance constraints between periods.

A typical monthly/weekly planning problem is to decide the catchment area allocated to each industry. A catchment area is a set of supply points that is to deliver a given assortment to each industry. Figure 2 illustrates catchment areas around a set of five mills. In this illustration only one assortment is used. Different assortments gives rise to different catchment areas. This type of a model appears in Carlsson and Rönnqvist (2005) and Frisk *et al.* (2006). In Gunnarsson *et al.* (2004) it is a part of a multi-period formulation.

In practice there are some issues regarding information that need to be addressed to develop a decision support tool. Information about supply and demand needs to be collected, stored and maintained. Information on supply

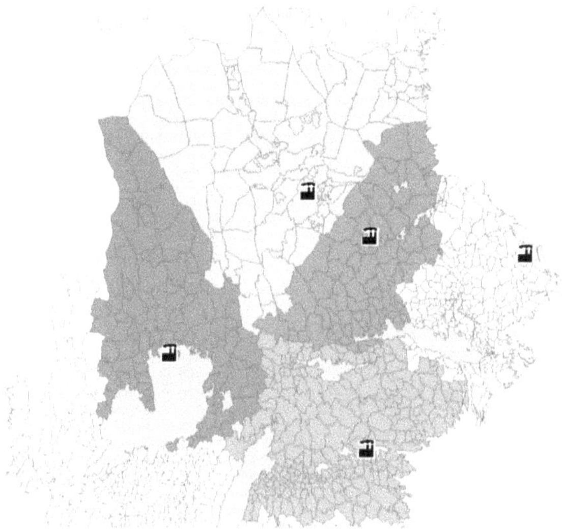

Figure 2. Illustration of catchment areas around a set of five mills.

can be collected online from harvest units or on a daily basis. This information also needs to be updated against what is actually transported by trucks. The costs are based on tariffs and agreements and these must be modelled in detail. Distances and travel times are important in this perspective and determining them is often not trivial. Use of GIS and road databases in this context is important. Models need to be robust and able to balance supply against demand, using penalties to prioritize deliveries when demand exceeds supply, or service remote supply points.

Harvesting is generally not in balance during the year. In the Nordic countries, the harvesting is often carried out during winter when harvesters and forwarders do not damage the soil. In other countries, like Chile it is the other way around and harvesting is concentrated in the summer. Because of this it is necessary to use storage terminals and several time periods. The extended model becomes a transshipment model and information on storage capacities, storage costs, handling costs, and so on etc must be included.

3 BACKHAULING

In the previous model [P1], the cost for each flow variable is based on the fact that each truck drives loaded from the supply to demand point and empty (unloaded) in the other direction. This gives an efficiency of just 50% as half of the travelled distance is loaded and the other half unloaded.

Efficiency would be improved if routes involving several loaded trips were used, that is, backhauling. Backhauling refers to cases where a truck that has carried one load between two points, carries another load on its return. By this, the unloaded proportion can decrease. In Fig. 3 we provide such an example. On the left part side of the figure, we have two direct flows where the corresponding routes are S1-D1-S1 and S2-D2-S2. On the right side, we have the backhaul route S1-D1-S2-D2-S1. The backhaulage trip clearly has a shorter unloaded distance.

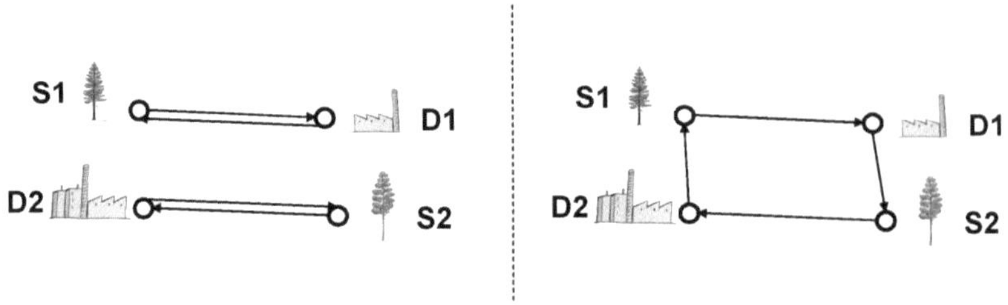

Figure 3. Illustration of two direct flows (*left*) and a backhaulage tour (*right*).

To use back-haulage flows may dramatically decrease the cost. Savings between 2% and 20% are reported in different analyses, see for example Carlsson and Rönnqvist (1998) and Forsberg *et al.* (2005). The possibilities for backhauling are dependent on the type of transportation and the geographical distribution of mills and harvest areas. There must exist wood flows going in opposite directions. This can be because of several assortments or because there are special restrictions between pairs of supply and demand points, resulting in each demand point not connected with the closest supply.

A model with backhauling flow can be expressed in the extended LP model [P2] below. This model is the same as [P1] but where variables y_k are added (from a set *K*). Each variable y_k denotes the flow in backhaulage route *k* and c_k the corresponding unit cost. The coefficients a_{ik} have value 1 if route *k* picks up at supply point *i* and b_{jk} has value 1 if route *k* delivers to demand point *j*. Otherwise the coefficients have values 0.

$$[\text{P2}] \quad \min \sum_{i \in I} \sum_{j \in J} c_{ij} x_{ij} + \sum_{k \in J} c_k y_k ,$$

$$\text{subject to} \sum_{j \in J} x_{ij} + \sum_{k \in K} a_{ik} y_k \leq s_i, \quad \forall \; i \in I, \tag{4}$$

$$\sum_{i \in I} x_{ij} + \sum_{k \in K} b_{jk} y_k = d_j, \quad \forall \; j \in J, \tag{5}$$

$$x_{ij}, y_k \geq 0, \quad \forall \; i, j, k. \tag{6}$$

The main problem with backhauling is the large increase in the number of variables. If a standard model using direct flows has, say, 100,000 variables, a model with backhauling may easily invlove 10–50 million potential variables. Because such numbers of variables cannot be explicitly used in the model, solution methods based on column generation are used. Here, the backhauling variables are dynamically generated in the solution process (Carlsson and Rönnqvist, 1998).

The use of backhauling is included in different systems. One example is the system FlowOpt developed by the Forestry Research Institute of Sweden (Forsberg *et al.*, 2005). In Carlgren *et al.* (2006) it is used integrated with harvest operations. For more operative purposes it is implemented in the web-based decision support system ÅkarWeb (Eriksson and Rönnqvist, 2003). In this system backhauling routes are generated each day and used as a basis in manual planning to find daily routes for trucks.

4 SCHEDULING AND DISPATCHING

The truck scheduling problem is to find a daily route for each truck in a fleet of trucks. This is a pickup and delivery vehicle routing problem (VRP) with time windows. In general there may be multiple pickups and deliveries but most often there is just one delivery. In addition, it is more complicating than a standard VRP as each location can be visited by several trucks; this as the supply or demand is generally more than one truckload. Figure 4 illustrate a route taken by a truck.

This problem can be formulated in several ways. One form often used in practice is based on representing routes as columns and is shown below. Here, we use binary variables to represent if routes are used or not. Each variable z_r is defined to be 1 if route r (from a set R) is carried out and 0

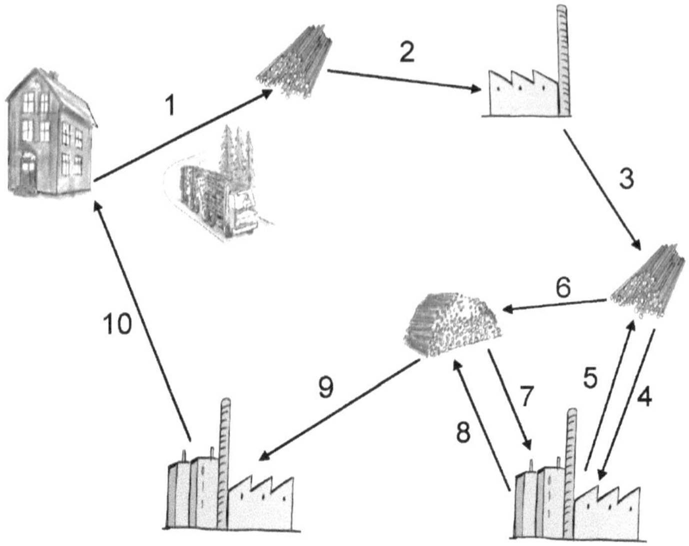

Figure 4. Illustration of a route in the scheduling problem.The numbers indicate the order in which the single parts of the trips are taken. The sequences 1–2 and 8–9 illustrate single pickups and sequence 5–6–7 illustrate a multiple pickup.

otherwise. The coefficient a_{ir} is the volume picked up at supply node i in route r and b_{jr} is the corresponding volume delivered to demand node j. The set R_t states which routes are linked to truck t (from a set T). The cost coefficient c_r gives the cost of an entire route. The basic formulation is

$$\text{[P3]} \min \sum_{r \in R} c_r z_r ,$$

$$\text{subject to} \sum_{r \in R_s} z_r = 1, \quad \forall \ t \in T, \tag{7}$$

$$\sum_{r \in J} a_{ir} z_r \leq s_i, \quad \forall \ i \in I, \tag{8}$$

$$\sum_{j \in J} b_{jr} z_r = d_j, \quad \forall \ j \in J, \tag{9}$$

$$z_r \in \{0,1\}, \quad \forall \ r \in R, \tag{10}$$

The constraint set (7) states that only one route can be used for each truck. Sets (8) and (9) restrict the supply and demand as in the previous models. The last set (10) requires each variable to be a binary variable. Model [P3] has many similarities with model [P2]. The main difference is the more detailed description of routes and the requirement to use only one route for each truck. Time restrictions on routes are handled during the generation process of columns. The number of potential columns or routes is much larger than the number of backhauling routes in [P2]. Solution methods are therefore often based on heuristics and column generation.

Since 1990 a computerized system called ASICAM, has been in use at the main forest companies in Argentina, Brazil, Chile, South Africa, Uruguay and Venezuela. It is a simulation system embedded with a heuristic solver, which produces a complete working schedule for 1 day for more than 100 trucks in a couple of minutes. The system also updates the schedules during the day to incorporate unforeseen events, to produce a real time solution. ASICAM has led to very significant improvements in transportation efficiency, with reduction in costs between 10% and 20% (Weintraub *et al.*, 1996; Epstein *et al.*, 1999). It has also been implemented for a large sawmill firm in Chile to schedule deliveries from plants to customers. In this case round trips are much longer and can take up to 5 days, given the distance from some cities to the forest region.

A system called EPO has been in use in Finland company; see Linnainmaa *et al.*, (1995). EPO is a system that deals with all stages from strategic to operative planning. The input data are collected on-line directly from the forests and the main output is a weekly schedule for each truck. The solution approach combines both heuristics and optimization. A solution approach based on branch and price is reported in Palmgren *et al.*, (2004). Here, the subproblem to generate routes becomes a very large constrained shortest path problem, due to time and volume discretization.

A system called RuttOpt was developed by the Forestry Research Institute of Sweden (Flisberg *et al.*, 2007). It is a system that integrated GIS with a road database. Different solvers can be used but for practical cases it uses a heuristic approach based on tabu search. In tests the system reports savings of 5%–20% compared to manual solutions. There are other systems available but many have not been reported in the research literature and it is difficult to evaluate the systems and their performance.

A real-time truck dispatch system where queuing is included for more than 100 trucks has been developed in New Zealand; see Rönnqvist and

Ryan (1995). Here, the entire daily schedule is not required and instead the objective is to generate one trip at a time for each truck that is available.

5 INTEGRATING TRAIN AND SHIP

Many forest companies operate over large areas and distances between supply and demand points are long. In these cases, substantial savings can be achieved using train or ship transportation for parts of the transportation. To use a train system requires a set of terminals where loading and unloading take place. At each terminal, there is a fixed cost for opening the terminal and a variable cost for handling, that is, loading, storage and unloading. The capacity of a train system is decided by the number of wagons in use and the frequency of the trips. There are also alternatives in how the operation is organized. One alternative is that the train is loaded directly when it arrives to a terminal and then leaves directly after a quick loading. Another is that the wagons are placed at the terminals and get loaded during several days and then locomotives arrive and take the wagons to unloading terminals. For each train system there is a fixed cost depending on the number of wagons and locomotives used, and different capacity levels lead to different fixed costs. The overall cost for a system is generally hard to estimate before an actual negotiation is done.

The introduction of trains may give a very different structure of the catchment areas. In the illustration shown in Fig. 5, we add two terminals and one train system (from Fig. 2) that can load at the terminals and unload at one of the mills. The impact after planning new catchment areas shows that the areas get a very different geographical structure.

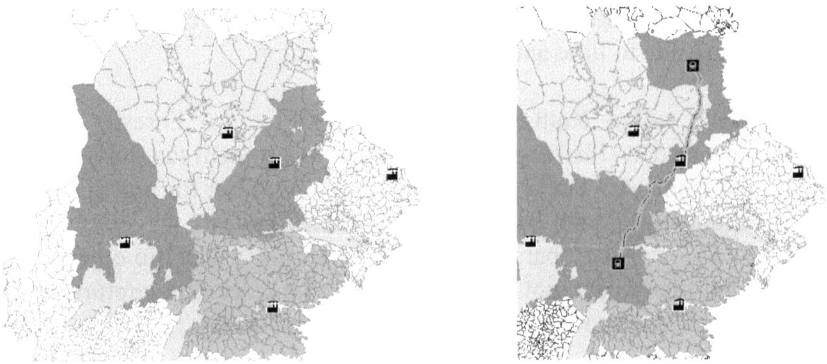

Figure 5. Illustration on catchment areas without trains (*left*) and when a train system is introduced (*right*).

Traditionally, truck and train transportation have been planned separately. Typically, the train transportation is decided in a first phase. Then, this plan provides a new set of supply and demand points, that is, terminals depending on whether loading or unloading take place. The second phase is then to make plans for the trucks. A two-phase approach leads to suboptimal solutions. Therefore, it is desired to integrate the two phases into one. To include train flows in the model requires variables describing at which terminals logs are loaded and unloaded. Constraints that must be added relate to the capacity of each link of the train system to the handling capacity of the terminals. Binary variables are normally needed to model the usage of terminals. Alternatively, if there are few alternatives these can be studied as different scenarios. To include ship transportation is relatively easy as the modelling properties are similar to that for train systems. In the system FlowOpt (Forsberg *et al.*, 2005) there is a possibility to integrate all three modes. In Broman *et al.* (2006) a case including truck, train and ship modes are used to establish a new logistic solution for a forest company.

6 ORGANISATIONAL ASPECTS

In addition to primary actors, representing the wood producers and wood consumers, there are also loggers and the transporters, harvesting and carrying the wood from forest to mill. The management of these operations can be centralized or decentralized. Even though all actors involved recognize the importance of co-operation and integration along the wood-flow chain it is easy to observe and explain why the different actors upon optimizing their individual short-term goals make decisions that can hinder integration and co-operation. There are for example problems with different planning systems, data capture, management and sharing sensitive information. However, savings between 5% and 15% are reported in the literature (Forsberg *et al.*, 2005) which acts as a motivational factor.

A strategic problem is how to optimize the timber supply scheduling when several companies operate in the same region. It is very common that transport distances, and costs, can be decreased if companies exchange wood, applying bartering between them. This issue is difficult, as planners do not want to reveal supply, demand and cost information to competitors. In practice this is solved by deciding on wood bartering of specific volumes. Today this is done in an ad hoc manner and is mostly dependent on personal relations. For more companies there needs to be a more general method to split costs and or profits among several companies. A number of different economic models for cost sharing are described in Frisk *et al.* (2006) where cooperation between eight companies are studied.

7 CONCLUSION

The chapter has discussed the important impact transportation has on operational decisions and costs. Transportation decisions are very complex, as they involve usually a large number of vehicles, multiple origins, destinations, time periods and assortments of types of products to be hauled from origins to destinations.

Most transportation is carried out via trucks, but other modes, in particular trains, can be attractive in some instances. Basic transportation decisions for which OR has been proposed are in the planning stages as well as in daily scheduling.

OR models can significantly improve decisions, as has been shown in concrete experiences, in particular for daily scheduling. Planning on dispatching for the next few weeks can also be improved through the use of models. As technology on communications improves and gets less expensive, it should drive the development of systems based on OR algorithms, which support decisions in real time.

Cooperation between organizations and companies is becoming more important and OR models play an important role in coming up with feasible and practical allocations of common costs.

ACKNOWLEDGEMENTS

The work of Rafael Epstein and Andres Weintraub was supported by several grants from Fondecyt and Fondef. The work by Mikael Ronnqvist was partially supported by Södra Skogsägarna and Holmen Skog.

8 REFERENCES

Broman, H, Frisk, M., and Rönnqvist, M. 2006. Usage of OR tools for logistics support in forest operations at Sveaskog after the storm Gudrun, *Proceedings of the International Conference on Information Systems, Logistics and Supply Chain*, May 15–17, Lyon, France.

Carlsson, D. and Rönnqvist, M. 2005. Supply chain management in forestry – case studies at Södra Cell AB, *European Journal of Operational Research*, 163, 589–616.
Carlsson, D. and Rönnqvist, M. 1998. Tactical planning of forestry transportation with respect to backhauling, Report LiTH-MAI-R-1998–18, Linköping University, Sweden.
Carlgren, C.-G., Carlsson, D., and Rönnqvist, M. 2006. Log sorting in harvest areas integrated with transportation planning using backhauling. Scandinavian Journal of Forest Research. 21(3): 260–271.
Eriksson, J. and Rönnqvist, M. 2003. Decision support system/tools: Transportation and route planning: Åkarweb – a web based planning system. *Proceedings of the Second Forest Engineering Conference*, Växjö, May 12–15, Sweden, 48–57.
Epstein, R., Morales, R., Serón, J., and Weintraub, A. 1999. Use of OR Systems in the Chilean Forest Industries, *Interfaces*, 29, 7–29.
Forsberg, M., Frisk, M., and Rönnqvist, M., 2005. FlowOpt – a decision support tool for strategic and tactical transportation planning in forestry, *International Journal of Forest Engineering* 16(2), 101–114.
Flisberg P., Liden B. and Rönnqvist M. 2007. A Hybrid Methods Based on Linear Programming and Tabu Search for Routing of Logging Trucks, Scandinavian Working Papers in Economics, NHH Discussion Paper 14/2007.
Frisk M., Jörnsten, K., Göthe-Lundgren, M. and Rönnqvist, M. 2006. Cost allocation in forestry operations, Proceedings of the 12th IFAC Symposium on Information Control Problems in Manufacturing, May 17–19, Saint-Etienne, France.
Gunnarsson H., Lundgren J.T. and Rönnqvist, M. 2004. Supply chain modelling of forest fuel. *European Journal of Operational Research* 158(1), 101–123.
Linnainmaa, S., Savola, J., and Jokinen O. 1995. EPO: A knowledge based system for wood procurement management, Paper presented at the 7th Annual Conference on Artificial Intelligence, Montreal, Canada.
Palmgren, M., Rönnqvist, M., and Värbrand, P. (2004). A near-exact method for solving the log-truck scheduling problem, International Transactions of Operations Research 11(4): 447–464.
Rönnqvist, M. and Ryan, D. 1995. Solving Truck Despatch Problem in Real Time, Proceedings of the 31st Annual Conference of the Operational Research Society of New Zealand, August 31-September 1, Wellington, New Zealand, 165–172.
Weintraub A., Epstein R., Morales R., Serón J., and Traverso P., 1996. A Truck Scheduling System Improves Efficiency in the Forest Industries, *Interfaces* 26(4), 1–12.

Chapter 21

OPTIMIZATION OF FOREST WILDLIFE OBJECTIVES

John Hof[1] and Robert Haight[2]

[1]*Rocky Mountain Research Station, USDA Forest Service, Fort Collins, CO;* [2]*North Central Research Station, USDA Forest Service, St. Paul, MN*

Abstract This chapter presents an overview of methods for optimizing wildlife-related objectives. These objectives hinge on landscape pattern, so we refer to these methods as "spatial optimization." It is currently possible to directly capture deterministic characterizations of the most basic spatial relationships: proximity relationships (including those that lead to edge effects), habitat connectivity/ fragmentation relationships, population growth and dispersal, and patch size/ habitat amount thresholds. More complex spatial relationships and stochastic relationships are currently best captured through heuristic manipulation of simulation models. General treatment of stochastic variables in spatial optimization is in its infancy.

Keywords: Habitat connectivity, landscape pattern, reaction-diffusion model, response-surface analysis, search heuristics, simulation optimization, spatial optimization, stochastic population model

1 INTRODUCTION

The objective of this chapter is to review emerging methods that allow analysts to make explicit recommendations (prescriptions) concerning the placement of different features in managed landscapes, so as to optimize wildlife-related objectives. We refer to this general set of methods as "spatial optimization." As used here, spatial optimization refers to methods that capture spatial relationships between different land areas in the process of maximizing or minimizing an objective function subject to resource constraints (we draw a distinction between this set of methods and "spatially-explicit optimization," which simply involves choice variables that are spatially defined and includes no spatial relationships).

2 STATE OF THE SCIENCE IN SPATIAL OPTIMIZATION

In our view, two basic approaches can be used to describe our current capabilities in spatial optimization of landscape pattern: direct spatial optimization approaches and heuristic manipulation of simulation models. We review each of these in turn.

2.1 Direct Approaches

The approach here is to directly include the spatial relationships of concern in a formulation that is focused on the optimization of landscape pattern, per se. We would characterize the approach as having a closed-form formulation with a formal solution method. We would include in this category recent augmentations of the basic "adjacency" formulation (discussed elsewhere in this book) to addressing ecological problems (see, e.g., Bettinger *et al.*, 1997; Barrett *et al.*, 1998; Falcao and Borges, 2001). We would also include recent contributions that have endeavored to add spatial relationships to the set-covering reserve selection model (some examples are Possingham *et al.*, 2000; Nalle *et al.*, 2002; Onal and Briers 2002, 2003; Fischer and Church, 2003). Other authors who have taken relatively direct approaches include Nevo and Garcia (1996), Farmer and Wiens (1999), and Loehle (1999).

Hof and Bevers (1998, 2002) explore a large number of direct spatial optimization formulations, including static models, dynamic models, models of spatial autocorrelation, and models of sustainability. As an example, wildlife habitat fragmentation (spatial division into disaggregated patches) is a common concern with regard to placement of timber harvests. The approach is to directly model the wildlife population growth and dispersal patterns that make habitat connectivity (nonfragmentation) important. This is a dynamic problem where management activities must be scheduled over time, wildlife habitat (determined by forest age) must be tracked as forest stands age and grow, and different wildlife species respond differently to those habitats. The method is related to the classic "reaction-diffusion models" (Skellum, 1951; Kierstead and Slobodkin, 1953; Allen, 1983).

First, the land is divided into cells, where the cell could be scaled to the ecology of the species (e.g., average home range or territory size) or could be scaled simply to provide adequate spatial resolution for the optimization problem. Then, a set of choice variables is defined for each cell, each of which represents a complete scheduled management prescription. For example, each prescription could define the time periods for harvesting the given cell, including a no-harvest option. Any harvest would reset the forest age class to 0 and would change the habitat for each wildlife species accordingly. Initial forest age classes are assigned to each cell, as well as initial population numbers for each wildlife species included.

The following definitions will be used:

i indexes species
k indexes the management prescription
h indexes the cells, as does n
t indexes the time period

qh = the number of potential management prescriptions for cell h

T = the number of time periods

X_{kh}= the area in cell h that is allocated to management prescription k

K_h= the total area in cell h

S_{iht}= the expected population of species i in cell h at time period t

a_{ihtk}= a coefficient set that gives the expected carrying capacity of animal species i in cell h at time period t, if management prescription k is implemented (based on forest age class)

N_{ih}= the initial population numbers for species i in cell h

g_{inh} = the probability that an animal of species i will disperse from cell n in any time period to cell h in the subsequent time period. This includes a probability for n=h, so that the g_{inh} sum to 1 for each combination of h and i

r_i = the "r value" population growth rate (net of mortality not related to dispersal) for species i, and

F_{it} = the total population for species i in time period t.

The simplest objective function would be to maximize a given species' expected total population:

Maximize $\sum_t F_{it}$ for a given i.

The minimum population over all time periods *(m),* for a given species could be maximized as follows:

Maximize m
Subject to $m \le F_{it}$, $t = 1,...,T$ for a given i.

Or, a weighted (V_i) sum of multiple species' populations could be maximized:

Maximize $\sum_i V_i \left(\sum_t F_{it}\right)$.

Many other objective functions are also possible.

The basic constraint set for such a model is

$$\sum_{k=1}^{q_k} X_{kh} = K_h, \forall h \tag{1}$$

$$S_{ih0} = N_{ih}, \begin{matrix}\forall i \\ \forall h\end{matrix} \tag{2}$$

$$S_{iht} \le \sum_k a_{ihtk} X_{kh}, \begin{matrix}\forall i \\ \forall h\end{matrix} \; t=1,\ldots,T \tag{3}$$

$$S_{iht} \le \sum_n g_{inh}\left[(1+r)S_{in(t-1)}\right], \begin{matrix}\forall i \\ \forall h\end{matrix} \; t=1,\ldots,T \tag{4}$$

$$F_{it} = \sum_h S_{iht}, \begin{matrix}\forall i \\ \forall t\end{matrix} \tag{5}$$

$$0 \le X_{kh} \le K_h, \begin{matrix}\forall k \\ \forall h\end{matrix}. \tag{6}$$

This model is linear, with continuous variables, and can be solved with the simplex algorithm. Equation 1 limits the total management prescription allocation to the area in each cell. The management prescriptions are defined with no action in the first time period (t=0), which is used simply to set initial conditions. Equation 2 sets the initial population numbers for each species, by cell. The S_{iht} (expected population by species by cell, for each time period) is determined by whichever of (3) or (4) is binding. Constraint set (3) limits each cell's population to the carrying capacity of the habitat in that cell, determined by forest age classes. Constraint set (4) limits each cell's population according to the growth and dispersal from other cells and itself in the previous time period. The growth and dispersal characteristics of each species are reflected by the parameters in constraint set (4). Constraint set (4) adds up the expected value of the population dispersing from all cells in the previous time period to the given cell in the given time period. It is important to note that whenever (3) is binding for a cell, some of the animals assumed to disperse into that cell are lost because of limited carrying capacity. Reaction-diffusion models assume that organisms dispersing into unsuitable regions will perish. This mechanism provides a probabilistic basis for the expectation that mortality (beyond the nondispersal-related mortality accounted for in the r value) occurs in proportion to the usage of inhospitable surroundings. Thus, actual population growth is determined by a combination of potential growth, dispersal, and spatially located carrying capacities determined by the management prescription allocations. Constraint set (5) defines the total population of each species, in each time period. And finally, constraint set (6) limits the choice variables to be between 0 and the area in each cell.

Hof and Bevers (1998, 2002) applied this basic type of model structure to problems of habitat placement for the black-footed ferret, which also accounted for release schedules of captive-born animals. As a follow-up, it was applied to the black-tailed prairie dog, accounting for population-dependent dispersal behavior. Hof and Bevers also modified this structure to account for water-borne seed dispersal and for habitat edge effects; converted the model to optimize the location of control measures in managing exotic pests; and applied these basic ideas to problems other than organism management (especially stormflow management and fire management).

The primary criticism that can be leveled at this type of approach is that there are limits to the complexity of ecological relationships that can be captured in a closed-form optimization formulation. An alternative explored in the next section is to start with a more complex ecological simulation model and use heuristic procedures to direct repeated predictions with different management regimes, hopefully converging on a near-optimum.

3 HEURISTIC MANIPULATION OF SIMULATION MODELS

The fields of wildlife management and conservation biology contain a long history in developing stochastic models of population viability, which are commonly used to inform wildlife management decisions (Boyce, 1992; Beissinger and Westphal, 1998). Demographic models predict the birth, death, and migration of individuals in one or more localized populations (e.g., Liu *et al.,* 1995). Incidence function models predict the extinction of local populations and colonization of empty habitat patches (Hanski, 1994). Both types of models incorporate uncertainty in one or more demographic parameters, and Monte Carlo methods are used to sample from the underlying distributions and simulate populations many times for different combinations of parameter values. Thus, stochastic population models yield probabilistic results, which are typically summarized by performance measures for the ending population such as mean patch occupancy or probability that population size exceeds a threshold. Stochastic population models are routinely used to determine the relative effects of habitat management options (e.g., Armbruster and Lande, 1993; Liu *et al.,* 1995). In addition, results of population models, such as relative growth rates of populations in potential reserve sites, are used in formulations of reserve selection models (e.g., Carroll *et al.,* 2003). Only a few studies combine optimization with stochastic population models to determine cost-effective habitat protection. Here we describe some of the basic ideas and approaches.

3.1 Stochastic Optimization with Heuristics

Stochastic population models can be optimized using search heuristics. Moilanen and Cabeza (2002) addressed the problem of selecting a subset of potential reserve sites to maximize the long-term persistence of a species living in a metapopulation given a limited budget for site protection. They formulated an incidence function model for the false heath fritillary butterfly, an endangered species in Finland, and applied the model with 125 potential reserve sites of varying size and isolation. Their site selection problem can be described with the following notation:

i, I = indices for individual sites and total number of sites,

b = upper bound on budget,

c_i = cost of protecting site i,

x_i = 0–1 variable for protecting site i; 1 if site i is protected, 0 otherwise,
$p(x_1,..., x_I)$ = the probability of extinction of the metapopulation in period T.

The optimization problem was formulated as follows:

$$\text{Maximize} \quad 1 - p(x_1, \ldots, x_I) \tag{7}$$

Subject to:

$$\sum_{i=1}^{I} c_i x_i \leq b \tag{8}$$

$$x_i \in [0,1], \quad \forall\, i \tag{9}$$

The objective of the optimization problem (Eq. 7) was to maximize the persistence of the metapopulation (one minus the probability of extinction) subject to a budget constraint (Eq. 8) and binary restrictions on the decision variables (Eq. 9). This is not a trivial optimization problem because the objective function value associated with each subset of sites must be estimated using a stochastic population model and the number of different subsets of sites (2^I) increases rapidly with I.

One way to estimate $p(x_1, \ldots, x_I)$ for a given subset of sites is to make N replicate simulations of the population and compute the proportion of simulations that go extinct before time T. The problem with this estimator is that N must be large to get a high level of precision if the probability of extinction is low, which will slow down the optimization considerably. Variance reduction techniques such as importance sampling can be used to increase precision of the estimator for a given sample size N (Haight and Travis, 1997). Importance sampling forces a larger number of rare events from the underlying distribution of the stochastic demographic parameter into the sample that is used for Monte Carlo simulation. Inference from the simulation results is done in a way to correct for sampling bias.

Another option is to define an objective function with properties similar to the probability of extinction $p(x_1, \ldots, x_I)$ and which requires a smaller sample size to obtain a given level of precision. Moilanen and Cabeza (2002) used the average one-step extinction probability of the meta-population calculated over time horizon T and N simulations:

$$\frac{1}{NT} \sum_{n=1}^{N} \sum_{t=1}^{T} \phi_{n,t+1}(x_1, \ldots, x_I; y_{n,t}; E_{n,t}) \tag{10}$$

where the function $\phi_{n,t+1}$ is the probability of extinction in period t+1 given the subset of sites protected ($x_1, \ldots, x_I$), the patch occupancy in period t ($y_{n,t}$), and a random environmental variable ($E_{n,t}$). Because their incidence function model provided a closed-form expression for the one-step extinction probability $\phi_{n,t+1}$, computation of the average (Eq. 10) was fast. Further, Moilanen and Cabeza (2002) found that the average one-step extinction probability was closely related to, but not always identical to, the extinction risk of the metapopulation during horizon T, and it does not require a large sample size to obtain a precise estimate.

Once an objective function is defined, a heuristic is needed to search for the best set of sites. The operations research community has developed a number of heuristics for optimizing stochastic systems using simulation (Andradottir, 1998; Goldsman and Nelson, 1998; Pichitlamken and Nelson, 2001). Moilanen and Cabeza (2002) used a genetic algorithm combined with a local search to optimize their incidence function model. A genetic algorithm operates on a set of alternative solutions, which are called individuals. Each individual is assigned a fitness value equal to the value of the objective function (e.g., Eq. 10). The fitness of an individual determines its probability of taking part in reproduction. Individuals with high fitness reproduce on average more often than those with low fitness. A genetic algorithm proceeds generation by generation with individuals in one generation combining and producing individuals in the next. Moilanen and Cabeza (2002) used 50 replicate simulations to estimate the fitness of each individual in a population. They found that problems with 125 candidate sites converged within 25 generations to almost identical site selections and objective function values.

Optimization of stochastic population models using search heuristics is in its infancy. Commercial simulation packages are beginning to incorporate heuristic optimization tools (e.g., April *et al.*, 2001), and it would be worthwhile to explore their value in optimizing stochastic population models.

3.1.1 Response-surface analysis

Another approach to simulator optimization involves response-surface analysis. Monte Carlo experiments with a stochastic population model can be used to create response surfaces of the relationships between measures of population performance and reserve design features. Then, regression equations representing those relationships can be included in an optimization model to determine the best reserve design features.

Haight *et al.* (2004) used elements of response-surface analysis to address a problem of allocating a fixed budget for habitat protection among disjunct populations of the endangered San Joaquin kit fox in California to maximize the expected number of surviving populations. A demographic model of population viability was used to quantify the risk of extinction of each population under different amounts of protected habitat. The predictions of the demographic model, in turn, were used to estimate risk-area curves for the populations. The risk-area curves and costs of habitat protection were incorporated into an optimization model to determine how best to allocate limited funds among the populations. The problem set up is as follows.

Suppose we have a set of disjunct populations of an endangered species and a limited budget to protect habitat. By disjunct we mean that each population is isolated enough that migration between populations is inconsequential. Further, assume that we have information for each population about the relationship between risk of population extinction and amount of habitat. Using these risk-area curves, we can formulate an optimization model for determining the amount of habitat to protect for each population that maximizes the expected number of populations that survive over the management horizon. The model has the following notation:

i, I = indices for individual populations and total number of populations,

a_i = amount of already-protected habitat for population i,

b = upper bound on budget,

c_i = unit cost of protecting additional habitat for population i,

d_i = upper bound on the amount of habitat available for protection for population i,

x_i = amount of habitat that is selected for protection for population i,

$p_i\,(a_i + x_i)$= function for the probability of extinction, population i.

The optimization problem is formulated as follows:

$$\text{Maximize} \sum_{i=1}^{I} 1 - p_i(a_i + x_i) \tag{11}$$

Subject to:

$$\sum_{i=1}^{I} c_i x_i \le b \tag{12}$$

$$0 \le x_i \le d_i \qquad i = 1, \ldots, I \tag{13}$$

The objective (Eq. 11) is to maximize the expected number of populations that survive over the management horizon. The probability of extinction of each population depends on the total amount of habitat protected, which is the sum of the already-protected habitat and the newly protected habitat. The risk function can be estimated using predictions from a demographic model of population viability under different amounts of protected habitat as described below. The first constraint (Eq. 12) requires that spending for habitat protection does not exceed the budget. The second set of constraints (Eq. 13) bounds the amount of habitat available for protection.

Haight *et al.* (2004) used a stochastic demographic model of a disjunct kit fox population to predict extinction risk in 100 years in habitat patches of increasing size. For each patch, the estimator of extinction risk was the percentage of 1,000 independent simulations in which population size was <10 individuals in 100 years. The predictions were used to estimate a relationship between extinction risk and patch area. The risk-area relationship was a logistic function estimated using a form of logistic regression called the minimum logit chi-squared method (Maddala, 1983). Logistic regression describes a binary response as a function of one or more explanatory variables. In this case, the binary response was extinction or persistence of a population in a habitat patch, and the explanatory variable was patch area. The minimum logit chi-squared method of estimation is appropriate when there are multiple observations of the binary response for each level of the explanatory variable. Let P_i be the proportion of the 1,000 observations in which the population became extinct in patch i and $P_i/(1-P_i)$ be the estimated odds of extinction. With the logistic model, the log of the odds of extinction is assumed to be a linear function of patch area. The model for San Joaquin kit fox was

$$\log \frac{\hat{p}_i}{1-\hat{p}_i} = b_0 + b_1 \frac{1}{y_i} + b_2 \log(y_i) + \mu_i, \tag{14}$$

where y_i is the area of patch i, b_0, b_1, and b_2 are the regression coefficients, and μ_i is the regression error. Because the log of the odds of extinction is a continuous variable without limit, ordinary or weighted least squares regression can be used to estimate the parameters of Eq. 14. Once the parameters of Eq. 14 were estimated, the equation was transformed into a risk-area relationship by solving for p_i on the left-hand-side. The risk-area curve for each of eight populations was incorporated into the optimization model (Eqs. 11–13) and solved using commercial nonlinear programming software. The results included priorities for reserve expansion under increasing upper bounds on funding and a cost curve showing funding required for incremental increases in population viability.

Hof and Raphael (1997) used elements of response-surface methodology to address a problem of locating habitat reserves for Northern Spotted Owl in the Olympic Peninsula, Washington State (USA), to maximize the overall size of the owl population (rather than the expected number of surviving populations in Eq. 7). They subdivided the landscape into 1681 cells and defined variables for the amount of habitat to be protected in each cell. The owl population in each cell depended on the amount of protected habitat in the cell and the total number of owls in adjacent cells. These functions were estimated using predictions of a stochastic model of spotted owl population viability. Although their equations for population size were nonlinear, Hof and Raphael (1997) approximated them as a series of linear line segments, which allowed them to formulate a linear programming model for habitat protection. Although their formulation had over 20,000 variables and 12,000 equations, it was easy to solve with off-the-shelf commercial software. Results of the optimization model were used to identify potential improvements in a proposed spotted owl habitat protection plan.

4 DISCUSSION AND CONCLUSION

The primary shortcoming of the direct approach is that the amount of ecological detail that can be captured is limited. The primary shortcoming of the Simulation Manipulation approach is, of course, that the outcome is only "the best" alternative from among the landscape layouts investigated. Even with a large number of layouts, near-optimality is not assured. To demonstrate the point, suppose we have 1001 and units (e.g., in a 10 x 10 grid). Even if we must consider only one action (v. none), with no scheduling component, there are still 2^{100} or 1.2676×10^{30} possible spatial layouts. Even if 99.9999999% (all but a trillionth) of the layouts can be eliminated as undesirable, we still have 1.2676×10^{21} options. Even if there are a trillion layouts that are acceptable, we only have a $7.886\times10^{-13}(1\times10^{9}\div1.2676\times10^{21}$ chance of hitting an acceptable solution if we randomly arrange our management actions. This suggests the need for optimization procedures in all but the simplest spatial problems. In addition, the implicit response surface may or may not be convex, such that a solution that appears to be near-optimal may actually not be at all.

Thus, the choice is between a precise optimum to a simplified model and an approximate optimum to a more precise model. In a given planning application, the answer may depend on the questions being asked and the circumstances surrounding the planning problem. For example, habitat placement choices may be limited because the pattern of land development has reduced the configuration possibilities (Saunders *et al.*, 1991). When habitat-placement options are restricted to a small set, simulation modeling may offer a very useful approach for ranking alternative configurations. If, however, placement choices are numerous, then formal spatial optimization may be more useful in determining a layout that really is "the best" given the objectives and constraints of the planning problem. Joint use of both strategies as in Hof and Raphael (1997) or Haight *et al.* (2002) might offer planners the opportunity to take advantage of both the ecological detail captured by simulation models and the analytical power of formal spatial optimization to select the best solution.

It should be fairly clear from this chapter that the state of the science in optimization of landscape pattern borrows heavily from the operations research or management sciences field. Thus, ecology in general and landscape ecology in particular could probably benefit from more utilization of management science methodology. It might be worth pointing out that the flow of knowledge also goes the other way. The study of natural processes has inspired several heuristic procedures in optimization. In particular, a class of heuristic solution algorithms called "genetic algorithms" solves mathematical programming problems by emulating evolutionary processes. In the heuristic search, new trial solutions are created by "mating" previous solutions so as to emphasize positive traits much like natural selection promotes evolution in natural systems. Another example is "simulated annealing" which was originally developed to simulate the annealing process of cooling metals, but which is now commonly used as an optimization search routine.

The most fundamental research need in optimizing landscape pattern is the ability to better capture the relevant ecological relationships in an optimization analysis. A natural reaction to the idea of optimizing spatial pattern across a landscape is that we simply do not know enough about ecological systems to actually optimize them. Indeed, we will probably never know as much about ecology as we would like to. Our reaction is that it is important to apply spatial optimization in the context of an adaptive learning process (as we have noted previously). We will probably never have a level of knowledge that is adequate to find a permanent optimal strategy for a managed ecosystem in a one-time optimization analysis. On the other hand, an adaptive management process that does not take advantage of optimization methods is much less likely to make progress either in learning about the ecological system or in managing it. Applied in a careful, learning process, spatial optimization of landscape pattern has the potential to illuminate new hypotheses for landscape ecology research as well as providing a mechanism to apply landscape ecology research in landscape management.

5 REFERENCES

Allen, L. J. S. 1983. Persistence and extinction in single-species reaction-diffusion models. *Bulletin of Mathematical Biology* 45:209–227.

Andradottir, S. 1998. A review of simulation optimization techniques. In: *Proceedings of the 1998 Winter Simulation Conference*, eds. D. J. Medeiros, E. F. Watson, J. S. Carson, and M. S. Manivannan. Institute of Electrical and Electronics Engineers, Piscataway, New Jersey. pp. 151–157.

April, J., F. Glover, J. Kelly, and M. Laguna. 2001. Simulation/optimization using "real-world" applications. In: *Proceedings of the 2001 Winter Simulation Conference*, eds. B. A. Peters, J. S. Smith, D. J. Medeiros, and M. W. Rohrer. Institute of Electrical and Electronics Engineers, Piscataway, New Jersey. pp. 134–138.

Armbruster, P. and R. Lande. 1993. A population viability analysis for African elephant (Loxodonta africana): how big should reserves be? *Conservation Biology* 7:602–610.

Barrett, T. M., J. K. Gilless, and L. S. Davis. 1998. Economic and fragmentation effects of clearcut restrictions. *Forest Science* 44:569–577.

Beissinger, S. R. and M. I. Westphal. 1998. On the use of demographic models of population viability in endangered species management. *Journal of Wildlife Management* 62:821–841.

Bettinger, P., J. Sessions, and K. Boston. 1997. Using tabu search to schedule timber harvests subject to spatial wildlife goals for big game. *Ecological Modelling* 42:111–123.

Boyce, M. S. 1992. Population viability analysis. *Annual Review of Ecology and Systematics* 23:481–506.

Carroll, C., R. F. Noss, P. C. Paquet, and N. H. Schumaker. 2003. Use of population viability analysis and reserve selection algorithms in regional conservation plans. *Ecological Applications* 13:1773–1789.

Falcao, A. O. and J. Borges. 2001. Combining random and systematic search heuristic procedures for solving spatially constrained forest management scheduling models. *Forest Science* 48:608–621.

Farmer, A. H. and J. A. Wiens. 1999. Models and reality: time-energy trade-offs in pectoral sandpiper (calidris melanotos) migration. *Ecology* 80:2566–2580.

Fischer, D. T. and R. L. Church. 2003. Clustering and compactness in reserve site selection: an extension of the biodiversity management area selection model. *Forest Science* 49: 555–565.

Haight, R. G., B. Cypher, P. A. Kelly, *et al.* 2002. Optimizing habitat protection using demographic models of population viability. *Conservation Biology* 16:1386–1397.

Haight, R. G., B. Cypher, P. A. Kelly, S. Phillips, K. Ralls, and H. P. Possingham. 2004. Optimizing reserve expansion for disjunct populations of San Joaquin kit fox. *Biological Conservation* 117:61–72.

Haight, R. G. and L. E. Travis. 1997. Wildlife conservation planning using stochastic optimization and importance sampling. *Forest Science* 43:129–139.

Hanski, I. 1994. A practical model of metapopulation dynamics. *Journal of Animal Ecology* 63:151–162.

Hof, J. and M. Bevers. 1998. *Spatial Optimization for Managed Ecosystems*. Columbia University Press, New York, p. 258.

Hof, J. and M. Bevers. 2002. *Spatial Optimization in Ecological Applications*. Columbia University Press, New York, p. 257.

Hof, J. G. and M. G. Raphael. 1997. Optimization of habitat placement: a case study of the Northern Spotted Owl in the Olympic Peninsula. *Ecological Applications* 7:1160–1169.

Goldsman, D. and B. L. Nelson. 1998. Comparing systems via simulation. *Handbook on Simulation*, ed. J. Banks, Chapter 8. Wiley, New York, NY.

Kierstead, H. and L. B. Slobodkin. 1953. The size of water masses containing plankton blooms. *Journal of Marine Research* 12:141–147.

Liu, J., J. B. Dunning, and H. R. Pulliam. 1995. Potential effects of a forest management plan on Bachman's sparrow (Aimophila aestivalis): linking a spatially explicit model with GIS. *Conservation Biology* 9:62–75.

Loehle, C. 1999. Optimizing wildlife habitat mitigation with a habitat defragmentation algorithm. *Forest Ecology and Management* 120:245–251.

Maddala, G. S. 1983. *Limited-dependent and Qualitative Variables in Econometrics.* Cambridge University Press, New York.

Moilanen, A. and M. Cabeza. 2002. Single-species dynamic site selection. *Ecological Applications* 12:913–926.

Nalle, D. J., J. L. Arthur, and J. Seesions. 2002. Designing compact and contiguous reserve networks with a hybrid heuristic algorithm. *Forest Science* 48:59–68.

Nevo, A. and L. Garcia. 1996. Spatial optimization of wildlife habitat. *Ecological Modelling* 91:271–281.

Onal, H. and R. A. Briers. 2002. Incorporating spatial criteria in optimum reserve selection. *Proceedings of the Royal Society of London B* 269:2437–2441.

Onal, H. and R. A. Briers. 2003. Selection of a minimum boundary reserve network using integer programming. *Proceedings of the Royal Society of London* B270, 1487–1491.

Pichitlamken, J. and B. L. Nelson. 2001. Selection-of-the-best procedures for optimization via simulation. In: *Proceedings of the 2001 Winter Simulation Conference*, eds. B. A. Peters, J. S. Smith, D. J. Medeiros, and M. W. Rohrer. Institute of Electrical and Electronics Engineers, Piscataway, New Jersey. pp. 401–407.

Possingham, H., I. Ball, and S. Andelman. 2000. Mathematical methods for identifying representative reserve networks. In: *Quantitative Methods for Conservation Biology*, eds. S. Ferson and M. Burgman. Springer-Verlag, New York. pp. 291–306.

Saunders, D. A., R. J. Hobbs, and C. R. Margules. 1991. Biological consequences of ecosystem fragmentation: a review. *Conservation Biology* 5:18–32.

Skellum, J. G. 1951. Random dispersal in theoretical populations. *Biometrika* 38:196–218.

Chapter 22

SPATIAL ENVIRONMENTAL CONCERNS

Alan T. Murray
Department of Geography, The Ohio State University, Columbus, USA

Abstract Balancing utilization and protection of the natural environment is a challenging task. Forest management in particular continues to deal with trade-offs inherent to responsible timber harvesting. This chapter focuses on harvest scheduling where one is interested in maximizing economic returns subject to maintaining a continued supply of timber in the future. This necessarily means managing resources in a sustainable manner. As such addressing spatial issues related to environmental concerns is critical. This chapter reviews approaches that have been relied upon to limit localized impacts of harvesting activity.

Keywords: Harvest scheduling, optimization, adjacency, green-up, area restrictions

1 INTRODUCTION

Human activities and their associated impacts on the environment continue to be recognized as a threat to the long-term sustainability of the Earth. Forestry represents an industry that relies on natural resources, forests and timber in particular, in a multiple-use context. Given demands for timber and wood products as well as the environmental impacts of harvest operations on flora and fauna, increasing attention has been directed to enhancing analysis and modeling detail in the management of natural resources. Of interest in this chapter is increased spatial and temporal specificity in harvest-scheduling analysis supported by optimization models.

Harvest-scheduling optimization models have characteristically focused on making decisions on how to treat standing timber over a horizon of several years to decades. Decision variables in these models relate to the sequencing of stands or blocks for harvesting to satisfy temporal timber demands and other constraining conditions. Given this, the orientation of harvest-scheduling models is to maximize economic returns in the management of a forest region.

This necessarily means minimizing management costs, such as operational overhead, transportation system development/maintenance, timber movement costs, and so on. Spatial environmental concerns arise when accounting for wildlife richness, creating habitat favorable to flora and fauna, promoting diversity, maintaining soil and water quality, preserving scenic beauty, and moving toward sustainability more generally. In order to address such concerns implicitly or explicitly, limiting spatial impacts is desired in harvest-scheduling models. Adjacency restrictions and green-up conditions have traditionally been relied upon to regulate localized activity.

2 ADJACENCY AND GREEN-UP

Avoiding concentrated harvest activity in any one area has been approached in optimization models by addressing adjacency relationships. Adjacency reflects spatial proximity of an area to another area. Typically, adjacency is defined as two areas sharing a common boarder or point, but certainly adjacency could be defined using distance between two areas as well. One way to limit localized harvest impacts is to prohibit any two adjacent areas from being simultaneously treated, as was the intent of Thompson *et al.* (1973). Consider a harvesting decision variable for management area *i*:

$$x_i = \begin{cases} 1 \text{ if area } i \text{ is harvested} \\ 0 \text{ otherwise} \end{cases}.$$

For two management areas *i* and *i'*, we can define a condition that would limit harvesting to at most one of these adjacent areas:

$$x_i + x_{i'} \leq 1. \tag{1}$$

Thus, restrictions would be imposed for all adjacent areas N_i to area *i*, and these conditions would be needed for each area *i*. Murray (1999) has referred to harvest-scheduling problems where adjacency between management units is imposed as the unit restriction model (URM). The assumption here is that the combined area of units *i* and *i'* exceeds an acceptable threshold. That is, $\alpha_i + \alpha_{i'} > A$, where α_i is the area of unit *i* ($\alpha_{i'}$ similarly defined) and *A* is the maximum permissible harvest disturbance area. Murray *et al.* (2004) indicate that maximum area limits of 16–49 ha are common in practice.

If we also take into account temporal aspects of spatial decision making, the earlier notation can be extended as follows:

$$x_{it} = \begin{cases} 1 \text{ if unit } i \text{ is harvested in time period } t \\ 0 \text{ otherwise} \end{cases}.$$

In the context of limiting localized impacts, condition (1) can be more broadly defined to include both spatial and temporal restrictions on harvest activity as follows:

$$\sum_{t'=t-p}^{t+p} \left(x_{it'} + x_{i't'} \right) \leq 1 . \tag{2}$$

where p is a pre-specified harvesting exclusion period. Condition (2) includes the so-called green-up requirement, where an area cannot be harvested if an adjacent unit has been harvested in a pre-specified interval of time before or after the current time period t. As such, condition (2) would be necessary for each units i and adjacent units i' in every each time period t.

3 AREA RESTRICTIONS

Current harvest-scheduling research increasingly focuses on a variant of the earlier problem, recognizing that management units may be defined such that two or more adjacent units do not necessarily violate the maximum area limitation (see Hokans, 1983; Lockwood and Moore, 1993; Murray, 1999). That is, it is possible that $\alpha_i + \alpha_{i'} < A$, representing a feasible management possibility. In this case, rather than adjacency constraints, one needs a maximum area restriction defined for sets of units if the intended condition is to be imposed in a harvest-scheduling optimization model. Murray (1999) has referred to harvest-scheduling problems where spatial limitations apply to sets of management units as the area restriction model (ARM).

While in the general case this is a particularly formidable problem to structure (and solve) using integer programming, under certain conditions it is possible to enumerate potential feasible harvesting blocks (or areas) a priori (see Murray *et al.*, 2004; Goycoolea *et al.*, 2005). As an example, consider the forest units shown in Fig. 1, assuming a maximum allowable impact area of A=49 ha. Given this, there are 17 potential combinations of feasible blockings for these individual management units: {1}, {2}, {3}, {4}, {5}, {1,2}, {1,3}, {1,4}, {2,4}, {2,5}, {3,4}, {4,5}, {1,3,4}, {1,4,5}, {2,3,4}, {2,4,5}, and {3,4,5}. A block, then, is an area comprising spatially connect, or contiguous, management units. As such, a feasible block would need to be identified using some enumerative scheme (see Goycoolea *et al.*, 2005).

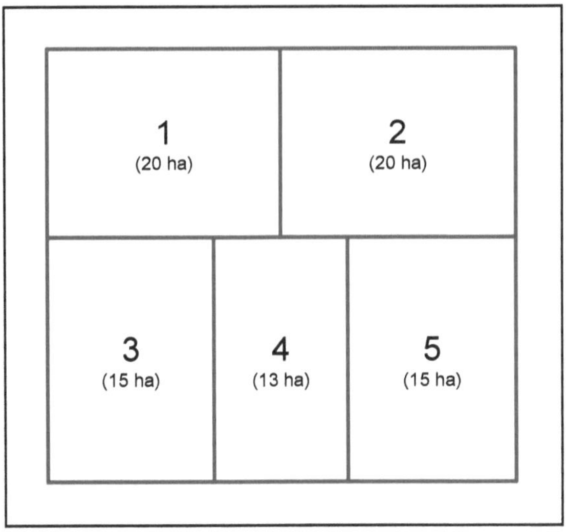

Figure 1. Forest management units.

From a modeling perspective, this somewhat changes how our problem is mathematically represented as we must account for these permissible spatial blocks. This can be done by introducing a new decision variable for each feasible block l:

$$y_l = \begin{cases} 1 \text{ if block } l \text{ is harvested} \\ 0 \text{ otherwise} \end{cases}.$$

Given this notation and a priori identified feasible blocks, there are two cases in which any two blocks cannot be simultaneously selected for harvest. First, if a unit i is common to both blocks, clearly both should not be allowed as a unit and cannot be harvested twice. Second, if two blocks are adjacent, then we assume that their combined area would result in a spatial violation. While it is true that their combined area may not actually exceed the stipulated maximum area restriction, if this combination of units is feasible it will be identified as a potential block (see Goycoolea *et al.*, 2005). Therefore, this harvesting option is present as a feasible block that can be selected. The implication of these two cases is that we can utilize an adjacency constraint to impose proscribed configurations of blocks as follows:

$$y_l + y_{l'} \leq 1 \qquad \forall l, l' \in \Omega_l, \tag{3}$$

where Ω_l is the set of blocks adjacent to block l and those blocks which share a common management unit with block l.

Returning to the example shown in Fig. 1, feasible blocks {1,4} and {2} would be prohibited based on adjacency given that their combined area (53 ha) exceeds the maximum allowed disturbance area of 49 ha. Mathematically, this can be imposed as follows:

$$y_{\{1,4\}} + y_{\{2\}} \leq 1.$$

Alternatively, feasible blocks {1,4} and {3,4,5} would be prohibited because they share a common area, unit 4. As such, the following additional constraint would also be needed, among others:

$$y_{\{1,4\}} + y_{\{3,4,5\}} \leq 1.$$

For this example, it is possible to encapsulate the harvesting decision variables and spatial constraints as a graph of nodes and arcs. The nodes in this graph represent feasible blocks to harvest and arcs depict adjacency or block overlap restrictions. Goycoolea *et al.* (2005) refer to this as a projected graph because it is derived from the forest region. Figure 2 illustrates the projected graph for the earlier forest example. In this case, it is nearly a complete graph, with no arcs between {1} and {5}, {2} and {3}, {3} and {2,5}, and {5} and {1,3}.

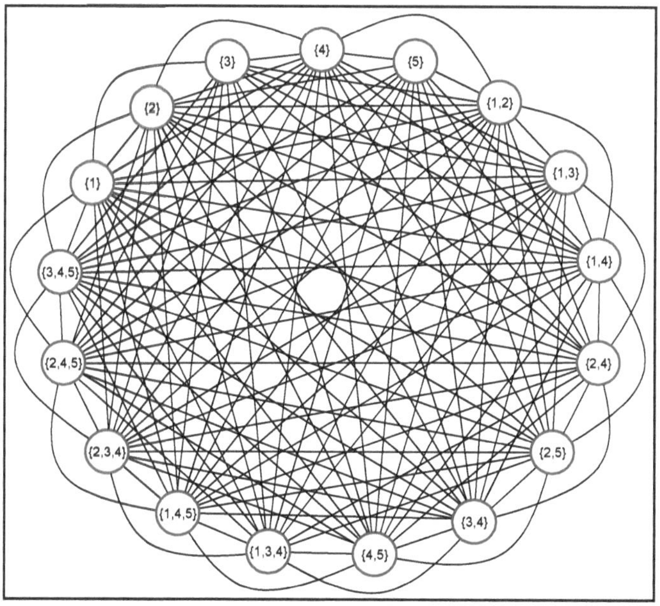

Figure 2. Graph depicting blocks and restrictions in the five unit-forest.

Given the projected graph, it is possible to structure the harvest-scheduling optimization problem as an integer program by restricting spatial impacts between blocks using the ARM.

4 SOLVING THE AREA RESTRICTION MODEL (ARM)

The remainder of this chapter will focus on the ARM, as the URM has been shown to be a special case of the ARM (see Murray, 1999). Murray and Weintraub (2002) provide an empirical examination of the relationship between the URM and ARM. The ARM can be formally stated as follows:

$$\text{Maximizes} \sum_{l}\sum_{t}\beta_{lt}y_{lt}, \tag{4}$$

subject to

$$\sum_{t'=t-p}^{t+p}\left(y_{lt'}+y_{l't'}\right)\le 1 \qquad \forall l,l'\in\Omega_l,t, \tag{5}$$

$$\sum_{l}v_{lt}y_{lt}\le U_t \qquad \forall t, \tag{6a}$$

$$\sum_{l}v_{lt}y_{lt}\ge L_t \qquad \forall t, \tag{6b}$$

$$\sum_{t}y_{lt}\le 1 \qquad \forall l, \tag{7}$$

$$y_{lt}=\{0,1\} \qquad \forall l,t, \tag{8}$$

where

β_{it} is the benefit of harvesting block l in period t,
v_{it} is the volume produced by harvesting block l in period t,
U_t is the upper bound on total volume harvested in period t, and
L_t is the lower bound on total volume harvested in period t.

The objective (4) maximizes net return in selecting blocks for harvest. Constraints (5) impose adjacency and incompatibility restrictions on the simultaneous selection of blocks. Constraints (6) establish upper and lower bounds on harvesting productivity in each time period. Constraints (7) allow a block to be harvested at most once over the planning horizon. Finally, constraints (8) indicate integer requirements on decision variables.

Common extensions to this basic model include road network construction and maintenance, minimum revenue requirements, age structure, and preservation of habitat (see Kirby *et al.*, 1986; Murray and Church, 1995; Caro *et al.*, 2003).

Solving the ARM has proven to be a challenge. Much of the initial work on solving the ARM utilized heuristic solution methods. Hokans (1983) detailed an artificial intelligence-based heuristic for the ARM. Following this were approaches based on simulated annealing and tabu search developed by Lockwood and Moore (1993), Clark *et al.* (2000), and Richards and Gunn (2000). Recent work in this area includes the evolutionary approach (genetic algorithm) of Falcao and Borges (2002) and the tabu 2-opt heuristic of Caro *et al.* (2003).

Murray *et al.* (2004) and Goycoolea *et al.* (2005) detail an approach for solving the ARM exactly using commercial integer-programming software. The idea behind the approach is to exploit properties of the projected graph. In particular, constraints (5) in the ARM are not particularly strong in the sense of inducing facets beneficial to integer-programming techniques. Specifically, integer-programming typically relies on linear programming (LP) coupled with branch-and-bound, where integer restrictions on decision variables are initially relaxed then systematically resolved in the branching phase. When constraints (5) are utilized, highly fractional LP solutions often result, if a relaxed solution can be obtained at all, requiring substantial effort to resolve fractions and prove optimality, again if this can even be done at all. To address this issue, Goycoolea *et al.* (2005) proposed higher-ordered cliques and other facet-defining constraints in the projected graph. A clique is a set whose members share a mutually exclusive relationship with all other members in the set. The cliques suggested in Goycoolea *et al.* (2005) are structurally similar to those developed for the URM by Murray and Church (1996).

Constraints (5) actually are low-ordered cliques, fundamentally containing two decision variables, for example, Eq. 3 and a right-hand side coefficient value of one. Interestingly, higher ordered cliques typically exist in projected graphs, making it possible to have many decision variables in one constraint while retaining a right-hand side coefficient of one. Thus, Eq. 3 can be generalized as follows:

$$\sum_{l \in C} y_l \leq 1, \tag{9}$$

where C is the set of blocks forming a clique (all blocks in the clique are mutually prohibited from being harvested together). Such a constraint in the ARM provides the facet-inducing property important for optimally solving integer-programming problems in practice.

For the forest example previously discussed, only three cliques are needed to impose all projected graph restrictions:

$$\begin{aligned} & y_{\{1\}} + y_{\{1,2\}} + y_{\{1,3\}} + y_{\{1,4\}} + y_{\{1,3,4\}} + y_{\{1,4,5\}} \\ & + y_{\{2\}} + y_{\{2,4\}} + y_{\{2,5\}} + y_{\{2,3,4\}} + y_{\{2,4,5\}} \\ & + y_{\{4\}} + y_{\{3,4\}} + y_{\{4,5\}} + y_{\{3,4,5\}} \leq 1 \end{aligned}, \tag{10a}$$

$$\begin{aligned} & y_{\{1\}} + y_{\{1,2\}} + y_{\{1,3\}} + y_{\{1,4\}} + y_{\{1,3,4\}} + y_{\{1,4,5\}} \\ & + y_{\{3\}} + y_{\{3,4\}} + y_{\{2,3,4\}} + y_{\{3,4,5\}} \\ & + y_{\{4\}} + y_{\{2,4\}} + y_{\{4,5\}} + y_{\{2,4,5\}} \leq 1 \end{aligned}, \tag{10b}$$

$$\begin{aligned} & y_{\{2\}} + y_{\{1,2\}} + y_{\{2,4\}} + y_{\{2,5\}} + y_{\{2,3,4\}} + y_{\{2,4,5\}} \\ & + y_{\{4\}} + y_{\{1,4\}} + y_{\{3,4\}} + y_{\{4,5\}} + y_{\{1,3,4\}} + y_{\{1,4,5\}} + y_{\{3,4,5\}} \\ & + y_{\{5\}} \leq 1 \end{aligned}, \tag{10c}$$

Assuming that an enumerative scheme is developed to identify all necessary cliques in a projected graph, a constraint for each clique k may be structured as follows:

$$\sum_{l \in C_k} \sum_{t'=t-p}^{t+p} y_{lt'} \leq 1 \qquad \forall k,t, \tag{11}$$

where k is the index of cliques. These constraints would replace constraints (5) in the ARM. The rationale for this replacement is that there will be substantially fewer constraints (11) than (5). Further, the facet inducing structure of constraints (11) is far superior to (5).

Goycoolea *et al.* (2005) provide computational experience using commercial integer-programming software to solve the ARM using constraints (11) for a range of harvest-scheduling problems. The largest problem solved had 1,363 management units and a planning horizon of 7 periods, resulting in some 9,500 scheduling decision variables alone. Extensions of the ARM to account for average area considerations were detailed in Murray *et al.* (2004), readily solving scheduling problems with 351 planning units. The point here is that the projected graph and higher-ordered cliques make it possible to solve fairly large ARM-based harvest-scheduling problems with modest computational effort.

5 TEMPORAL RESTRICTIONS

While much progress has been made in the development of optimization approaches to support harvest scheduling, a relatively under investigated area of research in modeling spatial environmental concerns is the impacts of temporal output requirements. This is not particularly surprising given that spatial restrictions have been challenging to represent and impose, and they have had substantial impact on model solvability (Murray and Church, 1996; Goycoolea *et al.*, 2005). Recent work by Vielma *et al.* (2007) has found that merely adding a temporal dimension to an ARM with a requirement on productivity in each time period greatly increased computational complexity. As an example, for a problem with 1363 management units and 15 time periods the addition of volume restrictions, constraints (6a) and (6b) increased computational effort by more than 200% just to find a solution within 1% of optimality. Thus, addressing both space and time presents difficulty, but is fundamentally important to responsible natural resource management practices (see Ware and Clutter, 1971; Bettinger *et al.*, 2003).

What is significant about the work of Vielma *et al.* (2007) is that the fraction-inducing behavior of temporal volume constraints becomes apparent. That is, temporal volume constraints do not tend to be integer-friendly. As a result, approaches for dealing with this aspect of modeling difficulty in harvest scheduling is necessary, which is precisely what was done in Vielma *et al.* (2007). Specially, Vielma *et al.* (2007) discussed approaches for constraint branching and relaxing strict volume constraints.

6 CONCLUSIONS

Spatial environmental issues are of central concern in forest management. Harvest scheduling has long been focused on using optimization models to support management and decision making. There has been an evolution of sorts in harvest-scheduling where greater spatial and temporal specificity is expected with increases in geographic data and a better understanding of ecological processes. To support this, harvest scheduling models have moved from unit-based to area-based approaches, such as the ARM. While many of the initial ARM applications made use of heuristic solution methods, recent work has demonstrated increased capabilities for optimally solving such problems. Improvements facilitated by the use of projected graphs and cliques necessarily exploit spatial problem structure. Along the temporal domain, advances are being made associated with the ARM, but there appears to be substantial opportunity for improvements based on space–time insights.

Future research addressing spatial environmental concerns will no doubt continue to push the envelop of computational capabilities for solving harvest-scheduling models. One can anticipate advances in both exact and heuristic approaches. It seems reasonable as well to expect research focusing on the impacts of temporal volume restrictions. Beyond this, extension of the basic ARM to address roading and other operational concerns is no doubt an important future area of work with real potential for significant contributions.

ACKNOWLEDGEMENTS

This research was supported by the National Science Foundation (Geography and Regional Science Program and the Decision, Risk, and Management Science Program) under grant BCS-0114362.

7 REFERENCES

Bettinger, P., D.L. Johnson and K.N. Johnson (2003). Spatial forest plan development with ecological and economic goals. *Ecological Modelling* 169: 215–236.

Caro, F., M. Constantino, I. Martins and A. Weintraub (2003). A 2-opt tabu search procedure for the multiperiod forest harvesting problem with adjacency, greenup, old growth and even flow constraints. *Forest Science* 49: 738–751.

Clark, M.M., R.D. Meller and T.P. McDonald (2000). A three-stage heuristic for harvest scheduling with access road network development. *Forest Science* 46: 204–218.

Falcao, A.O. and J. Borges (2002). Combining random and systematic search heuristic procedures for solving spatially constrained forest management scheduling models. *Forest Science* 48: 608–621.

Goycoolea, M., A.T. Murray, F. Barahona, R. Epstein and A. Weintraub (2005). Harvest scheduling subject to maximum area restrictions: exploring exact approaches. *Operations Research* 53: 490–500.

Hokans, R.H. (1983). Evaluating spatial feasibility of harvest schedules with simulated stand-selection decisions. *Journal of Forestry* 81: 601–603, 613.

Kirby, M., W. Hager and P. Wong (1986). Simultaneous planning of wildland management and transportation alternatives. *TIMS Studies in the Management Sciences* 21: 371–387.

Lockwood, C. and T. Moore (1993). Harvest scheduling with spatial constraints: a simulated annealing approach. *Canadian Journal of Forest Research* 23: 468–478.

Murray, A.T. 1999. Spatial restrictions in harvest scheduling. *Forest Science* 45: 45–52.

Murray, A.T. and R.L. Church (1995). Heuristic solution approaches to operational forest planning problems. *OR Spektrum* 17: 193–203.

Murray, A.T. and R.L. Church (1996). Analyzing cliques for imposing adjacency restrictions in forest models. *Forest Science* 42: 166–175.

Murray, A.T., M. Goycoolea and A. Weintraub (2004). Incorporating average and maximum area restrictions in harvest scheduling models. *Canadian Journal of Forest Research* 34: 456–464.

Murray, A.T. and A. Weintraub (2002). Scale and unit specification influences in harvest scheduling with maximum area restrictions. *Forest Science* 48: 779–789.

Richards, E.W. and E.A. Gunn (2000). A model and tabu search method to optimize stand harvest and road construction schedules. *Forest Science* 46: 188–203.

Thompson, E.F., B.G. Halterman, T.S. Lyon and R.L. Miller (1973). Integrating timber and wildlife management planning. *Forestry Chronicle* 47: 247–250.

Vielma, J.P., A.T. Murray, D.M. Ryan and A. Weintraub (2007). Improving computational capabilities for addressing volume constraints in forest harvest scheduling problems. *European Journal of Operational Research*, 176: 1246–1264.

Ware, G.O. and J.L. Clutter (1971). A mathematical programming system for the management of industrial forests. *Forest Science* 17: 428–445.

Chapter 23

HEURISTICS IN FOREST PLANNING

John Sessions[1], Pete Bettinger[2], and Glen Murphy[3]
[1]*Oregon State University, Department of Forest Engineering, USA;*
[2]*University of Georgia, Warnell School of Forest Resources, USA;*
[3]*Oregon State University, Department of Forest Engineering, USA*

Abstract Heuristics are often used in forest planning due to the size and nonlinear structure of many problems. Heuristics have been used at all levels of forest planning: strategic, tactical, and operational. An important strength of heuristics is their ability to capture the essence of the planning problem. The solution methods for forest-planning problems reflect the wide range of problems being solved, from rule-based systems to network-based algorithms, linear programming (LP)-heuristic combinations, as well as the more recent metaheuristics (simulated annealing, threshold accepting, tabu search, and genetic algorithms). The major barriers to solving planning problems have moved from hardware and software to costs of data capture, reliability, and uncertainty. Advances in data-capturing technologies will help. Trained and experienced people are important to the success of heuristic applications.

Keywords: Combinatorial optimization, resource scheduling, strategic planning, tactical planning, operational planning

1 INTRODUCTION

Heuristics used in forest planning can be divided into three broad applitions, strategic forest planning, tactical forest planning, and operations planning. Strategic planning involves the scheduling of activities across broad areas and over long time periods and is generally nonspatial in nature. Tactical planning involves scheduling activities over a shorter horizon, but uses higher detail and often includes spatial considerations. Operational planning is used for scheduling activities over short time periods and utilizes highly specific spatial information.

2 STRATEGIC PLANNING

Planning processes performed at the stand level and translated to the forest level can be viewed as unconstrained optimization of organizational goals. When this translation is made, higher level landscape goals do not affect the decisions made at the stand level. However, if higher-level goals affect the decisions made at the stand level, the forest-level planning process becomes constrained, and the translation of stand-level decisions to the forest-level is not necessarily transparent. In this case, the forest level decisions generally involve choosing among a number of alternatives for each stand.

2.1 Unconstrained Optimization – No Constraints Between Stands

The task of finding the optimal way to grow an individual stand of trees has been the focus of considerable research. The use of dynamic programming (Brodie *et al.*, 1978), optimal control (Haight *et al.*, 1985), and heuristics (Paredes *et al.*, 1987; Yoshimoto *et al.*, 1990) have all been used. In the early work in this area, the objectives were mainly economic, but recently there has been interest in ecological goals and multiple economic-ecological goals (Pukkala and Miina, 1997). The inherent stand-level optimization problem is a discrete, multiple period optimization problem. The decision variables are designed such that they represent the number of trees to be removed from each diameter class from each species in each time period to maximize a goal. Much of the recent work originated with Paredes *et al.* (1987) who used a one-state, one-stage dynamic programming approach to reduce the search region. This approach was extended by Yoshimoto *et al.* (1988), and applied by Cousar (1992) and Wedin (1999). Recent applications include developing prescriptions for fire-prone landscapes and obtaining structural conditions in a minimum amount of time (Graetz *et al.*, in press).

2.2 Optimization with Constraints Within Stands

Sometimes management decisions involve considering multiple goals within a stand. This has been approached in one of two ways, either one goal is used for the objective function and the remaining goals are represented as constraints, or goal-programming approach is used, where multiple goals are included in an objective function, and given weights to represent their importance. In first case, when dynamic programming is used, arcs leading to infeasible states are pruned from the network. What this suggests is that while one goal is optimized, the solutions can contain only those actions that do not exceed the bounds of other goals (expressed as constraints).

A considerable amount of discussion during planning meetings centers on the importance of a goal being represented by the objective or by a constraint, and how much influence each has on the final solution. In the latter case, the constraints are moved to the objective function and penalized (Haight *et al.*, 1992). Here, a trade-off (of sorts) among goals is optimized, yet the difficulty lies in scaling and weighting the goals in the objective function.

2.3 Optimization with Constraints Between Stands

Often, a number of stands must be managed to meet goals at a higher level. Heuristics have been used in strategic forest planning for many years. In Europe, early heuristics were developed to estimate the amount of volume that could be harvested during the time required to convert a forest from one condition to another. A classic text is Forest Regulation by Freiderich Judeich that appeared in 1871. The early methods involved simple formulas and discusses 18 approaches. Later, when digital computers became available, binary search procedures were developed to permit taking into account many types of stands and stand conditions to answer the same question. In the 1970s, linear programming (LP) replaced many of the heuristics, as the nature of the problem often resulted in a linear objective function (maximize volume or harvest value subject to constraints on land, stand condition, labor, and budgets). Normally the constraints were nonspatial. The harvest-scheduling process could usually be described as land classification, growth and yield projection of stands under differing levels of management intensity, and assignment of acres to various levels of management intensity.

The two main problem formulations for even-aged management are best described by Johnson and Scheurman (1977) who classify them into Model I and Model II. Model I is categorized by the description of activity inputs and outputs over the entire planning horizon, whereas a Model II activity defines inputs and outputs from the birth to the death of a stand. In both cases, the assignment of land to various types of prescriptions was performed with LP, but heuristic approaches were used to develop the management prescriptions eligible for each type of stand (strata). These heuristic approaches usually involve a trial and error approach or a "shot gun" approach designed to bracket acceptable prescriptions from which the LP could choose at solution time. More recently, there has been attention to more careful selection of prescriptions. Difficulty in the identification and development of "good candidate prescriptions" led to consideration of the intelligent search procedures from unconstrained optimization.

Beginning in the 1990s, concern over the size of harvest units began to appear in strategic plans. Lockwood and Moore (1993), for example, developed a four period, 1.6 million ha plan in eastern Canada using simulated annealing. By the end of the 1990s the USA-based National Council for Air and Stream Improvement had developed spatial forest-planning software to support the Sustainable Forestry Initiative using a simulated annealing heuristic (Van Deusen, 1999), and the state of Oregon was using a simulated annealing heuristic to develop strategic plans for the majority of its forests. The state of Oregon models combine the advantages of Model I and Model II by maintaining spatial control in a Model I sense while permitting acres to shift between prescriptions at regeneration harvest time (Model II). Most other land owners in North America continue to use strata-based LP or a strata-based binary search heuristic for strategic planning.

Heuristics have also recently been used for strategic planning that involves uneven-aged management (Bettinger *et al.*, in press). Here, the size of harvest units is not a concern, yet the location of treatments may be important, as this information may be transferred to a spatial fire behavior model. Integer decision variables were used to ensure that the exact location of uneven-aged stand-level management prescriptions were known, and the resulting forest structure influences the simulation of fire behavior across a broad area (100,000 ha or more).

3 TACTICAL PLANNING

While LP has been the dominant method for strategic forest planning in North America in the last three decades, as well as the dominant method for planning on large plantations of South America, New Zealand, and Australia, heuristics have dominated tactical-planning processes. However, in a few research applications mixed integer LP has been shown to be useful for tactical planning (e.g., Hof and Joyce, 1992). The difference in the choice of solution method for the type of planning process has been mainly due to spatial forest-planning considerations. The issues in tactical planning have generally revolved around how to implement the strategic plan, and incorporate constraints not considered at the strategic level, primarily spatial constraints. Embedded within tactical planning are harvest shape and size, delineation of riparian zones, and mitigation activities for fish and wildlife. These considerations invariably involve large numbers of integer variables. The most common approach is to try to find output levels corresponding to the strategic plan while meeting constraints imposed at the tactical level. Of course, if the additional constraints at the tactical level are binding, then the solution at the strategic level is to some extent infeasible. Heuristic algorithms used at the tactical level include pure Monte Carlo simulation, simulated annealing and other similar stochastic neighborhood searches, tabu search, and genetic algorithms. Where transportation is important, road systems are included in the tactical plan (Jones *et al.*, 1986; Bettinger *et al.*, 1998; McNaughton *et al.*, 1998; Clark *et al.*, 2000; Richards and Gunn 2000), but this has generally not been common. The research applications have involved tabu search, tabu search with strategic oscillation, simulated annealing, and other approaches. Specialized heuristic procedures for rounding and bounding linear-programming solutions have also been developed (Jones *et al.*, 1986; Weintraub *et al.*, 1994).

Because strategic planning omits constraints included at the tactical-planning level, it is difficult to interpret forest condition across the landscape where the metrics depend on not only how much (percentage) will occur, but where it will occur. This has prompted discussion of what is superior, a strategic plan that is "optimal" but may prove infeasible, or a strategic plan that is feasible, but may not be optimal. As an alternative, heuristics are being explored at the strategic level to directly incorporate spatial constraints. Examples include the heuristic to determine the sustainable harvest level for the majority of forests of the Washington Department of Natural Resources, USA and the majority of forests managed by Oregon Department of Forestry (ODF). In the former case, the heuristic involves primarily simulation and in the latter case simulated annealing is used.

4 OPERATIONAL PLANNING

Operational planning involves a wide variety of land management activities. At one end of the spectrum, it includes equipment assignment and development of logging site access (which is at the threshold of tactical planning), and at the other end, algorithms for real-time log-cutting decisions on harvesting equipment. Although LP and dynamic programming have had limited application in operational planning, the majority of the planning and decision-making applications have involved using heuristics. We discuss several of the categories of operational planning and decision making later.

4.1 Equipment Location and Road Access

One of the most basic operational-planning processes in forest management involves the decisions regarding where to install (or locate) equipment, the equipment type(s) to use, and the road access decisions. Approaches that have been used generally involve network-based heuristics. Some of these algorithms interact with a digital terrain model (DTM) to automatically determine the road network possibilities, such as automatically constructing the road network possibilities for various truck-turning radius requirements (Chung *et al.*, 2004; Epstein *et al.*, in press). Truck-routing problems have been described with rule-based algorithms (Weintraub *et al.*, 1996), and an LP master with a k-shortest path heuristic subproblem (Palmgren *et al.*, 2004).

4.2 Tree-Based Decisions

Perhaps the very finest level of planning in forest management involves what to do with individual trees. For example, forest managers are routinely confronted with decisions regarding the trees to mark for harvest, or the shortest path around a set of timber sampling points. One could also consider the decisions regarding the marking of trees to leave (as residual wildlife habitat) and the associated safety and logging efficiency concerns as similar tree-level decisions. The solution methods that have been used or proposed for these types of decisions involve integer decision variables. Tree-bucking decisions (the decision to cut a tree into logs) and market allocation decisions are other types of tree-based decisions that include as possible solution techniques LP with column generation using dynamic programming (Eng *et al.*, 1986; Epstein *et al.*, 1999) and heuristics such as tabu search (Laroze and Greber, 1997; Murphy 1998), as well as network-based heuristics (Sessions *et al.*, 1989). Individual tree harvesting and bucking decisions may also be considered as another class of integer decisions for either manual felling and processing or mechanized harvesting, and have been described with dynamic programming and rule-based algorithms (Kivinen and Uusitalo, 2002; Murphy *et al.*, 2004).

4.3 Location of Forestry Facilities

Sort yard location, a centralized place where forest products are sorted into classes and distributed to various processing facilities, has been described with network-based heuristics (Sessions and Paredes, 1987). In these types of problems, a planner is attempting to minimize costs or maximize net revenues by locating a temporary destination, along a path from the woods to the mill, to sort logs into different wood products. With a sort yard, a forest manager may be able to better meet the needs of various mills in their region. The decision choices in this type of planning process include determining the appropriate number and location of sort yards within a transportation network. Shortest path heuristics have been used to address the network problem.

4.4 Road Network Design

Foresters and forest engineers often attempt to develop transportation systems that have the lowest total road cost, while also protecting the associated soil and water resources (Akay, 2003). These land managers have to examine a large number of road design alternatives to ensure that the routes selected are cost-effective. These types of management decisions are very complex, and in the absence of computer-aided road design algorithms, could easily be made inefficiently. The decision to develop roads in areas void of a transportation system is guided by economic criteria as well as topographic and physiographic aspects of a landscape. Virtually all applications of road design involve manual trial and error processes, but there has been substantial recent activity in developing road design algorithms that use heuristics such as simulated annealing, tabu search, and genetic algorithms (Akay *et al.*, 2005; Aruga *et al.*, 2005).

5 CAPTURING THE ESSENCE OF PROBLEMS IN MODELS

One of the main limitations of most modeling efforts relates to the validation of the modeling process, which is inherently problematic. Some of the results of planning processes, such as projected harvest levels and spatial patterns of activities, can be compared against recent activity, however planning processes do not lend themselves well to validation procedures. When developing plans of action, a number of real-system dynamics are contingent on factors that may have unknown or uncertain distributions, such as climate change and human population growth (Carpenter, 2002). Therefore, the uncertainty surrounding plans is difficult to compute. Evaluating alternative plans of action have value, however. These force managers to think through decisions when accurate predictions of actions and consequences are not possible, and they broaden people's perspectives, and may tend to challenge conventional thinking (Carpenter, 2002).

As with many modeling efforts, there is considerable room for improvement of the modeling process. Linear and integer programming allow planners to represent management systems reasonably well, although when certain constraints are not represented, the results should be viewed as those from "relaxed" problems. Heuristics and simulation models can facilitate the modeling of an unlimited number of integer variables and complex resource assessments, although one loses the ability to say with confidence that the results obtained are the best one can hope to find for a specific problem. Ultimately, planners must acknowledge that they can never develop a complete and perfect representation of real management systems, since many uncertainties and unknown interactions make this task impossible. Finding the level of comfort (in quality of data, system dynamics acknowledged, etc.) for both the decision maker and planner may be the main challenge in capturing the essence of problems in forest-planning models.

6 DATA ACQUISITION, RELIABILITY, AND UNCERTAINTY

Data-development processes comprise perhaps 50–75% of the time required to develop forest plans. A typical industrial forestry organization may spend 2–3 months each year updating Geographic Information Systems and associated inventory databases before developing forest plans. A number of challenges face the integration of data and models for forest planning efforts. Next, we provide a few examples of issues related to forest-planning that should be considered, including those related to vegetation databases, topography, individual tree measurements, and product quality.

6.1 Vegetation Databases

Before 1960, one could speculate that the ability to solve forest-planning problems was limited by solution methods, not data. However, advances in data acquisition methods have not kept pace with the rapid advancements in solution capability. Many land managers, however, are still using inventory methods with which they have high confidence at the forest level, but have low confidence at the individual strata level. Others are using broad strata averages to represent the vegetation characteristics of individual stands. There is concern that allocating strata averages spatially across the landscape may provide the illusion of data adequacy for spatial planning, but this has not been quantified. A common method of gathering data is to sample photo-identified strata, or stand management record strata, with ground-based plot measurements. More recently, satellite images coupled with field plots and various heuristics have been used to classify the images, and provide estimates of vegetation data at the pixel level, with aggregation algorithms used to form stands (e.g., Ohmann and Gregory, 2002).

6.2 Topography

Forest-planning efforts since the 1960s have also used topographic maps to assist with the layout of harvest units and road systems. These have typically been derived from aerial photographs, combined with a sampling of ground elevations. The precision of contours was a function of the height of the vegetation at the time of estimation. With the advent of global positioning systems (GPS), airborne lasers (light detection and ranging, LIDAR) are becoming a practical way to delineate topographic features. LIDAR provides a surface profile from which vegetation and ground surface elevations are separated by data processing algorithms (Lefsky *et al.*, 1999; Drake *et al.*, 2002). The current accuracy of these devices in forested terrain is about 30 cm, but measurement and algorithms continue to improve. LIDAR promises to improve both vegetation measurement and ground profile measurement.

6.3 Projections of Individual Tree Measurements

Measurements of individual trees are used in algorithms that allow foresters to optimize the products derived from trees at the time of cutting (felling and bucking), and in stand- or forest-level planning processes that include constraints that are closely associated with forest structure (i.e., wildlife habitat models). Stem optimization requires accurate descriptions of tree form and wood quality. Stand- or forest-level optimization may require the development of growth regimes, which requires predicting tree size, stem taper, branch frequency, and size over time. Growth models have been developed for many important species, but uncertainty concerning genetics, precipitation, disturbance, reactions to disturbance, and nutrient availability cloud forecasting. Shorter-term projections are usually more accurate than longer term forecasts. Models such as ORGANON (Hann *et al.*, 1997) and the Forest Vegetation Simulator (Dixon *et al.*, 2004) can be used to provide the growth and yield projections necessary for stand- or forest-level planning.

6.4 Product Quality

For operational decisions such as tree bucking, the major obstacles are collecting stem form data and identifying interior defects. Manual measurement of stem properties in real time has not been successful, except for training and auditing purposes. The increase in mechanization of tree-felling operations has provided access to rapid survey of the stem during the delimbing operation. Mechanical methods that rely on measuring wheels to record length can have bark and weather-related challenges. Use of ultrasonic, x-ray, lasers, and other scanning technologies has the potential to provide surface quality and interior defect identification.

7 SOLUTION ALGORITHMS

The main types of solution algorithms used in stand-level planning have been dynamic programming, heuristics developed from dynamic programming, and simulation. In forest-level planning, linear programming, integer programming, heuristics, and simulation have all been used. In North America, a shift in planning from a dependence on LP to heuristics has been shown, although LP and other exact techniques continue to be used to illustrate strategic forest plans, plans without spatial components, or relaxed solutions to complex forest-planning problems (Bettinger and Chung, 2004). Mathematical-programming techniques have evolved to support the development of forest plans with complex non-timber goals, and spatial components within forest-planning processes have increased dramatically in the last decade (Bettinger and Sessions, 2003).

The initial papers on forest-level planning in North America include those related to linear and goal programming in forest management (Curtis, 1962; Loucks, 1964; Kidd, *et al.*, 1966; Kidd, 1969; Thompson and Haynes, 1971; Ware and Clutter, 1971; Field, 1973; Leuschner *et al.*, 1975). Of the heuristic methods, Monte Carlo programming (Nelson and Brodie 1990), simulated annealing (Lockwood and Moore, 1993; Murray and Church, 1995; Öhman and Eriksson, 1998; Sessions *et al.*, 2000), threshold accepting (Bettinger *et al.*, 2003), tabu search (Bettinger *et al.*, 1997, 1998, 1999; Boston and Bettinger, 1999; Caro *et al.*, 2003), and genetic algorithms (Falcǎo and Borges, 2001) have all been used in forest-level planning. The mathematical description of the forest management problem is found in earlier chapters. Readers interested in a general description of the primary heuristic methods in current use are referred to Glover and Kochenberger (2003).

The solution methods for operational forest problems reflect the wide range of problems being solved, from rule-based systems to network-based algorithms, LP heuristic combinations, as well as the more recent meta-heuristics (simulated annealing, threshold accepting, tabu search, and genetic algorithms). The wide acceptance of heuristics in forest operational problems generally reflects the difficulty with handling the large number of integer variables in mixed integer programming that is inherent with formulating these problems.

8 EXPERIENCE IN APPLICATIONS

A number of recent applications are available to illustrate the use of operations research techniques in natural resource management. We present three applications here: the first one applies operations research techniques to the planning needs of a state agency, the second applies the techniques to a large-scale forest landscape planning assessment, and the third illustrates how the techniques can be used in operational log-bucking decisions.

8.1 Oregon Department of Forestry

The Oregon Department of Forestry (ODF) uses simulated annealing to schedule harvests. The four main forest districts are 40,000 to 100,000 ha with average parcel size of about 1.0 ha. The forest goals include maximizing present net value, achieving a desired forest structure in a minimum amount of time, and controlling a log normal distribution of floating patches of complex forest on the landscape during each 5-year period of a 30-period-planning problem. Patches range from 20 to 1000 parcels. Harvest scheduling includes recognizing greenup constraints (48.5 ha) with each opening including 30–80 parcels. An explicit road system tree is recognized. Calculation of net present value includes construction, reconstruction, haul, and road maintenance. The scheduling model is a spatial Model II with approximately 100 thinning prescriptions available for each parcel, and a maximum of four regeneration harvests per parcel over the planning horizon. Eligible rotation length varies from 40 years to 150 years depending on land classification. Upslope and riparian prescriptions must be coordinated to recognize logging feasibility. The objective function is formulated as a goal-programming model with even one-way or two-way goal penalties for all goals except maximum clear-cut size which is a legal requirement and is represented as a constraint.

Each parcel has nine binary decision variables to describe regeneration times and intermediate silvicultural activities, so problem size is 300,000 to 900,000 variables. Solution time varies from 15 to 20 min on a 3.2 GHz computer. The planning team operates in real time using several machines and varies the goal weights to explore alternative solutions. The cost of data preparation including harvest and transportation system planning, prescription generation, and GIS support (excluding timber inventory cost) is approximately US$ 2–3 per ha. To ensure the harvest-scheduling model produces reliable answers, two auxiliary programs have been developed. One, a GIS-based tool, checks independently to see the spatial solution "follows the rules" and the other, a database tool, makes an independent check to see the solution is computationally correct. In a complex problem like this, independent checks have been found to be essential.

8.2 Coastal Landscape Analysis and Modeling Study (CLAMS)

The CLAMS project involves a large-scale analysis of management behavior and policies across a long time frame. A simulation model (LAMPS, Bettinger and Lennette, 2004) was developed as a strategic-planning model that can accommodate tactical-planning relationships. The time horizon used in the LAMPS model is 100 years, divided into twenty 5-year time periods. Each landowner group across the landscape is recognized, and a simulation of management behavior for the entire landscape is made. LAMPS includes Monte Carlo simulation processes as well as binary search, yet recognizes a hierarchy of spatial units. Both unit restriction and area restriction adjacency restrictions can be modeled as well as an unlimited array of green up periods. Harvest blocks are built dynamically as the simulation model progresses, transition probabilities determine the type of forest that returns after a regeneration harvest, and riparian and leave tree policies can be modeled. Interior habitat areas can be developed on state-managed land as well. The LAMPS model can simulate up to about 650,000 ha at one time, recognizing over 5 million decision variables; as a result, a single simulation could require as much as 2 h of processing time. The data input requirements are heavy, as over 300 Mb might be required. Output from the LAMPS model can be as great as 1 Gb, as a number of characteristics of each decision variable can be reported, including inventory tree lists used to describe the landscape over time.

8.3 Meeting Log Supplier Order Book Constraints

Modern mechanized harvesters are often fitted with sensors that measure stem dimensions and with computers that optimally buck each stem to maximize the value gained based on stem dimensions, qualities, log prices, and desired specifications. Optimally bucking individual stems, based on market prices, is unlikely to provide yields that meet order book constraints at the harvest unit or forest level.

An adaptive control heuristic was developed by imbedding an individual stem optimal bucking dynamic-programming procedure in a threshold accepting algorithm which adjusts relative prices and minimum small end diameter specifications to meet order book constraints (Murphy *et al.*, 2004). The heuristic was tested on four radiata pine plantation stands where the location and detailed stem description of every tree was known. Three of the stands were virtual stands designed specifically to test the adaptive control heuristic. The fourth stand was a real-world stand. Improvements in meeting order-book, target proportions were found for all four stands when the heuristic was used; 17% to 22% improvement with pre harvest inventory data and 19% to 26% improvement with stem information gathered as the harvester works its way through the unit. These results are similar to those reported by Kivinen and Uusitalo (2002) who found a 20% improvement in four Norway spruce stands when applying fuzzy logic to tree-bucking control where perfect information was available on all trees within the stand. Recent research using a genetic forest-level algorithm-based control system to generate stand-specific log-demand distributions has shown promise for improving bucking-to-order deliveries (Kivinen, 2006).

9 MAIN CHALLENGES

As with many types of systems that forestry organizations attempt to implement, forest planning, whether strategic, tactical, or operational, requires several important aspects. First, having access to the appropriate people is critical. This includes people to assist with the identification of reasonable forest management prescriptions (field foresters), to develop databases (inventory specialists, geographic information systems analysts), to develop and specify parameters of policies (managers, policy analysts), and to run the algorithms and report results in a manner than can be understood by decision makers. Second, having access to the appropriate type of databases is essential, and these include databases related to the vegetation, land ownership, stream systems, road systems, and other aspects of the landscape for which decisions will be based or effects measured. Third, technology, such as the type of computers and computer-programming languages used is critical. Investing in current computer technology could allow both larger models to be solved, and speed up the analytical cycle (i.e., development of scenario, running of model, analyzing results, re development of scenario, etc.). Of equal importance are algorithms that are flexible enough to change as problems or goals of an organization change. Finally, institutional acceptance and funding of the planning process is important. Too often organizations will under fund, discount, or under utilize the results of planning efforts. A limitation in one or more of these will likely reduce the success of the effort (Bettinger, 1999).

10 ROLE OF DEVELOPING TECHNOLOGIES

A number of the advances made in other sciences can be of value in optimizing decisions made at the strategic, tactical, or operational levels in natural resource management organizations. For example, some current pre harvest inventory systems are being used to match wood to markets based on optimal bucking of a sample of individual standing stems that have been described in detail. Mechanized harvesters, with their measurement technologies and on-board computer power, now provide an opportunity to augment or, in some cases replace, traditional pre harvest inventory systems and to reduce inventory costs. Recent studies of harvester-based inventory systems in Australia, and harvester measurements in New Zealand, Oregon, and Canada indicate that length and diameter measurement errors may be no worse than those associated with standing tree measurements for pre harvest inventory systems (Murphy *et al.*, 2004). The effects of speed, distance the observer is from the point of interest on the stem, operating conditions, and so on, need to be evaluated before we can say whether a better description of stem quality can be obtained from a harvester cab or from an on-the-ground cruiser.

GPS are being used to map forest conditions, track vehicle travel, and provide precise positioning for land-surveying efforts. GPS units fitted to a harvester are likely to be used for area measurements in the near term. Although research indicates that GPS may be as accurate as traditional methods for determining area at the stand or harvest unit level, more research is required to determine the effect of inventory plot, boundary location, GPS errors on yield predictions.

LIDAR offers a powerful new data source to improve operational planning. For large contiguous areas, DTMs can now be quickly produced for $1.00 to $3.00 per ha depending on resolution. The availability of high-quality digital data makes optimal road design in forested areas practical, by greatly reducing data costs over field data collection and greatly improving accuracy over previously available remote data acquisition techniques. The availability of high-quality LIDAR provides the data required for optimal road design, and optimal logging planning in steep terrain.

The availability of real-time data promises to reduce uncertainty in planning-allowing certain scheduling problems to be “measured and managed” in real time. Examples include the truck-routing heuristic ASICAM (Weintraub *et al.*, 1996), which is used to schedule truck trips before each day and has the potential to be rerun to reschedule trucks during the day as new information becomes available. Another example of real-time data is the measurement of road roughness by logging trucks for road maintenance scheduling. Individual trucks are equipped with recording devices that record road roughness. Road roughness measurements are downloaded along with GPS locations at the scale station and road maintenance vehicles are scheduled. Reported reductions in road management costs have been approximately 20%.

The increasing portability of computing devices also promises to be able to utilize real-time data more effectively. For example, during harvest plan implementation hidden rock walls, wet places, or other topographic obstacles unknown when the "optimal" road location was developed, can be resolved in the field to identify the next best solution. Replanning can be done when suitable stump anchors or intermediate supports are later identified as not available during field implementation. In addition, tree-bucking decisions can be modified in the field as either market or stand conditions change. Murphy *et al.* (2004) found that using a heuristic to adjust prices and specifications on a mechanized harvester, based on stem data from the previously harvested block, was better than solely relying on pre harvest inventory data to control product yields for the next block.

11 REFERENCES

Akay, A.E. 2003. Minimizing Total Cost of Construction, Maintenance, and Transportation Costs with Computer-Aided Forest Road Design, Ph.D. thesis. Oregon State University, Corvallis, OR, 229 pp.

Akay, A., K. Boston, and J. Sessions. 2005. The evolution of computer-aided road design systems. *Journal of Forest Engineering* 16(2):73–79.

Aruga, K., J. Sessions, and A. Akay. 2005. Heuristic planning techniques applied to forest road profiles. *Japan Journal of Forest Research* 10(2):83–92.

Bettinger, P. 1999. Distributing geographic information systems capabilities to field offices: benefits and challenges. *Journal of Forestry* 97(6):22–26.

Bettinger, P., K. Boston, Y.-H. Kim, and J. Zhu. In press. Landscape-level optimization using tabu search and stand density-related forest management prescriptions. *European Journal of Operational Research*. 176(2): 1265–1282.

Bettinger, P., K. Boston, and J. Sessions. 1999. Intensifying a heuristic forest harvest scheduling search procedure with 2-opt decision choices. *Canadian Journal of Forest Research* 29:1784–1792.

Bettinger, P., and W. Chung. 2004. The key literature of, and trends in, forest-level management planning in North America, 1950–2001. *The International Forestry Review* 6:40–50.

Bettinger, P., D.L. Johnson, and K.N. Johnson. 2003. Spatial forest plan development with ecological and economic goals. *Ecological Modelling* 169:215–236.

Bettinger, P., and M. Lennette. 2004. *Landscape Management Policy Simulator (LAMPS), Version 1.1 User's Guide*. Forest Research Laboratory, Oregon State University, Corvallis, OR. Research Contribution 43:117 pp.

Bettinger, P., and J. Sessions. 2003. Spatial forest planning: to adopt, or not to adopt? *Journal of Forestry* 101(2):24–29.
Bettinger, P., J. Sessions, and K. Boston. 1997. Using tabu search to schedule timber harvests subject to spatial wildlife goals for big game. *Ecological Modelling* 94:111–123.
Bettinger, P., J. Sessions, and K.N. Johnson. 1998. Ensuring the compatibility of aquatic habitat and commodity production goals in eastern Oregon with a Tabu search procedure. *Forest Science* 44:96–112.
Boston, K., and P. Bettinger. 1999. An analysis of Monte Carlo integer programming, simulated annealing, and tabu search heuristics for solving spatial harvest scheduling problems. *Forest Science* 45:292–301.
Brodie, J.D., D.M. Adams, and C. Kao. 1978. Analysis of economic impacts on thinning and rotation for Douglas-fir using dynamic programming. *Forest Science* 24:513–522.
Caro, F., M. Constantino, I. Martins, and A. Weintraub. 2003. A 2-opt tabu search procedure for the multiperiod forest harvest scheduling problem with adjacency, greenup, old growth, and even flow constraints. *Forest Science* 49:738–751.
Carpenter, S.R. 2002. Ecological futures: building an ecology of the long now. *Ecology* 83:2069–2083.
Chung, W., J. Sessions, and H. Heinimann. 2004. An application of a heuristic network algorithm to cable logging layout design. *Journal of Forest Engineering* 15(1):11–24.
Clark, M., R. Meller, and T. McDonald. 2000. A three stage heuristic for harvest scheduling with access road network development. *Forest Science* 46:204–218.
Cousar, P. 1992. A Multivariate Control Solution to the Mixed Species/Diameter Class Thinning and Final Rotation Problem, M.S. thesis. Department of Forest Resources, Oregon State University, Corvallis, OR, 49 pp.
Curtis, F.H. 1962. Linear programming the management of a forest property. *Journal of Forestry* 60(9):611–616.
Dixon, G.E., A.R. Stage, N.L. Crookston, R.A. Monserud, M. Moeur, D. Ferguson, W.R. Wykoff, and D.A. Hamilton, Jr. 2004. *Essential FVS: A User's Guide to the Forest Vegetation Simulator*. USDA Forest Service, Forest Management Service Center, Ft. Collins, CO, 196 pp.
Drake, J.B., R.O. Dubayah, D.B. Clark, R.G. Knox, J.B. Blair, M.A. Hofton, R.L. Chazdon, J.F. Weishampel, and S.D. Prince. 2002. Estimation of tropical forest structural characteristics using large-footprint lidar. *Remote Sensing of Environment* 79:305–319.
Eng, G., H. Daellenbach, and A. White. 1986. Bucking tree-length stems optimally. *Canadian Journal of Forest Research* 16:1030–1035.
Epstein R., E. Nieto, P. Chevalier, J. Gabarro, and A. Weintraub. 1999. A system for short term harvesting. *European Journal of Operations Research* 119:427–439.
Epstein, R., A. Weintraub, J.B. Sessions, J. Sessions, P. Sapunar, E. Neto, F. Bustamante, and H. Musante. 2006. A combinatorial heuristic approach for solving real size machinery location and road design problems in forestry planning. *Operations Research Applications.* 54(6): 1017–1027.
Falcão, A.O., and J.G. Borges. 2001. Designing an evolution program solving integer forest management scheduling models: an application in Portugal. *Forest Science* 47:158–168.
Field, D.B. 1973. Goal programming for forest management. *Forest Science* 19:125–135.
Glover, F., and G. Kochenberger. 2003. *Handbook of Metaheuristics*. Kluwer Academic Publishers, Dordrecht, 556 pp.
Graetz, D., J. Sessions, and S. Garman. In press. Using stand-level optimization to reduce crown fire risk in the Blue Mountains of eastern Oregon. *Landscape and Urban Planning.*
Haight, R.G., J.D. Brodie, and D.M. Adams. 1985. Optimizing the sequence of diameter distributions and selection harvests for uneven-aged management. *Forest Science* 31:451–462.
Haight, R.G., R.A. Monserud, and J.D. Chew. 1992. Optimal harvesting with stand density targets: managing Rocky Mountain conifer stands for multiple use outputs. *Forest Science* 38:554–574.

Hann, D.W., A.S. Hester, and C.L. Olsen. 1997. *ORGANON User's Manual: Edition 6.0.* Forest Resources Department, Oregon State University, Corvallis, OR.

Hof, J.G., and L.A. Joyce. 1992. Spatial optimization for wildlife and timber in managed forest ecosystems. *Forest Science* 38:489–508.

Johnson, K.N., and H.L. Scheurman. 1977. Techniques for prescribing optimal harvest and investment under different objectives – discussion and synthesis. *Forest Science Monograph* 18, 31 pp.

Jones, G., J. Hyde III, and M. Meacham. 1986. *Four Analytical Approaches for Integrating Land Management and Transportation Planning on Forest Lands.* USDA Forest Service, Intermountain Research Station, Ogden, UT. Research Paper INT-361, 33 pp.

Kidd, W.E. 1969. Forest regulation and the alternative rate – a case illustration. *Journal of Forestry* 67(12):885–887.

Kidd, W.E., E.F. Thompson, and P.H. Hoepner. 1966. Forest regulation by linear programming – a case study. *Journal of Forestry* 64(9):611–613.

Kivinen, V.-P. 2006. A forest-level genetic algorithm based control system for generating stand-specific log demand distributions. *Canadian Journal of Forest Research* 36(7): 1705–1722.

Kivinen, V.-P., and J. Uusitalo. 2002. Applying fuzzy logic to tree bucking control. *Forest Science* 48:673–684.

Laroze, A.J., and B.J. Greber. 1997. Using tabu search to generate stand-level, rule-based bucking patterns. *Forest Science* 43:157–169.

Lefsky, M.A., W.B. Cohen, S.A. Acker, G.G. Parker, T.A. Spies, and D. Harding. 1999. Lidar remote sensing of the canopy structure and biophysical properties of Douglas-fir western hemlock forests. *Remote Sensing of Environment* 70:339–361.

Leuschner, W.A., J.R. Porter, M.R. Reynolds, and H.E. Burkhart. 1975. A linear programming model for multiple-use planning. *Canadian Journal of Forest Research* 5: 485–491.

Lockwood, C., and T. Moore. 1993. Harvest scheduling with spatial constraints: a simulated annealing approach. *Canadian Journal of Forest Research* 23:468–478.

Loucks, D.P. 1964. The development of an optimal program for sustained-yield management. *Journal of Forestry* 62(7):485–490.

McNaughton, A., M. Ronnqvist, and D. Ryan. 1998. The scheduling of forest harvesting with adjacency constraints. *33rd ORSNZ Conference Online Proceedings*, Operations Research Society of New Zealand (accessed 28 November 2004: http://www.esc.auckland.ac.nz/Organisations/ORSNZ/conf33/papers/p75.pdf).

Murphy, G. 1998. Allocations of stands and cutting patterns to logging crews using a tabu search heuristic. *International Journal of Forest Engineering* 9(1):31–37.

Murphy, G., H. Marshall, and M.C. Bolding. 2004. Adaptive control of bucking on harvesters to meet order book constraints. *Forest Products Journal* 54(12):114–121.

Murphy, G.E., I. Wilson, and B. Barr. 2006. Developing methods for pre-harvest inventories which use the harvester as the sampling tool. *Australian Journal of Forestry* 69(1):9–15.

Murray, A.T., and R.L. Church. 1995. Heuristic solution approaches to operational forest planning problems. *OR Spektrum* [*Operations Research*] 17:193–203.

Nelson, J., and J.D. Brodie. 1990. Comparison of a random search algorithm and mixed integer programming for solving area-based forest plans. *Canadian Journal of Forest Research* 20:934–942.

Öhman, K., and L.O. Eriksson. 1998. The core area concept in forming contiguous areas for long-term forest planning. *Canadian Journal of Forest Research* 28:1032–1039.

Ohmann, J.L., and M.J. Gregory. 2002. Predictive mapping of forest composition and structure with direct gradient analysis and nearest-neighbor imputation in coastal Oregon, U.S.A. *Canadian Journal of Forest Research* 32:725–741.

Palmgren, M., M. Ronnqvist, and P. Varbrand. 2004. A near exact method for solving the log-truck scheduling problem. *International Transactions in Operational Research* 11:447–464.

Paredes V., L. Gonzalo, and J.D. Brodie. 1987. Efficient specification and solution of the even-aged rotation and thinning problem. *Forest Science* 33:14–29.

Pukkala, T., and J. Miina. 1997. A method for stochastic multiobjective optimization of stand management. *Forest Ecology and Management* 98:189–203.

Richards, E., and E. Gunn. 2000. A model and tabu search method to optimize stand harvest and road construction schedules. *Forest Science* 46:188–203.

Sessions, J., D. Johnson, J. Ross, and B. Sharer. 2000. The Blodgett Plan: an active-management approach to developing mature forest habitat. *Journal of Forestry* 98(12): 29–33.

Sessions, J., E. Olsen, and J. Garland. 1989. Tree bucking for optimal stand value with log allocation constraints. *Forest Science* 35:271–276.

Sessions, J., and G. Paredes. 1987. A solution procedure for the sort yard location problem in forest operations. *Forest Science* 33:750–762.

Thompson, E.F., and R.W. Haynes. 1971. A linear programming-probabilistic approach to decision making under uncertainty. *Forest Science* 17:224–229.

Van Deusen, P.C. 1999. Multiple solution harvest scheduling. *Silva Fennica* 33:207–216.

Ware, G.O., and J.L. Clutter. 1971. A mathematical programming system for the management of industrial forests. *Forest Science* 17:428–445.

Wedin, H. 1999. Stand Level Prescription Generation under Multiple Objectives, M.S. thesis. Department of Forest Resources, Oregon State University, Corvallis, OR, 178 pp.

Weintraub, A., R. Epstein, R. Morales, J. Seron, and P. Traverso. 1996. A truck scheduling system improves efficiency in the forest industries. *Interfaces* 26(4):1–12.

Weintraub, A., G. Jones, A. Madgenzo, M. Meacham, and M. Kirby. 1994. Heuristic system to solve mixed integer forest planning models. *Operations Research* 42:1010–1024.

Yoshimoto, A., R. Haight, and J.D. Brodie. 1990. A comparison of the pattern search algorithm and the modified PATH algorithm for optimizing an individual tree model. *Forest Science* 36:394–412.

Yoshimoto, A., G. Paredes, and J.D. Brodie. 1988. Efficient Optimization of an Individual Tree Growth Model. USDA Forest Service, Rocky Mountain Forest and Range Experiment Station, Ft. Collins, CO. General Technical Report RM-161, pp. 154–162.

Chapter 24

FORESTRY ECONOMICS: HISTORICAL BACKGROUND AND CURRENT ISSUES

Ronald Raunikar and Joseph Buongiorno
Department of Forest Ecology and Management, University of Wisconsin, 1630 Linden Drive, Madison, WI 53706

Abstract The Faustmann model embodies the application of fundamental economic principles to the choice of management methods and alternative land uses. The price of products is a key input in applications of this principle. For wood prices, forest sector models provide a link between forestry and the rest of the economy, and thus a means to predict wood prices consistent with expected demographic and economic growth. To include the price of non-wood forest outputs, Faustmann's principle needs to be expanded with modern benefit-cost analysis. Evaluation of forest amenities in the absence of markets poses conceptual challenges and requires special analytical techniques of contingent valuation. New concepts such as the environmental Kuznets curve and the Porter hypothesis provide the means to forecast future demand for forest amenities and better analyze the consequences of public policy choices.

...theory is only a tool for investigating practice, like a spade for digging up facts and converting them into an understandable system...

John R. Commons (1934)

Everybody thinks of economics whether he is aware of it or not. In joining a political party and in casting his ballot, the citizen implicitly takes a stand upon essential economic theories.

Ludwig Edler von Mises (1949)

Keywords: Harvest scheduling, optimization, adjacency, green-up, area restrictions

1 INTRODUCTION

Reducing forest ecosystems to money is understandably unappealing to those intimate with the full complexity and beauty of forested landscapes. However, economics encompasses much more than money. Economists study how people organize themselves to satisfy their various needs. As foresters, not only can we find many practical tools in economics, but we are also constantly choosing a particular economic theory by our actions.

The choices we make shape the capacity of forests to provide future generations with wood products, fresh water, and wildlife. Our actions determine the forests of the future and are shaped in large part by economics. The industrial revolution of Europe and the settlement of the Americas were economic watersheds that changed forever the nature and extent of forests. Presently, economic forces are largely deciding how much forest, if any, will remain in the Amazon and the Congo basin.

The purpose of this paper is to review briefly some major contributions of economic thinking to forestry, and to consider how modern economic concepts can help in the wise management of forests.

2 WOOD PRODUCTION AND EARLY FOREST ECONOMICS

Wood has been throughout human history an essential element of civilization, for fuel, construction, transport, and defence (Winters, 1974). The art of forest management for wood production has long been organized in a set of general principles founded on the scientific method. Over time, this has led to elaborate rules to optimize forest harvests in sustainable fashion. In the process, some foresters realized early on that economic principles were critical for good forest management.

Few ventures require as much time between investment and results as growing trees. Much of the rest of the economy evolves in a groping process. In relatively short order, firms either learn to operate efficiently enough to stay in business or they quickly disappear. They learn by themselves or by using the example of more successful firms. Although forestry enterprises do also partake of this selection process, the results are generally slower, and that other managers are good examples to follow is less obvious. Thus for foresters well-considered economic theory is of paramount importance.

This importance has been recognized for so long that solutions of some forest economic problems predate the discovery of similar results in general economics. As early as 1790 William Marshal noted that trees must be felled before they achieve their full growth or else both interest and the use of the land are lost by waiting. Richard Watson made the incomplete claim in 1794 that a tree should be cut when the increase in value due to growth is less than interest. Marshal subsequently completed Watson's solution by repeating that a tree must be cut when the value of growth is less than interest plus the rent on the land (Scorgie 1996).

More famous is Martin Faustmann's rigorous derivation of the value of forestland and of immature stands for forestry, and his attendant computation of the economic rotation (Faustmann, 1849). General economists largely missed Faustmann's insight, and it was not until the 1930s that, through the work of Irving Fisher and others, the general theory of investment was formulated as soundly as Faustmann's formula. Interestingly, Fisher, though possibly "the greatest single economic writer on interest and capital," gave the wrong solution to Faustmann's problem (Samuelson 1976).

In seeking the value of forestland, Faustmann recognized that it must be equal to the value of the net returns that one could expect from that land, if it were used in forestry. However, much of these returns would occur very far into the future, so that they would be worth less now, that is, they must be discounted at a suitable interest rate. Thus, the land value had to be equal to the net present value (NPV) of the full future stream of costs incurred and benefits derived from the forest. This was and remains a remarkable insight. It gives unambiguously the general principle to follow in choosing between forest management alternatives and different land uses: maximize the land expectation value.

By simplifying the forest management problem to only the value of wood harvested from an even-aged stand, Faustmann showed that the economic optimal rotation is less than the rotation that produces the maximum average annual biological yield. This conclusion seems to contradict the intuition that higher average annual production must also mean higher income. Faustmann's insight was to recognize that, besides the magnitude of the harvests over the rotation, their timing also matters.

Cutting and selling early gives income to either consume or to re-invest in forestry or alternative investments. The interest rate reflects this opportunity cost of postponing a harvest. One elegant aspect of Faustmann's method is that it recognizes the opportunity to plant a new stand of trees earlier when the rotation age is shortened. By summing the costs and revenues of an infinite series of replanted stands, Faustmann accounts for the opportunity cost of these future stands.

Alternatively, Faustmann could have maximized the present discounted value of a single rotation, and included the land rental value of the bare land left at the end of the rotation as Marshall suggested. However, the calculation of the appropriate market land rental rate is fraught with difficulties. It is one of the notable advantages of Faustmann's approach that the land rental value is not needed, but that instead it results directly from his formula. This result can then be compared with the land expectation value for alternative land uses (obtained again with the equivalent of Faustmann's formula, regardless of the land use). The highest land expectation value is, then, the land value for which the annual rental rate can be obtained as $EAI = LEV \cdot r$, where LEV is the land expectation value, r is the yearly interest rate, and EAI is the equivalent annual income, that is the constant annual rent whose discounted present value is equal to the land expectation value.

Faustmann extended his approach to compute the value of immature stands. The method has also been generalized to the case of selection forests, by recognizing that for an uneven-aged stand in the steady state, the land value would be equal to the NPV of expected returns minus the value of the residual stock of trees. Whichever plan leads to the greatest land value (i.e., the greatest return to the fixed input) is optimal. This allows, in particular, the calculation of the best economic cutting cycle in uneven-aged forests.

Despite controversies (Oderwald and Duerr, 1990), Faustmann's rotation appears to be valid for a regulated forest, that is, a forest yielding constant annual production (Chang, 1990), and in fact, it seems to be valid regardless of the initial condition of the forest (Buongiorno, 2001). Faustmann's formula can also be generalized to include benefits in addition to harvested wood. The difficulty, of course, is in determining this non-timber value of forests, a question to which we shall return, below. Another difficulty, not addressed here, lies in choosing a proper interest rate (see Fisher and Krutilla 1975, about the discount rate for natural or environmental resources in general; Harou, 1985 and Leslie, 1987 for forests in particular; Weitzman, 1998, 2001) for a reason to lower discount rates over time; Broome, 1993 for the distinction between discounting future commodities and future well being; and Partridge, 2003 for insight into "pure time preference").

3 MARKOV DECISION PROCESS

Risk of natural catastrophe and other uncertainties are ignored in the simplifying assumptions of the Faustmann analysis, but dealing with biological and economic risk are important to modern forest economic research (Perry and Maghembe, 1989). The Markov decision process (MDP) provides a generalization of Faustmann's model to cover risk. An MDP describes the forest stand and other variables (especially prices) with a matrix of probabilities: each being the probability of a future state given the current state.

Hool (1966) first suggested such a model for even-aged forests, but the first operational application was Lembersky and Johnson's (1975) work with Douglas fir plantations. The Markov model is very general; in particular it can account for correlated price changes (Taylor, 1984). The theory of MDPs is illuminating, in particular it demonstrates the stationarity of optimal decision rules (decisions depend only on system state, independently of its time), and as a result the independence of managing strategies with respect to initial conditions.

Although there have been few applications of MDP's to forestry, they have been shown to be adaptable to uneven-aged as well as even-aged forests (Kaya and Buongiorno, 1987). Powerful numerical solutions are available, based on linear programming, or successive approximation. As a result, it is possible to investigate, in a truly stochastic environment, management strategies with economic and ecological dimensions as objective functions or constraints (Lin and Buongiorno, 1998; Rollin *et al.*, 2005).

4 TIMBER PRODUCT MARKETS

Economic forest sector modeling has made much progress during the past thirty years. The models are used extensively to help set national forest policy (Adams *et al.*, 1996). Even at the international level multi-country models of production, consumption, trade, and prices of wood products (Zhu *et al.*, 1998; Buongiorno *et al.*, 2003) now help decide policy issues.

These models are typically based on equilibrium theory, whereby at every point in time there exists a unique set of prices that clear markets for all products in all countries. Their implementation often involves a combination of techniques: econometrics to estimate key relations, mathematical programming to compute equilibria, and systems dynamics to simulate changes in capacity and other constraints over time (Buongiorno, 1996).

Forest sector models represent a significant advance in how forest policy is decided, and forest decisions are made. In principle, the methods and assumptions are transparent, facilitating greatly the communication of ideas, their critique, and ultimate progress. Still, like all economic models, those of the forest sector lack accuracy. At best, they give an indication of possible direction of changes, given an internally consistent set of assumptions, but the future may turn out quite differently from what the models predict. Thus, in the foreseeable future, the timber prices that foresters should use even in the simplest Faustmann's formula will always be greatly uncertain. In addition, foresters must deal with the rising importance of the complex non-timber values of forests.

Key to these computations of financial performance based on Faustmann's principle is correct assessment of interest rate, prices, and costs. In particular, future prices depend naturally on the demand and supply conditions of the wood products markets. Here again, economics gives foresters useful tools to better understand what causes price to change, and help predict their future direction, if not their exact level.

Econometric models of forest product markets have a long history in forest economics (Buongiorno, 1990). The simplest market model would consist of two equations: one explaining demand from prices and demand shifters (population, income, etc.), the other explaining supply from prices, and supply shifters (resource stock, energy cost, money supply, etc.). The equations are estimated by statistical methods from regional, national, or international data, depending on the context. After calibration, the model can be used for forecasting and policy analysis. For example, given the necessary condition of demand-supply equality, a two-equation demand-supply system is solvable for price as a function of the demand and supply shifters (also called exogenous variables). This reduced-form price equation can then be applied to predict price, conditional on future exogenous variables such as income and population.

Presumably, income and population are themselves predicted by macro-economists and demographers, and in that way a linkage is established between the forest sector and the rest of the economy. Ultimately, the model gives a price projection essential for benefit-cost analysis, including calculations with Faustmann's formula to decide whether to begin, continue, or stop forest production activities.

5 NON-TIMBER VALUES

As the ecological value of forestland is increasingly recognized and understood by foresters and by the public, the non-timber value of forests that stems from their variety of life forms and functions is of growing interest. Economic theory can help define these values in monetary terms, and econometric techniques can be used to measure them.

Hartman (1976) reanalyzed Faustmann's optimal harvest age problem after including non-timber values of a mature forest such as flood control, recreation, and wildlife. He showed that if the services of the mature forest are valued more than the services of a newly planted forest, then it is optimal to extend the harvest age beyond the Faustmann solution computed with timber prices only.

Strang (1983) argued further that there might be situations where even though a finite local best rotation may exist; the true global optimum may be quite different. In particular, Strang showed that it might be preferable never to cut an existing old growth forest, due to the considerable non-timber values embedded in the old-growth forest.

6 EXTERNALITIES, PIGOVIAN TAXES, AND PROPERTY RIGHTS

Often time, values that derive from the presence of forest cover, such as flood and erosion control may benefit others than the forest owner, so the owner does not include such benefits in the NPV calculation. These values are externalities for the firm making the decision. Since, by definition, externalities do not profit the private firm, the amount of the externality is incidental to its decisions.

If the externality is positive like erosion control then society might optimally desire more than the private firm will provide spontaneously. The private firm may guard against excessive erosion during harvest to a degree because they want to preserve the land fertility, but a municipal water processing plant downstream will want more effective erosion control so they have less silt to remove.

One approach to achieving the socially best level of erosion control is direct regulation, decreeing, and enforcing standards to control erosion. Another is to institute Pigovian taxes. A tax would be waged on the forest owner for each ton of silt in the runoff thus internalizing the cost of the silt to the owner. If the tax is set at the right level, the forest owner will choose the socially optimal degree of erosion prevention. Discovering the correct Pigovian tax rate is the main difficulty in this approach.

A third approach is to establish clear property rights. If the water plant has the right to silt-free water, then the forest owners must pay the water plant for adding silt to its water supply. In this way, the externality of silting the water is internalized to the forest owners who incur an added expense for each ton of silt they add to the water, so they consider that cost in their management decision. For small externalities, the cost of enforcing the property right can make this approach impractical. Monitoring erosion on all of the watersheds and estimating how much purification costs increase for each incident of erosion might be more costly than ignoring the right. Coase (1960) described the conditions under which the optimum amount of an externality will be generated when property rights are well defined.

As another example, cedar rust is a disease that is incubated in red cedar and that attacks the leaves and fruit of apple trees. A 1914 law of the state of Virginia gave apple orchards the right to remove all red cedar trees within two miles of the orchard (Samuels, 1989). Though draconian, the law was an attempt to force cedar owners to internalize the externality they inflicted on orchards. The law meant to give the property right of cedar-free surroundings to orchard owners.

In some cases, property rights are impossible to establish. The beauty of the forest and the protection of threatened species are public goods that someone can enjoy freely without preventing others from doing the same. A public non-timber forest good, such as its beauty, may be valuable to the owners, yet it tends to be under-provided compared to the social optimum. The combined value of a public good for all citizens is greater than the value to each, so as forest owners optimize for themselves, they will provide less than what all citizens desire (Bergstrom *et al.*, 1986).

7 ASSESSMENT OF VALUE

The crucial step in Benefit-Cost Analysis (BCA) is assigning a monetary value to all costs and benefits. To bring some order to this complicated issue, benefit-cost analysts classify the value of the many qualities or outputs of forests as use-value option-value, bequest value, or existence-value. Use-value derives from a particular use of the forest or its products. For example, timber has a use-value, and so do hunting, grazing, and non-consumptive uses such as recreation and flood control.

Option-values pertain to forest resources that might have value in the future. A pharmaceutical use for a biochemical compound produced by an understory species might be discovered in the future. We maintain the option of collecting this value as long as forest conditions allow the species to survive.

Bequest-value is the value of maintaining a resource to pass on to future generations. The satisfaction we gain from the idea of passing a forest intact to future generations is its bequest-value to us. Last, the value that individuals and society derive from the forest merely being there now is its existence-value.

Existence-value may be, but is not necessarily related to some use values. An individual who greatly values the beauty of a forest is using the forest while viewing it, yet the greater part of the value of sightseeing might be affirming the existence of the forest. Others might value the existence of rare animal species harbored by the forest even though they might never see or in any other way use them. As suggested by the subtlety of their definitions, quantifying their value is no easy task.

8 MARKET PRICES

Economists have developed many techniques to do benefit-cost analysis. The easiest use-values to estimate in monetary terms are those that are bought and sold at a "market price."

Suppose the good is timber traded competitively in a region where there are many small private forests, and a large public forest. Under the current policy, the public forest produces nothing, so that the upward sloping line Supply_1 in Fig. 1 represents supply (totally private), while the downward sloping line represents total demand. Demand and Supply_1 cross at B, the quantity of private timber sold and bought is F m^3 per year, and the equilibrium price is A $\$/\text{m}^3$.

Suppose that the managers of the public forest consider producing timber, independently of price level. JG is the amount of public supply, regardless of price. This policy would result in a shift in the total regional supply curve

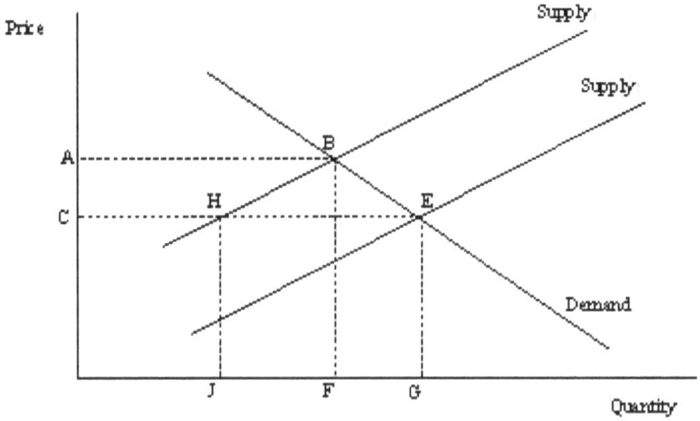

Figure 1. Welfare change due to the provision of a market good.

from Supply_1 to Supply_2. The new equilibrium would then be at a lower price, *C*, for a larger volume bought and sold, *G*. However, the amount sold by private forests would decrease from *F* to *J*. Consequently, there would be an increase in the welfare of timber buyers, measured by the area of the polygon ABEC, but a decrease in the welfare of private sellers, equal to the area of the polygon ABHC. There would then be an increase in total welfare of consumers and producers, equal to the area of the triange BEH.

This is an example of the most straightforward complete welfare accounting, with a full evaluation of benefits and costs, and of winners and losers. It still demands work to estimate the necessary demand and supply relations, and it is not possible without a market. As a result, for public goods, even those with use-value such as forest scenery, the assessment of value in monetary terms is difficult. If we knew the demand schedule for the good, we would calculate the value of, say, a change in supply as we did above, but there are usually no market data to estimate the demand schedule. We can count the number of sightseers and tabulate the time they spend viewing forest scenery, but short of a wall hiding the forest and a tollbooth we can neither limit sightseers nor charge them a fee. In this example, the ability to limit access to or charge for the public good is difficult or impossible; so most consumers will "free ride," that is benefit without paying to view the scenery. One can observe the quantity of a public good that free riders use at zero cost, but construction of the full demand curve requires the methods described later.

In another example, the sequestration of carbon by forests moderates atmospheric changes for everyone regardless of whether they pay. Without a way of charging for the use of the climate we cannot observe how use changes with price. As private forests photosynthesize carbon dioxide, the value of the carbon sequestration to others may not be fully considered by the owner. In choosing to build houses where there was forest, the value of the carbon storage is typically not considered by a private firm, so we cannot directly infer the value of this ecosystem service by observing the behavior of the firm.

For resources with option value such as the unknown medicinal value of forest flora, we might observe a value in an options market. Since the option-value exists because of a potential future market good, rights to that resource could be traded and a market value observed. For example, the pharmaceutical company Merck paid $1 million to bioprospect in Costa Rica under a two-year agreement that spelled out provisions for royalties (Columbia University, 1999). The biotech company Diversa gave the US National Park Service equipment, training, and an advance on royalties to do the same in the Yellowstone hot springs (ten Kate *et al.*, 1998).

To the extent that those deals were competitive, they should reflect some of the expected value of patentable biological discoveries over costs for the companies. Since these agreements include royalties, a claim on the profits generated by the biotechnology, the owners of the genetic resources retain part of the option value. Thus, only part of the expected value is reflected by the up-front fees. However, these two companies are searching for short-term value in the genetic material, which we have retained the option to use by maintaining the ecosystems with those life forms.

The uncertainties Merck and Diversa face in finding valuable biological material are small compared to the difficulty of evaluating longer-term option values. Effective markets for long-term option-value are hard to envision with the short profit horizon typically important to businesses. Some biotechnology companies are currently profitable (Kelly and Mbaria, 2004), but methods for tracing option-values from such market data are not clear.

The option-value of future non-consumptive goods is also not subject to market valuation. In the Midwest of the USA the existence and recreational value of traditional tall grass prairie ecosystems has been discovered recently after the destruction of all but small remnants. The option of re-creating prairies was preserved in the plants remaining in those remnants. Since barely one century ago few would have guessed that the prairie would some day have a non-use value they would have had a hard time guessing how much the present value of that option-value was worth.

The bequest-value and existence-value of forestlands are in part revealed when forestlands are bought and sold. These non-use values are part of the price paid. However, separating bequest values and existence values from the total price paid for the complex bundle of goods represented by the forest is not a simple task. Furthermore, if the price data come from private transactions, to obtain the total value of the forestland we must add its value to the rest of society to the private value.

9 NON-MARKET VALUATION

Economists use three kinds of methods to estimate the value of forest goods that are not traded (Braden *et al.*, 1991). The household production function approach uses the observed trade-off between forest attributes and markets goods to infer the value of the forest attribute. An example is the travel cost method (TCM). The cost of travel to the forest is used to infer the value of visiting the forest for an individual, and more generally the value of different characteristics of forests for different people.

For example, Scarpa and Thiene (2004) determined the value of climbing sites to various climbers in the Northeast Alps using TCM. With this method, Hesseln *et al.* (2004) found the net benefits of hiking in National Forests in Colorado ($12/trip) and Montana ($55/day). The value of the hikes was lower in forests recovering from crown fires, but higher in forests recovering from prescribed burns. Hanley and Ruffell (1992) used TCM to determine the value of the physical characteristics of forests in Canada. Fix and Loomis (1998) found that a mountain bike trip to Moab, Utah was worth $205.

Another approach is hedonic pricing in which a market good is viewed as a bundle of attributes. An implicit value of the attribute is inferred from the differences in prices of goods with various amounts of attributes. For example, the price of houses can be used to infer the amenity value of a neighboring forest. Along with pure housing attributes such as square footage, number of bedrooms, etc. is access to the forest, measured for example by its distance from the house. Given a sufficient number of houses bought and sold, we calculate by regression analysis the best equation to relate house price to the attributes. We can then infer how much more an otherwise equivalent house is worth for being near the forest. This difference is an estimate of the amenity value of the forest. With this technique, Bourassa *et al.*, (2004) found that a house in Aukland, New Zealand, is worth 59% more with a wide view of water. Li and Brown (1980) found that a house is worth $250 more near a conservation area, and $2,800 more next to a recreation area. Although these amenities are capitalized values, they can be transformed into an annual rent value of the forest amenity, with the analog of Faustmann's formula presented above.

Examples of application of hedonic pricing in forestry include Turner *et al.* (1991), and Roos (1995) who inferred the value of particular characteristics of forest estates, such as their location. Scarpa *et al.* (2000) estimated the amenity value of a stand of trees as the opportunity cost owners paid in the timber profit they could have gotten had they tried to maximize profits (according to Faustmann's rule), minus the profit from the timber they actually cut. By then regressing this non-timber value on stand data, they inferred the amenity value of trees of different species and size. They found that for most owners, the amenity value of trees was much larger than their timber value. This opportunity cost approach has also been applied to the value of even-aged forests (Lee, 1997) and mixed-aged mixed-species forests in the southern USA (Raunikar and Buongiorno, 2005a).

These two methods use market price information, for example the cost of travel or the price of a house, to infer non-market use-value. They are revealed-preference methods based on actual observed choices of people. In the contingent valuation method (CVM), instead, we ask individuals about how much they are willing to pay. One advantage of contingent valuation is that it can deal with non-use values such as the existence of a healthy forest, as well as use-values such as viewing that healthy forest (Pease and Holmes, 1993).

As a result, CVM is used extensively in benefit-cost analysis. For example, Xu, *et al.* (2003) found that urban households in the state were willing to pay $31.44 annually to improve biodiversity in the forests of Western Washington. Donovan and Nicholls (2003) found that consumers in Alaska would support a secondary wood products industry in the state. They were willing to pay an $82 premium for a wooden table manufactured locally. Mattson and Li (1993) used the CVM to quantify the value of on-site consumptive use (berry and mushroom picking), on-site non-consumptive use (hiking, and camping), and off-site visual experience. And, Crocker (1985) asked forest visitors their willingness to pay for a visit if the trees at the site showed slight, moderate, and severe damage from air pollution. With these data he estimated the willingness to pay function for air pollution damage to trees.

CVM is subject to biases, especially due to the hypothetical nature of surveys. If respondents do not expect to carry out the hypothetical transactions, they might respond strategically and they have less incentive to gather full information. Still, CVM has become a well-established valuation technique with methods that compensate for biases, despite doubt notably sown by litigants in the Exxon Valdez damages case (Boyle and Bergstrom, 1999). Thus, CVM is vital when revealed-preference methods based on actual observed choices of people are not possible.

10 ENVIRONMENT KUZNETS CURVE FOR FORESTS

The choice to enjoy environmental amenities is an economic choice. In the study of how economies provide themselves with environmental amenities, a characteristic pattern has emerged.

As first noted by Grossman and Krueger (1992) for emissions of sulfur dioxide, the emissions increased with national income among poor nations, but decreased for wealthier economies. This pattern has been dubbed the Environmental Kuznets Curve (EKC) in recognition of the analogous peaking relationship between income inequality and national income first observed by Simon Kuznets (1955).

Many other environmental attributes have been found to follow the EKC pattern such as wastewater, solid waste and other emissions in China (De Groot, *et al.*, 2004), carbon dioxide emissions (Lindmark, 2004; Martinez-Zarzoso and Bengochea-Morancho, 2004), toxic air, water and land emissions (Rupasingha *et al.*, 2004), land degradation (Rodriguez-Meza, *et al.*, 2004), and toxic emissions from new manufacturing facilities (Gleeson, 2004).

This EKC research has also dealt with forests. Bhattarai and Hammig (2001) found a strong EKC relationship between income and deforestation of tropical natural forests in Latin America, Africa, and Asia. They later found that improved agricultural technology and higher educational attainment reduce deforestation (Bhattarai and Hammig, 2004).

Ehrhardt-Martinez *et al.* (2002) supplemented the deforestation EKC for developing countries with effects of institutions and urbanization. Decreased reliance on agriculture is the main factor in tropical deforestation (Barbier and Burgess, 2001).

Cropper and Griffiths (1994) found that per capita national income affected deforestation in both Africa and Latin America, but not in Asia. Stern, *et al.*, (1996) used the published EKC deforestation data to predict an overall forest loss stabilization by 2025, but continued tropical deforestation at a constant rate. Panayotou (1993) using strictly cross-sectional international data reported a turning point in deforestation at $1275 of income per capita in 1985 prices, while Turner (2004) fixed it at $8,700. However, some studies have not found a deforestation EKC at all (Shafik and Bandyopadhyay, 1992; Koop and Tole, 1999; Lantz, 2002; Nguyen Van and Azomahou, 2003).

A few researchers have considered other attributes of forests rather than just their area. Skonhoft and Solem (2001) found no relation between the wilderness (measured by remoteness) of counties in Norway and their income per capita. However, with a more comprehensive measure of forest amenities, Raunikar and Buongiorno (2005b) did find a positive relationship between forest naturalness and income per capita of southern US forested counties.

Economic theory explains the EKC phenomenon as a tradeoff between the desire for consumption goods and environmental amenities in an optimal control context (Keeler *et al.*, 1972). The specific manner of this tradeoff is explained by abatement costs (Andreoni and Levinson, 2001), choice of technology (Stokey, 1998), choice of products (Jones and Manuelli, 2001), emissions as a factor of production (Chavas, 2004), relocation of polluting facilities (Copeland and Taylor, 1994), consumption choices (Rothman, 1998), or migration of household wealth (Gawande *et al.*, 2001).

According to these explanations, economic agents make choices on consumption, investment, production, and public policy that result in the aggregate economy moving along the growth curves in Fig. 2. Understanding this macroeconomic pattern allows foresters to forecast how much and what kind of forests will be needed in the future. This aggregate pattern involves a trade-off between income and forest amenities. More income would be provided along the dashed income line in Fig. 2 at the cost of poorer forest amenities, the dashed forest line. If the economy did not follow that pattern, it means that society, as a whole, values the forest amenities they would have lost more than the extra future income they would have gained. This suggests a macroeconomic method to value non-market forest amenities, based on income trends only.

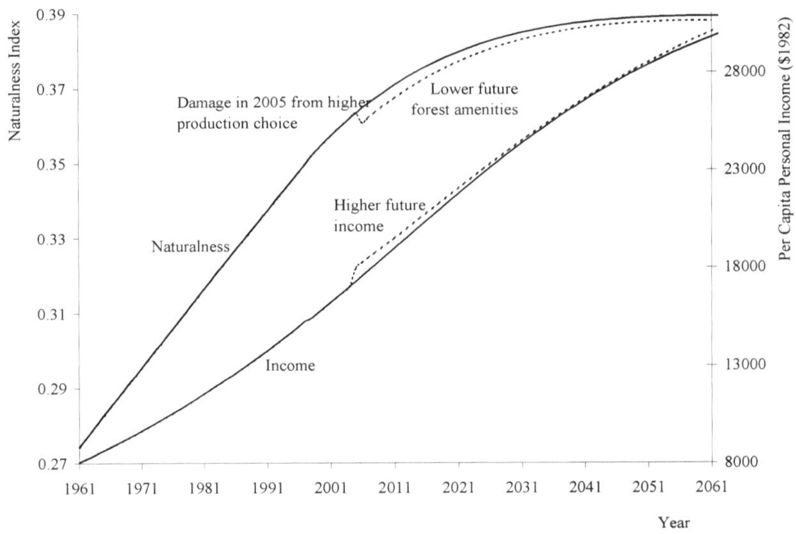

Figure 2. Personal Income growth and forest change in an economy

Another useful product of this approach would be to predict the future demand for natural forests. Such a forecast would help decide on forest policy. With a better grasp of how much forest with old growth characteristics would be needed in coming decades, management could be adjusted to promote the development of those characteristics on the correct amount of land. In using these projections as a guide, we would be recognizing an underlying trend that results from innumerable details of the many choices of consumers and producers, in the private and public sectors. Rather than trying to estimate the value of all goods and services, the EKC analysis leads to a macro relation in the same spirit as the aggregate consumption. It recognizes that values result from the choices of the many competing interests in the economy, some of which are made through the political process.

11 THE PORTER HYPOTHESIS IN FORESTRY?

As foresters like Faustmann have contributed to economics, so have many other disciplines. Business schools have long used case studies to deal with management issues. Based on case studies, Porter (1991) concludes that environmental regulation often stimulates innovation and thus improves manufacturing efficiency rather than hampers it. Further evidence of this phenomenon includes de Vries and Withagen (2005), Roediger-Schluga (2003), Horbach (2003), and Alpay *et al.* (2002).

A key point in the Porter hypothesis is that environmental regulation must be "properly constructed" to stimulate innovation. Roediger-Schluga (2004) describes the effect that technological opportunity, market conditions, patent law, and other factors have in the design of regulations that promote innovation in environmental technology. Ambec and Barla (2002) use an economic model to show that environmental regulation may both spur environmental innovation and increase profits by overcoming organizational inertia.

The stimulation of innovation by constraints has been a fundamental factor in the economic history of the USA. When labor was scarce and resources were abundant, the need for less labor-intensive methods stimulated automation, which greatly increased US productivity and competitive advantage. Using deliberate regulatory choices to recognize that resources have become much scarcer has the potential to stimulate development of efficient technology for the benefit of the USA and of the rest of the world.

Evidence for or against the Porter hypothesis is still scarce in the forest sector. Norberg-Bohm and Rossi (1998) found that the US pulp and paper industry has a strong preference for incremental change. Thus, they conclude that long-term goals for continual environmental improvement would be needed to stimulate radical innovation. Consistent with this conclusion, Marklund (2003) found no improvement in efficiency of Swedish pulp mills in response to tightened emission standards. Cashore and Vertinsky (2000) describe conditions that can deter innovation in sustainable forest management using the case of three North American forest companies. Understanding how and how much regulations stimulate innovation should improve as more cases are examined.

12 CONCLUSION

From Faustmann's classic valuation of forestland to the complex multidimensional choices in modern forest policy, economic principles and methods have contributed much to forest management decisions. Economics helps foresters grapple with the fundamental concept of opportunity cost, as it applies to time and alternative land uses. It gives us the framework and tools to handle risk objectively. Applied to the timber sector, economics is essential to predict the demand, supply, and prices of wood products. In the more difficult realm of amenity values, the methods of benefit-cost analysis are put to work constantly to measure the full social value of forests.

As economic theory and methods continue to develop, new opportunities open for their application to the management of forests. Modern macro-economic growth theory coupled with the EKC suggests a framework to overcome some difficulties of benefit-cost analysis, and finding more directly the value of forest amenities to all of society, and their future demand.

The amenity value of forests is likely to grow in importance in the day-to-day concerns of forest managers. Economics is helping in the assessment of these values. Nevertheless, there are definite limits to economics. Some forest policy issues such as the preservation of species far transcend the purely economics dimension and reach into the realm of ethics and religion. It may, then, be questioned whether economics methodology can truly give useful measures of value in those circumstances.

Conservation goals will most likely be set on broader grounds than purely economic considerations. Nevertheless, the means to reach conservation goals will certainly have an important economic dimension. They involve budgets, reallocation of resources, and sacrifices in current consumption. In sum, there is a very real opportunity cost to any forestry decision. It is the role of economics and its power to determine this cost exhaustively and accurately.

ACKNOWLEDGMENTS

We are grateful to the publishers of the Revue Forestière Française for permission to use some of the material published in the Special number of 1999, pp. 102-116. The preparation of this chapter was supported, in parts, by the USDA-CSREES NRI grant 2003-35400-13816, the PREISM program of the USDA-ERS, the USDA Forest Service Southern Forest Experiment Station, McIntire-Stennis Grant P879, and the School of Natural Resources, University of Wisconsin-Madison.

13 REFERENCES

Adams D.M., Alig R.J., McCarl B.A., Callaway J.M., Winnett S.M. An analysis of the impacts of public timber harvest policies on private forest management in the United States. Forest Science, 19, 1996, 343–358.

Alpay E., Buccola S., Kerkvliet J. Productivity growth and environmental regulation in Mexican and U.S. food manufacturing. American Journal of Agricultural Economics, 84(4), 2002, 887–901.

Ambec S., Barla P. A theoretical foundation of the Porter hypothesis. Economics Letters, 75(3), 2002, 355–360.
Andreoni J., Levinson A. The simple analytics of the environmental Kuznets curve. Journal of Public Economics, 80(2), 2001, 269–286.
Barbier E.B., Burgess J.C. The economics of tropical deforestation. Journal of Economic Surveys, 15(3), 2001, 413–433.
Bergstrom T., Blume L., Varian H. On the private provision of public goods. Journal of Public Economics, 29(1), 1986, 25–49.
Bhattarai M., Hammig M. Institutions and the environmental Kuznets curve for deforestation: a crosscountry analysis for Latin America, Africa and Asia. World Development 29(6), 2001, 95–1010.
Bhattarai M., Hammig M. Governance, economic policy, and the environmental Kuznets curve for natural tropical forests. Environment and Development Economics, 9(3), 2004, 367–382.
Bourassa S.C., Hoesli M., Sun J. What's in a view? Environment and Planning A, 36(8), 2004, 1427–1450.
Boyle K.J., Bergstrom J.C. Doubt, doubts, doubters: the genesis of a new research agenda? Chapter 7 of Valuing Environmental Preferences. Bateman I.J., Willis K.G., editors. Oxford University Press, 1999, p. 645.
Braden J.B., Kolstad C.D., Miltz D. Introduction. Chapter 1 of Measuring the Demand for Environmental Quality. Braden J.B., Kolstad C.D., editors. North-Holland, New York, 1991, p. 370.
Broome J. Discounting the future. Philosophy and Public Affairs, 23(2), 1993, 128–156.
Buongiorno J. Analyse econometrique du secteur forestier: un bilan de l'experience americaine. Cahiers d'economie et sociologie rurales, 15–16, 1990, 146–166.
Buongiorno J. Forest sector modeling: a synthesis of econometrics, mathematical programming, and system dynamics methods. International Journal of Forecasting, 12, 1996. 329–343.
Buongiorno J. Generalization of Faustmann's formula for stochastic forest growth and prices with Markov chain decision process models. Forest Science, 47(1), 2001, 466–474.
Buongiorno J., Zhu S., Zhang D., Turner J., Tomberlin D. The Global Forest Products Model: Structure, Estimation, and Applications. Academic Press, New York, 2003, p. 301.
Cashore B., Vertinsky I. Policy networks and firm behaviours: governance systems and firm responses to external demands for sustainable forest management. Policy Sciences, 33(1), 2000, 1–30.
Chang S.J. Comment on Oderwald and Duerr (1990). Forest Science 36(1), 1990, 177–179.
Chavas J.-P. On impatience, economic growth and the environmental Kuznets curve: a dynamic analysis of resource management. Environmental and Resource Economics, 28(2), 2004, 123–152.
Coase R.H. The problem of social cost. The Journal of Law and Economics, 3, 1960, 1–44.
Columbia University School of International and Public Affairs. Access to Genetic Resources: An Evaluation of the Development and Implementation of Recent Regulation and Access Agreements. Columbia University School of International and Public Affairs, Environmental Policy Studies, Working Paper #4, 1999.
Commons J.R. Institutional Economics; Its Place in Political Economy. The Macmillan Company, New York, 1934, p. 921.
Copeland B.R., Taylor M.S. North-South trade and the environment. Quarterly Journal of Economics, 109(3), 1994, 755–787.
Crocker T.D. On the value of the condition of a forest stock. Land Economics, 61(3), 1985, 244–254.

Cropper C., Griffiths M. The interaction of population growth and environmental quality. American Economic Review, 84, 1994, 1358–1393.
De Groot H.L.F., Withagen C.A., Minliang Z. Dynamics of China's regional development and pollution: an investigation into the environmental Kuznets curve. Environment and Development Economics, 9(4), 2004, 507–537.
de Vries F.P., Withagen C. Innovation and Environmental Stringency: The Case of Sulfur Dioxide Abatement. Tilburg University, Center for Economic Research, Discussion Paper 18, 2005.
Donovan G.H., Nicholls D.L. Estimating Consumer Willingness to Pay a Price Premium for Alaska Secondary Wood Products. Research paper PNW-RP-553. USDA Forest Service, 2003.
Ehrhardt-Martinez K., Crenshaw E.M., Jenkins J.C. Deforestation and the environmental Kuznets curve: a cross-national investigation of intervening mechanisms. Social Science Quarterly, 83(1), 2002, 226–243.
Faustmann M. Calculation of the value which forest land and immature forest stands possess for forestry. Allgemaine forst-un jagd-zeitung, 15, 1849, 441–455.
Fisher A.C., Krutilla J.V. Resource conservation, environmental preservation and the rate of discount. Quarterly Journal of Economics, August 1975, 358–370.
Fix P., Loomis J. Comparing the economic value of mountain biking estimated using revealed and stated preference. Journal of Environmental Planning and Management, 41(2), 1998, 227–236.
Gawande K., Bohara A.K., Berrens R.P., Wang P. A consumption-based theory of the environmental Kuznets curve. Ecological Economics, 37, 2001, 101–112.
Gleeson B. Manufacturing Industry Pollution and Community Characteristics. Ph.D. Dissertation, University of Wisconsin, Madison, 2004.
Grossman G.M., Krueger A.B. Environmental Impacts of a North American Free Trade Agreement. Discussion papers in economics/Woodrow Wilson School, Princeton University, vol. 158, 1992, pp. 1–39.
Hanley N., Ruffell R. The Valuation of Forest Characteristics. Queen's Institute for Economic Research, Discussion Paper: 849, 1992, p. 45.
Harou P.A. On a social discount rate for forestry. Canadian Journal of Forest Research, 15(5), 1985, 927–934.
Hartman R. The harvesting decision when a standing forest has value. Economic Inquiry, 14, 1976, 52–58.
Hesseln H., Loomis J.B., Gonzalez-Caban A. Comparing the economic effects of fire on hiking demand in Montana and Colorado. Journal of Forest Economics, 10(1), 2004, 21–35.
Hool J.N. A dynamic programming-Markov chain approach to forest production control. Forest Science Monograph, 12, 1966, 1–26.
Horbach J. Employment and Innovations in the Environmental Sector: Determinants and Econometrical Results for Germany. Fondazione Eni Enrico Mattei, Working Papers: 2003.47, 2003.
Jones L., Manuelli R. Endogenous policy choice: the case of pollution and growth. Review of Economic Dynamics, 4(2), 2001, 369–405.
Kaya I., Buongiorno J. Economic harvesting of even-aged northern hardwood stands under risk: a Markovian decision model. Forest Science, 33(4), 1987, 889–907.
Keeler E., Spence M., Zeckhauser R. The optimal control of pollution. Journal of Economic Theory, 4, 1972, 19–34.
Kelly K.J., Mbaria J. YES, WE GOT SPECIMENS FROM KENYA - GENENCOR. Global News Wire – Asia Africa Intelligence Wire, September 6, 2004.

Koop G., Tole L. Is there an environmental Kuznets curve for deforestation? Journal of Development Economics, 58(1), 1999, 231–244.

Kuznets S. Economic growth and income inequality. American Economic Review, 45, 1955, 1–28.

Lantz V. Is there an environmental Kuznets curve for clearcutting in Canadian forests? Journal of Forest Economics, 8, 2002, 199–212.

Lee K.J. Hedonic Estimation of Nonindustrial Private Forest Landowner Amenity Values. Unpublished Ph.D. Dissertation, North Carolina State University, 1997.

Lembersky M.R., Johnson K.N. Optimal policies for managed stands: an infinite horizon Markov decision process approach. Forest Science, 21(2), 1975, 109–122.

Leslie A. A second look at the economics of natural management systems in tropical mixed forests. Unasylva, 39(1), 1987, 46–58.

Li M.M., Brown H.J. Micro-neighborhood externalities and hedonic housing prices. Land Economics, 56(2), 1980, 125–140.

Lin C.R., Buongiorno J. Tree diversity, landscape diversity, and economics of maple-birch forests: implications of Markovian models. Management Science, 44(10), 1998, 1351–1366.

Lindmark M. Patterns of historical CO2 intensity transitions among high and low-income countries. Explorations in Economic History, 41(4), 2004, 426–47.

Mattson L., Li C. The non-timber value of Northern Swedish forests. Scandinavian Journal of Forest Research, 8, 1993, 426–434.

Marklund, P.-O. Environmental Regulation and Firm Efficiency: Studying the Porter Hypothesis using a Directional Output Distance Function. Umeå Economic Studies, Working Paper no. 619, 2003.

Martinez-Zarzoso I., Bengochea-Morancho A. Pooled mean group estimation of an environmental Kuznets curve for CO2. Economics Letters, 82(1), 2004, 121–126.

Nguyen Van P., Azomahou T. Deforestation croisance economique et population: Une etude sur donnees de panel (Deforestation, economic growth, and population: a panel data study). Revue Economique, vol. 54, no 4, 2003, pp. 835–855.

Norberg-Bohm V., Rossi M. The power of incrementalism: environmental regulation and technological change in pulp and paper bleaching in the US. Technology Analysis & Strategic Management, 10(2), 1998, 225–245.

Oderwald R.G., Duerr W.A. Koning–Faustmann: a critique. Forest Science, 36(1), 1990, 169–174.

Panayotou T. Empirical Tests and Policy Analysis of Environmental Degradation at Different Stages of Economic Development. Working Paper WP238, Technology and Employment Programme (Geneva: International Labor Office), 1993.

Partridge E. In search of sustainable values. The International Journal of Sustainable Development, 6(1), 2003, 25–41.

Pease P.H., Holmes T.P. Accounting for nonmarket benefits in southern forest management. Southern Journal of Applied Forestry, 17(2), 1993, 84–89.

Perry D.A., Maghembe J. Ecosystem concepts and current trends in forest management: time for reappraisal. Forest Ecology and Management, 26, 1989, 123–140.

Porter M.E. America's green strategy. Scientific American, 264(4), 1991, 168.

Raunikar R.P., Buongiorno J. 2006 Willingness to pay for forest amenities: the case of non-industrial owners in the south central United States. Ecological Economics. 56(1): 132–143.

Raunikar R.P., Buongiorno J. 2006 (Section 9) Environmental quality and economic growth: Some evidence from forested landscapes in the Southern United States. Department of Forest Ecology and Management, University of Wisconsin, Madison.

Rodriguez-Meza J., Southgate D., Gonzalez-Vega C. Rural poverty, household responses to shocks, and agricultural land use: panel results for El Salvador. Environment and Development Economics, 9(2), 2004, 225–239.
Roediger-Schluga T. Some micro-evidence on the "Porter hypothesis" from Austrian VOC emission standards. Growth and Change, 34(3), 2003, 359–379.
Roediger-Schluga T. The Porter hypothesis and the economic consequences of environmental regulation: a neo-Schumpeterian approach. Elgar, Cheltenham, U.K. and Northampton, Mass., 2004, pp. xi, 349.
Rollin F., Buongiorno J., Zhou M., Peyron J.-L. Management of mixed-species, uneven-aged forests in the French Jura: from stochastic growth and price models to decision tables. Forest Science, 51(1), 2005, 64–75.
Roos A. The price for forest land on combined forest estates. Scandinavian Journal of Forest Research, 10, 1995, 204–208.
Rothman D.S. Environmental Kuznets curves—real progress or passing the buck? A case for consumption-based approaches. Ecological Economics, 25(2), 1998, 177–194.
Rupasingha A., Goetz S.J., Debertin D.L., Pagoulatos A. The environmental Kuznets curve for US counties: a spatial econometric analysis with extensions, Papers in Regional Science, vol. 83 no 2, 2004, pp. 407–424.
Samuels W.J. Legal-economic Nexus. The George Washington Law Review, 57, 1989, 1556–1578.
Samuelson P.A. Economics of forestry in an evolving society. Economic Inquiry, 14, 1976, 466–491.
Scarpa R., Buongiorno J., Hseu J.S., Lee Abt K.J. Assessing the non-timber value of forests: a revealed-preference, Hedonic model. Journal of Forest Economics, 6(2), 2000, 83–108.
Scarpa R., Thiene M. Destination Choice Models for Rock Climbing in the Northeast Alps: A Latent-Class Approach Based on Intensity of Participation. Fondazione Eni Enrico Mattei, Working Papers: 2004.131, 2004.
Scorgie M., Kennedy J. Who discovered the Faustmann condition? History of Political Economy, 28(1), 1996, 77–80.
Shafik N., Bandyopadhyay S. Economic Growth and Environmental Quality: Time Series and Cross-country Evidence. Background Paper for the World Development Report 1992 (Washington D.C.: The World Bank), 1992.
Skonhoft A., Solem H. Economic growth and land-use changes: the declining amount of wilderness land in Norway. Ecological Economy, 37, 2001, 289–301.
Stern D.I., Common M.S., Barbier E.B. Economic growth and environmental degradation: the environmental Kuznets curve and sustainable development. World Development 24(7), 1996, 1151–1160.
Stokey N.L. Are there limits to growth? International Economic Review, 39(1), 1998, 1–31.
Strang W.J. On the optimal forest harvesting decision. Economic Inquiry, 21, 1983, 576–583.
Taylor C.R. Stochastic dynamic duality: theory and empirical, applicability. American Journal of Agricultural Economics, 66(3), 1984, 352–357.
ten Kate K., Touche L., Collis A. Benefit-Sharing Case Study: Yellowstone National Park and the Diversa Corporation. Report to the Executive Secretary of the Convention on Biological Diversity, 1998.
Turner J. Trade Liberalization and Forest Resources: A Global Modeling Approach. Ph.D. Dissertation, University of Wisconsin, Madison, 2004.
Turner R., Newton C.M., Dennis D.F. Economic relationships between parcel characteristics and price in the market for Vermont forestland. Forest Science, 37(4), 1991, 1150–1162.

Von Mises L.E. Human Action: A Treatise on Economics. Yale University Press, New Haven, 1949, p. 889.

Weitzman M.L. Why the far-distant future should be discounted at its lowest possible rate. Journal of Environmental Economics and Management, 36, 1998, 201–208.

Weitzman M.L. Gamma discounting. American Economic Review, 91(1), 2001, 260–271.

Winters, R.K. The Forest and Man. Vantage Press, New York, 1974, p. 393.

Xu W., Lippke B.R., Perez-Garcia J. Valuing biodiversity, aesthetics, and job losses associated with ecosystem management using stated preferences. Forest Science, 49(2), 2003, 247–257.

Zhu S., Tomberlin D., Buongiorno J. Global Forest Products Consumption, Production, Trade and Prices: Global Forest Products Model Projections to 2010. Working paper GFPOS/WP/01. Forest Policy and Planning Division, FAO, Rome, 1998.

Chapter 25

MULTIPLE CRITERIA DECISION-MAKING IN FOREST PLANNING: RECENT RESULTS AND CURRENT CHALLENGES

Luis Diaz-Balteiro and Carlos Romero
Technical University of Madrid, Department of Forest Economics and Management, Spain

Abstract Forest management is becoming a complex process that requires decision making involving economic, environmental and social criteria. This means that multiple criteria decision-making (MCDM) approaches need to be used in many forestry contexts. This chapter aims at assessing the efforts undertaken over the last 30 years towards formulating and solving forest management problems from an MCDM perspective. The goal of the chapter is not to compile an exhaustive list of MCDM applications in forestry but to detect the areas within forest management in which MCDM approaches have proven to be more productive or have significant future potential.

Keywords: Forestry, forest management, multiple criteria decision-making, goal programming

1 INTRODUCTION

Every decision taken in the management and planning of a forest resource generally affects several wide-ranging criteria, such as economic (timber, livestock, forage, hunting, etc.), environmental (biodiversity conservation, soil erosion control, carbon sequestration, etc.) and social (recreation, population settlement, level of employment, etc.). Therefore, for most forest management optimisation problems, especially concerning publicly owned forests, the decision maker (DM) has a preferences structure involving several criteria of very different nature. In short, in many cases, forest management models need to be formulated within the multiple criteria decision-making (MCDM) paradigm. For instance, the USDA Forest Service (1997) has devised a computer tool (SPECTRUM) for the management of public forests from a multi-criteria perspective.

This chapter aims to review the main contributions to the broad field of forest management made from a MCDM perspective during the last 30 years. It should be noted that the purpose of the chapter is not to present a detailed categorised bibliography of most of the MCDM applications in forestry reported in the literature. Readers interested in this type of information can consult a working paper by Diaz-Balteiro and Romero (2004), which categorises more than 220 references according to the type of problem analysed and the MCDM technique used. On the contrary, the purpose of this work is to explore the areas within forest management where the MCDM paradigm has proven to be most productive and fields where it has a lot of future potential. Previous review papers on MCDM in forestry taking a different approach are: Tarp and Helles (1995) and Kangas and Kangas (2002).

The main areas covered in the chapter are as follows: (i) harvest scheduling problems, (ii) forest biodiversity conservation, (iii) sustainability of forest management plans, and (iv) group decision-making problems. In all the areas studied, the main purpose will not be to exhaustively list contributions but to characterise the underlying MCDM problem, as well as analyse the pros and cons of the representative cases chosen. In this way, we attempt to assess the results derived from the use of MCDM approaches in forestry and to detect the main challenges to be addressed in the near future.

2 A BRIEF OVERVIEW OF MULTIPLE CRITERIA DECISION-MAKING APPROACHES MAINLY USED IN FOREST PLANNING

In this section the basic features of the main MCDM approaches used in forestry are reviewed to give readers unfamiliar with multi-criteria techniques an understanding and appreciation of the material presented in subsequent sections.

Let us divide the MCDM approaches into two groups: (i) methods devised to tackle continuous problems and (ii) methods for the purpose of addressing discrete problems. In the continuous case, we have a feasible set with infinite points usually defined by a set of linear and non-linear constraints. In the discrete case, we have a feasible set characterised by a finite and usually fairly small number of solutions (alternatives).

The chief approach worthy of note within the continuous case is goal programming (GP). GP links up with Simonian "satisficing" theories by attaching a target to each attribute under consideration. The target represents a desirable level of achievement for the respective attribute. By desirable we mean a target figure that the DM considers satisfactory and sufficient. The combination of the target and the attribute define the goal. Thus, for instance, to achieve a timber volume of at least 300,000 m^3 across the planning horizon is a goal.

The goals are included in the model by converting the inequalities into equalities by adding to each goal a negative and a positive deviation variable that measures possible underachievement and overachievement, respectively. GP minimises what the DM does not desire, that is, the unwanted deviation variables. Thus, the unwanted deviation variable is the negative one when the goal derives from an attribute "more is better", the positive one when the goal derives from an attribute "less is better" or both deviation variables when neither underachievement nor overachievement is desirable.

The formulation of the GP model implies the minimisation of a function of the former unwanted deviation variables. This function is called the achievement function. The most commonly used forms of achievement functions are as follows: (i) weighted goal programming, which minimises a composite objective function formed by a weighted sum of unwanted deviation variables, (ii) lexicographic goal programming, which attaches pre-emptive or exclusive priorities to the different goals for the purpose of minimising the unwanted deviation variables in a lexicographic order, (c) Chebyshev goal programming, which minimises the maximum deviation from the stated goals, and (iii) extended goal programming, which, encompassing the weighted and Chebyshev variants as particular cases, can establish compromises between these two options.

GP was introduced by Charnes and Cooper in 1961. We owe the first application of a GP model in forestry to Field (1973). For a classic presentation of the technical aspects of GP, see Ignizio (1976). Updated presentations of the GP approach are Ignizio & Romero (2003) and Romero (2004).

The second MCDM approach applicable to continuous problems to be reviewed is multi-objective programming, also called vectorial optimisation. This approach tackles the problem of simultaneously optimising several objectives subject to a set of constraints. Given a certain level of conflict among objectives, which is usual in most real problems, not all the objectives can be simultaneously optimised. Therefore, instead of searching for a non-existent optimum, the multi-objective optimisation approach seeks to find the set of efficient solutions, also known as non-dominated or Pareto optimal solutions. The elements of the Pareto set are feasible solutions, such that no other feasible solution can yield an improvement in one objective without causing a degradation in at least one other objective.

There are several methods for generating or at least approximating the Pareto set. The most operational are: the weighting method and the constraint method. The weighting method combines all the objectives into a single objective function by attaching a weight to each objective and then adding all the resulting components. The efficient set is then generated by weight parameterisation. In fact, under very general conditions, an extreme efficient point is obtained for each set of weights chosen. The constraint method is to some extent the dual of the weighting method. In fact, this technique involves optimising one of the objectives, while the other objectives are entered as constraints. Under very general conditions, an extreme or interior efficient point is generated for each set of values of the respective right-hand sides.

A detailed explanation of the multi-objective programming approach is to be found in Cohon (1978), and an updated version in Steuer (1989). Chang and Buongiorno (1981) describe one of the first applications of this approach to forest planning.

Compromise Programming was devised to help the DM to choose the "most suitable" efficient point or the "most suitable" portion of the efficient set. Compromise Programming starts by defining the ideal point as a vector whose components are given by the optimum values of the objectives considered. Given the usual conflict among objectives, the ideal point is not feasible, so the "most suitable" or "best compromise" solution is defined as the efficient solution closest to the ideal point within this approach. Depending upon the metric (measure of distance) used, a set of compromise solutions can be determined as the "most suitable solutions".

Compromise Programming was introduced by Yu (1973) and Zeleny (1974). It should be noted that there are strong links between GP and Compromise Programming (Romero, 2001). Field *et al.* (1980) describe the first application of this approach to forestry in the context of a harvest scheduling problem.

The three approaches that we have reviewed above are applicable to continuous, as well as to discrete problems. However, some specific approaches have been devised for the discrete case. Of these, the soundest from a theoretical point of view is multi-attribute utility theory (MAUT). The basic idea of this approach is to define a cardinal utility function comprising all the relevant criteria in the analysed decision problem. This multi-attribute utility function is optimised subject to the constraints of the problem.

The first step in the MAUT methodology is to elicit the individual utility functions for each of the criteria introduced in the decision problem. The second step is to amalgamate all the individual utility functions into a multi-attribute utility function. Both phases of the process require a very strong interaction with the DM by setting questions related to artificial random lotteries and asking the DM to give the respective certainty equivalent. Moreover, the second step calls for the acceptance of strong assumptions about DM preferences. Therefore, although it is very sound from a theoretical point of view, the MAUT approach is not widely used in forestry planning.

This approach was developed basically by Keeney and Raiffa (1976). Bell (1977) was the first to resort to MAUT within forestry in the context of quantifying stakeholder preferences for a forest region subject to an outbreak of a certain pest.

To avoid the above-mentioned problems associated with the MAUT approach, several less theoretically sound but more operational methods have been devised. Of these, the ELECTRE (Elimination and (et) Choice Translating Algorithm) and the AHP (Analytic Hierarchy Process) methods deserve a mention in a forestry context.

The basic idea of ELECTRE is to replace the preference-indifference relation underlying the MAUT approach by an "outranking" relationship. Thus, alternative A_i outranks alternative A_j, when "alternative A_i is at least as good as alternative A_j". The meaning of this statement is established according to the concept of concordance (i.e. the fact that for a relatively important number of criteria A_i is preferred to A_j) and discordance (i.e. the fact that there is no criterion for which alternative A_j is much better than alternative A_i). The set of alternatives is ranked using certain concordance and discordance thresholds.

ELECTRE was initially proposed by Roy (1968). For an updated version of the foundations of ELECTRE, see Roy (1991). Bertier and Montgolfier (1974) were the first to apply ELECTRE to forestry in the context of ranking a set of projects for a suburban motorway damaging a forest environment.

Saaty introduced the AHP in the late 1970s. This approach has had an enormous impact not only on forestry but also on many other applied areas.

The basic idea underlying the AHP approach is to represent the DM´s preferences through a "pairwise"comparison process among criteria and alternatives using a lingustic scale and within a certain hierarchical structure. From the "pairwise" comparison matrices, a set of weights coherent with the DM's responses are elicited to get the respective ranking of alternatives.

The foundations of AHP are to be found in Saaty (1977, 1980). An updated version of the AHP approach is described in Forman and Gass (2001). Mendoza and Sprouse (1989) were one of the first to apply AHP to a forestry context.

3 HARVEST SCHEDULING PROBLEMS

Timber harvest scheduling is the first area in the forestry field where the MCDM paradigm was widely applied. This can be primarily put down to the very nature of a harvest scheduling problem. In fact, this type of problem was initially formulated as a linear programming model with an objective function that maximises an economic criterion (e.g., net present value), subjected to a set of constraints covering at least the following aspects: (i) volume control (i.e. to get an even flow of timber volume harvested in each of the cutting periods considered, (ii) area control (i.e. each age-class must occupy the same area at the end of the planning horizon and (iii) ending forest inventory (i.e. there must be a sensible relationship between initial and final inventory in order to guarantee the persistence of the forest).

The above linear programming formulation is very sound from a forestry perspective; however, it is too rigid. Thus, there is no feasible solution to the proposed model in many cases; that is, there is not a harvest schedule satisfying the above forestry conditions expressed as constraints. In other cases, the feasible set is not empty but its size is very small, which makes the optimum net present value corresponding to the optimum solution so low that the solution is unacceptable to the decision-maker.

One possible way of handling the overly rigid specifications of the linear programming models within which the timber harvest scheduling problems were formulated is to treat the right-hand sides of the above constraints as targets that may or may not be achieved. Operating in this way, the linear programming models turn into goal programming formulations, for which there are always feasible solutions.

The goal programming approach was initially proposed as an operational way of reconciling economic, even-flow and regulation criteria by Kao and Brodie (1979), for which purpose they formulated a lexicographic goal programming model. Extensions in the direction proposed by Kao and Brodie, also resorting to lexicographic goal programming are the works by Field *et al.* (1980), Hotvedt *et al.* (1982) and Hotvedt (1983).

Along the same lines of the papers commented above, Riiters *et al.* (1982) use multi-objective programming techniques to determine the trade-offs among harvested timber volume, economic return and stand diameter within a context of thinning regimes. Diaz-Balteiro and Romero (1998) show how goal programming and compromise programming can be used to get sensible harvest schedules that represent good compromises between economic return, an even-flow policy, area control (regulation) and ending inventory. The preferential weights to be attached to each criterion are elicited through "pairwise" comparison matrices in an AHP fashion. These authors have demonstrated (Diaz-Balteiro and Romero, 2003) how this type of approach can easily accommodate other environmental criteria, like the net carbon captured across the planning horizon.

All the above-mentioned cases correspond to a strategic level of planning. In fact, tactical planning including spatial considerations in harvest scheduling have usually been dealt with using integer programming and more recently by metaheuristics approaches. However, multi-criteria tools have also been applied in an incipient but promising way in this field (e.g., Snyder and Revelle, 1997). Finally, other studies focus on hierarchical planning models that combine spatial landscape-level goals with owner-specific goals using multi-criteria models. For example, in Kurttila and Pukkala (2003), a MAUT methodology is modelled to achieve objectives at the landscape level that are compatible with small utility losses at the forest-holding level.

4 FOREST BIODIVERSITY CONSERVATION

The management of forest biodiversity from the perspective of species and habitat diversity has been recently addressed within a MCDM perspective. The main efforts in this direction can be summarised as follows.

The first idea was to formulate the strategic planning of a forest area within an analytic hierarchy structure (AHP). The hierarchy includes biodiversity as a whole at the level of the objectives. At lower levels of the hierarchy the biodiversity criterion is broken down into different components, such as richness, rarity and vulnerability of species. The outcome of the process is a priority index for each feasible forest plan. Some works in this direction, all of them formulated within an AHP framework, are Kangas and Kuusipalo (1993), Kuusipalo and Kangas (1994) and Mendoza and Prabhu (2001).

Another line of research, closely related to the one mentioned above, consists of making biodiversity operational by decomposing it into diversity indicators that measure the characteristics of individual stands. Some examples of biodiversity indicators are: proportion of old forest, mean volume of deciduous trees, volume of deadwood, and so on. These indicators are treated as decision-making criteria and the respective individual utility functions are elicited according to the MAUT methodology mentioned in Section 2. Later on, the individual utility functions are amalgamated into a multiattribute utility function. This aggregate utility function can output a ranking of forest plans feasible from a biodiversity perspective. Some applications of this approach are Kangas and Pukkala (1996) and Kangas *et al.* (1998).

An alternative way of dealing with forest biodiversity is based on the right management of the structural diversity of a forest stand. The structural diversity of a stand is described by means of the distribution of trees by species-size classes, and the classic Shannon index is used to measure the relative abundance by species-size class. Technically a multi-objective programming model is formulated, where an economic objective, such as the net present value, is maximised subject to a parametric constraint that measures structural diversity. This establishes the efficient Pareto frontier between economic returns and biodiversity and determines the trade-offs between both criteria. The key papers in this direction are Buongiorno *et al.* (1994, 1995) and Önal (1997a, 1997b).

Some works, like Carter *et al.* (1997) and Bevers and Hof (1999), among others, address the biodiversity problem by focusing on the optimisation of the spatial arrangement of forest stand age classes, provided that the habitat requirements of species with respect to the number of edges are previously known. Along these lines, Bertomeu and Romero (2001, 2002) proposed the integration of the maximisation of the edge contrast as an operational measure of habitat diversity with other relevant forest management criteria, such as volume control across the planning horizon and the ending forest volume inventory. The exercise is successfully undertaken by formulating several goal programming models (a weighted and a Chebyshev formulation).

Linked with the biodiversity problem is the management of national parks, reserves and any type of protected land. In these cases, the selection of activities to achieve the management objectives lead to a multi-criteria decision-making problem. Several works deal with this question. For example, Bojórquez-Tapia *et al.* (2004) utilise the AHP in a study to determine the best design of nature reserves inside a park in order to maximise their conservation value.

5 SUSTAINABILITY OF FOREST MANAGEMENT PLANS

In the few last years sustainable forest management, where sustainability is understood in a broad sense, has come to be of paramount importance. The current view of sustainability comprises not only timber production persistence, but also sustainability (i.e. persistence over time) of several attributes demanded by society and produced by the forest systems. These attributes involve the economic, ecological, as well as the sociological aspects of the forest systems.

Taking into account the above considerations, it is rather obvious that the concept and measurement of the sustainability of a forest system is a very complex problem, and there is no consensus about how to address it. In this respect, one of the most widely used orientations to measure the sustainability of a system is the so-called "indicators approach". Within this perspective, a key question is to aggregate the different indicators used into a single index that measures the sustainability of the forest system as a whole.

Analytically speaking, the above-stated problem of aggregation fits in very well with a MCDM approach. In fact, we only need to interpret each indicator as a criterion function in order to establish the respective equivalence, and the analogy is complete.

Efforts to connect the forest sustainability issue with the MCDM paradigm are very recent and can be summarised as follows. Ducey and Larson (1999) resort to fuzzy multi-objective programming to evaluate a discrete set of forest management plans. In a similar direction, Mendoza and Prahbu (2000, 2003) and Mendoza *et al.* (2002) demonstrate how to use some qualitative soft multi-criteria methodologies for the assessment of indicators of forest sustainability. The use of discrete multi-criteria methods, based upon outranking relationships, like ELECTRE, to select the most favourable kernel of alternatives from a sustainable perspective is well illustrated by Bousson (2001).

Diaz-Balteiro and Romero (2004) have proposed a procedure, based upon binary extended goal programming, that can establish the forest system with a higher level of achievement with respect to the targets that an expert or a panel of experts have attached to each indicator of sustainability. A natural extension of this procedure can determine a complete ranking of all the forest systems considered.

Phua and Minowa (2004) have applied several criteria and indicators for sustainable forest management to deal with a forest conservation plan at landscape level in a national park. Three criteria and eight indicators have been integrated in a GIS-based multi-criteria decision-making approach. Criteria and indicators are evaluated using GIS and remote sensing techniques. All the scores are normalised and introduced into an AHP model.

The concept and measurement of the sustainability of forest systems is still an open problem. Similarly, the incipient methods for addressing this type of problem derived from the MCDM field are still very tentative. However, the development of further efforts to articulate new methods based upon the MCDM paradigm to characterise and measure forest sustainability appears to be especially promising.

6 GROUP DECISION-MAKING IN FOREST PLANNING

Nowadays there is increasing awareness that the complexity associated with the management and protection of a public forest is due not only to the multiplicity of very different criteria involved in the process but also to the manner in which different segments of society or social groups perceive these criteria. Shields *et al.* (1999) offer a thorough presentation of the main models for addressing group decision-making problems from a single-criterion perspective and within an ecosystem management context.

In short, public forests planning is actually a decision-making problem, involving several criteria as well as several social groups. To address this type of problem properly, a crucial matter will be to tackle the problem of how to aggregate the preferences that the members of each social group have revealed for each criterion considered in the decision-making process. Accordingly, forestry planning problems strongly connect with what operational research terms the group decision-making discipline or with what economists refer to as the social choice discipline.

We owe one of the first attempts of addressing forestry planning problems from a multi-criteria as well as multiple decision-maker perspective to Tecle *et al.* (1998). These authors formulate a problem with five objectives in a group decision-making framework, by using two methodologies: compromise programming and cooperative games. By resorting to an interactive software the DM can provide different vector of weights in the compro mise programming model, or to change their utility vectors in the cooperative game approach.

Schmoldt and Petersen (2000, 2001) prioritise projects in national parks in the USA using the AHP methodology to arrive at a consensus between different subjective judgements using the geometric averages of the different judgements, as the AHP methodology suggests within a group decision-making context. A similar procedure was developed by Bantayan and Bishop (1998) to allocate land use in a forest reserve in the Philippines. This area was divided into 10 compartments and 8 different criteria were defined for each one. A choice was made from four alternatives. In this case, the median was used to represent group response, assuming that there were no intra-group differences in the weightings.

The integration of tactical plans for private forests into a more aggregated scale (landscape level) was solved by Pykäläinen *et al.* (2001) by applying an optimisation model with an interface to be used in group decision-making problems. This hierarchical model incorporates a MAUT method called "HERO" (Kangas and Pukkala, 1993) and presents four different ways to formulate a landscape-level forest planning model for group planning. Before applying this MAUT method, a goal programming model was used to achieve the landscape optimum. Finally, an interface enables an interactive planning process.

Kangas and Kangas (2003), Laukkanen *et al.* (2002) and Laukkanen *et al.* (2004) apply the multi-criteria approval (MA) method, suggested by Fraser and Hauge (1998) to group decision-making problems in forestry. The MA approach is an ordinal method that does not require too much preference information from the social groups/decision-makers involved. In Laukanen *et al.* (2004), the MA approach is applied to a case study in Finland, where nine timber harvesting alternatives are ranked according to seven criteria and seven social groups. In Laukanen *et al.* (2002) this methodology is applied in a small case study regarding the choice of the best tactical plan in a consortium consisting of three private forest owners. Five criteria and 20 alternatives were considered. Other discrete MCDM approaches were applied to achieve different rankings for the 20 forest plans considered. Finally, a compromise programming methodology is applied in Phua and Minowa (2004) to integrate the forest conservation priorities of several decision-makers.

The application of group decision-making/social choice methods with a multi-criteria perspective in forestry is a new area of research. Nowadays, this line of development represents more a promise of success than a productive reality. However, the adaptation of the enormous collection of ordinal and cardinal methods for group decision-making decisions to a forestry context appears to be an extremely attractive research area.

7 CONCLUDING REMARKS

The main purpose of this chapter was not to compile an exhaustive list of applications of MCDM techniques in forestry. On the contrary, we aimed at providing an overall judgement of the suitability of this paradigm to several forestry areas of application.

The main finding of this chapter is that MCDM has played a relatively important role in forestry over the last 30 years. Moreover, this importance is increasing as shown by the growth rate of papers published on forest management with a MCDM perspective. It should also be noted that there are areas, like harvest scheduling and biodiversity conservation, where the MCDM applications have reached a certain level of maturity. However, there are other areas, like sustainability and group decision-making, where, although there are interesting applications, the main interest of MCDM lies in potential future developments.

Regarding the use of the different approaches, there has been a clear reliance on operational approaches, like goal programming or AHP, with respect to more sophisticated approaches like MAUT that require a strong interaction with the DM, as well as the acceptance of very demanding behavioural assumptions.

It is not bold to state that the most appealing aspect of MCDM for forest researchers rests upon the idea of finding "satisficing" solutions among very different goals and/or finding "best compromise" solutions among conflicting objectives as underlies the goal programming approach. The success and acceptability of these theoretical orientations is reinforced by the fact that at the practical level of modelling, only a basic mathematical programming knowledge is required. And this type of knowledge has been in the tool-kit of forest researchers for decades.

ACKNOWLEDGEMENTS

The support of our colleagues from the Spanish Multi-Criteria Network is appreciated. This research was supported by the Spanish Ministry of Education and Science under project SEJ 2005-04392 and by the "Fundación BBVA" project Timber Sustainable Production: Economic and Environmental Aspects. We would like to thank Mrs. Rachel Elliott for editing the-language.

8 REFERENCES

Bantayan, N. C., and Bishop, I. D., 1998, Linking objective and subjective modelling for landuse decision-making. *Landscape and Urban Planning* 43: 35–48.

Bell, D. E., 1977, A decision analysis of objectives for a forest pest problem, in: *Conflicting Objectives in Decisions*, D. D. Bell, R. L. Keeney, and H. E. Raiffa, ed., Wiley, New York, pp. 389–418.

Bertier, P., and Montgolfier, J., 1974, On multicriteria analysis: an application to a forest management problem. *Metra* 13: 33–45.

Bertomeu, M., and Romero, C., 2001, Managing forest biodiversity: a zero-one goal programming approach. *Agricultural Systems* 68: 197–213.

Bertomeu, M., and Romero, C., 2002, Forest management optimisation models and habitat diversity: a goal programming approach. *Journal of the Operational Research Society* 53: 1175–1184.

Bevers, M., and Hof, J., 1999, Spatially optimizing wildlife habitat edge effects in forest management linear and mixed-integer programs. *Forest Science* 45: 249–258.

Bojórquez-Tapia, L. A., De La Cueva, H., Díaz, S., Melgarejo, D., Alcantar, G., Solares, M. J., Grobet, G., and Cruz-Bello, G., 2004, Environmental conflicts and nature reserves: redesigning Sierra San Pedro Mártir National Park, Mexico. *Biological Conservation* 117: 111–126.

Bousson, E., 2001, Development of a multicriteria decision support system adapted to multiple-use forest management: application to forest management at the management unit level in Southern Belgium, in: *Criteria and Indicators for Sustainable Forest Management at the Forest Management Unit Level*, A. Franc, O. Laroussinie, and T. Karjalainen, eds., EFI Proceedings, Joensuu, Finland, pp. 151–164.

Buongiorno, J., Dahir, S., Chih Lu, H., and Ching-Rong, L., 1994, Tree size diversity and economic returns in uneven-aged forest stands. *Forest Science* 40: 83–103.

Buongiorno, J., Peyron, J. L., Houllier, F., and Bruciamacchie, M., 1995, Growth and management of mixed-species, uneven-aged forests in the French Jura: implications for economic returns and tree diversity. *Forest Science* 41: 397–429.

Carter, D. R., Vogiatzis, M., Moss, C. B., and Arvanitis, L. G., 1997, Ecosystem management or infeasible guidelines? Implications of adjacency constraints for wildlife habitat and timber production. *Canadian Journal of Forest Research* 27: 1302–1310.

Chang, S., and Buongiorno, J., 1981, A programming for multiple use forestry. *Journal of Environmental Management* 13: 45–58.

Charnes, A., and Cooper, W. W., 1961, *Management Models and Industrial Applications of Linear Programming,* John Wiley and Sons, New York.

Cohon, J. L., 1978, *Multiobjective Programming and Planning*, Academic Press, New York.

Diaz-Balteiro, L., and Romero, C., 1998, Modeling timber harvest scheduling problems with multiple criteria: an application in Spain. *Forest Science* 44: 47–57.

Diaz-Balteiro, L., and Romero, C., 2003, Forest management optimisation models when carbon captured is considered: a goal programming approach. *Forest Ecology and Management* 174: 447–457.

Diaz-Balteiro, L., and Romero, C., 2004, Sustainability of forest management plans: a discrete goal programming approach. *Journal of Environmental Management* 71: 351–359.

Diaz-Balteiro, L., and Romero, C., 2004, MCDM and forest management: a review. Working Paper, Technical University of Madrid.

Ducey, M. J., and Larson, B. C., 1999, A fuzzy set approach to the problem of sustainability. *Forest Ecology and Management* 115: 29–40.

Field, D. B., 1973, Goal programming for forest management. *Forest Science* 19: 125–135.

Field, R. C., Dress, P. E., and Fortson, J. C., 1980, Complementary linear and goal programming procedures for timber harvest scheduling. *Forest Science* 26: 121–133.

Forman, E. H., and Gass, S. I., 2001, The analytic hierarchy process-an exposition. *Operations Research* 49: 469–486.

Fraser, N. M., and Hauge, J. W., 1998, Multicriteria approval: application of approval voting concepts to MCDM problems. *Journal of Multi-Criteria Decision Analysis* 7: 263–272.

Hotvedt, J. E., 1983, Application of linear goal programming to forest harvest scheduling. *Southern Journal Agricultural Economics* 15: 103–108.

Hotvedt, J. E., Leuschner, W. A., and Buhyoff, G. J., 1982, A heuristic weight determination procedure for goal programs used for harvest scheduling models. *Canadian Journal of Forest Research* 12: 292–298.

Ignizio, J. P., 1976, *Goal Programming and Extensions*, Lexington Books, Lexington, MA.

Ignizio, J. P., and Romero, C., 2003, Goal programming, in: *Encyclopedia of Information Systems*, Vol. 2, H. Bidgoli, ed., Academic Press, San Diego, CA, pp. 489–500.

Kangas, J., Alho, J. M., Kolehmainen, O., and Mononen, A., 1998, Analyzing consistency of experts' judgements-case of assessing forest biodiversity. *Forest Science* 44: 610–617.

Kangas, J., and Kangas, A., 2002, Multiple criteria decision support methods in forest management: an overview and comparative analyses, in: *Multi-Objective Forest Planning,* E. T. Pukkala, ed., Kluwer Academic Publishers, pp. 37–70.

Kangas, J., and Kangas, A., 2003, Multicriteria approval and Smaa-O in natural resources decision analysis with both ordinal and cardinal criteria. *Journal of Multi-Criteria Decision Analysis* 12: 3–15.

Kangas, J., and Kuusipalo, J., 1993, Integrating biodiversity into forest management planning and decision-making. *Forest Ecology and Management* 61: 1–15.

Kangas, J., and Pukkala, T., 1996, Operationalization of biodiversity as a decision objective in tactical forest planning. *Canadian Journal of Forest Research* 26: 103–111.

Kao, C., and Brodie, J. D., 1979, Goal programming for reconciling economic, even flow, and regulation objectives in forest harvest scheduling. *Canadian Journal of Forest Research* 9: 525–531.

Keeney, R. L., and Raiffa, H., 1976, *Decision with Multiple Objectives: Preferences and Value Trade-Offs*, John Wiley and Sons, New York.

Kuusipalo, J., and Kangas, J., 1994, Managing biodiversity in a forest environment. *Conservation Biology* 8: 450–460.

Laukkanen, S., Kangas, A., and Kangas, J., 2002, Applying voting theory in natural resource management: a case of multiple-criteria group decision support. *Journal of Environmental Management* 64: 127–137.

Laukkanen, S., Palander, T., and Kangas, J., 2004, Applying voting theory in participatory decision support for sustainable timber harvesting. *Canadian Journal of Forest Research* 34: 1511–1524.

Mendoza, G. A., and Prabhu, R., 2001, A fuzzy analytic hierarchy process for assessing biodiversity conservation, in: *The Analytic Hierarchy Process in Natural Resource and Environmental Decision Making*., D. L. Schmoldt, J. Kangas, G. A. Mendoza, and M. Pesonen, ed., Kluwer Academic Publishers, Dordrecht, pp. 219–233.

Mendoza, G. A., and Prabhu, R., 2000, Multiple criteria decision making approaches to assessing forest sustainability using criteria and indicators: a case study. *Forest Ecology and Management* 131: 107–126.

Mendoza, G. A., and Prabhu, R., 2003, Qualitative multi-criteria approaches to assessing indicators of sustainable forest resource management. *Forest Ecology and Management* 174: 329–343.

Mendoza, G. A., and Sprouse, W., 1989, Forest planning and decision making under fuzzy environments: an overview and illustration. *Forest Science* 35: 481–502.

Mendoza, G. A., Hartanto, H., Prabhu, R., and Villanueva, T., 2002, Multicriteria and critical threshold value analysis in assessing sustainable forestry: model development and application. *Journal of Sustainable Forestry* 15: 25–62.

Önal, H., 1997, Trade-off between structural diversity and economic objectives in forest management diversity. *American Journal of Agricultural Economics* 79: 1001–1012.

Önal, H., 1997, A computationally convenient diversity measure: theory and application. *Environmental and Resource Economics* 9: 409–427.

Phua, M., and Minowa, M., 2004, A GIS-based multi-criteria decision making approach to forest conservation planning at a landscape scale: a case study in the Kinabalu Area, Sabah, Malaysia. *Landscape and Urban Planning*, 71: 207–222.

Pukkala, T., and Kangas, J., 1993, A heuristic optimisation method for forest planning and decision-making. *Scandinavian Journal of Forest Research* 8: 560–570.

Pykalainen, J., Pukkala, T., and Kangas, J., 2001, Alternative priority models for forest planning on the landscape level involving multiple ownership. *Forest Policy and Economics* 2: 293–306.

Riitters, K., Brodie, J. D., and Kao, C., 1982, Volume versus value maximization illustrated for Douglas-fir with thinning. *Journal of Forestry* 80: 83–89, 107.

Romero, C., 2001, Extended lexicographic goal programming: a unifying approach. *Omega. The International Journal of Management Science* 29: 63–71.

Romero, C., 2004, A general structure of achievement function for a goal programming model. *European Journal of Operational Research* 153: 675–686.

Roy, B., 1968, Classement Et Choix En Presence De Points De Vue Multiples (La Methode ELECTRE). *Revue Francaise d'Informatique et de Recherche Operationnelle* 8: 57–75.

Roy, B., 1991, The outranking approach and the foundations of ELECTRE methods. *Theory and Decision* 31: 49–73.

Saaty, T. L., 1977, A scaling method for priorities in hierachical structures. *Journal of Mathematical Psychology* 15: 234–281.

Saaty, T. L., 1980, *The Analytic Hierachy Process: Planning, Priority Setting, and Resource Allocation*, McGraw-Hill, New York.

Schmoldt, D. L., and Peterson, D. L., 2000, Analytical group decision making in natural resources: methodology and application. *Forest Science* 46: 62–75.

Schmoldt, D. L., and Peterson, D. L., 2001, Efficient group decision making in workshop settings, in: *The Analytic Hierarchy Process in Natural Resource and Environmental Decision Making*, D. L. Schmoldt, J. Kangas, G. A. Mendoza, and M. Pesonen, ed., Kluwer Academic Publishers, Dordrecht, pp. 97–114.

Shields, D. J., Tolwinski, B., and Kent, B. M., 1999, Models for conflict resolution in ecosystem management. *Socio-Economic Planning Sciences* 33: 61–84.

Snyder, S., and Revelle, C., 1997, Multiobjective grid packing model: an application in forest management. *Location Science* 5: 165–180.

Steuer, R. E., 1989, *Multiple Criteria Optimization: Theory, Computation and Application*, John Wiley and Sons, New York.

Tarp, P., and Helles, F., 1995, Multi-criteria decision-making in forest management planning – an overview. *Journal of Forest Economics* 1: 273–306.

Tecle, A., Shrestha, B. P., and Duckstein, L., 1998, A multiobjective decision support system for multiresource forest management. *Group Decision and Negotiation* 7: 23–40.

USDA Forest Service, 1997, *Spectrum Users Guide*, Ecosystem Management Analysis Center, Fort Collins, CO.

Yu, P. L., 1973, A class of solutions for group decision problems. *Management Science* 19: 936–946.

Zeleny, M., 1974, A concept of compromise solutions and the method of the displaced ideal. *Computers & Operations Research* 1: 479–496.

Chapter 26

FOREST FIRE MANAGEMENT

Current practices and new challenges for operational researchers

David L. Martell
Faculty of Forestry, University of Toronto, Toronto, Ontario, Canada

Abstract Forest fire management systems share much in common with urban fire, police and ambulance systems, but the spatial and temporal variability of forest fire occurrence processes and the comparatively long distances over which forest fire management takes place pose special challenges to operational researchers. This chapter describes the basic structure of a forest fire management system and the decision-making problems faced by fire managers. It describes how operations research (OR) has been applied to forest fire prevention, detection, deployment and initial attack dispatch decision-making problems; large fire management, strategic planning and fuel management, and it identifies new challenges that are amenable to OR approaches.

Keywords: Wildland fire, wildfire, fire suppression, ecosystem management, emergency response systems, natural disturbance, operations research

1 INTRODUCTION

Forest fire management agencies are responsible for dealing with forest or wildland fire and its social, economic and ecological impacts on people and forest ecosystems.[i] Fire managers *predict* when and where fires might occur; invoke *prevention* measures to reduce the incidence of arson and accidental fires; attempt to *detect* fires while they are small; *acquire* and *deploy* fire fighters, aircraft and other suppression resources close to areas where fires are expected to occur to minimize response times; initiate *initial attack* control action to contain fires that are reported while they are small; and deal with the large *escaped fires* that are not controlled by the initial attack system. They also work with other land managers and use *prescribed fire* to enhance wildlife habitat and produce other beneficial impacts of fire; *modify fuels* to mitigate the damage that results if and when wildfires do occur; and develop *integrated fire/forest management strategies* designed to minimize the detrimental impact of fire on timber production and other forest resources.

Fire is a natural component of many forest ecosystems[ii] but forest or wildland fires can and often do pose significant threats to public safety, property and forest resources. Most North American forest management agencies were formed in response to large fires [some of which are described by Holbrook (1943), Lambert and Pross (1967) and Pyne (1982)] that burned across large forested areas, engulfed many communities, and killed many people in the early decades of the twentieth century. Although forest fire managers have long recognized the ecological benefits of fire and most North American fire management agencies have evolved from fire suppression to fire management organizations, fire has, until recently, been viewed primarily as a destructive force that should, for the most part, be suppressed or excluded from the forest.

In this chapter, I approach forest fire management from a decision-making perspective and attempt to address the diverse needs of forest fire managers, fire management systems researchers and operations research practitioners and researchers. I describe some of the research and development that has been carried out in the past and some of the many important challenges that remain in the hope that fire managers that read this chapter will gain a better understanding of what operations research (OR) has to offer and some OR specialists will decide to tackle some of the many important challenges that remain. Fire management practices vary around the globe due in part to variations in climate, vegetation and societal needs. I chose to focus on forest fire management in Canada, and to a lesser extent, the USA, in part because those are the jurisdictions with which I am most familiar and in part because OR has, I believe, had its most significant impact on forest fire management in those two countries. OR has, of course, been applied to fire management problems in other countries including Australia (e.g. Loane and Gould, 1986), Russia (e.g. Kurbatskii and Tsvetkov, 1976) and parts of Europe (e.g. Dimopoulou and Giannikos, 2004), but a comprehensive review of global practices would lengthen this chapter unreasonably and I would no doubt overlook important contributions in those other areas.

This chapter is not nor is it intended to be a comprehensive literature review. Readers interested in the early contributions to this subject should consult Martell (1982) who reviewed most of the research (much of which appeared in technical reports and theses) that was carried out during the period 1961 through 1981 and Martell *et al.* (1998), which updated some aspects of that literature review.

2 FIRE AND FOREST LAND MANAGEMENT

Fire is but one of many factors that forest land managers must consider. Fire managers should view forest managers as clients and attempt to manage fire in ways that contribute to their forest management objectives. The development of an integrated fire/forest management strategy should begin by having forest managers partition their land into compartments and specifying "how much" of "what type" of fire is required (i.e. the preferred fire regime) in each compartment.[iii] This preferred fire regime should then be transmitted to the fire managers who can respond with a proposed fire management programme designed to minimize the cost of achieving those fire regime objectives along with an assessment of the potential impact of their proposed fire programme on public safety, property and ecosystem processes.[iv] Forest managers must ultimately finance the fire management programme and given the very slim likelihood that the first fire management programme proposal will meet the preferred land management fire regime objectives at a cost that is acceptable to the land managers, should initiate an iterative planning process during which land managers successively refine their fire regime targets and fire managers refine their proposed plans to ultimately converge on a fire management programme where a "preferred" fire regime is produced at a cost acceptable to the forest managers.

Perhaps the best way to describe forest fire management to an OR audience is to borrow from the definition of supply chain management (see Simchi-Levi *et al.*, 2003) and define *forest fire management* as *getting the right amount of fire to the right place at the right time at the right cost*, a definition that conveys the importance of achieving an appropriate balance of the beneficial and detrimental impacts of fire on people and forest ecosystems at a reasonable cost to society. It is therefore not surprising that OR specialists, beginning with Shephard and Jewell (1961), were attracted to fire and developed and implemented many wildland fire management decision support systems that have the "look and feel" of their urban counterparts.[v]

3 FIRE OCCURRENCE PREDICTION

In order to model a forest fire management system one needs to model basic fire occurrence processes. Fire managers typically classify fires into one of two broad categories – people-caused fires and natural fires, the most common of which are those ignited by lightning. In the province of Ontario, in Canada, for example, lightning ignites roughly 45% of the fires but lightning-caused fires burn roughly 80% of the area burned (Wotton and Martell, 2005).[vi]

Cunningham and Martell (1973) studied daily people-caused fire occurrence and showed it is reasonable to assume the probability distribution of the number of fires that occur in a district each day is Poisson with an expected value that varies with the weather, the state of which is described using fire danger rating systems that are based on daily weather observations.[vii] The Poisson model is used to predict daily people-caused fire occurrence because of the underlying theoretical rationale for its structure; the fact that it fits historical fire occurrence reasonably well; and perhaps even more importantly, because it can serve as the basis of Poisson fire occurrence processes, the Markovian properties of which simplify fire management systems modelling (e.g. queueing models of initial attack systems) enormously. Significant effort has therefore been devoted to developing and testing statistical methods for predicting the expected number of fires in a designated area each day which can, when needed, be aggregated over time to produce weekly, monthly and annual predictions; and over space to produce district, regional, provincial and national fire occurrence estimates.

The Poisson distribution is also used to model lightning fire occurrence, which is simplified by the availability of lightning counter technology that records when and where (with some uncertainty) lightning strokes hit the earth. Kourtz and his colleagues pioneered the prediction of lightning-caused fire occurrence (see for example, Kourtz and Todd, 1992) and Wotton and Martell (2005) describe a model they recently developed for use in the province of Ontario.

4 FIRE PREVENTION

Fire prevention is seldom given the attention it deserves and with few exceptions [see for example, Heineke and Weissenberger (1974), Nickey and Chapman (1979), and Iliadis and Spartalis (2005)] has been largely ignored by the OR community. Prevention is essentially a marketing activity[viii] and although there are marketing specialists in the OR community, a keyword search of Marketing Science, a leading journal targeted to reach Operational Researchers working in marketing did not identify any papers that deal with either forest or urban fire management.

Although most forest fire management agencies do devote some resources to prevention activities, a lack of understanding of the impact of prevention measures on people-caused fire occurrence makes it difficult for fire prevention specialists to develop and rationalize their programmes. Comprehensive public education fire prevention programmes are particularly difficult to assess. Consider, for example, an agency that administers an intensive elementary school prevention programme and also places prevention advertising in local media. School programmes instill messages and habits that may persist for many years such that even if both such programmes were to be suspended for a number of years, I doubt that would precipitate a rise in fire occurrence in most jurisdictions.

There remains however, an important prevention measure that can be rigorously assessed – local forest closures. Most forest fire management agencies restrict travel or the use of campfires in designated areas when fire hazard is extreme. Such measures can have a number of impacts, three of which are a reduction in people-caused fire occurrence, inconvenience to forest users and a loss in revenue to local tourist operators. Clearly, such impacts should be balanced but I am not aware of any published efforts to do so. Given the availability of GIS technology, which can be used to archive when and where such measures are invoked, a sound decision analysis of forest closures should provide fire and forest managers with valuable decision support.

5 FIRE DETECTION

Forest fires are detected and reported by the organized detection system (e.g. detection patrol aircraft and lookout tower observers) and the public, what fire managers refer to as the unorganized detection system. The sooner a fire is detected and reported the sooner initial attack can begin and the smaller the fire will be when suppression action begins. Since the probability of initial attack success is a decreasing function of size at initial attack, a well-managed detection system can contribute to significant reductions in the number of fires that escape initial attack, fire suppression costs and losses.

Some forest fire management agencies use fixed towers or lookouts that provide continuous coverage of designated areas, some use detection patrol aircraft that provide intermittent coverage, and others uses mixes of such resources. In North America, fire management agencies in the mountainous western states and provinces tend to use towers more than their eastern counterparts. Kourtz and O'Regan (1968) showed that high value areas should generally be protected using fixed lookouts while less valuable areas should be protected by patrol aircraft.

The management of fixed detection sensor systems (e.g., lookout towers) poses interesting strategic and tactical decision-making problems – for example, how many lookout towers or sensors should be constructed or purchased, where they should be located and when the lookout towers should be staffed or the sensors turned on. Given the relatively high capital cost of towers and sensors and the annual salaries of observers and the relatively low daily cost of sensor operation and observer overtime wages, strategic management of fixed detection system resources provides more opportunities for detection system improvement than daily tactical detection planning.

Deciding where to locate fixed sensors, be they towers, cameras or smoke detectors, can be modelled as a simple location problem. Mees (1976) developed a simulation model which combined digital "area seen" maps and historical fire occurrence rates and used an explicit enumeration algorithm to evaluate potential tower sites with respect to the number of historical fires "seen" by alternative tower system configurations. That problem can, of course, be formulated as a simple coverage problem (e.g. maximize the number of historical fires covered with a specified budget or minimize the cost of covering a designated proportion of fires covered) and solved using mathematical programming methods, but I am not aware of any published accounts of efforts to do so.

Detection aircraft management poses much more difficult problems. Each year a fire management agency must decide upon the number and type of aircraft to charter and the terms of those charters. Detection aircraft charter contracts may, for example, stipulate some guaranteed minimum number of flying hours at a specified rate and the provision of additional flying hours at higher rates. Detection managers must also decide if an observer will accompany the pilot on some or all flights. Each day he or she must decide how many aircraft will fly what patrol routes keeping in mind the expected number and location of undetected fires, the damage that might result if their detection is delayed given the current and forecast weather, the values at risk in the areas where fires are thought to be burning, and the potential demand for detection flying hours throughout the remainder of the fire season.

Peter Kourtz of the Canadian Forest Service developed many strategic and tactical detection system models which are described in Martell (1982), beginning with Kourtz (1967), a simulation model of a hypothetical detection system comprising towers, patrol aircraft and the public. Kourtz (1971) developed a simulation model to investigate operating policies for airborne infrared detection aircraft after which O'Regan *et al.* (1975) developed a quadratic programming model to maximize detection system effectiveness.

Kourtz later extended his patrol route planning research to develop novel approaches that exploit the travelling salesperson structure of such problems. Suppose a detection system manager has partitioned his or her protected area into a large number of cells and fire occurrence prediction models are used to predict how many fires are burning undetected in each cell. He or she will subjectively combine that information with his or her knowledge of the predicted weather, fuel, and values at risk in each cell, and identify a set of cells to be visited and searched by detection patrol aircraft. The detection patrol route planning problem then becomes one of finding optimal routes for one or more aircraft that originate and end at one or more designated airports. The problem is compounded by the need to search some cells before others due to potential losses. It is, in short, not unlike the complex stochastic vehicle routing with time windows problems studied by transportation researchers (e.g. see Bramel and Simchi-Levi, 1996). The need to visit designated cells can be relaxed to a need to fly within some designated distance of designated cells, a variant of what Hodgson *et al.* (1998) describes as the tour covering problem. Given such decisions, the development and implementation of detection patrol route planning models is a potentially very rich source of interesting challenges for the OR community. The emergence of new technology (e.g. see San-Miguel-Ayanz and Ravail, 2005) such as tower mounted gas sensors and infrared cameras, satellite imagery,[ix] infrared sensor equipped drones will hopefully stimulate a revival of interest in this important area.

6 INITIAL ATTACK RESOURCE DEPLOYMENT

The rate at which fires are reported in the area surrounding an initial attack base can vary significantly from day to day and hour to hour posing significant challenges to fire managers that must deploy their scarce resources at bases located close to fires that have yet to be reported in order to minimize initial attack response times. Analysis of historical fire patterns reveals that most initial attack bases experience many days with little or no fire activity but from time to time, lightning and/or human activity coupled with prolonged drought and extreme fire weather conditions initiate short periods of intense fire activity or "fire flaps" characterized by large numbers of challenging fires that are reported over short periods of time. Fire management agencies do not attempt to equip all their attack bases to levels sufficient to cope with these intense, infrequent fire flaps but rather, they attempt to predict when and where such events will occur and re-deploy their resources from bases where they are not expected to be required to bases where they are expected to be needed.

Daily initial attack deployment poses many interesting challenges for fire managers and OR specialists. Consider, for example, the daily deployment of airtankers, which can be viewed as a "design and control of queueing systems" problem with fires as customers and airtankers as servers. In the case of a single base, the duty officer must decide how many airtankers to place on initial attack standby at that base each day to achieve specified initial attack response time objectives that may vary with fire weather severity. The problem is complicated by the fact that fire arrival rates and service times may vary throughout the day in response to diurnal variation in weather and human behaviour. Service times will not be exponential due to the need to include travel time to and from the fire in the service time and to complicate matters even more, service time depends upon waiting time as fires grow while they wait in the initial attack queue. The optimal solution may call for the number of airtankers on standby to vary throughout the day and, as was the case with police cars (see Green and Kolesar, 2004), the number of airtankers to be dispatched to each fire may be a random variable. Daily airtanker deployment is further complicated by the need to account for the fact that a fire region may have a number of airtanker bases with partially overlapping response zones. When an airtanker finishes serving a fire it is therefore not necessarily optimal for it to return to the airport from which it was dispatched.

Greulich (1976) formulated the multiple base daily airtanker deployment problem as a chance constrained linear programming model, which Greulich and O'Regan (1975) applied to the California Division of Forestry's District 1. Hodgson and Newstead (1978) formulated the daily airtanker deployment problem as a location-allocation problem. Lee later incorporated a variant of that model in his spatial fire management system (Lee *et al.*, 2002). Bookbinder and Martell (1979) appear to have been the first to use a queueing approach to daily deployment when they developed a queueing model that can be used to allocate initial attack transport helicopters to independent initial attack bases. Martell and Tithecott (1991) later extended that approach to airtankers but when they field-tested it in the Northwestern region of Ontario, the fire managers involved expressed concern that they had not modelled the interaction of bases. Islam (1998) extended Larson's hypercube queueing model to deal with interacting bases, time-dependant fire arrival rates and Erlang service times and by using numerical methods to solve the differential equations that describe the behaviour of such systems. Furthermore, he showed how the dynamic re-deployment of airtankers (how they should be re-deployed from base to base as the day progresses) can be modelled as an intractable complex stochastic dynamic programming problem that he did not attempt to solve.

Daily initial attack deployment is clearly an important problem that has been studied but not yet solved. Such problems have grown in importance with increased centralization of fire management organizations and that trend is expected to continue as fire agencies attempt to "do more with less" in the future. Fire management agencies will, much like their urban emergency response system counterparts, have to develop an appropriate mix of stochastic queueing and deterministic location allocation type models to satisfy their needs. Given the needs and what has been accomplished to date, daily deployment problems should prove to be a rich source of interesting problems for OR specialists.

7 INITIAL ATTACK DISPATCHING

Forest fire managers use the term *initial attack* to refer to the first suppression action taken on a wildfire. Most forest fire management agencies attempt to initiate suppression action, while fires are small in the hope that they can be contained at a small size in short period of time. The United States Forest Service (USFS) was one of the first agencies to formalize their initial attack objectives when in 1935, they developed what is referred to as the "10:00 A.M. rule", which called for fires to be controlled by 10:00 A.M. on the day following the day the fire is first reported or, failing that, by 10:00 A.M. the next day, ad infinitum (see Pyne, 1997, p. 195). Most forest fire management agencies use variants of that rule that call for fast, aggressive initial attack.

Initial attack dispatchers must decide what resources (e.g., fire fighters and airtankers) will be dispatched (by ground and/or air) to each fire that is reported, and when more than one fire is burning out of control, they must prioritize them and decide which will be attacked first or which will receive most of the scarce resources. Initial attack dispatching decisions must be resolved quickly and often with very limited information concerning current and potential future fire behaviour and values at risk. Fire managers typically develop initial attack dispatch guidelines or rules that stipulate what suppression resources will be dispatched to each fire and vary with respect to fire weather conditions and zone to reflect the need to respond more quickly and aggressively to fast-spreading fires in high value areas.

Parks (1964) developed an analytical model of initial attack using crews that construct fire line. That was followed by many subsequent studies, most of which were simulation models, including Stade (1967) who simulated airtanker effectiveness; Simard (1979), who's simulation model was designed to help identify the best use of airtankers and ground forces on individual fires and many others described by Martell (1982).

Despite the importance of initial attack dispatching and the amount of work that has been carried out by OR specialists, the peer reviewed literature contains little evidence that OR specialists have had any significant impact on initial attack dispatching. Most agencies appear to rely on well defined "run card" dispatching rules that are subjectively modified by dispatchers if and when they perceive the need to do so, and although the development and testing of such rules may have been influenced by OR models, there is little documented evident that has occurred. This is perhaps not surprising given Green and Kolesar's (2004) discussion of the limited use of the dispatch research that was carried out as part of the New York City Rand Fire Project in the 1970s. They believe the lack of "real time" implementation of that research was due in part to the fact that complex models produced insight that could be incorporated in simple dispatch rules and that senior managers were more interested in strategic issues such as the siting of fire stations rather than tactical issues like dispatch.

The apparent lack of success in implementing tactical dispatching models in both urban and forest fire organizations notwithstanding, I feel the OR community has a great deal to offer forest fire managers that must resolve dispatch-related decision-making problems.

8 LARGE FIRE MANAGEMENT

Large fires that result from initial attack "failures"[x] usually materialize under extreme burning conditions when small, intense, fast spreading wind-driven fires that cannot be contained by initial attack forces rage out of control or when lightning storms ignite large clusters of fires that simply overwhelm the initial attack system. Although fuel, topography and diurnal variation in weather and burning conditions often enable fire managers to contain many "escaped" fires at relatively small sizes, large fires that burn tens of thousands of hectares are common in many areas, particularly the boreal forest region of Canada.[xi] Although large fires are relatively uncommon, some of those that do occur may persist for weeks or even months during which they draw upon the organization's suppression capabilities and undermine the initial attack system. This can lead to even more large escaped fires and result in expenditures of millions of dollars, to say nothing of their impact on public safety and property.

Large fires typically progress through several distinct phases. Most fires escape initial attack during severe burning conditions that preclude safe and fire suppression. Most of the large fires that occur in the province of Ontario experience at most one such "burning" day but multiple burning days are not uncommon (Podur, 2006). Fire managers typically draw back from such fires and wait for favourable weather conditions to materialize.[xii] During such "runs", fire managers consult weather forecasters, whereas fire behaviour analysts study the fuel and topography in the vicinity of the fire and predict when and where the fire is expected to slow its progress. The Incident Management Team responsible for the fire[xiii] develops a strategy for dealing with the fire, requests the resources required to implement their strategy and begins to assemble those resources at a staging area near the fire. As soon as the fire slows to a point where it can be fought safely, they implement their plan and quickly deploy fire crews and other suppressions resources around the perimeter of the fire where they work to quickly establish control lines before the weather deteriorates and blows the fire out of control yet again. Once they establish a preliminary control line they solidify it and then gradually work their way from the perimeter towards the interior of the fire extinguishing flames and mopping up smoldering fuels farther and farther from the fire's edge. Once the fire is declared under control, mop up activity may proceed for days or weeks until the fire is judged to have been extinguished and it is declared out. Resources commitments escalate very rapidly during the early stages of a large fire suppression operation and then tend to decrease gradually as the fire becomes more controlled to a point where in the later stages of a large fire, only a small number of crews may remain to patrol the fire on the ground or by air. Some large fires in the boreal forest region of Ontario are not formally declared out until winter sets in.

Despite the early interest in large fire management (e.g. see Shephard and Jewell, 1961), there have been few efforts to bring OR to bear on large fire management problems. Bratten (1969) developed a non-linear mathematical model that could be used to optimize the location of control line segments subject to constraints on the availability of suppression resources. Hafterson (1979) described how the US Forest Service used decision analysis methods to enhance the Escaped Fire Situation Analysis (EFSA) methods that were used to structure the evaluation of alternative strategies for managing large fires. Saporta (1995) showed how timber harvest scheduling models can be incorporated in EFSA frameworks so that fire managers could assess large fire management strategies in terms of their potential impacts on forest resources as well as suppression costs and areas burned. Mees *et al.* (1994) developed a stochastic large fire suppression model. More recently, Hof *et al.* (2000) developed an integer programming model of large fire containment. Lastly, Donovan and Noordijk (2005) studied how well fire mangers assessed the potential final size suppression costs of the escaped fires they were managing.

Large fire management operations are very costly and the Incident Management Teams that are responsible for managing them are well trained and equipped with modern information systems technology, communications resources and fire behaviour prediction technology. Modern supply chain logistics tools, personnel scheduling models and other OR applications could, I expect, be exploited to enhance the management of large fires.

9 STRATEGIC PLANNING

Forest fire managers must decide how many and what type of airtankers and transport aircraft to acquire, how many fire fighters to hire, how many trucks and what fire suppression equipment to lease or purchase and where all those resources should be based to minimize the cost of quickly satisfying local demands that vary throughout the course of a fire season. Such decision-making is complicated by the fact that fire occurrence and the need for fire suppression resources is highly variable over both time and space. Most fire management organizations re-deploy their resources to meet daily demands and share their resources with other agencies, sometimes on continental scales. It is therefore not surprising that many OR specialists have studied strategic fire management planning problems and given the capital cost of aircraft, that many of them have focused on airtanker fleet composition decision-making problems.

One of the earliest contributions of OR to forest fire management was Stade (1967) who developed a simulation model that contributed to the development of Canadair's (now Bombardier's) CL-215 airtanker. The USFS developed the FOCUS simulation model (see Davis and Irwin, 1976) and used it to evaluate initial attack strategies for relatively small independent planning units (e.g. national forests). Martell *et al.* (1984) developed a strategic initial attack simulation model that the Ontario Ministry of Natural Resources used to help decide how many and what type of airtankers it should acquire to satisfy its initial attack needs. Their Interfaces publication appears to be one of the first documented applications of OR to forest fire management that appeared in the open literature. Their so called "initial attack model" was subsequently modified by McAlpine and others[xiv] to produce what is now referred to as the LEOPARDS model (see McAlpine and Hirsch, 1999), which the OMNR has used on a number of occasions to help resolve other airtanker fleet management decision-making problems and to explore strategies for dealing with the potential impact of climate change on fire management in Ontario.

Other important contributions in this area include Fried and Gilless's CFEES2 simulation model that was initially developed for use in the state of California (a recent application of which is described in Fried *et al.*, 2006) and the USFS's National Fire Management Analysis System (a recent assessment of which is described by Dimitrakopoulos and Omi, 2003).

Strategic fire management planning remains an important area for further research. Boychuk's enhancement of LEOPARDS to deal with daily deployment decision-making constitutes an important advance but there is a need for further research to develop a more fully integrated hierarchical planning system that links long-term strategic planning with daily deployment and initial attack operations. Fire management agencies now share resources much more frequently than in the past and the economies of scale and continent-wide resource sharing should be considered when individual agencies decide what resources to acquire and where to base them. It is, for example, important to assess the likelihood that an agency can borrow airtankers from another agency across the continent, the possibility of which will depend upon the probability that both agencies will need such resources at the same time. There is perhaps, a need for a mix of planning models that can be used to explore how fire management agencies can collaborate effectively across regional, provincial state and national borders.

Most strategic fire management planning models use relatively simple measures of effectiveness (e.g. predicted initial attack response times or flying costs) that reflect suppression cost effectiveness. Fire management agencies are now developing and implementing new fire management policies that address a broader range of fire suppression and fire use activities, and when they implement such policies they often specify complex objectives that vary spatially across their jurisdiction (e.g. the need for aggressive suppression in some areas and the need to allow fire to play a more natural role in some parks and protected areas). Strategic planning models need to be enhanced to respond to such evolving demands.

10 FUEL MANAGEMENT

Fuel management poses a special class of strategic planning problems that have received very little effort from the OR community. Fire behaviour is influenced by fuel, weather and topography and although people can influence climate in the long run, only fuel can be managed in the short run. Fire managers who once treated fuel as a "given" must now deal with the fact that in some areas, their "successes" have contributed to fuel buildups, some of which are located in or near Wildland Urban Interface (WUI) areas where people build residences in or near very flammable fuel complexes. WUI problems contributed to the challenges faced by fire managers in Australia, British Columbia and California in 2003 and in southern Europe in 2003 and 2005.

Tremendous amounts of resources have been devoted to fuel management research in recent years and much of it has been directed at developing a better understanding of fire behaviour in natural and modified fuel complexes. Fire specialists are relying on what is currently known about fire behaviour and developing guidelines for fuel management, most of which are directed at community protection, but with very few exceptions [e.g. see Omi (1979) or Hof *et al.* (2000)] there have been very few attempts to bring OR to bear on such problems.

Fuel management represents rich untapped sources of interesting problems for OR specialists, particularly those interested in spatially explicit stochastic integer programming under uncertainty. Consider for example, the problem faced by a fire manager who is faced with deciding how to reduce the flammability of an inhabited flammable forest landscape. He or she must, for example, decide what fuel patches will be modified using what fuel treatment strategy subject to a fuel treatment budget constraint. When doing so he or she must consider uncertainty concerning fire ignition and spread, initial attack success and large fire growth. There is a very rich source of interesting problems that will push even the best stochastic integer programming methodologies to their limits.

11 DISCUSSION

Most North American forest fire management agencies were, as noted above, established to counter the destructive impact of fire on people, property and forest resources. Their initial focus was on community protection but timber protection grew in importance as lumber and pulp and paper companies and other resource industries grew and spread across the heavily forested regions of the continent. Logging slash was sometimes burned to reduce the fire hazard it posed, to facilitate post-harvest forest regeneration, and to manipulate forest vegetation to achieve wildlife management objecttives. It was not until the mid-1970s that park managers and others in the USA and to a lesser extent Canada, began to develop true "fire management" strategies that called for more extensive use of prescribed fire and the modified management of "wild" fire to help achieve ecological objectives.

Many North American fire managers are now faced with the difficult tasks of restoring fire to parks and protected areas and using fire to reduce fuel buildups, some of which are located near populated WUI areas that contain expensive homes and other structures. Smoke emissions are problematic and in some areas, fuel loads are so volatile they simply cannot be burned because of the risk to people and property.

Many North American fire management agencies are gradually moving into an era during which timber protection will diminish in importance, public safety and property protection will grow in importance and the ability of fire mangers to cope with such problems will be challenged by increasing pressure on fire management budgets due to growing health care costs and climate change. This of course will create many opportunities for OR specialists.

OR specialists have worked with fire managers in the past to develop decision support systems that can enhance fire management, but the number of OR specialists still working in the area is substantially less than the numbers involved in the 1960s and 1970s and there remain very substantial gaps between fire managers' needs and the decision support systems currently being used.

In their retrospective article, Green and Kolesar (2004) describe differences between police work and traditional OR. They concluded that "meaningful models of emergency systems cannot be developed without intimate knowledge of the organization, its operations, and its objectives" and pointed out that members of the New York City Road Fire Project Team "went to fires with the firefighters, slept over in firehouses, [and] sat at the shoulders of the dispatchers in the communications center" to acquire such knowledge. The importance of such practices, which might best be described as embedding, (given the current widespread use of that term by journalists) has perhaps been most eloquently articulated by Woolsey in his many popular Interfaces articles (e.g. see Woolsey, 1995). What Green and Kolesar found in urban settings is no different in the forest fire sector. In order to contribute, OR specialists need to get out into the woods to observe what is really happening. The time commitment required to do so will be more than made up for by the fact that fire management poses many interesting challenges and, as I have illustrated above, is a rich source of novel problems that call out for basic and applied research in stochastic modelling and optimization.

ACKNOWLEDGEMENTS

Many of the ideas presented here emerged from collaboration with former graduate students and discussions with OMNR fire managers who took time to show me and explain to me what they did, discussed their decision-making, provided data and suggestions concerning how to model fire management systems, and tested many of the models my graduate students and I developed.

12 REFERENCES

Beverly, J.L., and D.L. Martell. 2004. Modeling prescribed surface-fire regimes for *Pinus strobus* conservation. *Conservation Biology* 18(6): 1541–1552.

Bond W.J., and J.E. Keeley. 2005. Fire as a global 'herbivore': the ecology and evolution of flammable ecosystems. *Trends in Ecology and Evolution* 20(7): 387–394.

Bookbinder, J.H., and D.L. Martell. 1979. Time-dependent queuing approach to helicopter allocation for forest fire initial attack. *INFOR J. (Can. J. Oper. Res. Inf. Process.)* 17(1): 58–69.

Bratten, F.W. 1969. A mathematical model for computer allocations of firefighting resources on large wildfires. U.S. For. Serv. Pac. Southwest For. Range Exp. Stn., Berkeley, CA.

Bramel J., and D. Simchi-Levi. 1996. Probabilistic analyses and practical algorithms for vehicle routing problem with time windows. *Operations Research Society of America* 44(3): 501–509.

Chase, R.A. (1980). FIRESCOPE: A new concept in multiagency fire suppression coordination. General Technical Report PSW-40. Berkeley, CA: USDA Forest Service Pacific Southwest Experiment Station.

Cunningham, A.A., and D.L. Martell. 1973. A stochastic model for the occurrence of man-caused forest fires. *Canadian Journal of Forest Research* 3(2): 282–287.

Davis, J.B., and R.L. Irwin. 1976. FOCUS: a computerized approach to fire management planning. *Journal of Forest Research*. 74(9): 615–618.

Dimitrakopoulos, A.P, and P.N. Omi. 2003. Evaluation of the fire simulation processes of the national fire management system's initial attack analysis processor. *Environmental Management* 31(1): 147–156.

Dimopoulou M, and I. Giannikos. 2004. Towards an integrated framework for forest fire control. European *Journal of Forest Research* 152(2): 476–486.

Donovan G.H., and P. Noordijk. 2005. Assessing the accuracy of wildland fire situation analysis (WFSA) fire size and suppression cost estimates. *Journal of Forest Research* 103(1): 10–13.

Fried J.S., J.K. Gilless, and J. Spero. 2006. Analysing initial attack on wildland fires using stochastic simulation. *International Journal of Wildland Fire* 15(1): 137–146.

Gorte, J.K., and R.W. Gorte. 1979. Application of economic techniques to fire management – a status review and evaluation. Gen. Tech. Rep. INT-53. USDA Forest Service. Intermountain Forest and Range Experiment Station, Ogden, Utah. 26 pp.

Green, L.V. and P. J. Kolesar. 2004. Improving emergency responsiveness with management science. *Management Science* 50(8): 1001–1014.

Greulich, F.E. 1976. A model for the seasonal assignment of airtankers to home bases under optimal daily transfer rules. Ph.D. dissertation, University of California, Berkeley, California.

Greulich, F.E., and W.G. O'Regan. 1975. Allocation model for air tanker initial attack in firefighting. U.S. For. Serv. Res. Note PSW-301.

Hafterson, J. 1979. Escaped fire situation analysis. In *Proceedings. Fire Control in the 80's*. Edited by R.J. Barney. Intermountain Fire Council and the Fire Working Group. Society of American Foresters, Missoula, MT. pp. 220–226.

Heineke, J.M., and S. Weissenberger. 1974. An analysis of wildfire prevention. IEEE *IEEE Transactions on Systems, Man, and Cybernetics* 4(6): 513–522.

Hodgson, M.J., and R.G. Newstead. 1978. Location-allocation models for one-strike initial attack of forest fires by airtankers. *Canadian Journal of Forest Research* 8: 145–154.

Hodgson M.J., G. Laporte, and F. Semet. 1998. A covering tour model for planning mobile health care facilities in Suhum district, Ghana. *Journal of Regional Science*. 38(4): 621–638.

Hof, J., P.N. Omi, M. Bevers, and R.D. Laven. 2000. A timing-oriented approach to spatial allocation of fire management effort. *Forest Science* 46(3): 442–451.

Holbrook, S.H. 1943. *Burning An Empire*. The Macmillan Company. New York.

Iliadis, L.S., and S.I. Spartalis. 2005. Fundamental fuzzy relation concepts of a D.S.S. for the estimation of natural disasters' risk (The case of a trapezoidal membership function). *Mathematical and Computer Modelling* 42(7–8): 747–758.

Islam, K.M.S. 1998. Spatial dynamic queueing models for the daily deployment of airtankers for forest fire control. Ph.D. Thesis, Department of Mechanical and Industrial Engineering, University of Toronto.

Kourtz, P.H. 1967. Computers and forest fire detection. *Forest Chronicle*. 44(2): 22–25.

Kourtz, P.H. 1971.Analysis of an infrared forest fire detection system. IUFRO Cong., Gainesville, FL.

Kourtz, P.H., and W.G. O'Regan. 1968. A cost-effectiveness analysis of simulated forest fire detection systems. *Hilgardia* 39(12): 341–366.

Kourtz, P.H., and B. Todd. 1992. Predicting the daily occurrence of lightning-caused forest fires. Forestry Canada, Information Report PI-X-112. Petawawa National Forestry Institute, Chalk River, Ontario.

Kurbatskii, N.B., and P.A. Tsvetkov. 1976. Integrated optimization in forest fire control management. Lesn. Khoz. 9: 66–70. *In Russian; English translation, Environ. Can. Transl. OOENV Tr-1476, Ottawa, ON, 1978.

Lambert, R.S., and P. Pross. 1967. *Renewing Natures Wealth*. Ontario Department of Lands and Forests. Toronto.

Lee, B.S., M.E. Alexander, B.C. Hawkes, T.J. Lynham, B.J. Stocks, and P. Englefield. 2002. Information systems in support of wildland fire management decision making in Canada. *Computers and Electronics in Agriculture* 37(1–3): 185–198.

Loane, I.T. and J.S. Gould. 1986. Project Aquarius. Aerial Suppression of Bushfires. Cost-Benefit Study for Victoria. National Bushfire Research Unit. CSIRO Division of Forest Research. Canberra ACT. April 1986.

Larson, R.C., and A.R. Odoni. 1981. *Urban Operations Research*. Prentice Hall. Englewood Cliffs, NJ.

Martell, D.L. 1980. The optimal rotation of a flammable forest stand. *Canadian Journal of Forest Research* 10(1): 30–34.

Martell, D.L. 1982. A review of operational research studies in forest fire management. *Canadian Journal of Forest Research* 12(2): 119–140.

Martell, D.L., and A. Tithecott. 1991. Development of daily airtanker deployment models. In *Proceedings of the 1991 Symposium on Systems Analysis in Forest Resources*. Charleston, S.C. March 3–6, 1991. Compiled by M.A. Buford.

Martell, D.L., R.J. Drysdale, G.E. Doan, and D. Boychuk. 1984. An evaluation of forest fire initial attack resources. *Interfaces* 14(5): 20–32.

Martell, D.L., E.A. Gunn, and A. Weintraub. 1998. Forest management challenges for operational researchers. *European Journal of Operations Research* 104: 1–17.

McAlpine, R.S., and K.G. Hirsch. 1999. An overview of LEOPARDS: The level of protection analysis system. *Forest Chronicle* 75(4): 615–621.

Mees, R.M. 1976. Computer evaluation of existing and proposed fire lookouts. U.S. For. Serv. Gen. Tech. Rep. PSW-19.

Mees R., D. Strauss, and R. Chase. 1994. Minimizing the cost of wildland fire suppression – a model with uncertainty in predicted flame length and fire-line width produced. *Canadian Journal of Forest Research* 24(6): 1253–1259.

Nickey, B.B., and C.B. Chapman. 1979. Evaluating fire prevention effectiveness through a probability model. *Fire Technology* 15: 291–306.

Omi, P.N. 1979. Planning future fuelbreak strategies using mathematical modeling techniques. *Environment Management* 3(1): 73–80.

O'Regan, W.G., P.H. Kourtz, and S. Nozaki. 1975. Patrol route planning for an airborne infrared forest fire detection system. *Forest Science* 21(4): 382–389.

Parks, G.M. 1964. Development and application of a model for suppression of forest fires. *Management Science* 10(4): 760–766.

Podur, J. 2006. Weather, forest vegetation, and fire suppression influences on area burned by forest fires in Ontario. Ph.D. Thesis, University of Toronto.

Pyne, S.J. 1982. *Fire in America: A Cultural History of Wildland and Rural Fire*. Princeton University Press. Princeton, NJ, 654 pp.

Pyne, S.J. 1997. *World Fire: The Culture of Fire on Earth*. Paperback edition published by the University of Washington Press. 384 pp.

Reed, W.J., and D. Errico. 1986. Optimal harvest scheduling at the forest level in the presence of the risk of fire. *Canadian Journal of Forest Research* 16: 266–278.

Richards S.A., H.P. Possingham, and J. Tizard. 1999. Optimal fire management for maintaining community diversity. *Ecological Applications* 9(3): 880–892.

San-Miguel-Ayanz, J., N. Ravail, V. Kelha, and A. Ollero. 2005. Active fire detection for fire emergency management: potential and limitations for the operational use of remote sensing. *Natural Hazards* 35(3): 361–376.

Saporta, R. 1995. Escaped fire situation analysis – a forest level perspective. M.Sc.F. Thesis, Faculty of Forestry, University of Toronto.

Shephard, R.W., and W.S. Jewell. 1961. Operations research in forest fire problems. *In* A study of fire problems. *National Academy of Sciences-National Research Council, Commitee on Fire*, pp. 145–165.

Simard, A.J. 1979. A computer simulation model of forest fire suppression with air tankers. *Canadian Journal of Forest Research* 9: 390–398.

Simchi-Levi, D., P. Kaminsky and E. Simchi-Levi. 2003. *Designing and Managing the Supply Chain: Concepts, Strategies, and Case Studies*. 2nd ed. McGraw-Hill. New York, 354 pp.

Stade, M. 1967. Cost effectiveness of water bombers in forest fire control. Journal of *Canadian Operational Research Society* 5(1): 1–18.

Stocks, B.J. B.D. Lawson, M.E, Alexander, C.E. Van Wagner, R.S. McAlpine, T.J. Lynham and D.E. Dube. 1989. The Canadian forest fire danger rating system - an overview. Forest. Chron. 65 (6): 450–457.

Walker, W.E., J.M. Chaiken, and E.J. Ignall. 1979. *Fire Department Deployment Analysis: A Public Policy Analysis Case Study*. The Rand fire project. Elsevier North Holland, Inc., New York, NY.

Ward, P.C, A.G. Tithecott and B.M. Wotton. 2001. Reply – a re-examination of the effects of fire suppression in the boreal forest. *Canadian Journal of Forest Research* 31: 1467–1480.

Whalen, R.J. 1995. *The Ecology of Fire*. Cambridge University Press. Cambridge. 346 pp.

Woolsey, R.E.D and R.A. Maurer. 1995. The fifth column: on the proper training of future management II. *Interfaces* 25(2): 74–80.

Wotton, B.M., and D.L. Martell, 2005. A lightning fire occurrence model for Ontario. *Canadian Journal of Forest Research* 35(6): 1389–1401.

Endnotes

---

[i] I use the term "forest fire", which is commonly used to refer to fires that burn in the forested areas of Canada and many other countries. The term "wildland fire" is often used to refer to such fires in the USA, some of which burn areas dominated by grass, brush and other forms of vegetation. Both "forest fire" and "brush fire" are used in Europe, depending upon the predominant fuel type being burned, and the term "bushfire" is commonly used in Australia.

[ii] Bond and Keeley (2005) discuss the importance of forest and wildland fire from a global perspective.

---

iii The attributes of a fire regime (e.g. see Whalen, 1995) and how fire managers might assess the impact of their programmes on those attributes are beyond the scope of this chapter.

iv The social, economic and ecological impact of fire and fire management's role in forest land management are often classed as "fire economics" and are beyond the scope of this chapter. For an introduction to the fire economics literature see Gorte and Gorte (1979). Most of the OR applications in fire economics deal with the impact of fire on timber supply and the timber harvest scheduling under uncertainty. Examples include Martell (1980) and Reed and Errico (1986). Although the ecological aspects of fire are growing in importance, there have been few applications in this area to date. Examples include Richards *et al.* (1999) and Beverly and Martell (2004).

v Although forest fire managers share much in common with their urban counterparts (see for example, the important OR contributions described in Walker *et al.*, 1979 and Larson and Odoni, 1981), the naturalness of fire is a very significant distinguishing factor. Structural fires in urban areas almost always pose threats to public safety and property but natural fires that burn in remote areas can, and sometimes are allowed to burn relatively freely to reduce suppression costs and produce ecological benefits.

vi Lightning-caused fires burn a disproportionate area because they generally occur in remote areas and are not detected as early as people-caused fires and they tend to arrive in spatial and temporal clusters that can overwhelm the initial attack system.

vii Forest fire managers use fire danger rating systems, indices based on daily observed weather that are used to predict fire occurrence, behaviour and impact. Canadian forest fire managers use the Canadian Forest Fire Danger Rating System (CFFDRS), which is described by Stocks *et al.* (1989).

viii Forest fire prevention specialists often describe the different elements of their programmes as the "3 E's", Education, Engineering and Enforcement.

ix Satellite imagery is used to monitor the progress of large fires in some remote areas of northern Canada but is not currently used to detect fires in intensively protected areas as the technology currently available to fire management agencies cannot detect fires at the small sizes required for effective initial attack. As higher resolution technology becomes available, the "routing" of satellite scanners (i.e. deciding where to focus high-resolution scanners) will produce a rich source of new routing problems analogous to detection aircraft routing.

x Fire management agencies sometimes classify large fires as initial attack failures but it is important to note that such fires may result from prevention or detection failures or simply from the fact that fuel, weather and topography conditions are producing such intense and erratic fire behaviour that the fire could not be controlled by any initial attack force and that such fires are not a result of initial attack or any other system failures.

xi Ward *et al.* (2001) reported that during the 25-year period from 1976 through 2000, an average of 1,580 fires were reported per year in the 49,281,000 ha Intensive and Measured (I & M) protection zone of Ontario, the portion of the province in which most forest fire management activity takes place. Those fires burned an average of 81,544 ha or 0.17% of the I & M zone per year.

xii There is one important exception to this general rule. Burnout teams equipped with torches slung from helicopters sometimes carry out "burning out" operations under severe burning conditions when there is no other suppression action taking place. They may, for example, set "backfires" along rivers ahead of the fire to reduce the likelihood that the main fire will "spot" across such barriers or create other burned buffer zones that fires crews can work from when they ultimately do initiate suppression action on the ground.

[xiii] Most forest fire management agencies and many other North American emergency response systems use the Incident Command System (ICS) to structure their emergency response teams. The ICS was first developed in response to wildfire problems in southern California (see Chase, 1980). For a recent assessment of ICS see "The Incident Command System: a 25-year evaluation by California practitioners" by Dana Cole, Assistant Chief, California Department of Forestry and Fire Protection, St. Helena, California. An applied research project submitted to the National Fire Academy as part of the Executive Fire Officer Programme. February 2000. Accessed at http://www.usfa.fema.gov/downloads/pdf/tr_00dc.pdf on May 16, 2006.

[xiv] Dennis Boychuk, one of the developers of the original Ontario initial attack model, has enhanced LEOPARDS by incorporating daily deployment decision-making to capture the important variability of fire activity over time and space.

Chapter 27

A MODEL FOR THE SPACE–TIME SPREAD OF PINE SHOOT MOTH

Roberto Cominetti and Jaime San Martín
Departamento de Ingeniería Matemática, Centro de Modelamiento Matemático, Universidad de Chile, UMI-CNRS

Abstract We develop a diffusion-type model for the space–time spread of the pine shoot moth in the south of Chile. The model is stated as a discrete time evolution with non-overlapping generations and combines a drift term determined by the dominant winds with a filter term that accounts for the fact that the moth is specific to pine trees. We fit the model using the captures on pheromone traps on the two consecutive seasons 1991–1992 as well as the observed average spread of the plague during the period 1984–1999.

Keywords: Diffusion, plague, shoot moth

1 INTRODUCTION

In 1984, the presence of pine shoot moth (*Rhyacionia buoliana*) was detected in the south of Chile (see Shoeder (1986)). Since then the plague has infested virtually all the pine forest run by the Chilean forest industry. The damage produced by this moth on the pine trees occurs mainly at the larvae stage, especially during the spring after surviving the winter in a bud. When a larvae feeds from the main apex of a tree, it produces deformations that may range from small deviations in the stem to fork-type anomalies with an important economic impact. For details on the biology of the pine shoot moth and the damages it produces on pine tree we refer to Robredo (1970) and Robredo (1975).

This work describes a model to predict the space–time evolution of this plague and may be used to design and evaluate different control policies. The main explanatory variable for the spatial spread of the moth is the flux of predominant winds, which implies that certain geographical areas are more exposed to immigration of the moth than others. In this way, the model can be used to produce a risk map of pine plantations as well as to identify the upstream regions where control must be applied to protect a given area.

A standard tool for modelling the spread of a disease or an insect is based on diffusion equations which are well suited to model short-range interactions of species. More precisely, if $u(t,x)$ represents the density of insects located at x by time t, a general evolution model may be described by

$$u(t+e,x)=\sum_{y\approx x}u(t,y)p(y,x,e),$$

where $p(y,x,e)$ represents the chances that an insect located at y (a neighbour of x) jumps over x in e units of time. The system is described by modelling p and the neighbourhood of x at the microscopic level. Under appropriate assumptions, a limit process where the vicinity $y \approx x$ shrinks towards x yields a diffusion model for $u(t,x)$. For example, taking $p = 1/4$ and choosing for the neighbourhood of x a rectangular planar grid of width $\sqrt{\delta}$, the limit when $\delta\to 0$ yields the standard two-dimensional Brownian motion. More generally, in a two-dimensional setting the limit process leads to diffusion equations of the form

$$\frac{\partial u}{\partial t}=\sum_{i,j}a_{ij}(x)\frac{\partial^2 u}{\partial x_i \partial x_j}+\sum_i b_i\frac{\partial u}{\partial x_i}+c(x)u,$$

in which the left-hand side represents the rate of variation of the density of insects, while the right-hand side contains three terms of different nature: a pure diffusion term associated to the random short-range movements of insects with coefficients $a_{ij}(x)$ that represent spatial asymmetries, a second term called the drift which accounts for the movement of the population due to an external force such as wind, and a third term representing the reproduction rate of the population.

In Banks (1994), a diffusion model with exponential growth was used to study the dispersal of the gypsy moth throughout New England at the beginning of the twentieth century; and a similar model was used for the spread of muskrats in central Europe during the period 1905–1927. Among the rich literature on the theoretical aspects of diffusions, we mention the excellent books by Protter (1990) and Oksendal (1992). For a background on the mathematical modelling of biological dispersal, we recommend the books by Murray (2003), Okubo (1980) and Leibhold *et al.* (1995), as well as the papers by Kendall (1948), Long (1977), Polymenopoulos and Long (1990) and Skellam (1951).

The model considered in this chapter represents a departure from this standard diffusion model. A first difference comes from the fact that the pine shoot moth has non-overlapping generations with one generation per year, so that a discrete-time model is more appropriate. Secondly, the long-range effects dominate over the local diffusion phenomena as the plague covers a long distance each year moving along with the flux determined by the dominant winds. As a matter of fact, starting from the south of Chile in 1984, the plague has moved northbound covering a distance of 850 km in 15 years for an average speed of 57 km/year which cannot be explained simply by local random movements. Finally, the available data to fit the model are not based on direct density measures but rather on captures of males in pheromone traps at different locations, once again each one spaced 1 year in time.

2 THE MODEL

Let us begin by stating some preliminary definitions required to write down the model. The territory is divided into cells which we assume for simplicity to have equal surface S. We denote by C the set of cells and for each $x \in C$ we let $a(x)$ represent the fraction of surface covered by pine plantations, which is assumed constant over time (a simplification justified since we restrict our study to a period of a few years). For each year t we let $u(t,x)$ denote the number of female moth at cell x per unit of forest surface, so that an empty cell in that year is represented by the equality $u(t,x) = 0$.

For each cell x we also denote $F(x)$ the upstream cells from which an insect can reach x by following the flux of dominant winds, that is to say, y is in $F(x)$ if and only if an insect starting at y and suspended in the wind reaches the cell x by following the wind streamlines. By convention, we assume x is not in $F(x)$. Clearly far-away cells in $F(x)$ should not contribute to cell x in the same way as the nearby ones. Our model reflects this fact in two different and complementary ways. First we restrict to the cells $F(x,T) \subset F(x)$ from which it takes less than T units of time to reach x following the wind flow. The value of T is one of the parameters of the model and will be estimated from the data. Secondly, we observe that for a moth in a cell $y \in F(x,T)$ it should be harder to reach x if along the path connecting the two cells there is a large amount of pine forest where it can land. In this way, the pine forest acts like a filter. For $y \in F(x,T)$, we denote

$$a(y,x) = \sum_{z} a(z),$$

where the sum extends over those $z \in F(x,T)$ that are between y and x along the flux, which is the same as saying that the path starting from y to x crosses z. We notice that $Sa(y,x)$ is the total surface of forest which is encountered when travelling from y to x following the dominant wind flow.

As already mentioned, the observations indicate that the average distance covered by the plague per season is around 57 km, a fact which is hard to explain on the ground of a pure diffusion model. Thus, we assume that a fraction α of females leaves the home cell and is dispersed by the wind. Alternatively, one may assume that all females are dispersed but a fraction $1-\alpha$ of the eggs is put at the home cell. The observed phenomena in both cases will be the same. Due to saturation effects, it is conceivable that the parameter α may depend on the population density. However, since we are going to fit our model for small population densities (first or second generation) we take α as a constant to be estimated from the data. The larger this parameter α is, the faster the plague will move. The case $\alpha = 0$ corresponds to a pure diffusion model whereas $\alpha > 0$ is a drifted model. We estimate α to fit the observed mean speed of 57 km/year.

With these preliminaries, we may now state the model, which is simply a balance of mass and takes the form

$$\begin{aligned} Sa(x)u(t+1,x) = {} & \tfrac{1}{2}h(1-\alpha)\,Sa(x)u(t,x) \\ & +\tfrac{1}{2}h\alpha \sum_{y\in F(x,T)} Sa(y)u(t,y)\exp(-\eta a(y,x))[1-\exp(-\eta a(x))]. \end{aligned} \tag{1}$$

This model reads as follows: the total number of females at cell x in the next period $t+1$ is equal to the number of females staying at x plus the females arriving from other cells driven by the wind, multiplied by the average number of female descendants which will survive to the adult stage (here h represents the average number of eggs per female that will reach the adult stage, while the factor 1/2 is explained by a sexual ratio of 1:1). The factor $\exp(-\eta a(y, x))[1-\exp(-\eta a(x))]$ represents the probability that a female born at y reaches x. The justification for this term is as follows. Let $a = a(y, x)$ be the total amount of forest between y and x. Divide this quantity into m equal pieces that are aligned one after the other. Let $p = p(m)$ the probability that a moth is trapped in the first of these pieces. Assuming that each experiment is realized independently (which clearly is a simplification) then the probability that a female trespasses all these traps is $(1-p(m))^m$. The value $p(m)$ should be a decreasing function of m and in first approximation it should have the form $p(m) \approx 1/m$. In addition, $p(m)$ should scale with $a(y, x)$ so that we postulate $p(m) = \eta a(y,x)/m$ for some constant $\eta > 0$. In this way, the probability that a female born at y passes over x is $(1-\eta a(y,x)/m)^m$ which converges to $\exp(-\eta a(y,x))$ as $m \to \infty$.

A difficulty for fitting the model (1) is that in the available data the moth population is not measured directly but through captures on pheromone traps. In the Appendix we derive the model

$$N = N_\infty \left(1 - \kappa^{u\theta R_0^2}\right),$$

which relates the number of captures N in a trap with the density u of females in the surroundings. The parameter N_∞ is interpreted as the saturation level of a trap whereas κ is a function of the relative strength between the trap and a female. Inverting this formula and replacing into Eq. 1, we get the following relation between captures $N(t,x)$ at site x in two consecutive seasons:

$$\log\left(1 - \tfrac{N(t+1,x)}{N_\infty}\right) = \tfrac{1}{2} h(1-\alpha)\log\left(1 - \tfrac{N(t,x)}{N_\infty}\right)$$
$$+R(x) \sum_{y \in F(x,T)} a(y) \log\left(1 - \tfrac{N(t,y)}{N_\infty}\right)\exp(-\eta a(y,x)),$$

where $R(x) = \frac{1}{2} h\alpha \, [1 - \exp(-\eta a(x))]/a(x)$.

If we consider a cell x located beyond the northern boundary of the region colonized in season t, we have $N(t,x) = 0$ and therefore

$$\log\left(1 - \tfrac{N(t+1,x)}{N_\infty}\right) = R(x) \sum_{y \in F(x,T)} a(y) \log\left(1 - \tfrac{N(t,y)}{N_\infty}\right)\exp(-\eta a(y,x)), \qquad (2)$$

from which we get the following explicit model in terms of captures

$$N(t+1,x) = N_\infty\left(1 - \exp\left[R(x) \sum_{y \in F(x,T)} a(y) \log\left(1 - \tfrac{N(t,y)}{N_\infty}\right)\exp(\eta a(y,x))\right]\right). \qquad (3)$$

We use this last model to calibrate the parameters N_∞, α, h, η and T by using least squares estimation. We observe that we can only estimate the product $h\alpha$ and not each one separately. In the language of statistics, we say that in this model h and α are not estimable but their product is. Moreover, it turns out that the parameter η is small so that $1-\exp(-\eta a(x)) \approx \eta a(x)$ and $\exp(-\eta a(y,x)) \approx 1$ and therefore the model we are fitting is close to

$$N(t+1,x)=N_{\infty}\left(1-\exp\left[\tfrac{1}{2}h\alpha\eta\sum_{y\in F(x,T)}a(y)\log\left(1-\tfrac{N(t,y)}{N_{\infty}}\right)\right]\right), \tag{4}$$

on which the identifiable parameters are N_{∞} and the product $h\alpha\eta$. In order to estimate them separately we tested different combinations of these parameters to simulate the advance of the plague during the period 1984–1999, comparing these simulations to the observed advance in terms of the northern boundary for the period 1991–1992 which was located around Concepción; and the fact that in the year 1999 the plague reached Valparaíso, the northern Chilean territory with pine plantations.

3 DATA AND ESTIMATION

The available data to fit the model are based on two different sources. The first is the geographical data, which provide two main variables for each cell: the fraction $a(x)$ covered by pine trees and the direction/magnitude of the dominant wind $v(x)$. The second source is the location of pheromone traps and the number of captures on each for the whole season.

It is known that the moth is carried out by the wind near the sunset. Since the plague is active as adults only during late spring and summer (November through March in Chile), the relevant data were taken close to 8 P.M. in that period. Although wind data present day-by-day variations in direction and strength, we only considered the average values. Using data obtained from different climate stations, we produced a wind map that associates to each point z a vector $v(z)$ representing the dominant wind at z. Numerical integration was then used to solve the equation

$$\frac{\mathrm{d}z}{\mathrm{d}t}=\mathbf{v}(z(t)),$$

backward in time to determine the sets $F(x,T)$.

The basic data for captures on pheromone traps correspond to the years 1991 and 1992. Each year the Chilean forest service CONAF, in collaboration with forest companies, distributed traps within a window of several kilometres around the northern border of the plague and keeping a certain overlapping from year to year. These observations provide the position and total number of captures for each trap on these two consecutive years. From these data, we estimated the northern boundary at year $t = 1991$ and then used model (3) for the period $t + 1 = 1992$ at the cells x located to the north of this boundary to estimate $N_{\infty}, h\alpha, \eta, T$. Since some of the cells contain many traps whereas others contain no traps, we smoothed the number of captures for 1991 and 1992 for all the cells containing pines, by averaging the captures in nearby traps with a weight proportional to exp($-d$) where d is the distance from the trap to the centre of the cell.

Since the model (3) is highly non-linear in the parameters, it is not possible to use the standard linear model theory to obtain confidence intervals for the parameters. In addition, because of spatial correlation between the traps, we may not assume the observations to be independent and a sophisticated analysis would be required to get asymptotic distributions for the parameters. We content ourselves by performing a sensibility in the parameter estimation through variations in the correlation coefficient r^2 between observations and predictions.

As mentioned earlier, since η is small there is a further estimation problem, which is reflected by the fact that there is a curve on the plane $(h\alpha, \eta)$ where the parameters produce essentially the same fit. This curve is approximately given by $h\alpha, \eta = 0.18$ (obtained empirically). The natural range for h is from a few eggs to a maximum of 180 observed in laboratory, so that η should be found on the interval [0.02, 0.0002]. Table 1 shows the best fit obtained for different combinations of these parameters.

Table 1. Parameters and r^2.

N_∞	$h\alpha$	η	T	r^2	N_∞	$h\alpha$	η	T	r^2
20	3.9	0.20	14.7	0.58	20	4.9	0.05	9.6	0.67
50	4.2	0.20	14.7	0.57	50	5.9	0.05	9.3	0.69
100	4.3	0.20	14.7	0.57	100	5.9	0.05	9.3	0.68
200	4.3	0.20	14.7	0.56	200	5.9	0.05	9.3	0.68

N_∞	$h\alpha$	η	T	r^2	N_∞	$h\alpha$	η	T	r^2
20	2.4	0.10	9.3	0.64	20	9.8	0.02	9.3	0.69
50	2.7	0.10	9.3	0.64	50	10.7	0.02	9.3	0.71
100	2.8	0.10	9.3	0.64	100	10.8	0.02	9.3	0.71
200	2.8	0.10	9.3	0.64	200	10.8	0.02	9.3	0.71

N_∞	$h\alpha$	η	T	r^2	N_∞	$h\alpha$	η	T	r^2
20	17.7	0.01	9.3	0.70	20	33.5	0.005	9.3	0.70
50	19.2	0.01	9.3	0.71	50	36.5	0.005	9.3	0.72
100	19.5	0.01	9.3	0.71	100	37.0	0.005	9.3	0.72
200	19.6	0.01	9.3	0.72	200	37.2	0.005	9.3	0.72

The data used to estimate the model correspond to a first or second year of infestation with low population densities and small captures (the maximum observed capture was 18), so we are far from the saturation level of a trap and therefore the estimation of N_∞ is not precise. On the other hand, assuming that N_∞ is large enough we can linearize Eq. 2 obtaining a model which does not depend on this parameter, this also explains why the fitting of N_∞ is not precise. Fortunately, since our ultimate interest is to fit the model (1), a precise value of N_∞ is not needed.

On the contrary, we observe that the estimation of T is very stable. Since a moth flies on the average around 1–2 h each day, the estimated value of $T = 9.3$ h corresponds to 4–9 days of flight, and in terms of distance it gives 80–100 km/year, which is in the order of magnitude of the observed mean advance of 57 km (notice also that the first is the maximum advance and the second is the average displacement).

Once the parameters $N_\infty, h\alpha\eta, T$ are estimated by using least squares, we estimate h, α, η by fitting the observed mean advance of 57 km/year and the observed 1991 boundary. To this end, we run simulations with model (1) for different values of the parameters and then compared the average advance of the northern boundary in successive years.

Figure 1 shows the 1991–1992 boundary as determined by the northern captures during the season 1991. This boundary may be slightly sub-estimated since at low population densities there may be no captures at all.

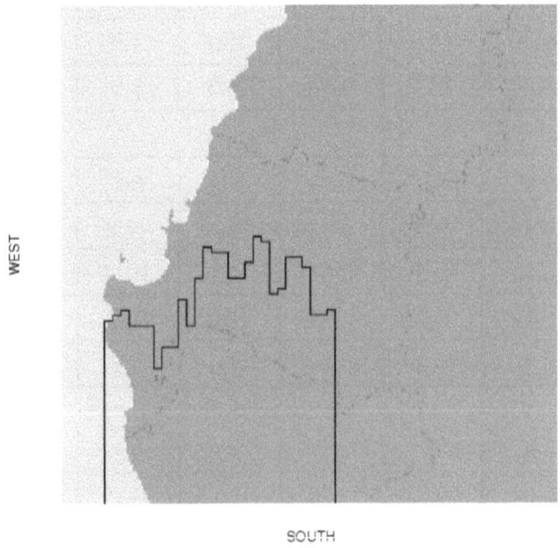

Figure 1. Boundary 1991–1992.

Figures 2 and 3 show two simulations with different values for the parameters. The figures show the south of Chile (Puerto Montt-Santiago) with pine plantations represented by small green dots and main cities represented by circles. We plot the simulation on a (logarithmic) colour scale ranging from blue = 5 moth/ha to red = 90,000 moth/ha. This range is divided into 18 classes. The fifth class (light blue) corresponds to 50 moth/ha and was considered as the threshold for positive captures, so the northern boundary of the plague predicted by the model should be around light blue and light green. For each set of parameters, we show three figures. The first one is common to all the simulations and shows the initial condition with a low density concentrated in a small region on the south. The second one is the simulation for the period 1991–1992 and the third one is the situation on 1999.

Using such simulations, we found that the best combination of parameters was: $h = 27$, $\alpha = 0.4$ and $\eta = 0.02$. This means that at low densities, the population (females) is amplified on average by a factor of 13.5 from 1 year to the next, and 40% of them emigrate from their home cell.

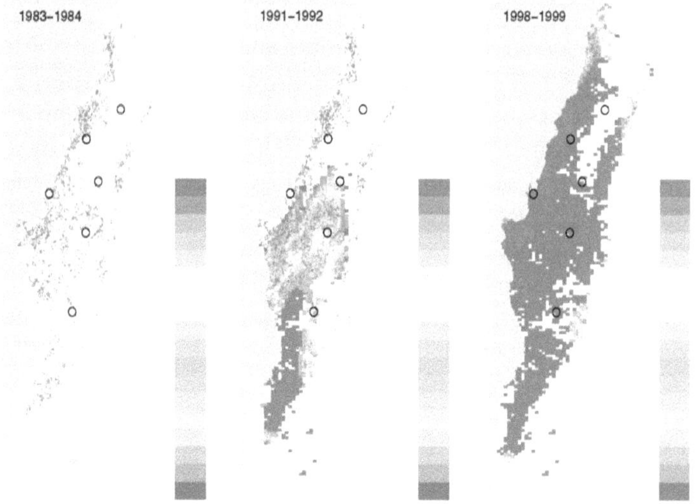

Figure 2. $h = 27$, $\alpha = 0.4$ and $\eta = 0.02$.

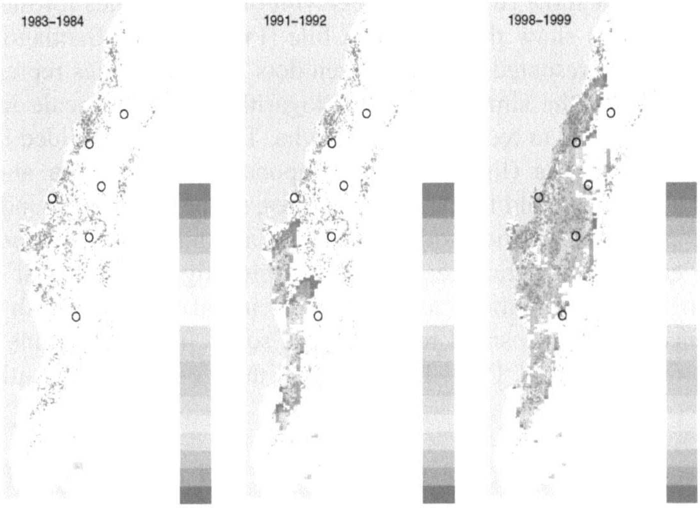

Figure 3. $h = 4.9$, $\alpha = 0.6$ and $\eta = 0.1$.

4 CONCLUSIONS

We described a model for the evolution and dispersion of the pine shoot moth in southern Chile. The model combines a drift term determined by the dominant wind flows, with a filter scheme that depends on the area of pine forest found along the fly path of a moth. The model depends on a few parameters with physical or biological meaning. Despite the fact the model is highly non-linear, these parameters were successfully calibrated from field observations, attaining a high correlation coefficient between the observations and the model predictions. The model is also consistent with the average growth and temporal distribution of the plague, replicating the northern boundaries during the period 1984–1999.

5 APPENDIX

In this appendix, we justify the model

$$N = N_{\infty}\left(1 - \kappa^{u\theta R_0^2}\right), \tag{5}$$

for predicting the number of captures N in a pheromone trap as a function of the density of females u in the surroundings. To this end we assume that a trap and a female attract a male moth in a similar way, except that the range of action and strength of the trap are larger. For a description of the mechanisms in pheromone traps see Daterman (1974). As a simple geometrical approximation, we consider that the basin of attraction of a trap (female) is a cone with vertex at the trap and angle θ with respect to the wind direction. Since traps are designed to deliver pheromone at a constant rate, a reasonable assumption is that the attraction exerted by a trap decreases as $1/d$, where d is the distance from the male to the trap. Correspondingly, the attraction of a female at the same distance is γ/d where $\gamma < 1$ represents the relative power of a female as compared with a trap.

The typical path followed by a male towards a trap is described in Birch and Haynes (1982) and is illustrated in Fig. 4. If a male and a female are dropped at random in the cone of attraction of a trap, then the male will miss the female and will be captured at the trap if and only if at every point $m(s)$ along this path the following inequality holds $\gamma\,\text{dist}(m(s),\text{trap}) \leq \text{dist}(m(s),\text{female})$.

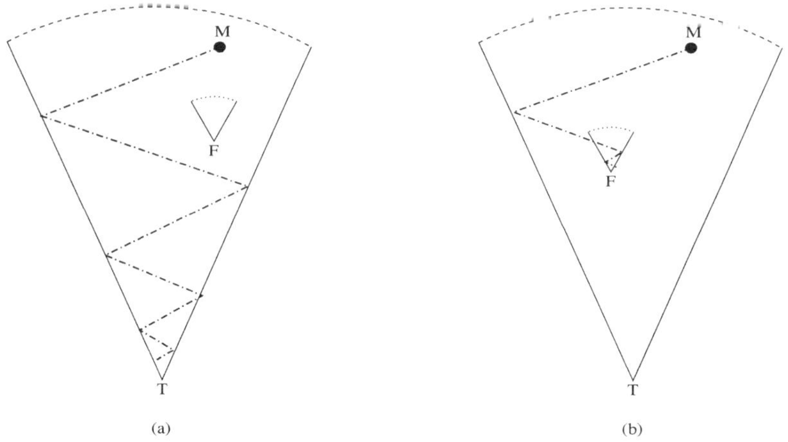

Figure 4. (a) A path to the trap. (b) A path intercepted by a female.

Suppose the male is located initially at the position (R,β_0) measured in polar coordinates from T, and consider an infinitesimal displacement along the male's path from point $A = (r,\beta)$ to point $B = (r-\Delta r,\beta + \Delta\beta)$ (see Fig. 5). Let μ be the angle of this segment to the vertical and denote $AB = \Delta l$, $AT = r$, $AC = \gamma r$. The male can arrive from A to B if no female is found inside the parallelogram P defined by the points A, B, C, D. The area of this parallelogram is $\gamma r\Delta l\cos(\theta-\mu)$ while up to first order we have

$$\Delta l \approx \Delta r\sqrt{1+r^2\left(\mathrm{d}\beta/\mathrm{d}r\right)^2}.$$

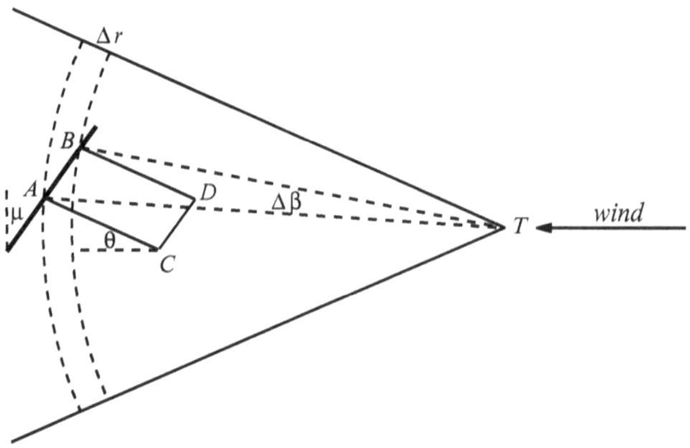

Figure 5. Infinitesimal displacement of a moth.

Integrating from $r = 0$ to $r = R$ (the initial distance from the male to the trap), we get a total area

$$A^{+}(R,\beta_0) = \gamma\cos(\theta-\mu)\int_0^R \sqrt{1+r^2(\mathrm{d}\beta/\mathrm{d}r)^2}\, r\mathrm{d}r.$$

Since $r(\mathrm{d}\beta/\mathrm{d}r) = \tan(\alpha)$ with $\alpha = \pi + \beta - \mu$, using the change of variables $r - Rw$, we obtain

$$A^{+}(R,\beta_0) = R^2\gamma\cos(\theta-\mu)\int_0^1 \sqrt{1+\tan(\alpha(w,\beta_0))^2}\, w\mathrm{d}w.$$

The function $\alpha(w,\beta_0)$ can be made explicit if we assume that the male initially moves upwards. A similar computation gives the area A^{-} for an initial downward movement. With these areas we estimate the probability that a male at (R,β_0) reaches the trap given that there is a female in the cone of attraction at a distance R from T, by the following expression

$$P((R,\beta_0)\,|\,1) = \frac{1}{2}\left[1 - A^{+}/\theta R^2\right] + \frac{1}{2}\left[1 - A^{-}/\theta R^2\right] = \kappa(\beta_0).$$

If we now compute this probability assuming there are n females we get $P((R,\beta_0)\,|\,n) = \kappa(\beta_0)^n$, and since in terms of the density u of females we have $n \approx u\theta R^2$ we obtain

$$P((R,\beta_0)\mid n)=\kappa(\beta_0)^{u\theta R^2}.$$

Integrating this function we deduce that the probability that a male dropped at random in the zone of action of the trap finally reaches it, within an environment with a density u of females, is

$$\frac{1}{\theta R_0^2}\int_{-\theta}^{\theta}\int_0^{R_0}\kappa(\beta_0)^{u\theta R^2}R\mathrm{d}R\mathrm{d}\beta_0=\frac{1}{u\theta R_0^2}\frac{1}{2\theta}\int_{-\theta}^{\theta}\frac{\kappa(\beta_0)^{u\theta R_0^2}-1}{\ln(\kappa(\beta_0))}\mathrm{d}\beta_0.$$

Finally, since the expected number of males in the zone of action for the trap is also $u\theta R_0^2$, the expected number of males caught at the trap for a given density u of females is

$$N=\frac{1}{2\theta}\int_{-\theta}^{\theta}\frac{\kappa(\beta_0)^{u\theta R_0^2}-1}{\ln(\kappa(\beta_0))}\mathrm{d}\beta_0.$$

An approximation of this expression using the mean value theorem gives

$$N=\frac{\left[\kappa^{u\theta R_0^2}-1\right]}{\ln(\kappa)}.$$

where $\kappa<1$ is a measure of the relative attraction of a female as compared with a trap, and letting $N_\infty=-1/\ln(\kappa)$ we finally obtain the model (5). Notice that this model presents a saturation level, with the number of captures approaching a finite limit N_∞ as u grows to infinity.

ACKNOWLEDGEMENTS

This work was partially funded by FONDEF projects. Jaime San Martín also thanks the support of the programme Millennium Nucleus in Information and Randomness.

6 REFERENCES

Banks R (1994) Growth and diffusion phenomena. *Springer-Verlag TAM* 14.

Birch M, Haynes K (1982) *Insect Pheromones*. The Institute of Biology Studies in Biology 47, London.

Daterman G (1974) *Synthetic Sex Pheromones for Detection Survey of European Pine Shoot Moth*. USDA Forest Service, Research Paper PNW-180.

Kendall DG (1948) A form of wave propagation associated with the equation of heat conduction. *Proceedings of the Cambridge Phil. Society* 44:591–594.

Leibhold AM, MacDonald WL, Bergdahl D, Mastro VC (1995) Invasion by exotic pests: a threat to forest ecosystems. *Forest Science Monograph* 30.

Long GE (1977) Spatial dispersion in a biological control model for larch casebearer. *Environmental Entomology* 6(6):843–852.

Murray JD (2003) *Mathematical Biology, Interdisciplinary Applied Mathematics*, Vol. 18, Springer-Verlag, Heidelberg.

Oksendal B (1992) *Stochastic Differential Equations*. Springer-Verlag, Heidelberg.

Okubo A (1980) *Diffusions and Ecological Problems: Mathematical Models*. Springer-Verlag, Heidelberg.

Polymenopoulos AD, Long G (1990) Estimation and evaluation methods for population growth models with spatial diffusion: dynamics of mountain pine beetle. *Ecological Modelling* 51:97–121.

Protter P (1990) Stochastic integration and differential equations. *Applications of Mathematics*, Vol. 21. Springer, Heidelberg.

Robredo F (1970) Contribución al conocimiento de la bioecología de *Rhyacionia buoliana* Den. et Schiff y sus daños. *Boletín del Servicio de plagas forestales* XIII 26:181–186.

Robredo F (1975) Contribución al conocimiento de la bilogía de *Rhyacionia buoliana* Den. et Schiff (Lepidoptera: Tortricidae) I. Estudio del adulto. *Boletín del Servicio de Defensa contra plagas e Inspección Fitopatológica* 1(1):69–81.

Shoeder D (1986) Consultacy on the Biological Control of European Pine Shoot Moth, *Rhyacionia buoliana* in Chile, IICA.

Skellam JG (1951) Random dispersal in theoretical populations. *Biometrika* 38:196–218.

Chapter 28

ADAPTIVE OPTIMIZATION OF FOREST MANAGEMENT IN A STOCHASTIC WORLD

Peter Lohmander
Department of Forest Economics, SLU, Sweden

Abstract Management decisions should be based on the sequentially revealed information concerning prices, growth, physical damages etc. Future flexibility is valuable in a stochastic world and should be optimized. Stochastic dynamic programming, stochastic scenario tree optimization, and optimization of adaptive control functions with stochastic simulation of the objective function are relevant alternatives.

Keywords: Stochastic dynamic optimization, forest management

1 INTRODUCTION

Economic optimization of forest management is a highly interesting area, which covers theories from several quantitative fields and optimization methods of all kinds. One reason for this fact is that trees grow. Trees represent interesting biological units that form stands and forests. Growth can be modelled at the tree level and at higher levels. The stock may be observed and controlled over time via activities such as thinning and fertilization. Growth, in several dimensions, is a function of the measurable state. Hence, we may view the forest as a controllable Markov process.

Future growth is affected not only by the present properties of the trees. Wind throws, insect damages, fungi damages, rodents, moose, elephants and other animals, changes in climate, air pollution, forest fires, and many other factors affect forest damages and growth and the future state of the forest. In most cases, such factors may not be perfectly predicted. Storms, fires, and different forms of other damages may usually be regarded as stochastic. It should also be clear that there are many different types of "stochastic disturbances" of this nature. Different types of disturbances may have very different effects on the forest state. Some disturbances will only influence some species and leave other species unaffected. Other disturbances will affect some parts of the trees, such as the roots or some branches. Some damages are local and affect only one tree. Other types of damages have spatial dimensions and influence the development of several square kilometres, such as forest fire damages and wind throws.

In most places, it is very expensive to go to the forest to harvest and to collect just one tree. We have to consider the different alternatives on a larger scale: Is it economically optimal to harvest one or several forest stands in the area in this period via clear felling or should we perform partial harvesting in the form of continuous cover forestry? We conclude that there are mostly considerable economies of scale in harvesting and other operations. For this reason, it is seldom optimal to perform "continuous" harvesting. During a time interval when you harvest in an area, you harvest much more than the growth during the same time interval. It would in most cases be far too expensive to keep the harvester, the forwarder, and the labour in the forest stand for ever. Hence, irrespective of whether you make a clear felling or make a thinning, it is economically optimal to wait a considerable time, usually many years, before you visit the site again in order to harvest.

When we consider stochastic events of importance to forest management and economic optimization, we must not forget the markets. The market value of a tree is a function of the general market conditions. If we want to determine the economic values of trees, we have to consider the prices of logs of different qualities and dimensions in different market places, the prices of pulpwood and fuel wood and several other alternative wood products. We also have to consider the costs of harvesting and forwarding, the capacity of the forest road network at different points in time, and the availability and prices of trucks and labour. Clearly, in most cases the timber producers do not control the market prices. Inventions and innovations in process technology will change future relative prices and can, by definition, not be perfectly predicted. If you can predict the exact properties of a future invention, it has already taken place! Furthermore, political changes at the national and international scales may change the markets in ways that are impossible to predict. As a confirmation, econometrics research has not yet been able to give us perfect price predictions covering long horizons.

Hence, if we want to address the relevant forest management problems, we have to admit that future prices cannot be perfectly predicted. This has not been possible in the past and there is no reason to hope for such options in the future. As a result, we may conclude the following:

We may view the forest management problem this way: There is a controllable Markov process. The state of the forest is affected by growth and damages that have to be regarded as stochastic. The stochastic disturbances may have many different properties and influence the forest in different scales. The market prices and other external conditions usually change over time. These changes cannot be perfectly predicted and may be regarded as stochastic processes with many different properties. There are considerable set-up costs in harvesting and other operations. Each time you undertake some operation, you have to move harvesters, forwarders and labour considerable distances between objects. Hence, you visit each stand or forest area only rarely. In the mean time, between these visits, the forest grows and may be affected by different kinds of stochastic damages. Furthermore, between these visits, considerable market changes may take place.

Now, it is time to take a close look at earlier attempts and new ways to handle the forest management decisions in an economically optimal fashion. Of course, this text cannot cover all kinds of relevant problems and applications in the area. Hence, just a small number of typical and economically important examples from forest management will be analyed in detail. The reference list contains studies that discuss many more applications of the presented methods and may be consulted by the reader at a later point in time.

2 GENERAL MATHEMATICAL TOOLS AND METHODS

Economic forest management decision problems have been addressed by all kinds of mathematical optimization methods during the years. Faustmann (1849) defined the present value of an infinite series of identical forest generations. It was implicitly assumed that everything was known with certainty. A number of authors have continued in the same tradition. Johansson and Löfgren (1985) give a survey of this field. They use a number of different methods in different chapters, all but one essentially based on deterministic derivations and optimization. During more than one century, the dominating forest economic decision problem has been the determination of the optimal harvest year, based on alternative deterministic assumptions. Typically, the decision problems were solved stand by stand via one-dimensional present value maximization. The options to simultaneously optimize the number of seedlings per hectare, the number of thinnings, the timing and intensity of thinnings, and the year of the final felling, were mostly neglected. As one typical example, Johansson and Löfgren (1985) present a very detailed one-dimensional analysis of the optimal harvest year decision problem, assuming that all future forest generations will be identical to the present.

Linear programming made optimization and coordination at higher levels possible. George Danzig developed an efficient method for linear programming problems many years before he published his well-known book, Danzig (1963). Linear programming rapidly became a widely used forest planning tool, partly thanks to the development of computers and easily available standard software. Mostly, the forest management model assumptions included perfect information, large numbers of forest compartments, and long horizons. Much later, it became known that linear programming also is a very useful tool when it is necessary to solve a completely different type of problem with stochastic disturbances of many different kinds. See later sections.

Bellman (1957) presented a conceptually new method: dynamic programming. In many of the typical applications, dynamic programming is used to optimize decisions over time under the assumption that future parameters are known with certainty. However, dynamic programming can also be used for many other purposes. It turns out that you may handle set-up costs, sequential information, and adaptive decisions in a very simple and consistent way. Ross (1983) concentrates on the very important and relevant field stochastic dynamic programming. Among other things, Ross (1983) shows how you can find an optimal stationary policy for stochastic dynamic programming problems based on Markov chains via linear programming. Winston (2004), Chapter 19, gives a very convenient summary of this approach and related methods.

It has often been considered more elegant and sometimes more relevant to deal with continuous time formulations and solutions. Pontryagin (1961) is often regarded as the founder of optimal control theory. Fleming and Rishel (1975) present most of the general theory and proofs in a complete fashion. Sethi and Thompson (2000) give a very good description of the area and derive the central proofs using dynamic programming in an efficient way. In fact, several authors have used continuous time optimal control theory to derive optimal solutions in forestry. Clark (1976) derives the optimal thinning policy in forestry using deterministic optimal control in continuous time. Sethi and Thompson (2000) develop the model from Clark

(1976) in a similar way. Clark (1976) and Sethi and Thompson (2000) refer to Kilkki and Vaisanen (1969), who developed the original optimization model in discrete time using dynamic programming. The later authors then transformed the model into continuous time.

The author of this paper is not convinced that the later continuous time versions of the Kilkki and Vaisanen (1969) model are more relevant or better in any way. The optimal stock-level function is smooth and elegant in the continuous time version. The discrete time model however has the very important advantage that the harvest cost function in a very simple way can include the set-up cost, the cost of moving harvesters, forwarders and labour to the site, and many other parameters that may vary in many different ways between periods. Hence, the discrete time version may come as close as you want to the true optimal solution. It is never optimal to "continuously harvest" the forest. The prices and variable harvest costs per unit are assumed constant over time in the Sethi and Thompson (2000) continuous time version. Furthermore, there are unfortunate errors in the Hamiltonian function analysis in the otherwise very well-written Sethi and Thompson (2000) book, since the discounting factor was forgotten. The original dynamic programming version is much more easily described to the reader and the dynamic programming version can also easily be extended to the really relevant case where you may have stochastic disturbances of many different kinds. Then, we simply go to stochastic dynamic programming in discrete time. A final argument is that very few, if any, real things are continuous. In the time dimension, as one example, we note that growth and harvesting conditions normally change over the year. Some seasons are warm, others are cold, some are wet, and some are dry.

Furthermore, as the discrete time intervals approach zero, we should not neglect the existence of day and night, since the light conditions usually affect biological growth in forests and elsewhere and most connected activities in the economies. As a result, it may be better to handle the real and relevant periods via discrete time dynamic programming than to assume that they do not exist, using continuous time optimal control. Continuous time optimal control can be extended to stochastic continuous time optimal control. This approach has found many applications. Sethi and Thompson (2000) give a good summary with typical applications from different fields. The reader should be aware that the underlying process assumptions often are very restrictive. If the assumptions are relevant to the application at hand, this may not be a problem. Gleit (1978) investigates a stochastic optimal control problem in continuous time and derives an optimal harvest function. The mathematical analysis is well performed and the analytical difficulties are openly demonstrated. In order to be able to derive some explicit results, several highly restrictive assumptions have to be introduced. One of these is the assumption that the expected growth is proportional to the stock level. From a biological perspective, such assumptions are seldom relevant. Even such a very special analysis with simplifying assumptions required 13 pages of advanced formulae. The author of this paper is convinced that it is far from easy to handle the types of stochastic disturbances described in the introduction via continuous time optimal control theory. Discrete time stochastic dynamic programming is mostly a more relevant approach, in particular since this makes it possible to include set-up costs, most types of functional forms (or at least discrete approximations), and periods with all kinds of different properties in a convenient way.

3 STOCHASTIC DYNAMIC PROGRAMMING

Here, a very general definition of the economic management problem is given along the lines found in Winston (2004) and many earlier publications. $W(i,t)$ is the expected present value at time t, i is the state, and r is the rate of interest. $h(i,t)$ denotes the control, such as the harvest, as a function of state and time. We use stars to indicate optimal values. $W^*(i,t) = W(i,t,h^*(i,t)$. $R(i,t,h)$ is the profit at time t. T is the horizon. $W^*(i,T+1) - 0\ \forall i$.

$\tau(j|i,t,h)$ denotes the conditional probability of reaching state j in period $t+1$ and $\sum_{j=1}^{J}\tau(j\mid i,t,h)$ is the expected value of the entering state in period $t+1$.

For $t\in\{1,2,\ldots T\}$ and $i\in\{1,2,\ldots I\}$, we determine the optimal harvest (and other) decisions.

$$W^*(i,t) = \max_{\substack{h \\ h\in H(i)}}\left(R(i,t,h) + e^{-r}\sum_{j=1}^{J}\tau(j\mid i,t,h)W^*(j,t+1)\right),$$

where $H(i)$ is the feasible control set and may sometimes also contain a time argument, which may be important if harvest capacity is changed over time.

If the planning horizon is infinite and functions do not change over time, we can simplify the problem by dropping the time index, using the following definition:

$W^*(i,t) = W^*(i) = W_i^*$. In the same manner, we write:

$$R(i,t,h) = R(i,h) = R_{i,h}.$$

Since each W_i should be optimal and independent of t, the following inequalities must hold.

$$W_i \geq R_{i,h} + e^{-r} \sum_{j=1}^{J} \tau(j \mid i,h) W_j, \quad \forall i,h \mid h \in H(i).$$

Furthermore, there is an upper bound on each W_i. W_i cannot exceed the value obtained if the best decision is selected.

$$W_i^* = R_{i,h^*} + e^{-r} \sum_{j=1}^{J} \tau(j \mid i,h^*) W_j^*.$$

Hence, the optimal values can be determined from this linear programming formulation:

$$\min Z = \sum_{i=1}^{I} W_i$$

s.t.

$$W_i - e^{-r} \sum_{j=1}^{J} \tau(j \mid i,h) W_j \geq R_{i,h} \quad \forall i,h \mid h \in H(i).$$

4 STOCHASTIC SCENARIO TREE OPTIMIZATION

The approach of multi-period stochastic programming used below has been well described by Birge and Louveaux (1997). The particular forest management problem was suggested by the author and first presented at the MODFOR conference in 2002. Let us denote adaptive multi period stochastic programming with scenarios, A1, and stochastic dynamic programming, A2. A1 and A2 have different advantages and disadvantages in practical applications. The author has used A2 in many applications with very good results but thinks that A1 may be an approach which is more easily used as a "default value" standard tool within some application areas. Here, we describe A1: We maximize the expected present value, W, of the net profits from all periods, 1, 2, …, n.

$$W = \sum_{d}\sum_{s_1} k_1\theta(s_1)(p_{ds_1}v_{ds_1} + L_d)h_{ds_1} +$$

$$\sum_{d}\sum_{s_{12}} k_2\theta(s_{12})(p_{ds_{12}}v_{ds_{12}} + L_d)h_{ds_{21}} +$$

$$\ldots$$

$$+\sum_{d}\sum_{s_{12\ldots n}} k_n\theta(s_{12\ldots n})(p_{ds_{12\ldots n}}v_{ds_{12\ldots n}} + L_d)h_{ds_{12\ldots n}}$$

d = Forest department

$s_{12\ldots t}$ = One scenario defining the states of the exogenous stochastic process(es) from period 1 until period t

$\theta(.)$ = Probability (of a scenario) estimated before the exogenous stochastic process outcomes have been observed

r = Rate of interest in the capital market

m = Time length of a period

K = Discounting factor. We assume that all results from a period occur in the middle of that period. Time zero is the point in time where the first period starts. The discounting factors of the different periods, t, are k_t:

$$k_1 = e^{-(1-\frac{1}{2})mr}$$

$$k_2 = e^{-(2-\frac{1}{2})mr}$$

$$\ldots$$

$$k_n = e^{-(n-\frac{1}{2})mr}$$

$p_{ds_{12\ldots t}}$ = Net price (price - harvesting costs per volume unit) in forest department d in period t if the scenario which has been followed until period t is $s_{12\ldots t}$

$v_{ds_{12\ldots t}}$ = Volume per area unit (density) in department d in period t if the scenario which has been followed until period t is $s_{12\ldots t}$

L_d = Land value (of bare land) in department d

$h_{ds_{12...t}}$ = Harvest area in department d in period t if the scenario which has been followed until period t is $s_{12...t}$

d = Forest department

$s_{12...t}$ = One scenario defining the states of the exogenous stochastic process(es) from period 1 until period t

$\theta(.)$ = Probability (of a scenario) estimated before the exogenous stochastic process outcomes have been observed

r = Rate of interest in the capital market

m = Time length of a period

k = Discounting factor. We assume that all results from a period occur in the middle of that period. Time zero is the point in time where the first period starts. The discounting factors of the different periods, t, are k_t:

$$k_1 = e^{-(1-\frac{1}{2})mr}$$

$$k_2 = e^{-(2-\frac{1}{2})mr}$$

$$\dots$$

$$k_n = e^{-(n-\frac{1}{2})mr}$$

$p_{ds_{12...t}}$ = Net price (price - harvesting costs per volume unit) in forest department d in period t if the scenario which has been followed until period t is $s_{12...t}$

$v_{ds_{12...t}}$ = Volume per area unit (density) in department d in period t if the scenario which has been followed until period t is $s_{12...t}$

L_d = Land value (of bare land) in department d

$h_{ds_{12...t}}$ = Harvest area in department d in period t if the scenario which has been followed until period t is $s_{12...t}$

Constraints and equations:

The probabilities of the different scenarios $s_{12...t}$ should be calculated via the (possibly time dependent) state transition probabilities for each t such that $2 \le t \le n$. The definitions of the exogenous stochastic process(es) may be different. The exogenous stochastic processes usually influence the variables in the problem, the objective function coefficients, the set of feasible decisions, or something else. All of these effects must be calculated for each time period and scenario. The parameter vectors of different scenarios may be identical in the first periods. In such cases, constraints must exist that make sure that also the decisions are the same. As long as the parameter vectors are the same, the "true" scenario cannot be identified. This has been discussed by Lohmander (1994).

If there are physical constraints, such that a particular forest department area only can be harvested once during the defined time horizon, these constraints must be defined for each scenario covering the complete time horizon. The dynamic timber volume developments in the different departments should be calculated through relevant difference equations. There may be other constraints, connecting the possible decisions in different forest departments. Perhaps the total harvest volume has to stay within some interval which is feasible because of delivery contracts? Perhaps there is a local pulp mill which needs a more or less constant flow of pulpwood? Then, constraints have to be defined which make sure that, for each scenario, the total harvest volumes stay within the feasible sets in the different periods. In some cases, there are harvest area constraints because of forest act regulations. In Sweden, such rules define dynamic sets of feasible harvest activities at forest properties. One Swedish forest property application including the forest act regulations is found here:

http://www.lohmander.com/kurser/MODFOR02/M DFOR02.htm

A forest logistics application of the stochastic scenario method is found in Lohmander and Olsson (2005).

5 OPTIMIZATION OF ADAPTIVE CONTROL FUNCTIONS WITH STOCHASTIC SIMULATION OF THE OBJECTIVE FUNCTION

This approach is very flexible. You can in principle handle all kinds of functions and constraints. You just define the complete model as a stochastic simulation model. You determine your adaptive strategy, the principles that give the decisions as a function of the revealed state of the system, time, etc. Then, when you simulate the system, you let the system decide your decisions according to your suggested adaptive strategy and calculate your objective function value, for instance the expected present value. You simulate the complete system a large number of times, let us say 1,000 times, and divide the sum of the objective function values by the same number, for instance 1,000. That way, you get an estimate of the expected present value. Then you may use different numerical methods in order to search a strategy that is close to the optimum. One approach which has been tested in typical forestry applications and turned out to come close to the optimum rather rapidly is the following: Let us assume that the decision, such as the harvest level, is an adaptive function of the state of the system including information about the stock level and the price. We assume that this function has two parameters defined by the point (X,Y). (X,Y) should be optimized. An example of such an adaptive harvest function could be $h(V, P, X, Y)$ where h denotes harvest as a function of stock, V, and price, P. The parameters are X and Y.

Initially, you guess the optimal parameter combination (X,Y). You investigate, via a large number of stochastic simulations, the expected objective function value, F, in this position. Then, you take steps, $\mathrm{d}X$ and $\mathrm{d}Y$, in each direction, and visit points $(X + \mathrm{d}X,Y)$ and $(X,Y + \mathrm{d}Y)$. In these coordinates, you investigate the expected objective function values. Now, you have useful information and can determine approximations to the partial derivatives of the expected objective function value $\partial F/\partial X$ and $\partial F/\partial Y$. With this information, you determine (an approximation of) the direction of the steepest path. You take steps of equal size in that direction and investigate F each time. As long as F increases, you continue in the same direction. When you have passed the peak, you reduce the step size and go back. You continue to go back and forward, reducing the step size, until you are "satisfied". Seen from the "peak" in the first selected direction, it is possible that the derivatives $\partial F/\partial X$ and $\partial F/\partial Y$ are different. Hence, you check these derivatives again and go in the "locally best" direction until you find a new peak. The process is repeated until you are satisfied. This approach is conceptually simple and flexible. One problem is that we do not know if we find a global maximum. Another problem is that stochastic deviations from the locally optimal solutions are likely, because we cannot "afford" to spend too much computing time with very large numbers of system simulations in each position. The step size is another issue that deserves some attention. Still, despite all the numerical problems with this approach, it may serve as a good alternative in some cases. The studies by Lohmander (1992a, 1993) are two such examples.

6 ADAPTIVE STAND-LEVEL OPTIMIZATION WITH FINAL HARVESTS

Forest management has been optimized at the stand level via stochastic dynamic programming in discrete time with continuous probability density functions of stochastic prices by Lohmander (1983, 1987, 1988). Other authors have later published the same type of models and results. This approach makes it possible to derive optimal reservation prices and expected optimal values of the forest stand and land explicitly via recursion and analytical methods. A reservation price is the price which makes you indifferent between the alternatives to harvest directly and to wait longer. If the price offer is higher than the reservation price, you should harvest directly. If the price offer is lower than the reservation price, you should wait longer. Here, the main structure of the model will be presented. W_t is the expected present value in period t before the price p in the same period has been observed. Prices are stochastic and have the probability density function $f(p)$. $f(p) > 0 \forall p$. In each period, new stochastic prices are revealed in the market. This type of price process is a special case of a stationary stochastic process. If we consider time periods of 5 years, this is a model that is not easy to reject on statistical grounds in the Swedish market and many other markets. The prices really denote "real net prices", nominal prices – harvesting costs per volume unit deflated by consumer price indices. The reservation price in a particular period is denoted by q_t. If $q_t > p_t$, then you should wait at least one period more for a new price offer. If $q_t = p_t$, the probability of which is almost zero, you are indifferent between the alternatives to harvest at once and to wait at least one more period. If $q_t < p_t$, then you should harvest directly. V is the volume per area unit and L is the value of the land released after harvest. We select to start the calculations far in the future, at the horizon, T. It turns out that the particular choice of T is not critical to the results as long as T is sufficiently large, for instance three times the age of the optimal forest rotation age in a deterministic analysis.

$$W_T^* = 0$$

$$W_t = \int_{-\infty}^{q_t} W_{t+1}^* f(p)dp + \int_{q_t}^{\infty} e^{-rt}(pV(t) + L)f(p)dp, \quad \forall t \big| t < T.$$

The reservation prices are optimized for each period t such that $t < T$.

$$\frac{\mathrm{d}W_t}{\mathrm{d}q_t} = f(q_t)\left(W_{t+1}^* - e^{-rt}(q_t V(t) + L)\right) = 0$$

$$f(q_t) > 0 \Rightarrow \left(W_{t+1}^* - e^{-rt}(q_t V(t) + L)\right) = 0$$

$$\left[\frac{\mathrm{d}W_t}{\mathrm{d}q_t} = 0\right] \Rightarrow \left[e^{-rt}(q_t V(t) + L) = W_{t+1}^*\right].$$

The present value of harvesting, given that the price is equal to the reservation price, is the same as the expected present value in case you wait at least one more period and take optimal decisions in future periods based on the revealed outcomes of the stochastic prices. In optimum,

$\left(W_{t+1}^* - e^{-rt}(q_t V(t) + L)\right) = 0$, which implies that

$$\frac{\mathrm{d}^2 W_t}{\mathrm{d}q_t^2} = -f(q_t)e^{-rt}V(t) < 0.$$

We find that the solution is a unique maximum in each period. The optimal reservation prices and expected present values are determined recursively, starting from T via the backward algorithm of dynamic programming.

$$q_t^* = \frac{e^{rt}W_{t+1}^* - L}{V(t)}.$$

You may state the optimization problem as:

$$W_t^* = \int_{-\infty}^{\infty} \max\left\{W_{t+1}^*, e^{-rt}\left(pV(t) + L\right)\right\} f(p)\mathrm{d}p, \quad \forall t \middle| t < T.$$

As a result, it is clear that the expected present value is a nonstrictly decreasing function of time. $W_t^* \geq W_{t+1}^*$. In most empirically relevant cases, $q_t < \infty$. $f(p) > 0 \forall p$ and $W_t^* > W_{t+1}^*$. As a result, the expected present value is a strictly decreasing function of time. This is not really surprising: If we move forward in the time dimension and still have not harvested, this indicates that we have not experienced prices above the optimal reservation prices yet. The longer we have to wait before we experience such a good price, the worse. Before we knew that this would happen, we had many valuable options ahead of us. That is why the expected present value is a decreasing function of time. The reader should be aware that we should not decide the harvest year in advance if the market prices are stochastic. We should wait and see what happens in the market before we decide to harvest or to wait longer. The reservation prices can however be optimized along the lines suggested in this section. Often, the optimal reservation prices are decreasing functions of time. However, this does not always have to be the case. Furthermore, if we make a more detailed analysis, we have to consider the changing dimensions of the trees, proportions of possible timber and pulpwood harvest levels, and quality-dependent price lists. If we look at the decision problems using shorter periods, such as months or years, we usually have to consider autocorrelation issues in detail. Different kinds of stochastic price and growth processes may lead to different results. Such detailed analyses have been done by, for instance, Lohmander (1987). General findings in this area have been reported by Lohmander (2000).

7 ADAPTIVE CONTINUOUS COVER OPTIMIZATION AT THE STAND LEVEL

Now, we move on to discrete time optimization of continuous cover forest management using dynamic programming. The analysis will start with a deterministic model that becomes transformed to a stochastic version. The analysis below was originally presented by the author at EURO/Informs, Istanbul, 2003. The volume stock levels, *S*, are defined in such a way that the stock moves up one level per 5-year period. Volume is denoted by *V*. The volume growth is assumed to follow this process $G(V) - \alpha V + \beta V^2$, which is consistent with the classical logistic growth assumption in natural resource economics. We may rewrite the function in this way: $G(V) = sV(1 - V/K)$, where *s* is the "intrinsic growth rate" and *K* is the "carrying capacity" of the environment. In the analysis, we assume that $s = 5\%$ and $K = 400$ cubic metres per hectare, which are typical parameters in some Swedish forests. The particular numerical values are however not important to the qualitative discussions. Of course, other numerical values will be the results if other growth parameters are used. The qualitative results are however the same. In the cases without price risk, it is assumed that the real price per cubic metre is 300 SEK, which is close to the average value in Sweden. The real variable harvesting cost is 100 SEK per cubic metre. (7 SEK≈$US 1). When there are set-up costs, representing the costs of moving harvesters, forwarders, and labour to the objects, these are assumed to be 500 SEK per hectare and occasion in real terms, which is typical in Sweden. When stochastic prices are assumed, then the prices per cubic metre are 220, 260, 300, 340, and 380 when the "business states" are 1, 2, 3, 4, or 5 respectively. Then, all business states are assumed to have equal probability. This degree of price variation is typical in Sweden and it is hard to statistically reject the hypothesis that a more or less uniform net price probability distribution is relevant, using historical data. All of the analysis in this section concerns optimization with infinite horizon. In the examples, a 3% real rate of interest is used. The stock level is constrained to simplify the illustration: $0 < S < 13$ (Table 1).

Table 1. The "classical" and most simple case with a constant price of 300 SEK per cubic metre and without set-up costs: the table shows the optimal harvest volumes per 5-year period as a function of the entering stock level.

Entering stock (cubic metres per hectare)	Optimal harvest volume (cubic metres per hectare)
30	0
37	0
45	0
55	0
67	0
81	14
97	30
116	49
136	69
159	92
183	116
207	140

In the relevant case, demonstrated in Table 2, the set-up costs and stochastic prices are treated consistently and simultaneously. The optimal harvest is an increasing function of the stock level and the price level. We also note that "low volume harvesting" should be avoided when possible, which is natural since we have set-up costs.

Table 2. The relevant case with set-up costs and price risk: The table shows the optimal harvest volumes (cubic metres per hectare) per 5-year period as a function of the entering stock level and the price level.

Entering stock (cubic metres per hectare)	Price (SEK per cubic metre)				
	220	**260**	**300**	**340**	**380**
30	0	0	0	0	0
37	0	0	0	0	0
45	0	0	0	0	0
55	0	0	0	0	0
67	0	0	0	0	37
81	0	0	0	0	51
97	0	0	0	0	67
116	0	0	0	49	86
136	0	0	0	69	106
159	0	0	0	92	129
183	0	0	46	116	153
207	25	25	71	140	178

8 ADAPTIVE COMPANY-LEVEL FOREST MANAGEMENT OPTIMIZATION

In forest companies, you often face global constraints, such as harvest capacity constraints, delivery contracts, etc. Optimal harvesting decisions in forestry under the influence of stochastic prices have mostly been studied under the assumption of complete stand separability. In situations when the harvester capacity is a binding constraint, the optimal stands to harvest cannot be determined without explicit consideration of this constraint, which means that stand separability assumptions are not relevant. Then, it may be optimal to harvest a particular stand if the timber price is low but not if the price is high. The profitability of harvesting may be more sensitive to the timber price in some other stand. Hence, it may be more important to use the machines in another stand if the timber price is very high. The expected shadow price of harvester capacity is an increasing function of the degree of timber price variation, indicating that the optimal harvester capacity is higher under risk than under certainty in some cases. Many other types of "global constraints" usually exist in forest sector enterprises. Many of them are expected to give "disturbances" to the classical "optimal economic stand management" decision rules. You may think that dynamic programming cannot be applied to relevant problems because of the curse of dimensionality. Maybe this is not always quite true. Optimal adaptive decisions can be determined at the forest company level. Some examples are found in the reference list: Lohmander (1992b, 1993, 1997a). Optimal infrastructure investments are also important. Compare Olsson and Lohmander (2005).

The expected present value increases if the risk of the stochastic prices increases and there are options to wait for the best harvesting occasions. The positive effects of increased price risk are reduced in case the harvesting capacity is low. If the harvesting capacity increases, you have a more flexible system and can take advantage of sudden price increases in a better way. The expected shadow price of harvest capacity is an increasing function of the degree of timber price variation. The optimal harvest capacity is higher under risk than under certainty in some cases. The traditional deterministic analysis does not give the correct shadow prices. This, in turn, leads to too low-capacity investments. The approach in this paper gives the correct expected shadow prices and can be used to optimize harvest capacity investments.

9 DISCUSSION

No paper is complete in the sense that all relevant issues are mentioned. The area of adaptive optimization of forest management contains many different kinds of special topics and if all of these should be discussed, you would not have been able to go into details in any particular direction. Hence, a selection of some of the most important problems had to be made. This paper contains a treatment of the final harvesting problem with adaptive optimization. We have also analysed the continuous cover forestry problems with and without price risk and set-up costs. Some other problems were mentioned and different types of global constraints were discussed in connection to the stochastic programming formulation and typical methodological tools were described. Now, the reader is advised to search for relevant applications in other directions. Optimal forest sector logistics is one area where new and relevant results can be obtained, in particular since roads and other parts of the logistics network may be disturbed by unexpected rains, snow and ice, and other problems that cannot be perfectly predicted. If you cannot deliver a sufficient flow of wood to the pulp mills, you may have to stop production, which may be very costly. Hence, there is an optimal combination of road capacity, trucks, pulpwood storage, and locations, which is not always easy to optimize. Some efforts in this direction have however been made in recent years and can be found in the reference list. Stochastic damages of many kinds have been analysed and the optimal adaptive strategies have been determined. Some of the recently investigated areas in this class concern species-selective damages caused by moose in Sweden and the optimal mix of species and selective thinnings in plantations. In large parts of Northern Sweden, moose damages to Scots pine cause severe problems and mixed species plantations are sometimes the economically best solution. Compare Lu and Lohmander (2005).

Another topic with reported optimal results is the spatial and temporal management of forest areas where stochastic winds randomly cause windthrows. In Sweden, the windthrow topic has been quite dominating during the spring of 2005 because a hurricane, named Gudrun, destroyed very large forest areas in southern Sweden completely. Research results existed much earlier, indicating that the optimal harvest ages are lower in stormy areas, that one should keep large areas together without partial harvests since the stands protect each other from the wind and that one should modify the spacing and thinning intensity. Compare Lohmander and Helles (1987) and Lohmander (1987b). However, the Swedish forest act did not take such things into account and the forest owners could not deviate from the detailed forestry regulations. With some luck, the forest act may be modified in the near future and consider these problems more carefully. So far, we have not mentioned the fact that the markets sometimes may be described as dynamic games. When the number of players is small, which is sometimes the case, at least locally, we may use deterministic or stochastic differential or difference games to study the optimal decisions. Compare Lohmander (1997b). This is a very large field that deserves much more efforts in the future. The area of stochastic difference games may be regarded as a very natural extension of adaptive optimization. Of course, this is highly relevant in the forest sector.

ACKNOWLEDGEMENTS

The author is grateful to Professor Andres Weintraub for highly construc tive editorial suggestions. Furthermore, valuable economic support from FORMAS made it possible to present and discuss the results at IFORS 2005.

10 REFERENCES

Bellman, R., *Dynamic Programming*, Princeton University Press, Princeton, NJ, 1957.

Birge, J.R., Louveaux, F., *Introduction to Stochastic Programming*, Springer, Berlin Heidelberg New York, 1997.

Clark, C., *Mathematical Bioeconomics, The Optimal Management of Renewable Resources*, Pure and Applied Mathematics, Wiley-Interscience, 1976.

Danzig, G., *Linear Programming and Extensions*, Princeton University Press, Princeton, NJ, 1963.

Faustmann, M., Berechnung des Wertes welchen Waldboden sowie noch nicht haubare Holzbestände fur die Waldwirtschaft besitzen, *Allgemeine Forst und Jagd-Zeitung*, 25, 441–455, 1849.

Fleming, W.H., Rishel, R.W., *Deterministic and Stochastic Optimal Control*, Springer, Berlin Heidelberg New York, 1975.

Gleit, A., Optimal harvesting in continuous time with stochastic growth, *Mathematical Biosciences*, 41, 111–123, 1978.

Johansson, P.O., Löfgren, K.G., *The Economics of Forestry and Natural Resources*, Blackwell, Oxford, 1985.

Kilkki, P., Vaisanen, U., Determination of optimal policy for forest stands by means of dynamic programming, *Acta Forestalica Fennica*, 102, 100–112, 1969.

Lohmander, P., Optimal harvest policy under the influence of imperfections and uncertainty, Swedish University of Agricultural Sciences, Dept. of Forest Economics, WP 22, 1983.

Lohmander, P., The economics of forest management under risk, Swedish University of Agricultural Sciences, Dept. of Forest Economics, Report 79 (Doctoral dissertation), p. 311, 1987.

Lohmander, P., Pulse extraction under risk and a numerical forestry application, *Systems Analysis – Modelling – Simulation*, 5(4), 339–354, 1988.

Lohmander, P., Continuous harvesting with a nonlinear stock dependent growth function and stochastic prices: Optimization of the adaptive stock control function via a stochastic quasi-gradient method, in: Hagner, M. (editor), *Silvicultural Alternatives, Proceedings from an Internordic workshop*, June 22–25, 1992, Swedish University of Agricultural Sciences, Dept. of Silviculture, No. 35, pp. 198–214, 1992a.

Lohmander, P., The optimal dynamic production and stock levels under the influence of stochastic demand and production cost functions: theory and application to the pulp industry enterprise, *Systems Analysis – Modelling – Simulation*, 10(2), 1992b.

Lohmander, P., Decision optimization with stochastic simulation subroutines: relation to analytical optimization of capacity investment and production, presented at the "4th International Conference of Systems Analysis and Simulation 1992", *Systems Analysis – Modelling – Simulation*, 10, 279–314, 1993.

Lohmander, P., Qualitative conclusions in dynamic resource economics and the consequences of alternative models, in: Helles, F., Linddal, M. (editors), *Scandinavian Forest Economics*, Proceedings from the biennial meeting of the Scandinavian Society of Forest Economics, Denmark, November, 1993, No. 35, 153–164, 1994.

Lohmander, P., Optimization of decentralized adaptive truck decision rules: a spatial dynamic stochastic forest company problem, in: Saastamoinen, O., Tikka, S. (editors), *Scandinavian Forest Economics*, 36, 73–87, 1997, Proceedings of the Biennial meeting of the Scandinavian Society of Forest Economics, Mekrijärvi, Finland, March 1996, University of Joensuce, Joensuce, Finland, pp. 73–87, 1997a.

Lohmander, P., The constrained probability orbit of mixed strategy games with marginal adjustment: General theory and timber market application, *Systems Analysis – Modelling – Simulation*, 29, 27 55, 1997b.

Lohmander, P., Optimal sequential forestry decisions under risk, *Annals Of Operations Research*, 95, 217–228, 2000.

Lohmander, P., Helles, F., Windthrow probability as a function of stand characteristics and shelter, *Scandinavian Journal Of Forest Research*, 2(2), 227–238, 1987.

Lohmander, P., Olsson, L., Adaptive optimisation in the roundwood supply chain, in Olsson L., Optimisation of forest road investments and the roundwood supply chain, Doctoral Thesis, Acta Universitatis Agriculturae Sueciae. Silvestria nr 310, 2004.

Lu, F., Lohmander, P., Optimal mixed stand management under risk, in Lu, F., Optimization of forest management decision making under conditions of risk, Doctoral Thesis, Acta Universitatis Agriculturae Sueciae, Swedish University of Agricultural Sciences, No. 333, 2004.

Olsson, L., Lohmander, P., Optimal forest transportation with respect to road investments, *Forest Policy and Economics*, 7, 369–379, 2005.

Pontryagin, L.S., Optimal regulation processes, *American Mathematical Society Translations, Series 2*, 18, 321–339, 1961.

Ross, S., *Introduction to Stochastic Dynamic Programming*, Academic Press, London, 1983.

Sethi, S., Thompson, G.L., *Optimal Control Theory, Applications to Management Science and Economics*, 2 ed., Kluwer Academic Publishers, Dordrecht, 2000.

Winston, W.L., *Operations Research, Applications and Algorithms*, 4 ed., Thomson, 2004.

PART IV

MINING

The articles presented in mining aim to show how OR has affected mine planning, both for open pit and underground mines. There are obvious significant differences in the extraction methods for both cases.

In "Applications of optimization techniques in open pit mining" Caccetta discuses how OR techniques have been used successfully to decide on basic issues, such as the design of the open pit, investments and mine production scheduling. In the planning process, constraints commercial packages with approximate approaches as well as MIP models have been used.

In "Optimization in underground mining" Alford, Brazil and Lee make a similar analysis of underground mining, which has more complexities than open pits, with different forms of carrying out the mining process. Again, main problems are the design, including in some cases the determination of an economic cut-off grade and scheduling of the mine production. The authors discuss the different systems developed and heuristic approximations to MIP models.

A specific case is presented by Newman, Kuchta and Martinez in "Long- and short-term production scheduling at LKAB's Kirune mine", an iron mine located in Sweden, where MIP models are used for different levels of resolution to schedule the extraction of the blocks and the machine placement in the mine to satisfy production targets.

In "Modeling the production chain in open pit and underground mines: A challenge for OR" by Caro, Epstein, Santibanez and Weintraub, the authors discuss the possible advantages of modelling the integrated production chain, from copper extraction to final processing in plants instead of following the traditional sequential planning and how OR models can be used for decisions in investments and in all stages of the extraction and production process.

Chapter 29

APPLICATION OF OPTIMISATION TECHNIQUES IN OPEN PIT MINING

Louis Caccetta
Western Australian Centre of Excellence in Industrial Optimisation, Department of Mathematics & Statistics, Curtin University of Technology, Western Australia

Abstract The economic viability of the modern day mine is highly dependent on careful planning and management. Declining trends in average ore grades, increasing mining costs and environmental considerations will ensure that this situation will remain in the foreseeable future. The operation and management of a large open pit mine having a life of several (5 50) years is an enormous and complex task. Optimisation techniques can be successfully applied to resolve a number of important problems that arise in the planning and management of a mine. These applications include ore-body modelling and ore-reserve estimation; the design of optimum pits; the determination of optimal production schedules; the determination of optimal operating layouts; the determination of optimal blends; the determination of equipment maintenance and replacement policies; the determination of an efficient mine site rehabilitation program; the determination of the best choice of equipment (trucks, loaders) for the mining operations and a range of logistics problems such as the design and efficient operation of a transport and logistic network to support the mining operations. This chapter discusses some of these applications. In particular, we focus on: the design of optimum pits and the mine-scheduling problem.

Keywords: Mine planning, mine scheduling, open pit design, mixed integer linear programming

1 INTRODUCTION

In today's highly competitive global market environment, careful planning and management are crucial for the survival of individual businesses and industries. These tasks are made more difficult by a number of factors including the conglomeration of industry, which leads to a larger scale of operation; increasing responsibilities taken on by governments, particularly in a regulatory sense; environmental and safety restrictions in which business and industry must operate; technological advances which provide an ever-increasing volume of data that needs careful consideration and analysis at all levels of operation of business, industry and government. This growing complexity has accelerated the need for the development of sophisticated mathematical techniques for improving efficiency, effective planning and decision making. Over the past decade, this challenge has been met by significantly improved computational models as well as huge advances in technology.

The design and efficient operation of industrial systems first requires a careful mathematical analysis capturing all their essential features. The resulting mathematical models are then used to investigate system behaviour and identify the crucial parameters influencing system performance. With this knowledge, one can focus on the problem of optimising system performance. Optimisation techniques have been successfully applied to improve the performance of many industrial systems in such diverse areas as rural land-use planning, natural resource allocation, planning of telecommunications and urban transportation systems, mine design and management, human resources planning, and planning for the agricultural and forestry industry. Internationally, there is a growing awareness that business and industrial organisations which use optimisation techniques have a distinct competitive advantage (Caccetta, 2003). This chapter focuses on the contribution of optimisation to the open pit mining industry.

The mining industry presents an excellent source of challenging optimisation problems covering a very wide range of applications as noted earlier. The problems arising in these applications are difficult for a number of reasons, including the problem size in terms of the number of variables and the number of constraints; multi-criteria objectives; non-linearity and the need for a comprehensive sensitivity analysis to establish robustness. The mining industry is very important globally and is valued at over 4% of the world's GDP. Consequently, improved solutions can affect profitability significantly.

Since the 1960s, numerous publications have appeared in the literature concerned with the application of optimisation technology in the mining industry. For example, the APCOM symposium series has been a major forum for the discussion of the application of computers and operations research methods in the mining industries since 1961. The APCOM publications,[1] in particular its proceedings volumes, represent a major body of

[1]http://www.smenet.org/education/apcom/index.cfm

research in the mining industry. The book, edited by Weiss (1979), is also a good reference detailing some of the early work in computing, and operations research in the mining industry. Caccetta and Giannini (1990) provide a comprehensive account on the applications of operations research techniques in open pit mining. Considerable progress has been achieved since this publication. We will report on some of this progress in this chapter.

Modelling and optimisation plays an important role at all stages in the life of an open pit mine. At the planning and feasibility stage, one needs to develop an accurate model of the ore body. This model, based on geological exploration and drill-hole data, must capture the structure of the ore body and provide accurate economic, metallurgical, grade and geotechnical data that allow mining engineers to establish the economic potential of the ore body and the viability of mining the ore body.

An important input into the economic viability of a mine is the determination of the ultimate pit limit of the ore body. That is, that contour which is the result of extracting the volume of material which provides the total maximum profit whilst satisfying the operational requirements of safe wall slopes. The ultimate pit limit gives the shape of the mine at the end of its life. Usually this contour is smoothed to produce the final pit outline. Optimum pit design is not only important in establishing the feasibility of the mine, but plays a crucial role in all stages of the life of the mine should the operation go ahead. Now, if the feasibility phase establishes viability then we enter the operating and management phase where a large range of optimisation problems arise. A typical mining operation is represented in the flow diagram of Fig. 1. This flow diagram shows the movement of material (waste and ore) from local pits as well as more distant sites to the waste dump for disposal or the plant for further processing. Usually local ore goes to the run of mill (ROM) pad or crusher or stockpile. From there, the ore is blended into products and may undergo further processing. The ultimate output of the plant is products and waste. Products are transported to markets whereas waste materials (which may be highly toxic) need to be correctly disposed. The overall management objective is to develop and implement operational plans that optimise the net present value of the ore body over the life of the mine, which satisfy a range of operational, contractual and legal constraints.

Effective mine management and planning requires serious consideration of a range of important optimisation problems. The flow diagram of Fig. 1 facilitates the development of a high-level systems model that optimises the whole mining operations. Indeed, we have developed mixed integer linear programming (MILP) models for operating mines. Obvious constraints in these models include production requirements; mine extraction sequences; milling and crusher capacity; capacity of ROM pad; stockpile; material movement capacities; etc. The high-level systems model identifies a number of important sub-problems involving optimisation that need to be resolved. Some of these are detailed in the following paragraphs.

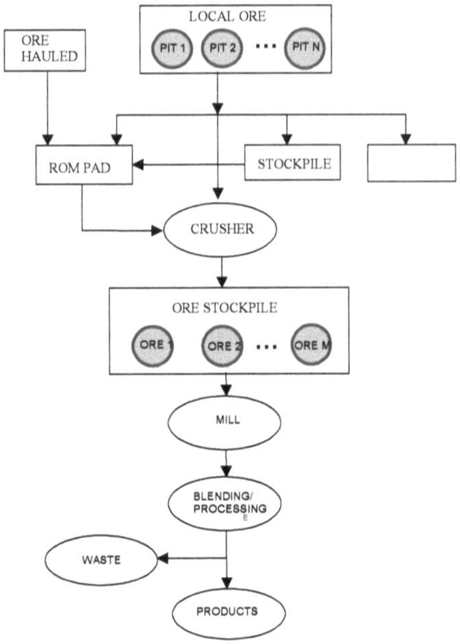

Figure 1. Mining operation.

An important problem in mine management and planning is the determination of effective long-range mine plans. These plans are crucial in that they establish the initial investment and surface facilities requirements to develop the mine and detail the quantity and quality of the available ore to be mined. The investment plan needs to consider the cost of developing the mine and requires the solution of a number of sub-problems. These sub-problems include determining: a whole of life pit design with a full sensitivity analysis; the best selection of equipment (loaders, trucks, crushers, blasting equipment, etc.) to use; strategies for the location and operation of surface facilities and many more. Surface facilities include treatment plants; waste dumps; tailing ponds; a haulage network for mining the ore/waste from mine site to plant/waste dump and a transportation network for moving ore/products from plant to distribution centres and for moving resources such as fuel and equipment to the mine sites (Burt *et al.*, 2005).

The long term plans provide the framework for developing a range of operational plans and strategies at the operational phase of the mine. These include: the generation of a spectrum of optimal pit designs that respond to changes in the economic data (metal prices, costs) and the geotechnical data (ore reserves, ore grades, wall slopes); the determination of the optimal sequence of production schedules over the life of the mine; the determination of the best equipment to achieve the objectives of the mine plan (Burt *et al.*, 2005); the design and efficient operation of a transport and logistic network to support the mining operations (Caccetta and Hill, 2004), and the establishment of an effective mine site rehabilitation plan. In addition, towards the end of a mine's life, one needs to determine an economic termination of the project.

The earlier discussion briefly identifies many problems in modelling and optimisation that are of considerable interest to the mining industry. We cannot cover in detail all problems in this chapter and so we restrict ourselves to just a few. More specifically, we focus on two fundamental problems: the design of optimum pits and mine scheduling. The remainder of the chapter is organised as follows. Section 2 discusses some basic modelling tools including the block model of an ore body. Section 3 considers the design of optimum pits. Section 4 details some very recent progress on the mine-scheduling problem.

2 ORE-BODY MODELLING

We have already identified that an early task in mine management is the establishment of an accurate model for the deposit. Since, in a real application it is difficult to analytically express either the economic functions or the pit shape functions, we must utilise numerical methods to solve the optimisation problems that arise. This involves the discretisation of the ore body. Though a number of models are available, the regular 3D fixed-block model is the most commonly used as it is well suited to describing the mining operations. This model is based on the ore body being divided into fixed size blocks. The block dimensions are dependent on the physical characteristics of the mine, such as pit slopes, dip of deposit and grade variability as well as the equipment used. The centre of each block is assigned, based on drill-hole data and a numerical technique, a grade representative of the whole block. The numerical technique used is some grade extension method such as distance-weighted interpolations, repression analysis, weighted moving averages and kriging (David, 1988). Using the financial (mining and processing costs as well as market prices for the products), metallurgical (processing and recovery data) and geotechnical (pit slopes) data, the net profit of each block is determined. The block model can be produced in a variety of ways depending on the structure of the ore body (Giannini *et al.*, 1991).

The wall slope requirements for each block are described by a set (typically 4 to 8) of azimuth-dip pairs. From these, we can identify for each block x the set of blocks S_x which must be removed before block x can be mined. This collection of blocks, $x \cup S_x$, is usually referred to as a "cone". This cone can be easily generated using the minimum search pattern algorithm described by Caccetta and Giannini (1988b). The key assumptions in the block model are the cost of mining each block does not depend on the sequence of mining and the desired wall slopes and pit shape can be approximated by the removed blocks.

The discretisation of the ore body means that many of the optimisation problems that arise in mining are combinatorial optimisation problems. Graph theory and network concepts and methods feature prominently in the area of combinatorial optimisation. The block model of an ore body can be represented as a weighted directed graph in which the vertices represent the blocks and the arcs represent the mining restrictions. More precisely, our graph contains the arc (x, y) if blocks x and y are "adjacent" and the mining of block x is dependent on the removal of block y. The profit resulting from the mining of a block is represented as a block weight. We note that if the mining of block x requires the removal of block y and the mining of block y requires the removal of block z, then there is no need to have the arc (x, z) in our graph, as it is redundant. Computationally it is important that redundant arcs are excluded from the model. The minimum search pattern algorithm for generating cones generates a graph with no redundant arcs.

In the context of mining an important graph theoretic concept is that *closure*. The *closure* of a weighted directed graph is defined as a set of vertices C such that if $x \in C$ and (x, y) is an arc in the graph, then $y \in C$. The weight of the closure is the sum of weights of vertices in the closure. Observe that in a mining context, a *closure* represents a feasible pit contour and its weight is the value of mining the blocks in the closure. Thus, the problem of determining the pit contour that satisfies the safe wall slope restrictions and which maximises the net economic return translates into a graph optimisation problem of determining, in a weighted directed graph a closure of maximum weight. Alternatively, this problem can be easily formulated as an integer linear programming (LP) problem:

$$\text{Maximise} \sum_{i=1}^{N} p_i x_j$$

$$\text{subject to} \sum_{i=1}^{N} x_i - x_j \leq 0 \text{ for all arcs } (i, j) \text{ and } x_i = 0, 1 \text{ for all } i.$$

Here the block are numbered 1, 2, …, N; p_i is the profit of mining block i and $x_i = 1$, if block i is mined and $x_i = 0$, otherwise. Despite the simplicity in formulation, this integer programming problem is well outside the capability of commercial packages because in most practical applications N is too large. In the next section, we will discuss the solution of this problem.

We remark that in some applications, the mining operations are described in terms of "faces". A face is a generalisation of a block having the following characteristics: specified quantities of ore and waste; a depth specification; faces that must be removed before the start of mining on this face; a face may be partially mined in a period.

3 OPTIMUM PIT DESIGN

The ultimate pit limit problem is easy to formulate mathematically, but not so easy to solve because of its size. A consequence of this is that much of the early optimisation research focused on heuristic/approximation algorithms (Caccetta and Giannini, 1986). Thomas (1996) has investigated the use of artificial intelligence methods such as genetic algorithms and simulated annealing. The Lerchs–Grossmann (Lerchs and Grossmann, 1965) graph theory method was not implemented until the mid-1980s (Caccetta *et al.*, 1986, 1991, 1994; Caccetta and Giannini, 1988a; Whittle,[2] 1990; Giannini *et al.*, 1991; Hochbaum and Chen, 2000). Powerful commercial integer programming packages such as CPLEX[3] are ineffective when applied to large ore bodies.

The Lerchs–Grossmann algorithm (LGA) constructs a sequence of "normalised trees", T^0, T^1, T^2,…, T^n, terminating when the set of "strong vertices" of T^n forms a closure of the graph representing the ore body. Each normalised tree T^i is constructed from T^{i-1} by following certain well-defined rules, which can be implemented by a clever labelling scheme (Caccetta and Giannini, 1988a). In applying the algorithm, the digraph D representing the

---

[2]http://www.whittle.com.au
[3]http://www.cplex.com

ore body is augmented by the addition of root vertex x_0 and joining it to every vertex of D. The vertex x_0 is given a negative weight and so will never be part of any minimum closure.

An alternative to the LGA is the network flow formulation of Picard (1976). Here a network N is obtained from the graph model D by adding a source node *s* and a sink node *t*. Node *s* is joined to every vertex *x* of D having a positive weight by an arc with capacity equal to the weight of vertex *x*. Vertices of D having non-positive weight are joined to *t* by an arc having a capacity equal to the absolute value of the weight of the vertex. Arcs of D are given an unrestricted capacity. Picard (1976) showed that a maximum closure of D corresponds to a minimum cut of N. Consequently, the maximum closure of a graph (optimum pit contour) can be determined by the application of any network flow algorithm.

The PITOPTIM package developed by Giannini *et al.* (1991) effectively produces the maximum closure. The package incorporates both the LGA and a network flow method and has been extensively tested on producing mines (Caccetta *et al.*, 1991, 1994; Giannini *et al.*, 1991). Pits with up to 3, 175, 200 blocks have been optimally designed with the network flow method requiring 20 min of CPU time. Speedups were achieved through effective graph constructions and effective data-reducing algorithms based on the theory of bounding (Caccetta and Giannini, 1985). The network flow algorithm used is a modified Dinic's method that exploits the structure of the problem. The software provides a sensitivity analysis for the cost and wall slope parameters and allows the generation of incremental pits. A comparative analysis of the methods is given in Caccetta *et al.* (1994) and Hochbaum and Chen (2000). An interesting outcome is that the LGA is computationally superior only for the case of "dense" (in terms of arcs) networks.

4 MINE SCHEDULING

The open pit mine production scheduling problem can be defined as specifying the sequence in which "blocks" should be removed from the mine to maximise the total discounted profit from the mine subject to a variety of physical and economic constraints. Typically, the constraints relate to the mining extraction sequence; mining, milling and refining capacities; grades of mill feed and concentrates; stockpile-related restrictions; a range of logistics issues and various operational requirements such as minimum pit bottom width and maximum vertical depth. The scheduling problem can be easily formulated as a mixed integer linear program (MILP) (Caccetta and Hill, 2003). This approach caters for mining operations consisting of multiple products and a number of sites each with a number of pits. In addition, cut-off grade can be optimised by allowing the model to decide whether or not extracted ore is milled or treated as waste. Further, the modelling can easily incorporate the maximisation of a user-defined weighted function of the life of the operation and the net present value (NPV). However, in real applications the resulting formulation is too large, in terms of both the number of variables and the number of constraints, to solve by a direct application of any available commercial MILP software packages. Thus the options available to the mining industry are either to consider simpler sub-problems or to develop special solution methods exploiting the structure of the MILPs. We consider each of these options.

The complexity of the problem has led the mining industry to focus on the easier sub-problems. In particular, manual procedures that generate schedules through a series of refinements, usually starting with the final pit contour are quite common. Several heuristic approaches have appeared in the literature including methods based on Lagrangian relaxation (Caccetta *et al.*, 1998); parameterisation (Matheron, 1975; Francois-Bongarcon and Guibal, 1984; Dagdelen and Johnson, 1986); dynamic programming (Tolwinski and Underwood, 1996); MILP (Gershon, 1983; Dagdelen and Johnson, 1986; Caccetta *et al.*, 1998; Ramazan *et al.*, 2005); simulated annealing and genetic algorithms (Denby and Schofield, 1995) and neural networks (Denby *et al.*, 1991). All these approaches suffer from one or more of the following limitations: cannot cater for most constraints that arise; yield only suboptimal solutions and in most cases without a quality measure (in fact generated schedule may not even be feasible); can only handle small-sized problems; time and other variable factors not well catered for.

The most commonly used method is parameterisation, initially introduced by Lerchs and Grossmann (1965). A set of nested pits is generated, starting with the final pit contour, by varying the economic parameters (the value of each block *i* is reduced by a specified amount). For each parameter value, the LGA is applied to generate the optimum pit. The most widely used software packages that use this method include: Whittle's Four-D and Four-X[4] and Earthworks NPV Scheduler (Ver. 3.2.5).[5] The latter product has a restricted tree search procedure that is used to re-sequence the "pushbacks" in an effort to improve the NPV.

---

[4]http://www.whittle.com.au

[5]http://www.earthworks.com.au/index1.htm

We now turn our attention to the exact MILP models. The attractiveness of these models has already been mentioned at the beginning of this section. A further attraction is the sensitivity analysis capability. The difficulty in solving the MILPs that arise in mining is their size. An effective way of solving a large-scale problem is to partition the problem into a number of smaller problems that are easier to solve. The basic idea of the branch and bound method is to subdivide the feasible solution set into successively smaller subsets, placing bounds on the objective function value over each subset and using these bounds to discard subsets from further consideration and to select the next subset to further subdivide. In our case the relaxed sub-problems are linear programming problems with some variables fixed. These sub-problems can easily be solved using a commercial linear programming package such as CLPEX. The difficulty that arises is that the LP relaxations produce weak bounds and so the search tree becomes excessively large. This difficulty can be overcome by using the method of branch and cut (Caccetta, 2000). The branch and cut method adds constraints (cuts) to strengthen the upper bound at each node within a branch and bound procedure. In recent years, the method has emerged as a very powerful technique for solving large-scale MILPs. Much of the computational advancement has been achieved with respect to specific applications, where the structure has been exploited. Motivated by the success of the method in real industrial applications in the airline industry and transport/distribution networks, Caccetta and Hill (2003) took up the challenge of developing an effective global optimisation method suitable for large-scale mine scheduling. The result is an innovative branch and cut approach which heavily exploits the structure of the problem. An important feature of our method is that it explicitly incorporates all constraints in the optimisation as well as time considerations. Some features of our method are detailed in Caccetta and Hill (2003), but commercial issues prevent a full disclosure.

A software package implemented in C^{++} and containing some 30,000 lines of code has been developed and extensively tested on operating mines. Our software can work with the usual block model or with benches (pre-defined chunks of resource). On single pit gold mines (block model), our methodology produced solutions that yield an increase of at least 15% on NPV profit (Caccetta and Hill, 2003). Often the life of the mine is increased.

The package has also been developed for multiple site mines. One test data set concerning an iron ore operation involved the concurrent scheduling of benches from 68 pits from 6 different sites over a 16-year time horizon. Some sites had individual characteristics that had to be built into the model such as sharing a rail line and only one site having a concentrator. The ore was broken down into different product types that had to satisfy strict assay grades. Quality schedules were produced. Competing products (that do not globally optimise) failed to produce a 16-year period schedule.

Having established the feasibility of using MILP to schedule the mining of large open pit mines, the current task is to extend the work and develop a general mine-scheduling package that can be easily applied to any mine operation and has the full set of features required by the industry.

MILP problems can also be approached using approximation techniques. Caccetta *et al.* (1998) proposed a Lagrangian relaxation method for solving the MILP. At each step a problem, similar to the ultimate pit limit problem, is solved using the PITOPTIM package described earlier. The additional constraints are dualised and sub-gradient optimisation is used to reduce duality gaps. Solutions within 5% of optimality were generated using data from producing mines. The same idea was used by Caccetta and Kelsey (2001) to tackle the land surface reshaping problem that arises in mine rehabilitation. In both these applications the speed of PITOPTIM is crucial.

5 CONCLUDING REMARKS

The economy of many nations is highly dependent upon the earnings derived from the mining industry. The economic viability of the mining industry is highly dependent on efficient practices. This chapter has demonstrated that accurate mathematical modelling, along with effective optimisation tools, provides the opportunity to achieve productivity gains. We have detailed how the fundamental problems of optimal mine design and mine scheduling can be effectively solved by cutting-edge optimisation technology. The current challenge is to develop a holistic modelling and optimisation approach that integrates all aspects of the mining operations.

6 REFERENCES

Burt, C, Caccetta, L, Hill, SP and Welgama, P: Models for mining equipment selection, *Proc. MODSIM 2005 International Congress on Modelling and Simulation* (Andre Zerger, Editor), Townsville, Queensland (2005), pp. 170–176.

Caccetta, L: Branch and cut methods for mixed integer linear programming problems, *Progress in Optimization II* (XQ Yang *et al.*, Editors), Kluwer Publishers, Dordrecht, Applied Optimization Series 39 (2000), pp. 21–44.

Caccetta, L: optimisation techniques in industrial modelling, *East–West Mathematics Journal* (2003), pp. 1–20.

Caccetta, L and Giannini, LM: On bounding techniques for the optimum pit limit problem, *Proceedings of Australasian Institute of Mining and Metallurgy* 290 (1985), pp. 87–92.

Caccetta, L and Giannini, LM: Optimization techniques in open pit mining, *Proceedings of Australasian Institute of Mining and Metallurgy* 291(8) (1986), pp. 57–63.

Caccetta, L, Giannini, LM and Kelsey, P: Application of Optimization Techniques in Open Pit Mining, Proc. 4th International Conference on Optimization Techniques and Applications (L. Caccetta et al, Editors) 1 (1998), pp. 414–422.

Caccetta, L and Giannini, LM: Applications of operations research techniques in open pit mining, *Proc. APORS '88* (Assoc. of Asian-Pacific Operational Research Societies) (Byong-Hun Ahn, Editor), Elsevier Science Publishers, Amsterdam (1990), pp. 707–724.

Caccetta, L, Giannini, LM and Carras, S: The optimum design of large open pit mines, *Australasian Institute of Mining and Metallurgy Symposium* 48 (1986), pp. 195–200.

Caccetta, L, Giannini, LM and Kelsey, P: Optimum open pit design: a case study, *Asia-Pacific Journal of Operational Research* 8 (1991), pp. 166–178.

Caccetta, L, Giannini, LM and Kelsey, P: On the implementation of exact optimisation techniques for open pit design, *Asia-Pacific Journal of Operational Research* 11 (1994), pp. 155–170.

Caccetta, L, Giannini, LM and Kelsey, P: Application of optimization techniques in open pit mining, *Proc. 4th International Conference on Optimization Techniques and Applications* (L. Caccetta *et al.*, Editors), Vol. 1 (1998), pp. 414–422.

Caccetta, L, Kelsey, P and Giannini, LM: Open Pit Mine Production Scheduling, *Technical Proceedings of Third Regional APCOM, Computer Applications in the Mineral Industries International Symposium* (AJ Basu *et al.*, Editors), The Australasian Institute of Mining and Metallurgy Publication Series No. 5/98 (1998b), pp. 65–72.

Caccetta, L and Hill, SP: An application of branch and cut to open pit mine scheduling, *Journal of Global Optimization* 27 (2003), pp. 349–365.

Caccetta, L and Hill, SP: Modelling distribution networks, *Proc. 6th International Conference on Optimisation Techniques and Applications (ICOTA6)* (A Rubinov and M Sniedovich, Editors) (2004), Paper No. 235, CD-Rom, ISBN 18768115.

Caccetta, L, and Kelsey, P: Models for mine site rehabilitation, *Proc. MODSIM 2001 International Congress on Modelling and Simulation* (F. Ghassemi *et al.*, Editors) ANU, Canberra (2001), pp. 2155–2160.

Dagdelen, K and Johnson, T.B: Optimum open pit mine production scheduling by Lagrangian parameterization, *Proc. 19th APCOM Symposium of the Society of Mining Engineers*, AIME, New York (1986), pp. 127–142.

David, M: *Handbook of Applied Advanced Geostatistical Ore Reserve Estimation*, Elsvier Science, Amsterdam (1988), pp. 1–216.

Denby, B and Schofield, D: The use of genetic algorithms in underground mine scheduling, *Proc. 25th APCOM Symposium of the Society of Mining Engineers*, AIME, New York (1995), pp. 389–394.

Denby, B, Schofield, D and Bradford, S: *Neural Network Applications in Mining Engineering*, Department of Mineral Resources Engineering Magazine, University of Nottingham, UK (1991), pp. 13–23.

Francois-Bongarcon, DM and Guibal, D: Parametization of optimal designs of an open pit beginning of a new phase of research, *Trans. SME, AIME* 274 (1984), pp. 1801–1805.

Gershon, M: Mine scheduling optimisation with mixed integer programming, *Mining Engineering* 35 (1983), pp. 351–354.

Giannini, LM, Caccetta, L, Kelsey, P and Carras, S: PITOPTIM: a new high speed network flow technique for optimum pit design facilitating rapid sensitivity analysis. *Proceedings o Australasian Institute of Mining and Metallurgy* 296 (1991), pp. 57–62.

Lerchs, H and Grossmann, IF: Optimum design of open pit mines, *Canadian Institute Of Mining Bulletin* 58 (1965), pp. 47–54.

Matheron, G: *Le Parametrage des Contours Optimaux*, Technical Report 403 (1975), Centre de Geostatistiques, Fontainebleau, France.

Picard, J.C: Maximum closure of a graph and applications to combinatorial problems, *Management Science* 22 (1976), pp. 1268–1272.

Ramazan, S, Dagdelen, K and Johnson, TB: Fundamental tree algorithm in optimising production scheduling for open pit mine design. *Mining Technology (Trans Inst Min Metall. A)* 114 (March 2005), A45.

Thomas, G: Pit optimization and mine production scheduling – the way ahead, *Proc. 26th APCOM Symposium of the Society of Mining Engineers*, AIME, New York (1996), pp. 221–228.

Tolwinski, B and Underwood, R: A scheduling algorithm for open pit mines, *IMA Journal of Mathematics Applied in Business and Industry* 7 (1996), pp. 247–270.

Weiss, A: *Computer Methods for the 80's in the Mineral Industry*, AIME, New York (1979).

Chapter 30

OPTIMISATION IN UNDERGROUND MINING

Christopher Alford[1], Marcus Brazil[2], and David H. Lee[3]
[1]*WH Bryan Mining Geology Research Centre, Sustainable Minerals Institute, The University of Queensland, Queensland 4072 Australia;* [2]*ARC Special Research Centre for Ultra-Broadband Information Networks (CUBIN), Department of Electrical and Electronic Engineering, The University of Melbourne, Victoria 3010 Australia;* [3]*Department of Mathematics and Statistics, The University of Melbourne, Victoria 3010 Australia*

Abstract Efficient methods to model and optimise the design of open pit mines have been known for many years. Although the underground mine design problem is conceptually more difficult it has a similar potential for optimisation. Recent research demonstrates some useful progress in this topic. Here we provide an overview of some of this research.

Keywords: Underground mining, infill drilling, cut-off grade, stope optimisation, mine design, network optimisation

1 INTRODUCTION

One of the great successes in the industrial applications of operations research has been the use of the Lerchs–Grossman algorithm, and its subsequent improvements, in the optimisation of the design of open pit mines. This technique has been successfully implemented in a number of widely used software systems. Most open pit designs are developed in an optimisation framework traceable back to the method launched by Lerchs and Grossman (1965), although more recently a number of authors, including Barbaro and Ramani (1986), Akaike and Dagdelen (1999) and Caccetta and Hill (2003), have reported integer programming models for this task.

The underground mine design problem is conceptually more difficult and less constrained than the open pit problem. However recent research demonstrates some progress in the topic. This chapter reviews some of this progress with emphasis on the optimisation of the design of underground mines.

Underground mining is the planned extraction and transportation of a mineral resource from its underground location to a mill or processing plant on the surface.

This deceptively simple statement disguises the complex set of development and operational phases needed to achieve this transformation but it is a good starting point for an operational research analysis. Many different types of minerals occur in many different styles of mineralisation in nature. In this chapter we concentrate on optimisation issues in hard rock mines – gold, silver, lead, zinc, copper or multi-metal deposits. We typically assume that the mineral resource is represented in a computer model as a three dimensional assembly of blocks each with grade, geology and metallurgical characteristics.

Different methods of extraction of the economic material have been devised depending on the geometry of the orebody and the geotechnical stability of excavation volumes and the surrounding rock. This review is focussed on two of the most common extraction techniques in large modern underground mining, sublevel open stopping and sublevel caving. In the first the excavation shape is drilled and blasted and the broken rock is extracted by load-haul-dump equipment and the void is then filled with a cemented slurry, which sets and restores the geotechnical stability of the region and allows adjacent stopes to be mined. This sequence of operations immediately imposes an important scheduling constraint on mining operations, which is amenable to operations research analysis (see later sections). Sublevel caving extracts material by relying on drilled and blasted material progresssively collapsing into excavation tunnels under a controlled draw sequence with overlying broken rock caving into the void. Other techniques such as block-caving, cut and fill and room-and-pillar are also in use.

Once the blocks to be mined are aggregated into mineable stope shapes or caving zones, the dominant working structure of an underground mine is a set of interconnected tunnels or mine development (called drives and ramps or declines), orepasses, which are near vertical passages down which ore is dropped, and vertical haulage shafts which provide access to and conduits for the transport of ore from the stopes to the mill. This set of interconnected tunnels forms a network. The layout of this network – both its geometry and topology – is a key to achieving the low cost operations needed to make a mine profitable.

A classical decision for underground mine design is whether to use a vertical shaft for haulage of mined material or rely on a fleet of trucks hauling material up the access network. A vertical shaft incurs a large fixed cost investment at the start of a project and relies on the subsequent low unit cost movement of materials to the surface for its economic justification. Typically deep, long-life mines warrant shaft haulage. Truck haulage, while more expensive per unit of material transported to the surface, has advantages of earlier recovery of ore in the life of a mining project and requires only progressive capital expenditure matched to material flow. While many of the methods described in this chapter relate to both shaft and decline-haulage mine operations, the emphasis is on decline operated mines.

2 OVERVIEW OF MINE DESIGN

A typical development sequence for an underground mining project follows.

The location of mineralised zones is determined by geological observations and surface drilling. This is followed by a more detailed delineation of the mineralisation via an infill drilling programme from underground development. The size of the reserves, their depth, geology, geotechnical characteristics and the market for the embedded minerals are all factors in determining how the deposit will be mined. Lane (1988) and others emphasise the importance of decisions on the cut-off grade, used to distinguish between economically mineable ore and waste. The cut-off grade strategy for a mine must ultimately include the amortisation of all fixed costs, including mineral processing costs as well as mine development and mine operating costs.

Once an engineering decision on the most appropriate mining method has been made, the locations and geometries of the stopes and the contained tonnages of ore can be determined (to some degree of accuracy) by analysing the computer resource model developed from infill drilling core data. Orebody simulation techniques can be used to provide robust models of the ore zone indicated by the drilling programme data (Grieco and Dimitrakopoulos, 2004). When the stopes are located in space the set of access and draw points (from whence the ore is drawn) can be determined.

A major design task is to determine the mine development network (horizontal drives, inclined ramps, orepasses and vertical shafts) needed to provide access to these draw points and provide material movement paths to transport the excavated ore to the processing mill at the surface. A key design consideration in underground mine design is that all development in the network providing underground access must be navigable by trucks and mining equipment. This means all ramps and drives have large cross-sections, typically larger than 4m × 4m, to accommodate large capacity haulage trucks; they have an upper limit on vertical gradient – typically modern equipment can haul up to a grade of 1:7; and they are navigable – turning radii must be large enough to accommodate all operating vehicles with typical minimum design radii in the range 15–30 m. Mines must be ventilated for human and machine operations. Generally mine ventilation is handled as a secondary system after the development and extraction sequnce is defined.

In recent mines (particularly in Australia) the underground development is an extension of an existing mined-out open pit mine or an old underground mine where new ore bodies have been discovered. In these cases the location of the surface portal or breakout point from the existing mine infrastructure is likely to be fixed or at least strongly constrained.

3 OPTIMISATION OBJECTIVES AND METHODS

A significant problem in developing a general framework for the optimisation of an underground mine is that there is a wide range of mining strategies, so that each deposit has a comparatively specialised solution. This makes a general approach to the optimal design of underground mines a complex task, and one that is currently supported by a fairly small amount of research.

Inevitably, the optimisation problem has been decomposed into tractable subproblems. We now identify a number of important optimisation subproblems in underground mine design, and discuss the models and solution techniques.

3.1 Infill Drilling Optimisation

The development of an underground mine, or an extension of a mine to a new zone, frequently includes the requirement to host an infill drilling programme. The aim of an infill drilling programme is to gather sufficient drill core data to characterise the mineralisation of a target zone in enough detail to establish reserves and develop an economic mining plan for the region. This activity is critical to the information gathering process and generally represents a substantial cost investment. In a case study of the Vera South project within the Normandy Mining Limited (now Newmont) managed Pajingo field in Queensland, Australia, the infill drilling programme amounted to more than 30% of the total project cost.

Two important ways in which the process can be optimised are through optimisation of drillhole spacing and optimisation of the physical infill drilling programme design.

The first of these optimisation problems involves balancing cost savings arising from reduction in drilling density against the costs associated with mineralisation misclassification. At some stage the cost of obtaining additional and more accurate information exceeds the benefit of such information. A geostatistical approach to this problem based on the geometry of the drilling configuration has been suggested (Goovaerts, 1997) but without reference to other factors such as local grade variability or an economic cost analysis. More recent work on this problem has tended to focus on open pit mining, but has relevance to underground mining. Boucher *et al.* (2005) have advocated the use of a stochastic simulation framework making the use of conditional simulation with maximum/minimum autocorrelation factors. Another related conditional simulation approach is suggested by Froyland *et al.* (2004). In each of these approaches, the aim is to realistically link the optimal expected net present value (NPV) of the project with the amount of drilling information. Whether this is really possible in underground mining remains to be seen.

The second optimisation problem is that of optimising the physical design of the infill drilling programme for a given drilling density. For deep deposits a very expensive infrastructure is often required for the infill drilling programme. This may require constructing a system of drill drives (often breaking out from existing declines) along which drill stations are set up to conduct the drilling. A drill fan is produced from each drill station for gathering information about the ore body at the required density. Brazil *et al.* (2000) have observed that this can be treated as a network optimisation problem where the objective is to optimise the cost of drilling combined with the cost of drives and infrastructure to support the drill stations. In particular, one can apply the physical network design methods discussed in Sections 3.4 and 3.5. Research to date has shown that a dynamic programming approach to this problem is very effective. For the case study at Pajingo, the approach demonstrated savings of over 10% over the cost of the proposed drilling programme.

3.2 Determination of Cut-Off Grade

The characterisation of mineralisation obtained from geological exploration allows a mining company to estimate the "mineral resources" available. A widely accepted definition of a Mineral Resource is as follows (quoted from Rendu, 2004):

A "Mineral Resource" is a concentration or occurrence of material of economic interest in or on the Earth's crust in such a form, quality and quantity that there are reasonable prospects for eventual economic extraction.

The certainty with which a mineral resource can be delineated depends on the accuracy and sample spacing in the exploration process.

The *ore bodies* or *ore reserves* of interest to the mining company are those parts of the mineral resource deemed to be economically valuable. A key step in determining the ore reserves is to establish a *cut-off grade*, which is the lowest grade of mineralisation that qualifies as ore. The estimation of optimum cut-off grades is a difficult problem in operations research, and one that deserves more attention. Many operations use a cut-off that is essentially a break-even grade, that is, one that ensures that every tonne of ore that is mined "pays for itself". Such a policy, however, will generally not lead to value maximisation.

An important step to understanding this problem has been the influential work of Lane (1988). He explores methods for balancing short-term considerations against long-term economic consequences of the cut-off grade while taking into consideration a number of constraints on the production system, such as mine, mill or market limited output. His theories have proved useful in the design of open pit mines but are more difficult to apply to large-scale underground mining.

Poniewierski *et al.* (2003) have further illustrated the important economic consequences of optimally choosing the cut-off grade. They describe the application of a rule-based scheduling and discounted cash flow technique to evaluate the cut-off for each ore body, applying the principles of Lane. There is still much work to be done in this area. Poniewierski *et al.* point out that a key to determining an optimum cut-off grade is the ability to rapidly perform complex optimal mine layout designs combined with rapid output of multiple potential schedules. Thus, this step ultimately relies on the ability to rapidly optimise all stages in underground mine design. Further industry experience is reported in Hall and Stewart (2004).

3.3 Stope Optimisation

The selection of a stoping method will take into account the size and orientation of the ore body, ground conditions that will affect the size of any excavation opening, and the percentage of waste material in the planned dilution. The method used to drill and blast the ore in advance of excavation will also dictate an acceptable stope dimension.

For narrow ore bodies a single stope will mine the full ore width. For wide orebodies multiple stopes may be needed to mine the full ore width. Waste pillars are required between open stopes for support before the stopes being filled with crushed rock or cemented sand fill.

In stope optimisation these factors can be reduced to dimensional constraints on the minimum and maximum stope size, acceptable stope shapes and orientations, and pillar widths. The metal contained in the stope must be sufficient to cover mining, haulage, processing and marketing costs and the associated mine development required to access the stope.

For narrow and steeply dipping deposits, the primary decision is the width of ore to be mined. This reduces the stope optimisation problem from a three-dimensional to a one-dimensional optimisation problem.

Gershon and Murphy (1987, 1989) analyse the selection of mining intervals for layered sedimentary deposits for an open pit mining operation. The single dimension of primary interest is the economic depth of ore in a vertical drill hole. A dynamic programming solution is outlined, but an extension to three dimensions is not advanced.

Rendu (1982) outlines procedures to determine the optimal position of the hanging wall and foot wall of ore zones intersected by drillholes, as a precursor to determining a mining scheme. The constraints considered are minimum mining width, minimum internal waste width, restrictions on the number of geological zones, and the maximum mining width. The minimum and maximum widths are usually specified vertically, horizontally or normal to the direction of mineralisation (Rendu, 1982, p. 2) so the first stage is to calculate the limiting lengths in the direction of the drill hole. While no solution to the general problem is advanced, several special cases are investigated. The continuity of ore and waste between drill holes is left to visual and manual interpretation.

For the full three-dimensional stope optimisation problem a number of approaches have been reported in the literature, or implemented in commercial mining software packages.

Deraisme *et al.* (1984) have constructed two-dimensional sectional models for a number of different stoping methods. Image transformation techniques found in mathematical morphology (Serra, 1982) are used to transform the image of ore blocks above cut-off grade to another image satisfying the stope geometry constraints. Similar work has also been reported by Muge *et al.* (1995).

Cheimanoff *et al.* (1989) describe a method of generating mineable volumes based on an octree model for the shape geometry (i.e. a model built up as a tree of octants). The first phase "Object Manipulator" gathers mineralised veins into convex blocks distinguishing "large" veins that justify a mineable block by themselves, merges those "close enough" into a single block, and separates those "too far from one another" into two separate blocks. A second phase, "Shape Generator", progressively subdivides a bounding volume in an octree till the smallest subdivision matches the smallest mining unit.

The Floating Stope is a technique implemented in the DATAMINE mining software package (Mineral Industries Computing Limited) to determine the optimal (boundary) limit for mineable ore, that may be economically extracted by underground stoping methods (Alford, 1995).

The term Floating Stope is derived from the technique of floating a minimum stope shape through the ore body to evaluate the stope grades for any stope position. Two envelopes are created. The maximum envelope is the union of all possible economic stope positions. The minimum envelope is found by taking the union of all best grade stope positions for every ore block in the ore body. The envelopes provide a limit for the engineer to design final stope positions, with the recommendation that the minimum envelope be used as the guide in the first instance.

Ataee-pour (2000) has introduced a concept of "Maximum Value Neighbourhood" for the stope optimisation. This method proceeds in a similar fashion to the Floating Stope method but differs in the approach to defining the envelope.

The Stopesizor is a mining software package used internally by Snowden Consultants (Thomas and Earl, 1999). Stopesizor produces a single mining outline for a selected cutoff grade. This is done by constructing a number of selective mining blocks (SMB). Each SMB comprises a single, contiguous group of resource blocks that honours the minimum mining width and dip angle in each dimension. Stope dimensions must be defined in whole increments of the resource model block dimensions. Stopesizor identifies the highest grade SMB, and then the next until all economic SMB's are exhausted.

Each resource block in the final outline is assigned the mean value of its SMB. If a range of cutoff values are supplied in decreasing order, then a single output model can be produced that represents the optimum mining outlines at these cutoff values. Using the identified value sequence and a discount rate they claim to optimise the NPV of the stope outline and the mining sequence using a supplied production rate.

A fundamental characteristic of previous approaches to stope optimisation is that the location and size of the final stope shape is not properly characterised. An engineer can use the result to guide a final manual design. A preferred approach would be for the stope optimisation procedure to generate the optimal stope design without manual adjustment, including both the stope and waste pillar positions for the one- and three- dimensional cases. Recently, several new formulations of the stope optimisation problem, with proven optimality, have been successfully applied by Alford in 2006.

In stope optimisation the development and haulage costs must be anticipated or averaged at any location in the mine before optimisation. The final mine development layout is generated using location and layout of optimised stopes. For many mine designs this requires an iterative mine design procedure. Recent research at the University of Melbourne (Alford, 2006) has focussed on the problem not of defining the minimal cost network to mine a predefined set of stopes but what mine development layout is required to most profitably mine a subset of all possible stopes.

Dealing with uncertainty in ore body modelling is of increasing interest in the mining industry. Greico and Dimitrakopoulos (2004) report on a case study using the Floating Stope method on conditionally simulated ore bodies to quantify the grade risk associated with stope designs.

3.4 Mine Development Network Design

One of the principal differences between the modelling of open pit and underground mining operations is the complexity of realistically modelling the costs associated with access to the ore. This is a much more significant problem in an underground mine. Understanding the space of feasible solutions and then optimising a cost function over such a space is a highly complex problem. The first serious analytic solution method is found in the work of Brazil *et al.* (2000). Further investigations of this approach and details of the underlying mathematics have also appeared in Brazil *et al.* 2001, 2002, 2004, 2005.

The key to finding a tractable solution technique for this problem has been to model it as a network optimisation problem, and then develop a theory of three-dimensional Steiner networks in a suitable metric space. The mine is represented using a weighted network model. The network can be treated as being embedded in Euclidean 3-space, and coordinated according to the coordinates of the mine. In this model, the given draw points and surface portal correspond to fixed nodes of the network known as *terminals*. The ramps in the mine are represented by *links* in the network whose embeddings correspond to the centre lines of the ramps. Finally, the junctions at which three or more ramps meet are represented by variable nodes in the network, known as *Steiner points*.

The main assumptions in this network model are that the locations of all draw points at the stopes are given, together with the expected tonnage for each given draw point. The surface portal is also assumed to be fixed (or at least strongly constrained in its location). The cost of each link in the network is modelled as a combination of construction and haulage costs. The variable component of this cost can be assumed to be proportional to the Euclidean length of the link. This proportion, however, will be different for each link, depending on the tonnage of ore to be hauled along the corresponding part of the mine.

The principal constraint is that ramp components are constrained to a maximum allowable slope. This slope is measured as an absolute gradient m (i.e. m is the absolute value of the ratio of change in horizontal displacement over change in height). In underground mining problems m is generally in the range 1/9 to 1/7 depending on mining equipment specifications. Other constraints such as navigability of the drives by mine equipment (defining a minimum turning circle) and obstacle avoidance (to prevent sterilisation of the ore, for example), are also important but can often be treated as secondary constraints, particularly in large-scale designs. The navigability constraint is discussed in more detail in Section 3.5.

The optimisation problem can now be formulated as follows:

1. GIVEN: A set of points N in Euclidean 3-space, and a gradient bound $m > 0$.

2. FIND: A network T interconnecting N embedded in Euclidean 3- space, such that
   (a) The embedded links are piecewise smooth curves whose absolute gradient at each differentiable point is at most m,
   (b) The total construction plus haulage costs are minimized

This is effectively a variation of the well-known Steiner network problem – the problem of constructing a minimum cost network interconnecting a given set of points in a metric space (Hwang *et al.*, 1992). The control variables here are: the topology (i.e. underlying graph structure) of the network; the locations of the variable nodes (Steiner points), corresponding to the junctions in the mine; and the embedding of each of the links in Euclidean 3-space.

A key to finding good approximation algorithms for solving such an optimisation problem is to appreciate and exploit the geometric properties of its exact solutions. Initial work in this direction has appeared in Brazil *et al.* (1998) and Brazil *et al.* (2001). The process of finding good solutions is also assisted by the existence of convexity properties, under certain conditions, for a fixed topology (Brazil *et al.*, 2002, 2005).

3.5 Decline Design

The approach described in Section 3.4 tends to be very effective in large-scale mine design projects involving multiple ore bodies. Suppose, however, there is a single ore-zone for a proposed new underground mine or an extension to an existing mine described by its outline and either cross-cut entry or level access points on a sequence of levels. Here the topology of the network is no longer an issue, since the main network of haulage ramps (known as a *decline*) forms a single path. In this case, however, the navigability and obstacle avoidance constraints are likely to be significant factors in the optimal solution and can no longer be treated as secondary optimisation objectives. For example, there is generally a requirement that a decline approach an ore-zone no nearer than some stand-off distance to avoid possible sterilisation of the ore and to allow a minimum working length in the cross-cuts (i.e. the ramps connecting the level access points to the main decline).

In modelling this problem, the surface portal or breakout point of the decline is fixed or strongly restricted in position and the decline is modelled as a concatenation of straight and curved ramp-links with variable length cross-cut links attached at points which are the Steiner vertices in the model. One can assume that the cross-cuts are perpendicular to the decline to within a given angle tolerance and that they access the ore body at a fixed point (or one of a group of fixed points) on each given level. The key constraints are curvature, gradient and "no-go" regions. The optimisation problem can be formulated as follows:

1. GIVEN: A set of points N in Euclidean 3-space with an ordering placed upon them, a gradient bound $m > 0$, a minimum radius of curvature bound, and a set of "no-go" regions.

2. FIND: A network T interconnecting N embedded in Euclidean 3-space, such that a given cost function of T is minimised, where T satisfies the following constraints:
   (a) T contains a smooth path (the decline) interconnecting the first and last terminals. The decline contains all Steiner points, and the terminals are connected to the decline in the given order via straight horizontal links perpendicular to the decline (corresponding to cross-cuts).
   (b) Each link has gradient at most m.
   (c) The decline satisfies the minimum radius of curvature bound.
   (d) T avoids the specified "no-go" regions.

Designing such a network so that it has optimal cost is an extremely difficult problem. In order to make the problem mathematically tractable, the first step towards a solution is to simplify it to one in which (a) is replaced by:

(á) T joins two given points, s and t, in three-dimensional space, and has two given direction vectors at s and t.

Once a solution method has been developed for the modified problem, one can proceed with a dynamic programming methodology to solve the original problem, visiting the specified points and amalgamating the path entering a point and the one leaving it provided the two paths have the same start and finish directions. Analytic techniques for solving the modified problem have been outlined in Brazil *et al.* (2003), building on the geometric methods of Dubins (1957) for the problem in the horizontal plane. These principles have been incorporated into an algorithm and successfully implemented as a Decline Optimisation Tool, DOT (Brazil *et al.*, 2003).

3.6 Stope Scheduling

A long-term mine production schedule specifies the mining sequence for economic stopes and the associated mine development required to achieve production targets over a 2–5 year time frame subject to equipment and other resource constraints.

In the past decade three University research projects have sought to apply mixed integer programming techniques to long-term mine production scheduling.

Earlier work by Gershon (1982), and Barbaro and Ramani (1986) outlines attempts to apply mixed integer programming to mine scheduling optimisation.

Trout (1995) has developed a general formulation for ore extraction and stope backfill operations. Stope production is modelled in four distinct phases: preparation, extraction, void and backfilling. The model has been applied to base metal mines at Mt. Isa Mines and BHP Cannington.

Kuchta *et al.* (2003) have described a formulation suitable for sublevel caving operations at the Kiruna Mine, Sweden. The model was focused on loader machine (LHD) placement in production blocks, with horizontal and vertical sequencing constraints between production blocks, and blending of ores for the three main raw ore products. The primary data included information for 1,173 production blocks but due to the ore body geometry and sequencing considerations for sublevel caving the data was reduced to 56 machine placements. Backfill requirements are not applicable for this mining method. The objective function chosen was to minimize deviation from the production target in each time period. Judicious preprocessing of the data led to tighter limits on early and late start values for machine placements. A model with 36 time periods and 56 machine placements could be solved in minutes. Further research is reported by Newman and Kuchta (2003). Aggregation of production blocks was used, and solutions from these smaller models was used to guide the search in the original model. The model reported had 60 time periods (5 years) and 56 machine placements.

Smith *et al.* (2003) have developed a model for lead-zinc production at Mt. Isa Mines. Approximately 1,500 stopes have been aggregated into 32 mining blocks to cover 13 years of planned production. A detailed model formulation is not provided in the published paper. The model was developed with a fixed cutoff grade, and the inclusion of a dynamic cutoff grade (through a tonnage-grade curve) is part of their continuing research.

Each of these approaches implements a variation of the resource constrained project scheduling problem, but the number of periods, activities and resource constraints provides a real challenge in modelling and optimisation with mixed integer programming. Recent commercial applications are reported by Whittle (2004) using alternative metaheuristic techniques based on simulated annealing.

3.7 An Integrated Approach

Given the developments in optimisation of stope definition and infrastructure described earlier in this chapter, it is not surprising that some work on broadening the scope of underground mine optimisation is emerging. The ultimate goal is to embrace the design of the drilling programme, cut-off grade strategies, stope definition, infrastructure development and mine scheduling in one integrated optimisation model.

Poniewierski *et al.* (2003) have described the application of an optimisation approach taking into account cut-off grade, stope definition and scheduling to maximise NPV of the Enterprise mine at Mount Isa.

Carter *et al.* (2004) have used a similar approach to determine the maximum NPV design for a tabular ore body but here the work includes explicit optimisation of the infrastructure needed to support models of the mine. Existing stope definition and scheduling software was used in this integrated approach. The optimisation process was used to decide between an open stope and sublevel caving mining method for the ore body in question.

4 CONCLUSIONS

The complexity of the underground mine design problem and the unique mine design solutions sought for each ore body suggest that there will never be an elegant solution method analogous to that which exists for open pit mining. By decomposing the design problem into tractable subproblems, such as infill drilling design, stope definition, topological network design, decline design, and soon highly effective though non-globally-optimal solutions can be found. While the short-term emphasis will remain on optimising components of the overall design, there is emerging evidence of the potential for research, some of which is described in Section 3.7, to guide more comprehensive and integrated optimisation capability. As optimisation techniques become automated via the resulting software tools like Stopesizor and DOT, designers can explore alternative designs much more efficiently than traditional design methods allowed.

ACKNOWLEDGEMENTS

This research has been conducted with financial support from the Australian Research Council and Newmont Australia Limited via a Linkage collaborative research grant, as well as additional financial support from AMC Consultants and Datamine Software Ltd.

5 REFERENCES

A. Akaike and K. Dagdelen, "A strategic production method for an open pit mine", 28th APCOM Symposium, Golden, Colorado, 1999, 729–738.

C. Alford "Optimization in underground mine design", 25th APCOM AusIMM, 1995.

C. Alford, Optimization in underground mine design. PhD, Department of Mathematics and Statistics, The University of Melbourne, 2006.

M. Ataee-pour, A heuristic algorithm to optimize stope boundaries. PhD, Faculty of Engineering, University of Wollongong, 2000.

R.W. Barbaro and R.V. Ramani, "Generalized multi-period MIP model for production scheduling and processing facilities selection", *Mining Engineering*, 38 (1986), 107–114.

A. Boucher, R. Dimitrakopoulos and J.A. Vargas-Guzman, "Joint simulations, optimal drillhole spacing and the role of the stockpile", in Geostatistics Banff 2004 (Quantitative Geology and Geostatistics, 14), Springer, Berlin, 2005.

M. Brazil, D. A. Thomas and J. F. Weng, "Gradient constrained minimal Steiner trees", in Network Design: Connectivity and Facilities Location (DIMACS Series in Discrete Mathematics and Theoretical Computer Science, Vol. 40), American Mathematical Society, Providence, 1998, pp. 23–38.

M. Brazil, J. H. Rubinstein, D. A. Thomas, D. Lee, J. F. Weng and N. C. Wormald, "Network optimization of underground mine design". *The Australasian Institute for Mining and Metallurgy Proceedings*, Vol. 305, No.1, pp. 57–65, 2000.

M. Brazil, J. H. Rubinstein, D. A. Thomas, J. F.Weng and N. C. Wormald, "Gradient-constrained minimal Steiner trees (I). Fundamentals", *Journal of Global Optimization,* 21, (2001), 139–155,.

M. Brazil, D.H. Lee, J.H. Rubinstein, D.A. Thomas, J.F. Weng, N.C. Wormald, "A network model to optimize cost in underground mine design", *Trans. South African Institute of Electrical Engineers*, 93 (2002), 97–103.

M. Brazil, D.H. Lee, M. Van Leuven, J.H. Rubinstein, D.A. Thomas, N.C. Wormald, "Optimizing declines in underground mines", *Mining Technology: Trans. of the Institution of Mining and Metallurgy, Section A*, 112 (2003), 164–170.

M. Brazil, D.H. Lee, J.H. Rubinstein, D.A. Thomas, J.F. Weng, N.C. Wormald, "Optimization in the design of underground mine access", *Proceedings Orebody Modelling and Strategic Mine Planning*, 2004, pp. 3–6.

M. Brazil, D.A. Thomas, J.F. Weng, D.H. Lee, J.H. Rubinstein, "Cost optimization for underground mining networks", *Optimization and Engineering*, 6 (2005), 241–256.

L. Caccetta and S. Hill, "An application of branch and cut to open pit mine scheduling", *Journal of Global Optimization*, 27 (2003), 349–365,.

P.G. Carter, D.H. Lee and H. Baarsma, "Optimization methods for the selection of an underground mining method", *Proceedings Orebody Modelling and Strategic Mine Planning*, 2004, pp. 7–12.

N.M. Cheimanoff, E.P. Deliac and J.L. Mallet, "GEOCAD: An alternative CAD artificial intelligence tool that helps moving from geological resources to mineable reserves", 21st APCOM, 1989.

J. Deraisme, C. de Fouquent and H. Fraisse, "Geostatistical orebody model for computer optimization of profits from different underground mining methods", 18th APCOM IMM, 1984.

L.E. Dubins, "On curves of minimal length with a constraint on average curvature, and with prescribed initial and terminal positions and tangents". *American Journal of Mathematics*, 79 (3) (1957), 497–516.

G. Froyland, M. Menabde, P. Stone, and D. Hodson, "The value of additional drilling to open pit mining projects", *Proceedings Orebody Modelling and Strategic Mine Planning*, 2004, pp. 169–176.

M.E. Gershon, "Mine scheduling optimization with mixed integer programming", *Society of Mining Engineers of AIME*, Preprint No 82–324, 1982.

M. Gershon and F.H. Murphy "Using dynamic programming for aggregating cuts in a single drillhole", *International Journal of Surface Mining*, 1 (1987), 35–40.

M. Gershon and F.H. Murphy "Optimizing single hole mine cuts by dynamic programming", *European Journal of Operational Research*, 38 (1989), 56–62.

P. Goovaerts, *Geostatistics for Natural Resources Evaluation*. New York, Oxford University Press, 1997.

N. Grieco and R. Dimitrakopoulos, "Grade uncertainty in stope design: Improving the optimization process", *Proceedings Orebody Modelling and Strategic Mine Planning*, 2004, pp. 249–255.

B. Hall and C. Stewart, "Optimising the strategic mine plan – methodologies, findings, successes and failures", *Proceedings Orebody Modelling and Strategic Mine Planning*, 2004, pp. 49–58.

F. K. Hwang, D. S. Richards, and P. Winter, The Steiner Tree Problem (Annals of Discrete Mathematics 53). Elsevier, Amsterdam, 1992.

M. Kuchta, A. Newman and E. Topal, "Production scheduling at LKAB's Kiruna Mine using mixed integer programming", *Mining Engineering* 55 (2003), 35–40.

K.F. Lane, The Economic Definition of Ore. London, Mining Journal Books Limited, 1988.

H. Lerchs, and I. F. Grossmann, "Optimum design of open-pit mines," *Trans. Canadian Institute of Mining and Metallurgy*, Vol. LXVIII (1965) 17–24.

Mineral Industries Computing Limited, Datamine Studio V2 User Documentation, 2005.

F. Muge, P. Pina and J.L. Vieira, "Some applications of image analysis techniques to mine planning", 24th APCOM IMM, 1995.

A. Newman and M. Kuchta, "Eliminating variables and using aggregation to improve the performance of an integer programming production scheduling model for an underground mine", preprint, 2003.

J. Poniewierski, G. MacSporran and I. Sheppard, "Optimization of cut-off grade at Mount Isa Mines Limited's Enterprise Mine", *Conference Proceedings – Twelfth International Symposium on Mine Planning and Equipment Selection (MPES)*, 2003, pp. 531–538.

J-M. Rendu, "Computerized estimation of ore and waste zones in complex mineral deposits", Society of Mining Engineers of AIME, Preprint No 82–95, 1982.

J-M. Rendu, "Orebody Modelling, mine planning, reserve evaluation and the regulatory environment", *Proceedings Orebody Modelling and Strategic Mine Planning*, 2004, pp. 201–208.

J. Serra, *Image Analysis and Mathematical Morphology*. New York, Academic Press, 1982.

M.L. Smith, I. Sheppard and G. Karunatillake "Using MIP for strategic life-of-mine planning of the lead/zinc stream at Mount Isa Mines", APCOM 2003 Conference, 2003, pp. 1–10.

G. Thomas and A. Earl, "The application of second-generation stope optimization tools in underground cut-off grade analysis", Whittle Strategic Mine Planning Conference, 1999.

P. Trout, "Underground mine production scheduling using mixed integer programming", APCOM XXV 1995 Conference, 1995, pp. 395–400.

G. Whittle, "Global asset optimization", *Proceedings Orebody Modelling and Strategic Mine Planning*, 2004, pp. 261–266.

Chapter 31

LONG- AND SHORT-TERM PRODUCTION SCHEDULING AT LKAB'S KIRUNA MINE

Alexandra M. Newman[1], Mark Kuchta[2], and Michael Martinez[3]
[1]*Division of Economics and Business, Colorado School of Mines, 1500 Illinois Street, Golden, Colorado, 80401 USA;* [2]*Mining Engineering Departament, Colorado School of Mines, 1500 Illinois Street, Golden Colorado, 80401 USA;* [3]*Departament of Mathematical Sciences, United States Air Force Academy, Colorado Springs, Colorado, 80840 USA*

Abstract LKAB's Kiruna mine is an underground sublevel caving mine located above the Arctic circle in northern Sweden. The iron ore mine currently uses a long-term production scheduling model to strategically plan its ore extraction sequence. In this chapter, we describe how we modify this model to consider several different levels of time resolution in the short- versus long-term, and provide guidance for increasing model tractability. We demonstrate numerically the increase in schedule quality and model tractability as a result of these modifications.

Keywords: Integer programming, production scheduling, underground mining, applications

1 INTRODUCTION AND BACKGROUND

LKAB's Kiruna mine, located above the Arctic circle in northern Sweden, has been operational for more than a century. The mine currently employs about 600 workers and produces approximately 24 million tons of iron ore per year in the form of three raw iron ore products: *B1*, *B2*, and *D3*. These raw products are used to supply planned production quantities at four ore postprocessing plants or mills. The ore products are classified according to their phosphorus content, and are processed into fines and pellets, both of which are used as raw materials in the manufacture of steel end-products.

For about half a century, iron ore at Kiruna was extracted exclusively via surface methods, but about 1960, the pit deepened to such an extent that it became more cost-effective to mine underground. The underground mining method Kiruna currently employs is known as large-scale sublevel caving. This method is used for extracting ore from vertically positioned, fairly pure, large, vein-like deposits. The mine is divided into ten main production areas, about 400 meters (m) to 500 m in length. Each production area has its own group of ore passes; such a group is also known as a shaft group. A shaft group is located at the center of each production area, and extends down to the main level. Each production area consists of about 10 sublevels, and entry to these sublevels is gained via access ramps. One or two 25-ton-capacity electric Load Haul Dump Units (LHDs) operating on a sublevel within each production area transport the ore from the crosscuts (from which the ore is extracted) to the ore passes, where loaded trains haul the ore to a crusher. At the crusher, the ore is broken into pieces small enough to be hoisted to the surface via vertical shafts. Up to 18 LHDs can operate daily throughout the mine; however, the allowable number of LHDs within each shaft group is restricted to about two or three to prevent LHD drivers from driving over and damaging LHD cables.

The site on which each LHD operates is also referred to as a machine placement. The number of machine placements that can be started in a given time period is restricted due to the availability of the crew that prepares the machine placement to start to be mined. The number of active machine placements, i.e., machine placements currently being mined, is also restricted due to LHD availability. Each machine placement belongs to a unique shaft group. A machine placement averages 200 to 500 m in length and contains from one to three million tons of ore and waste rock. A machine placement possesses the same height as the mining sublevel and extends from the hangingwall to the footwall. Between one and five smaller (100 m) entities known as production blocks constitute a machine placement. About one month is required to mine each production block. If a machine placement is left partially mined, old explosives (which only have a life of about 30 days) must be replaced to reblast the solidified cave rock. This requirement, coupled with the aggravation of tracking partially-mined machine placements, results in operational restrictions that require continuous production within a machine placement until all available ore has been removed. Whether a machine placement can (or must) be mined depends on the relative position of machine placements where mining has already begun. Specifically, certain machine placements beneath a given machine placement cannot start to be mined until some portion of the given machine placement has been mined, and machine placements to the right and left of a given machine placement must start to be mined after a specified portion of the given machine placement has been mined (to prevent blast damage on adjacent machine placements). These operational constraints are referred to as vertical and horizontal sequencing constraints, respectively. Each machine placement possesses a series of notional drawdown lines, consisting of several production blocks each. Within a machine placement, the order in which production blocks must be mined is regulated by this series of drawdown lines, which also helps to enforce continuous mining of a machine placement. These drawdown lines cut horizontally or at a 45 degree angle though several blocks within the machine placement and preclude production blocks in a drawdown line underneath a given drawdown line from being extracted until all ore in a given drawdown line is extracted. This mining pattern is necessary to correctly execute the sublevel caving method so that the mined out areas do not collapse on top of ore that is yet to be retrieved. Minimum and maximum production levels per month govern the rate at which the blocks within a machine placement are mined. These rates ensure continuous mining of machine placements, as discussed above, as well as adherence to production capacity restrictions. Because of vertical and horizontal sequencing constraints and the relative positions of machine placements and the production blocks within them, there are only certain time periods in which these can be mined.

Figure 1 illustrates the relationship between the ore body, machine placements, production areas, ore passes, levels and sublevels, vertical shafts, shaft groups, the crusher, crosscuts (i.e., production drifts), the hangingwall

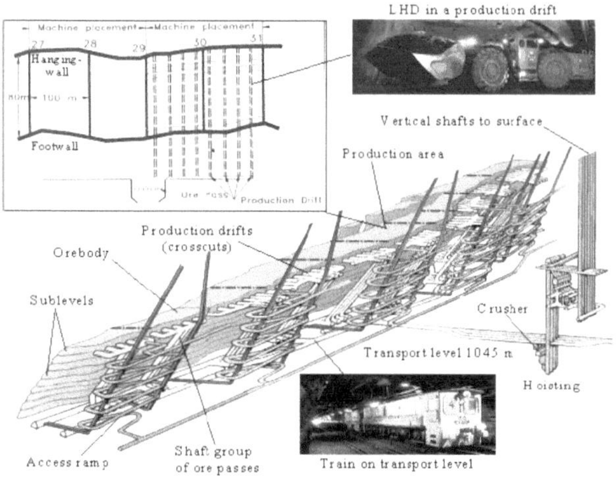

Figure 1. The main body in the figure depicts the relationship between the production areas, ore passes, shaft groups, sublevels, and vertical shafts to the surface. The enlargement in the upper left hand corner shows a plan view of a machine placement, consisting of various production drifts, within a production area. The inset in the upper right hand corner shows a load haul dump unit, while the inset in the lower right hand corner depicts a train hauling ore to the crusher.

(H.W.), and footwall (F.W.). Figure 2 shows the relationship between production blocks, machine placements, and drawdown lines. See Topal (2003) for a more detailed description of the Kiruna mine and its characteristics.

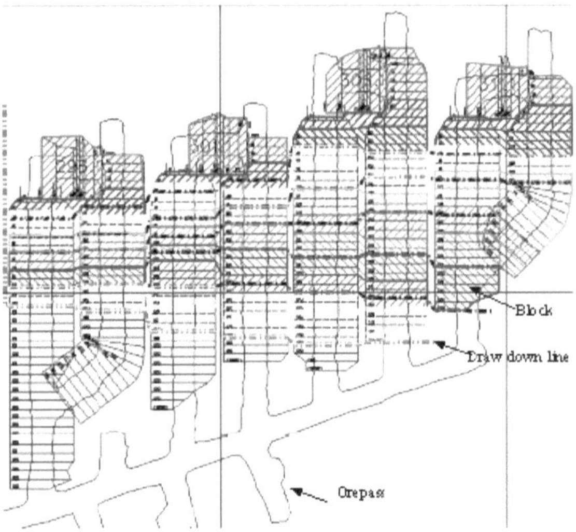

Figure 2. Machine placement MP849_29-30 consists of eight columns of production blocks. Notional drawdown lines pass through a "row" of blocks. The ore body is mapped onto an underlying network of production drifts, which connects to the ore pass.

2 LITERATURE REVIEW

Winkler (1996), among others, identifies the importance of using integer programming models to determine discrete mine production schedules. However, many researchers are unable to solve realistic mining scenarios, for example Trout (1995), Smith (1998), Smith *et al.* (2003). Earlier attempts at the Kiruna mine failed to yield a production schedule of requisite length in a reasonable amount of time, for example Almgren (1994), Topal (1998), Dagdelen *et al.* (2002). Rather than determining an optimal schedule, these authors resort to shortening the schedule time horizon and/or to sacrificing schedule quality. Carlyle and Eaves (2001) present a tractable model that maximizes revenue from Stillwater's platinum and palladium mine, as do Sarin and West-Hansen (2005) for an underground coal mine; however, both models differ from ours in their objective and, because of the mining method (sublevel stopping, and long-wall, room-and-pillar, and retreat mining, respectively), in the constraints. Kuchta *et al.* (2003) provide a more comprehensive list of references.

3 SCHEDULING MODEL

Kiruna currently uses a long-term production scheduling model (Kuchta *et al.*, 2004) to strategically plan its ore extraction sequence. Before the existence of the model, a mine planner would have to develop by hand over the course of a week or more a schedule that would determine when to start mining each machine placement. Because of the vast number of choices and all the scheduling rules, a planner could easily narrow his options unacceptably by choosing to mine certain machine placements that would later preclude sequencing constraints from being satisfied and/or that would result in unacceptably high deviations from demands. As a result, LKAB opted to employ optimization techniques to more quickly generate better-quality production plans. The model is a mixed-integer program that contains thousands of binary variables representing whether or not to mine a specific machine placement in each month of the planning horizon. The model considers the physical limitations of the mine, while meeting as closely as possible the planned production quantities of each raw ore product.

We have modified the original model to comprise several levels of detail. At the coarser (original) level of detail, decision variables consist of which machine placements to start mining each month. For machine placements that are already being mined, we model decisions at a finer level of detail with variables that represent the amount to mine from each production block in each month. Correspondingly, the model also tracks which drawdown lines have been completely mined. This extra level of detail allows the mine planner to more closely control the amount and types of ore that are extracted from each machine placement in each time period. We show in the numerical results section how this extra flexibility helps us to achieve better-quality schedules than those using the long-term model.

A formulation of our combined long- and short-term model follows:

PARAMETERS

◊ LHD_t = number of machine placements that can start in time period t

◊ LHD_v = maximum number of active machine placements in shaft group v

◊ d_{kt} = demand for ore type k in time period t (kilotons)

◊ $r_{at'tk}$ = reserves of ore type k available at time t in machine placement a given that the machine placement started to be mined at time t' (kilotons)

◊ R_{bk} = reserves of ore type k contained in block b (kilotons)

◊ $\overline{C}_{at}$ = maximum production rate of machine placement a in time period t (kilotons per time period)

◊ $\underline{C}_{at}$ = minimum production rate of machine placement a in time period t (kilotons per time period).

◊ $$\rho_{at't} = \begin{cases} 1 & \text{if machine placement } a \text{ is being mined at} \\ & \text{time } t \text{ given that it started to be mined at time } t' \\ 0 & \text{otherwise.} \end{cases}$$

VARIABLES

◊ $\overline{z}_{kt}$ = amount mined above the demand for ore type k in time period t (kilotons)

◊ $\underline{z}_{kt}$ = amount below the demand for ore type k in time period t (kilotons)

◊ x_{bt} = amount mined from production block b in time period t (kilotons)

◊ $$w_{lt} = \begin{cases} 1 & \text{if we finish mining all blocks contained in drawdown line } l \\ & \text{by time period } t \\ 0 & \text{otherwise.} \end{cases}$$

◊ $$y_{at} = \begin{cases} 1 & \text{if we start mining machine placement } a \text{ at time period } t \\ 0 & \text{otherwise.} \end{cases}$$

FORMULATION

$$\text{(P) Min} \sum_{k,t} \underline{z}_{kt} + \sum_{k,t} \overline{z}_{kt},$$

subject to

$$\sum_a \sum_{t' \le t} r_{at'tk}\, y_{at'} + \sum_b \frac{R_{bk}}{\sum_{\hat{k}} R_{b\hat{k}}} x_{bt} + \underline{z}_{kt} - \bar{z}_{kt} = d_{kt} \quad \forall\, k,t, \tag{1}$$

$$\sum_a \sum_k \sum_{t' \le t} r_{at'tk}\, y_{at'} + \sum_b x_{bt} = \sum_k d_{kt} \quad \forall\, t, \tag{2}$$

$$\sum_a \sum_{t' \le t} \rho_{at't}\, y_{at'} + \sum_l (1 - w_{l\hat{t}}) \le LHD_v \quad \forall\, v, \hat{t}, t, \tag{3}$$

$$\sum_a y_{at} \le LHD_t \quad \forall\, t, \tag{4}$$

$$\sum_t x_{bt} \le \sum_k R_{bk} \quad \forall\, b, \tag{5}$$

$$w_{lt} \le w_{l,t+1} \quad \forall\, l,t, \tag{6}$$

$$\sum_b \sum_{u \le t} x_{bu} \ge \sum_b \sum_k R_{bk} w_{lt} \quad \forall\, l,t, \tag{7}$$

$$\sum_{u \le t} x_{bu} \le \sum_k R_{bk} w_{lt} \quad \forall\, b,l,t, \tag{8}$$

$$\sum_b x_{bt} \le \overline{C}_{at} \quad \forall\, a,t, \tag{9}$$

$$\sum_b x_{bt} \ge \underline{C}_{at} (1 - w_{lt}) \quad \forall\, a,l,t, \tag{10}$$

$$w_{lt} \ge y_{at} \quad \forall\, a,l,t, \tag{11}$$

$$\sum_{t \le \hat{t}} y_{at} \ge w_{l\hat{t}} \quad \forall\, a,l,\hat{t}, \tag{12}$$

$$\sum_{t \le t'} y_{at} \ge y_{a't'} \quad \forall\, a, a', t', a \ne a', \tag{13}$$

$$\sum_{t' \le t} y_{a't'} \ge y_{at} \quad \forall\, a, a', t, a \ne a', \tag{14}$$

$$\sum_{t} y_{at} \le 1 \quad \forall\, a, \tag{15}$$

$$w_{lt}, y_{at} \in \{0,1\} \quad \forall\, l, a, t. \quad \bar{z}_{kt}, \underline{z}_{kt}, x_{bt} \ge 0 \quad \forall b, k, t. \tag{16}$$

The objective function measures the tons of deviation. Note that we could weight this either by ore type, or by time period, or both. Constraints (1) record for each ore type and time period the amount in excess or deficiency of the required amount of ore. Constraints (2) require that for each time period in the short term, the total amount of ore required, regardless of ore type, is mined. This prevents the postprocessing mills from sitting idle. Constraints (3) limit the maximum number of active machine placements in each shaft group and time period. The index t belongs to the set of time periods in which drawdown line l can finish being mined, and $\hat{t}$ is the time period by which all blocks in drawdown line l must finish being mined. Constraints (4) constrain the number of long-term machine placements that can be started in a time period. Constraints (5) preclude mining more than the available reserves. Constraints (6) indicate that once a drawdown line has finished being mined, it has finished for the horizon. Constraints (7) relate the finish of mining a drawdown line to mining the blocks within that drawdown line. Constraints (8) preclude a block b in a drawdown line from starting to be mined unless all blocks in constraining drawdown lines (l) have been mined, and holds for all time periods in which drawdown line l can finish being mined. Constraints (9) and (10) enforce maximum and minimum production rates, respectively, for each machine placement and time period. Constraints (11) and (12) enforce vertical and horizontal sequencing, respectively, between machine placements in the short term. Note that in constraints (11), the drawdown line, l, in a constraining machine placement controls access to a constrained machine placement, a. (The relationship is reversed in constraints (12).) Constraints (13) and (14) enforce vertical and horizontal sequencing, respectively, between machine placements in the long term. In these two constraints, a' belongs to the set of machine placements whose access is restricted vertically, or forced by adjacency, respectively, to machine placement a. Constraints (15) allow a machine placement to start to be mined at most once during the time horizon. Finally, nonnegativity and integrality are enforced, as appropriate. In the interest of brevity, we have omitted the use of a large number of sets, specifying, for example, the machine placements in a shaft group or the eligible time periods in which a machine placement can start to be mined. Martinez *et al.* (2005) give a detailed formulation.

The model is unique in several respects: (i) it does not account for the difference in costs from mining various machine placements due, for example, to their location in the mine, (ii) the objective does not consider the net present value of ore, (iii) model instances are not necessarily solved for the life of the mine, which may result in undesirable end effects, and (iv) there is no allowance for holding inventory, or stockpiling. With respect to the first issue, we assume that all ore will be mined eventually, and hence, total mining costs are sunk. Therefore, we need not consider discrepancies in costs between mining various machine placements. The second aspect is explained by the difference between the markets for iron ore and precious metals. Precious metals such as gold and silver are traded on, for example, the Commodity Exchange of New York. These metals are bought and sold worldwide, and the strategy of mines extracting these metals is to maximize profits by producing as much as is economically viable given current market prices. By contrast, markets associated with base metals such as iron ore are regionalized, as transportation costs are high relative to the value of the commodity. Within these markets, steel companies enter into a contract with an iron ore producer, settling on a price commensurate with the chemical and physical characteristics of the iron ore. Large buyers tend to influence prices in contracts between other buyers and iron ore producers. The negotiated prices generally hold for about a year, and iron ore producers are obligated to supply a certain amount of iron ore to each buyer with whom they hold a contract. Therefore, iron ore mines like Kiruna are concerned with meeting contractual demands as closely as possible. With respect to the third point, given the future uncertainty in the iron ore composition of each block and the computational time currently required to solver smaller models, attempting to produce life-of-mine schedules is impractical. To mitigate end effects, we solve the model on a rolling horizon basis, updating the ore type composition of each machine placement as the information becomes available.

Finally, company policy does not allow LKAB to stockpile iron ore. A traditional inventory constraint would not apply in this setting at any rate. Specifically, there is physically no space in which to store more than about 50 kilotons of extracted iron ore. Furthermore, because LKAB's goal is to meet demand as closely as possible in each time period so as to regulate the amount of ore processed at the mills, a shortage in one time period cannot be compensated by a surplus in, say, the following time period. We could instead recommend using the results from our original model as follows. For each ore type and time period in which there exists production excess (positive deviation), we add to a stockpile until the limit of 50 kilotons has been reached. Correspondingly, for each ore type and time period in which there exists a shortage (negative deviation), we draw up to 50 kilotons from this stockpile, decrementing the total amount stored, as appropriate. However, too many successive periods of overproduction prevent a significant amount of the overproduced ore from being stored in the stockpile, and too many successive time periods of underproduction deplete the 50 kilotons buffer without overproducing to replenish it. We find that, for Kiruna's current scenarios, there are few instances of alternating excess and under production between time periods.

4 SOLUTION TECHNIQUES

As the model is large, we use techniques to eliminate all variables that would necessarily assume a value of zero in the optimal solution, and, in fact, in any feasible solution. We capitalize on a modified version of a resource-constrained critical path model to determine earliest and latest possible start dates for each machine placement in the scheduling horizon, allowing us to eliminate a portion of the y_{at} binary variables. The modification manifests itself in that not only are there vertical sequencing constraints that dictate the duration of an "activity," that is mining a machine placement, but there exist also horizontal sequencing constraints that require, rather than allow, a subsequent activity to be started only after a given activity has started (or has been completed). LHD availability is the resource. We can also assign a late start date to each machine placement. This late start date would eliminate decisions to start to mine a machine placement so late that adjacent and underlying machine placements eventually become "locked in," thereby increasing the amount of deviation between the actual and the planned production quantities beyond values otherwise obtained in an optimal solution. We use information about active machine placements together with horizontal sequencing constraints to determine the latest time period in which mining a subset of machine placements can start. We can then assign late start dates to machine placements whose start dates are affected by machine placements within this subset.

We can also eliminate variables associated with mining a drawdown line before an earliest finish date or after a late finish date. We can determine an early finish date for a drawdown line based on the principle of a critical path model by comparing the tonnage available in each drawdown line with the tonnage that can be mined in each time period. Similarly, the latest time at which the mining of a drawdown line could finish is the time at which the first drawdown line in the machine placement finishes being mined added to the longest amount of time it would take all drawdown lines overlying the given drawdown line to be mined. Note that we can use similar principles to establish early start and late finish dates for production blocks to eliminate x_{bt} variables corresponding to mining a production block before its earliest start date or after its latest finish date. However, because the variables associated with mining a production block are continuous, the direct benefit of eliminating such variables is small. However, an indirect benefit of an early start date for each production block is its use in establishing an early start date for a drawdown line, which is simply the earliest early start date among all blocks in a drawdown line. Early start dates for a drawdown line help to eliminate irrelevant terms in constraints (3). Martinez *et al.* (2005) provide details regarding the early and late start, and the early and late finish algorithms, as well as the early start algorithm itself.

In previous research addressing only the long-term model, we have used not only variable elimination based on early and late start and finish dates, but also an optimization-based heuristic, which we term the *aggregation procedure*, to eliminate all but a reasonably good set of starting times for each machine placement. This allows us to restrict the model to a subset of start date choices beyond the restrictions we determine with the early and late start algorithms. To date, we have found that this procedure is useful only for eliminating the y_{at} variables because the loss of fidelity inherent in the procedure would be unacceptable for short-term decisions. We refer the interested reader to Newman and Kuchta (2007) for more details.

5 NUMERICAL RESULTS

We demonstrate the benefits of our solution procedures, as well as the improvements we gain in the superiority of the solution by using the combined long- and short-term model over simply using the long-term model for production planning. We conduct our numerical experiments using the AMPL programming language (Fourer *et al.*, 2003; and AMPL Optimization LLC, 2001) and the CPLEX solver, Version 9.0 (ILOG Corporation, 2003), and use the CPLEX parameter setting, which applies its relaxation induced neighborhood search heuristic every 40 nodes. We run all model instances on a Sunblade 1,000 computer with 1 GB RAM. The scenario we use possesses current data from LKAB's Kiruna mine. The data set contains three ore types and spans 24 months.

We summarize the reduction in model size as a function of the number of continuous and binary variables, and the number of constraints when we employ each of the early start, late start, early finish and late finish algorithms *independently of each other*. We apply each algorithm to all relevant model entities: that is, the early start algorithm applies to machine placements, drawdown lines, and production blocks. The late start algorithm applies only to machine placements. The early finish algorithm applies to drawdown lines, while the late finish algorithm applies to both production blocks and drawdown lines.

The monolithic model contains over 4,000 binary variables and over 8,000 constraints. Applying the early start algorithm gives an approximately 25% reduction in both the number of binary variables and in the number of constraints. The number of continuous variables decreases by about 15%. Applying the late start algorithm gives about a 3% reduction in the number of binary variables, and about a 1% reduction in the number of constraints. Applying the early finish algorithm reduces the number of binary variables and the number of constraints by more than 10%. Applying the late finish algorithm reduces the number of binary variables and the number of constraints by about 40%, and the number of continuous variables by 15%. Using all four algorithms in conjunction with each other yields a model with about 20% of the original number of binary variables, continuous variables, and constraints as found in the monolith.

We also make comparisons regarding the quality of the solutions from the long-term and combined models. We apply the variable reduction techniques, mentioned earlier, to both models, as applicable. Additionally, we weight the objective functions both in the long-term and combined models so as to penalize deviations in earlier time periods more heavily. In contrast to the reduced combined model (with presolve), which has 672 binary variables, 579 continuous variables, and 1,702 constraints, the reduced version of the long-term model (including presolve) possesses 416 binary variables, 144 continuous variables, and 734 constraints. The long-term model solves to within 5% of optimality in 4 seconds, whereas the combined model requires 4,350 seconds of solution time to reach the same gap.

The extra fidelity in the combined model increases its size and decreases its tractability. However, the combined model yields a 69.6% reduction in deviation compared with that corresponding to the schedule generated by the long-term model. The long-term model only makes decisions at the machine placement level. Once a machine placement starts to be mined, the monthly production quantities within that machine placement are fixed, and mining must occur according to that fixed sequence. With the combined long- and short-term model, production rates are allowed to vary between set minimum and maximum values for each production block within a machine placement, thereby allowing partial mining of a monthly production block in order to more closely meet demands for the three ore types.

Figure 3 shows a comparison of total deviation (both under- and over-production for all three ore types) for each month in the horizon. Currently active machine placements, that is initial conditions, cause high deviations in the early time periods in the solutions of both models relative to the deviations in the later months of the horizon.

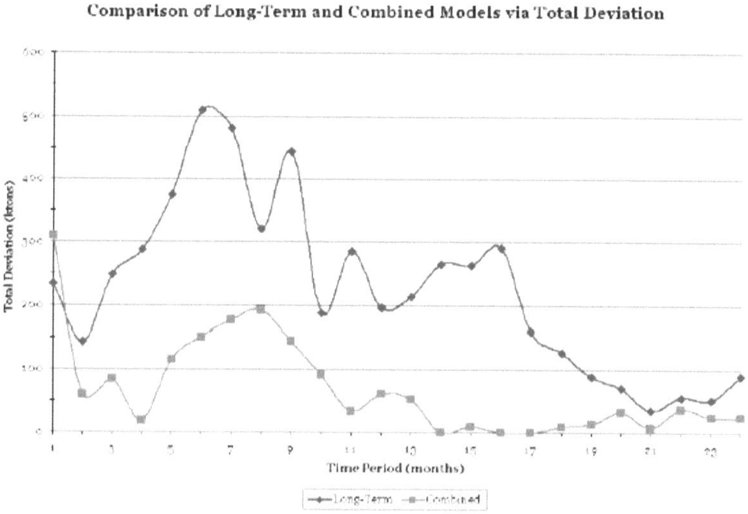

Figure 3. Depiction of total deviation (ktons of ore) as a function of the monthly time periods in the planning horizon for both the long-term and combined models. Although the graphs appear qualitatively similar, the amount of deviation in the former graph far exceeds that in the latter graph for all but one time period.

6 CONCLUSIONS

We present a model that considers both short- and long-term production scheduling for LKAB's Kiruna mine. The benefits of this combined model are its short-term fidelity in directing miners at an operational level which ore to extract and its long-term clairvoyance showing mine planners at a strategic level the mining areas to develop. Adding short-term fidelity to the original long-term model improves the objective of meeting demands for each ore type and time period. Because the model contains thousands of binary variables and constraints over just a 2-year horizon, we present methods for reducing the size of the model, hence increasing its tractability. Specifically, we develop several algorithms to determine eligible time periods in which a machine placement, a production block, and a drawdown line can be mined. This, in turn, allows us to eliminate variables whose values would equal zero in the optimal solution. Future research entails developing additional methods to enhance model tractability, to enable the generation of production schedules over, say, a 4- or 5- year horizon. Ultimately, life-of-mine schedules, though the varying availability of data means they can only serve as estimates, would be attractive to mine developers.

ACKNOWLEDGMENTS

The authors would like to thank LKAB for the opportunity to work on this challenging project and for permission to publish these results. The authors would specifically like to thank LKAB employees Hans Engberg, Anders Lindholm, and Jan-Olov Nilsson for providing information, assistance, and support. We also thank Professor Kevin Wood of the Naval Postgraduate School, Professor Candi Yano of University of California, Berkeley, and Professors F.S. and J.K. Newman of the University of Illinois at Urbana-Champaign for helpful comments on earlier drafts.

7 REFERENCES

Almgren, T. (1994). An Approach to Long Range Production and Development Planning with Application to the Kiruna Mine, University of Technology, Sweden, Luleå, Doctoral Thesis Number 1994:143D.

AMPL Optimization LLC. (2001). *AMPL*, Version 10.6.16. www.ampl.com.

Carlyle, M. and B.C. Eaves. (2001). Underground planning at Stillwater Mining Company, *Interfaces* 31(4): 50–60.

Dagdelen, K., M. Kuchta, and E. Topal. (2002). Linear programming model applied to scheduling of iron ore production at the Kiruna Mine, Kiruna, Sweden, *Transactions of the Society for Mining, Metallurgy, and Exploration* 312: 194–198.

Fourer, R., D. Gay, and B. W. Kernighan. (2003). *AMPL: A Modeling Language for Mathematical Programming*, Thompson Learning, Pacific Grove, CA.

ILOG Corporation. (2003). *CPLEX*, Version 9.0. www.ilog.com.

Kuchta, M., A. Newman, and E. Topal. (2003). Production scheduling at LKAB's Kiruna Mine using mixed integer programming, *Mining Engineering* April, 35–40.

Kuchta, M., A. Newman, and E. Topal. (2004). Implementing a production schedule at LKAB's Kiruna Mine, *Interfaces* 34(2): 124–134.

Martinez, M., A. Newman, and M. Kuchta. (2005). Using Decomposition to Optimize Long- and Short-Term Production Planning at an Underground Mine, Working Paper, Division of Economics and Business and Mining Engineering Department, Colorado School of Mines, Golden, CO, November.

Newman, A. and M. Kuchta. (2007). Using aggregation to optimize long-term production planning at an underground mine, *European Journal of Operational Research* 176(2): 1205–1218.

Sarin, S. and J. West-Hansen. (2005). The long-term mine production scheduling problem, *IIE Transactions* 37(2): 109–121.

Smith, M.L. (1998). Optimizing short-term production schedules in surface mining: Integrating mine modeling software with AMPL/CPLEX, *International Journal of Surface Mining* 12(4): 149–155.

Smith, M.L., Sheppard, I. and G. Karunatillake. (2003). Using MIP for strategic life-of-mine planning of the lead/zinc stream at Mount Isa Mines, in *Proceedings, 31st International Symposium on the Application of Computers in the Mineral Industry*, Capetown, South Africa: 465–474.

Topal, E. (1998). Long and Short Term Production Scheduling of the Kiruna Iron Ore Mine, Kiruna, Sweden, Master of Science Thesis, Colorado School of Mines, Golden, CO.

Topal, E. (2003). Advanced Underground Mine Scheduling using Mixed Integer Programming, Ph.D. Dissertation, Colorado School of Mines, Golden, CO.

Trout, L.P. (1995). Underground mine production scheduling using mixed integer programming, in *Proceedings, 25th International Symposium on the Application of Computers in the Mineral Industry*, Brisbane, Australia: 395–400.

Winkler, B.M. (1996). Using MILP to optimize period fix costs in complex mine sequencing and scheduling problems, in *Proceedings, 26th International Symposium on the Application of Computers in the Mineral Industry*, Pennsylvania State University, University Park, PA: 441–446.

Chapter 32

AN INTEGRATED APPROACH TO THE LONG-TERM PLANNING PROCESS IN THE COPPER MINING INDUSTRY

Rodrigo Caro, Rafael Epstein, Pablo Santibañez, and Andres Weintraub
Department of Industrial Engineering, University of Chile, Santiago, Chile

Abstract Long-term mine-metallurgy planning in the copper mining industry is a complex process that simultaneously establishes ore extraction policies for multiple mines, an investment plan, and an operating plan for processing plants such concentrators, smelters, and refineries. These strategic plans must specify extraction and processing decisions for each cubic meter of mine ore while maintaining their consistency with medium- and long-term objectives.

This chapter demonstrates the advantages and implications of integrated models that can simultaneously plan the entire chain of production from the extraction of ore to the final cathodes, subproducts, and inputs. Little work has been done in this area, which the authors believe is a promising field for development and application in operations research. We begin with a brief description of the long-term mine planning problem in the copper industry and then discuss the advances made in recent years. This is followed by an overview of the copper production process, and finally, a look at the challenges involved in integrated planning.

Keywords: Mine planning, mine scheduling, plant planning, plant scheduling, mixed integer linear programming

1 INTRODUCTION TO THE LONG-TERM MINE-METALLURGY PLANNING PROCESS IN THE COPPER INDUSTRY

The long-term mine-metallurgy planning problem in the copper industry is to establish a strategic plan for production and investment, which ensures a maximum return on investment over a mine's time horizon while remaining consistent with short- and medium factors, and is subject to multiple constraints whether technical (such as the mine's geomechanical stability), environmental, physico-chemical, process-related or of some other type. For large deposits the time horizon will typically be 20 years, but in exceptional cases may be as many as 50.

The planners charged with this task are multidisciplinary groups of interacting specialists who draw up a mine-metallurgy plan, which defines the investment and production schedules for the mines and plants over the entire evaluation horizon, including such items as technology, inputs, processing methods and the scheduled flow of each cubic meter of ore. The result is a document with sufficient detail to be reviewed and audited for consistency and feasibility. Any error at this strategic level could have drastic consequences for the future of the business. A wrongly scaled processing plant, for example, could limit production capacity, while poorly sequenced ore extraction could render unusable significant sections of the mine. In the intimate relationship between the various levels of decision-making, long-term planning is the backbone of medium-term planning, which in turn forms the foundation for short-term planning.

The long-term planning process must find solutions to three main problems:

- Ore deposit production
- Selection, design, and scheduling of investment projects
- Plant operating policy and strategies

Although these problems are simultaneous and highly interrelated, in practice they are solved sequentially and the solution to either of the first two is the input to the next one. This tactic is necessitated by the methodological difficulties involved in a more integrated approach and the fact that even taken individually, the three problems are very complex. Nevertheless, there is general awareness that partial solutions are suboptimal in terms of the overall problem and that their integration would offer major benefits.

In view of the foregoing, the various decisions involved in a mine-metallurgy plan may also be classified into three main categories:

Extraction:

- Which resources to extract and where to transport them. This leads to a definition of the reserves
- When to extract them
- Which technology to extract them with

Investment:

- How much to invest
- When to invest
- Mine extraction capacity and expansion potential
- Plant processing capacity and expansion potential

Processes:

- How each plant should be operated. This includes defining the operating variables
- How much to process in each plant
- Sale of byproducts and intermediate products

Among the technical factors that frame the planning process are sector sequencing, expansions, production smoothing in underground mine production plans, and open-pit slope angles. Other technical issues include observing production rate limits, handling pollutants, and the optimal blend of ore feed to the plants. As regards economic factors, various expected costs and prices over the time horizon including the value of time as given in the plan's discount rate must be considered, and the return on investment and income from products and subproducts such as molybdenum must be maximized.

The evaluation of investments must be approached in a systematic fashion. Examining each project individually is of little use; an investment plan makes sense only if it is harmonized with mine production and processing plans. As the list of projects grows, the difficulties in identifying the combination that maximizes profits while remaining consistent with the mine production plan become increasingly apparent. If we then add the time factor, that is, the scheduling of each of the projects, the decisions associated with both investment and production are clearly a major challenge for planners. All of this gives rise to a set of problems with multiple alternatives and constraints that is large, complex, and difficult to model.

1.1 The Traditional Planning Approach

The long-term planning process for copper mine production and investment begins with the geological exploration of a deposit. The data collected are used to build a geological model in which the deposit is divided conceptually into small ore blocks measuring 20 m^3, and key information such as tonnage, grades, and mineralogical attributes are then determined for each one. Meanwhile, teams of economic experts work on the definition of sales strategies. Specialists in open-pit or underground mine development then define the phases or expansions and extraction points. Multi-disciplinary teams create scenarios for evaluating different investment plan configurations, such as the enlargement of a concentrating plant. The scenarios vary as regards value of ore, planning horizon costs, performance of new technology and quantity and quality of deposits. One or two scenarios are chosen for detailed development in terms of both investments and the production plan.

It should be noted that the optimal operating policy for the processing stages such as crushing, concentrating, and leaching will depend on the volume and characteristics of the ore they are fed. In mine production plans the plant operating coefficients are generally assumed to be constants or linear, a simplification that implicitly supposes a given level of production and set of ore characteristics. However, when developing a detailed plan an optimal operating policy is sought using models that best reflect actual plant operation, whose processes are usually non-linear.

The plan is defined via an iterative process. The planning team attempts to find the best solution by sequentially modifying the investment, mine production and processing plans. Figure 1 illustrates this process and the relationships between each of these constituent plans.

Such an iterative approach leads to feasible solutions that are generally satisfactory, but they could be improved upon through the development of a methodology that effectively integrates the various decisions regarding investments, mine, and plant operation.

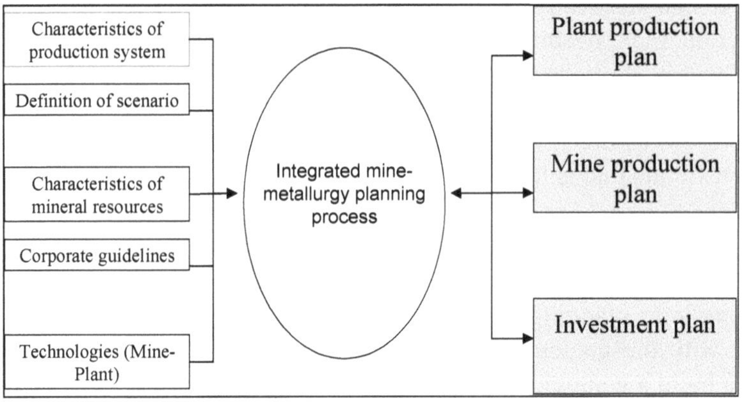

Figure 1. Planning process.

2 REVIEW OF THE LITERATURE

The most commonly cited work on mine planning is that of Lane (Lane, 1988). The author employs economic theory to argue for the advantages of using cut-off grades for defining mineral reserves. His study assumes a simplified operation that enables him to derive simple rules for solving the mining problem in accordance with economic logic. The results may be used as a basis for mine planning in real applications with their singularities and imperfections.

The definition of an optimal pit in an open-pit mine has also been extensively studied. The objective is to find the final pit that maximizes return subject to the physical design constraints. A number of algorithms for solving this problem have been developed over the years, including those suggested by Lerchs and Grossmann (1965), Robinson and Prenn (1977), Underwood (1998), Koenigsberg (1982), Dow and Onur (1992), Zhao and Kim (1992) and Huttagosol and Cameron (1992). Lerchs and Grossman propose an optimizing algorithm based on calculating the maximum flow of a network, Zhao and Kim's algorithm is a variant on Lerchs and Grossman, and Huttagosol and Cameron offer a transportation model approach to the problem. The others are heuristic algorithms that do not guarantee an optimal solution. Hochbaum and Chen (2000) describe some of these algorithms and comment on the most efficient implementations. Elsewhere in this handbook, Caccetta reports on optimization tools that have contributed to a solution.

The long-term mine planning problem has also been tackled using mathematical optimization models. This approach has been increasingly employed in recent years thanks to advances in modeling techniques, solution algorithms, and computer power that have made it possible to deal with complex problems formerly considered intractable. Noteworthy examples in underground mining are the work of Epstein *et al.* (2003), Brazil *et al.* (2002, 2005), Kuchta *et al.* (2003), and Carlyle *et al.* (2001). In the case of Epstein *et al.*, the optimization approach was extended to planning of underground and open-pit mines simultaneously.

Some mining companies, such as Codelco-Chile (Epstein *et al.*, 2003) and Kiruna (Kuchta *et al.*, 2004), have reported positive results using optimization techniques. Other cases have not been published in the literature, possible due to the companies' policy on confidentiality. Studies included in this handbook by Alford, Brazil, Lee and Newman, Kuchta and Martinez (2004) open the way to further progress in underground mining applications.

The mid-1980s saw the appearance of the first studies applying real options theory to the mining problem (Brennan and Schwartz, 1985). This approach attempts to maximize the net present value of future flows generated by a project. Explicit account is taken of stochastic behavior in the principal variables that shape mine planning such as metal prices. Real options theory also allows the incorporation of data from metal futures markets. Unlike more traditional approaches, real options theory does not try to determine a single plan for the planning horizon, but rather seeks to develop a strategy that can be adapted to the values taken on by the stochastic variables. The methodology's heavy demands in terms of calculations and computer power have, however, prevented its use in real-world planning cases, and the solutions found in the literature deal with highly simplified and ideal situations. The main studies in this area are Brennan and Schwartz (1985), Cortazar *et al.* (1998), Schwartz (1997) and Caldentey *et al.* (2006).

This bibliographic overview demonstrates how little, if any, effort has gone into integrating production and plant processes in mine planning. Indeed, we were not able to find a single study that attempted to deal with this issue. In our view, better planning and coordination of the various stages of the mine-metallurgy business has the potential to generate huge opportunities for its improvement.

3 THE COPPER INDUSTRY PRODUCTION PROCESS

Due to the highly specialized and interdependent nature of their assets and the need to be assured of a stable supply of ore for their processing plants, most copper mining companies are vertically integrated from the ore extraction stage to the production of cathodes.

The subproducts obtained will depend on the mineralogical characteristics of a mine's deposits, and the plant processes are calibrated in accordance with the specific conditions they face. There are two types of copper ore, sulfides and oxides. The production line for sulfide ores consists of six stages: mine extraction, crushing, grinding, flotation, smelting, and electrorefining. For oxide ores the grinding, flotation, smelting, and electrorefining stages are replaced by the chemical processes of leaching, solvent extraction (SX), and electrowinning (EW). Each of these stages is described in the following paragraphs.

Extraction: The object of this process is to extract the copper ore from the rock mass in the mine (which may be open-pit, underground, or a combination of the two) and send it for crushing. Low-grade material may be stockpiled for later processing, while very low-grade ore whose processing is not economically viable with current technologies is disposed of in dumps.

Crushing and grinding: The crushing and grinding stages reduce and homogenize the size of the material to particles of no more than 180μm. This is done using a variety of equipment types such as crushers, conventional mills (rod or ball), or SAG mills.

Flotation: The slurry produced at the grinding stage is mixed with frother, depressant, and collector reagents as well as other additives before being sent to the flotation cells. Air bubbles ascending from the cell bottoms attach themselves to the copper particles, which then float up to the top. This yields a marketable concentrate of approximately 31% copper.

Smelting: This stage involves a pyrometallurgical process that takes place in high-temperature furnaces where concentrates with 31% copper content are turned into metal containing 99% copper and separated from any other minerals present such as iron, sulfur and silica. The process has three sub-stages – fusion, conversion and refining – each of which yields marketable products and byproducts.

Electrorefining: Copper anodes are suspended alternately with pure copper cathodes known as starter sheets in electrolytic cells containing a sulfuric acid solution. A low-voltage direct current is applied, causing the anodes to dissolve and the copper to be deposited on the starter sheets. This produces copper cathodes that are 99.99% pure.

Leaching (Oxides): In the case of oxides the process is a chemical one, and larger-sized rock can therefore be used. The ore is arranged in heaps and exposed to sulfuric acid (H_2SO_4), thus generating a copper solution ($CuSO_4$).

Solvent extraction (Oxides): This stage uses selective solvents to separate the copper solution ($CuSO_4$) into one product rich in copper, which is sent for electrowinning and another product containing the impurities.

Electrowinning: This is the last stage in the oxides process. Electrochemical reactions involving a cathode and an anode deposit copper on the cathode.

4 OPPORTUNITIES CREATED BY INTEGRATED MINE-METALLURGY PLANNING

In this section we analyze the advantages accruing to integrated mine-metallurgy planning, based on the notions that the result of one process sets the conditions for the next one and that an investment plan must take into account the entire production chain if the results are to be optimized.

4.1 Mine-Concentrator Planning

Copper sulfide ore is processed in concentrators to yield 31% copper that is then smelted and refined. Since costs and output in the concentration stage depend on the characteristics of the extracted ore, mine planning models should include this process. Incorporating these parameters in mine production planning will enable the overall process results to be optimized. In what follows we describe the most important parameters.

4.1.1 Hardness

The hardness of the ore negatively impacts milling capacity, implying that plant treatment capacity also depends on the plan. Figure 2 is a simple graph of processing capacity (in tpd) as a function of the ore's work index (a hardness measure), illustrating the non-linear nature of the relationship.

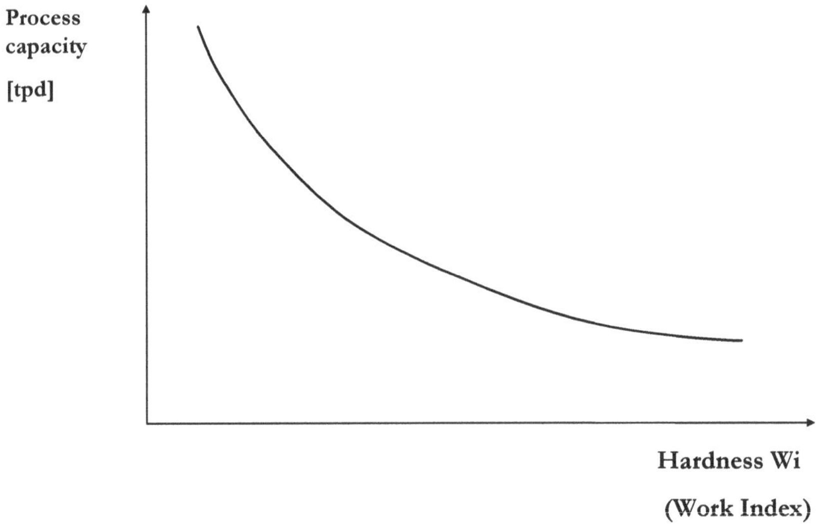

Figure 2. Processing capacity (in tpd) as a function of the work index for a given ore size.

The ore is reduced first by crushing and then by grinding. The mills are classified into conventional, which use steel balls or rods, and semiautogenous (SAG), in which reduction is obtained by collision between the rock particles themselves. The latter method consumes less energy.

4.1.2 Ore Size

Copper recovery at plant level is influenced by the size of the rock feed, which in turn is determined by ore lithology and the crushing and grinding processes. Recovery is greater with smaller rock sizes but size reduction has associated costs, and finding the optimal strategy is not an easy matter. Figure 3 illustrates the relationship; as with hardness, it is observed to be nonlinear.

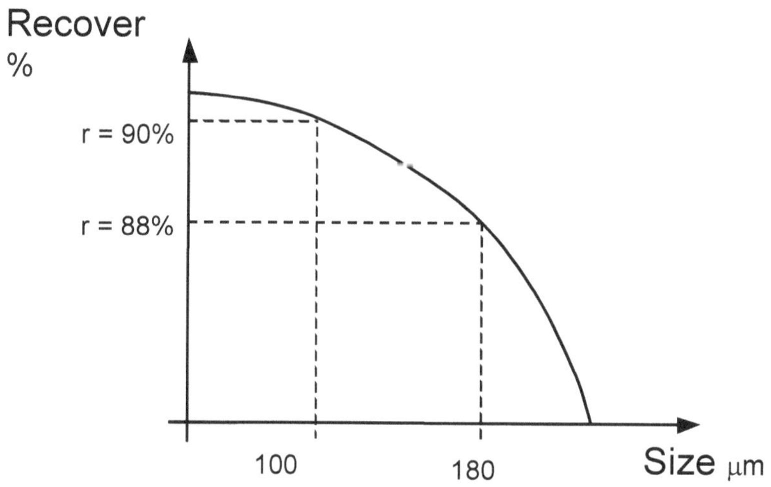

Figure 3. Copper recovery versus ore size.

4.1.3 Pollutants and Impurities

The production of pollutants such as arsenic in the concentration process may limit production capacity and elevate mitigation costs. Pollutant levels can be controlled through the use of an appropriate blend of plant feed to the concentrator, another advantage of integral mine and concentrator planning. Copper ore also contains impurities, many of which have commercial value (molybdenum, silver, gold, antimony, etc.), that should be incorporated into the value maximization procedure. Other impurities merely add costs to the process and their presence should be minimized. Here again we see how a range of variables affect decision-making, and only an integral vision of the entire process will enable company results to be maximized.

4.2 Smelter-Refinery Planning and Its Integration with Mine-Plant Planning

As noted earlier, the production process begins with the extraction of ore from a deposit. Copper may be found in combination with sulfur, forming sulfide ores, or with oxygen, in which case it forms oxide ores. In the case of sulphides, the process involves reducing the rock before sending it to the concentration plant. The output of the plant will depend on the ore size and characteristics. The plant produces concentrates of approximately 31% copper plus impurities and pollutants. Concentrates of molybdenum are also obtained. The planner has various mines or mine sectors to work with, each with its own mineralogical characteristics that strongly impact the result. Concentration plants may be operated in different ways, prioritizing variables such as copper production, metal recovery, minimization of pollutants, or some combination of these.

The copper concentrates are fed to smelters that produce mainly copper anodes with a purity of 99%. They also produce blister copper and white metal, intermediate products that may be marketed or further processed to obtain anodes. These are then sent to a refinery where they are purified to obtain cathodes of 99.99% copper. Anode bars containing commercially valuable impurities such as gold, silver, antimony, and bismuth are also produced. The gases given off by the furnaces are sent to a gas cleaning plant for dust abatement and production of sulfuric acid.

In the case of copper oxides, an alternative process known as leaching generates a single marketable product, which is copper in the form of cathodes. It is generally cheaper than the traditional concentration and smelting-refining process.

The processes described here are shown schematically in Figure 4.

4.2.1 Integration of Input Supply Logistics and Sale of Products in the Smelting-Refining Process

Given the high cost of ore transport, concentration plants and leaching facilities are located relatively close to deposits. Smelters and refineries, on the other hand, may be installed relatively far from the ore source. This is due to the higher value added of the inputs such as concentrates, white metal, blister copper, and anodes, making their transport relatively efficient. Some plants are situated near railway lines or seaports to facilitate the shipping of products and delivery of inputs.

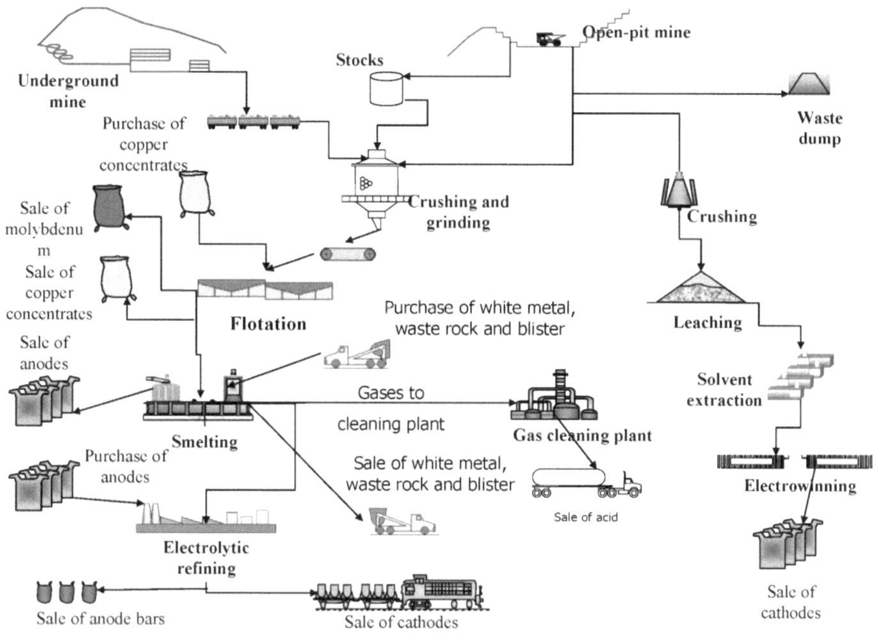

Figure 4. Schematic diagram of process.

Although the smelting-refining process is usually integrated with the overall mining business, particularly in large companies, in some cases it is a separate business managed independently of any particular ore deposit, and functioning as a business center that markets commercially valuable products and subproducts generated by the process such as sulfuric acid, anode bars, molybdenum concentrates, blister copper, and white metal.

The efficient running of a smelting-refining operation consists in optimizing input supply logistics and maximizing return on the portfolio of products. This includes skillful marketing management and plant planning that is also market-focused with a view to determining the optimal mix of products and prices. Needless to say, optimal operation of the physico-chemical plant processes is also necessary.

To optimize input supply to the plants it is essential that every alternative be evaluated, taking into account the cost of raw material, transport costs, and the metallurgical characteristics of the inputs. Similarly, to maximize returns on plant products we must evaluate the options in the various markets, considering price, transport cost, and market size.

An efficient way of carrying out such evaluations is through the use of optimization models that focus on maximizing returns from product sales less transport and production costs. These models generally have a linear structure and include the subproblem of input and product transport. Certain commercial constraints in addition to technical constraints on production capacity are also incorporated. These models are highly useful in ensuring efficient input and sales management that could form the basis for a strategy of running the plants as profitable business units.

4.2.2 Limiting Factors in Copper Production

Certain aspects of energy and water resources constitute limiting factors on copper production. The demand for these resources is directly related to mine and plant production levels. Tables 1 and 2 summarize relative water and energy consumption for the various stages of the production process.

Table 1. Relative consumption of water resources by stage.

Process	**Consumption (%)**
Crushing-grinding-flotation	53
Smelting	10
Hydrometallurgy	15
Other	22

Table 2. Relative consumption of energy resources by stage.

Process	**Consumption (%)**
Crushing-grinding-flotation	48
Fundition	19
Refinery	4
Others	29

Water is a scarce resource whose value is rising. Some major copper deposits are located far from any water source, as is the case in Chile where most production is found in the Atacama desert. In other cases, environmental regulations restrict the use of water resources for industrial purposes. In practice, water consumption is a limiting factor on copper production, and planners face a real challenge to correctly assign water resources in such a manner as to optimize the value of the business. A poorly conceived or shortsighted assignment could have a drastic impact on the business and limit production.

Energy inputs and their associated costs are also significant limiting factors in the planning process, and energy variables affect strategic decisions in both the design and operation of a mine. As an example, changing from open pit to underground mining is a decision that can be justified by energy savings given that as the pit becomes deeper the increase in truck fuel costs to haul the ore to the surface may render the operation noneconomic. Another example is the very energy-intensive grinding and crushing operations, whose energy consumption depends on the degree of size reduction required and the hardness of the ore. Planners can improve both variables by looking at the entire business when optimizing its value.

In the smelting-refining stage, the most serious limiting factors are those relating to environmental pollution. The main pollutants generated by these processes are arsenic and sulfur. In order to protect the environment and comply with legislation on the subject, mining companies are required to treat these residues and mitigate their negative effects. Treatment capacity and the associated costs are thus limiting factors in mine-metallurgy planning. If these constraints are active, planners can choose higher tonnages of lower-grade ore containing fewer pollutants in order to increase total refined production.

The three resource variables just discussed – water, energy and the environment – must be carefully considered in the planning process, as they can significantly limit the production derived from an ore deposit. In some operations, these resources are not properly valued, leading to their overexploittation or inefficient use and consequent harmful effects in the medium term.

5 CONCLUSIONS

Mine-metallurgy planning must deal with three principal issues: an investment plan, processing plant planning and mine production planning. Although they arise simultaneously and are closely interrelated, in practice they are dealt with sequentially and the solutions adopted for one is the starting point for solving the next.

This sequential procedure is adopted because of the methodological difficulties involved in a more integrated approach to what are already highly complex problems when considered individually. Nevertheless, there is general awareness that separate solutions are suboptimal in terms of the overall problem, and integration would yield great potential benefits.

Operations research in the copper industry has primarily focused on solving mine production planning problems, with good results for both underground and open-pit operations.

Among the principal challenges facing operations research in mine-metallurgy planning is the development of methodological approaches that make it possible to integrate investment, plant, and mine production decisions so that strategic plans create greater value and optimize the entire copper production chain.

Another significant challenge is an improved incorporation of the various risks inherent in a long-term mine-metallurgy plan.

ACKNOWLEDGMENTS

This work was partially supported by FONDEF, through project number 311064 and CODELCO for the Grant in Logistcs and Planning in Mining awarded to the Industrial Engineering Department of the University of Chile.

6 REFERENCES

Brazil, M., D. Lee, J.H Rubinstein, D.A Thomas, J.F Weng, N.C Wormald. 2002. A network model to optimise cost in underground mine design. *South African Institute of Electrical Engineers* 93, 97–103.

Brazil, M., D.Lee, J. H. Rubinstein, D. A. J. F. Thomas, Weng N. C. Wormald. 2005. Optimisation in the Design of Underground Mine Access", Orebody Modeling and Strategic Mine Planning. Uncertainty and Risk Management in Orebody Modeling and Strategic Mine Planning: Spectrum Series Volume 14, Melbourne, Australia: Australasian Institute of Mining and Metallurgy. 121–124.

Brennan M., E. Schwartz. 1985. Evaluating natural resources investments. *Journal of Business* 58, 135–157.

Carlyle, M., B.C. Eaves. 2001. Underground planning at Stillwater Mining Company. *Interfaces*, 31(4), 50–60.

Caldentey R., R. Epstein, D. Sauré. 2006. Real Option Approach for Valuating Long-term Planning in the Mining Industry. Working paper.

Cortazar G., E. Schwartz, M. Salinas. 1998. Evaluating environmental investments: A real options approach. *Management Science* 44(8), 1059–1070.

Dow, P. A., A. H. Onur. 1992. Optimising open pit design and sequencing. *Proceedings of the 23rd International. APCOM Symposium*, 411–422.

R. Epstein, Gaete S, Caro F, Weintraub A, Santibañez P, Catalan J. 2003. Optimizing Long Term Planning for Underground Copper Mines. *Internacional Copper Conference (2003)*, Copper 2003 – Cobre 2003.

Huttagosol, P., R. Cameron. 1992. A computer design of ultimate pit limit by using transportation algorithm. *Proceedings of the 23rd International. APCOM Symposium*, 443–460.

Hochbaum D, Chen S. 2000. Performance analysis and best implementations of old and new algorithms for the open-pit mining problem. *Operations Research* 48(6), 894–914.
Koenigsberg, E. 1982. The optimum contours of an open pit mine: An application of dynamic programming. *Proceedings of the 17th International. APCOM Symposium*, 274–287.
Kuchta, M., A. Newman, and E. Topal. 2004. Implementing a Production Schedule at LKAB's Kiruna Mine. *Interfaces* 34(2), 124–134.
Lane K.F. 1988. *The Economic Definition of Ore*. London, Mining Journal Books Limited.
Lerchs, H., I. F. Grossmann. 1965. Optimum design of open-pit mines. *Transactions*, C.I.M. LXVIII 17–24.
Robinson, R. H., N. B. Prenn. 1977. An open pit design model. *Proceedings of the 15th International. APCOM Symposium,* 1–9.
Schwartz E. 1997. The stochastic behavior of commodity prices: Implications for valuation and hedging. *The Journal of Finance* 52(3), 923–973.
Underwood R., B. Tolwinski. 1998. A mathematical programming viewpoint to solve the ultimate pit problem. *European Journal of Operational Research* 107, 96–107.
Zhao, Y., Y. C. Kim. 1992. A new optimum pit limit design algorithm. *Proceedings 23rd International. APCOM Symposium*, 423–434.

Early Titles in the
INTERNATIONAL SERIES IN
OPERATIONS RESEARCH & MANAGEMENT SCIENCE
Frederick S. Hillier, Series Editor, *Stanford University*

Saigal/ *A MODERN APPROACH TO LINEAR PROGRAMMING*
Nagurney/ *PROJECTED DYNAMICAL SYSTEMS & VARIATIONAL INEQUALITIES WITH APPLICATIONS*
Padberg & Rijal/ *LOCATION, SCHEDULING, DESIGN AND INTEGER PROGRAMMING*
Vanderbei/ *LINEAR PROGRAMMING*
Jaiswal/ *MILITARY OPERATIONS RESEARCH*
Gal & Greenberg/ *ADVANCES IN SENSITIVITY ANALYSIS & PARAMETRIC PROGRAMMING*
Prabhu/ *FOUNDATIONS OF QUEUEING THEORY*
Fang, Rajasekera & Tsao/ *ENTROPY OPTIMIZATION & MATHEMATICAL PROGRAMMING*
Yu/ *OR IN THE AIRLINE INDUSTRY*
Ho & Tang/ *PRODUCT VARIETY MANAGEMENT*
El-Taha & Stidham/ *SAMPLE-PATH ANALYSIS OF QUEUEING SYSTEMS*
Miettinen/ *NONLINEAR MULTIOBJECTIVE OPTIMIZATION*
Chao & Huntington/ *DESIGNING COMPETITIVE ELECTRICITY MARKETS*
Weglarz/ *PROJECT SCHEDULING: RECENT TRENDS & RESULTS*
Sahin & Polatoglu/ *QUALITY, WARRANTY AND PREVENTIVE MAINTENANCE*
Tavares/ *ADVANCES MODELS FOR PROJECT MANAGEMENT*
Tayur, Ganeshan & Magazine/ *QUANTITATIVE MODELS FOR SUPPLY CHAIN MANAGEMENT*
Weyant, J./ *ENERGY AND ENVIRONMENTAL POLICY MODELING*
Shanthikumar, J.G. & Sumita, U./ *APPLIED PROBABILITY AND STOCHASTIC PROCESSES*
Liu, B. & Esogbue, A.O./ *DECISION CRITERIA AND OPTIMAL INVENTORY PROCESSES*
Gal, T., Stewart, T.J., Hanne, T. / *MULTICRITERIA DECISION MAKING: Advances in MCDM Models, Algorithms, Theory, and Applications*
Fox, B.L. / *STRATEGIES FOR QUASI-MONTE CARLO*
Hall, R.W. / *HANDBOOK OF TRANSPORTATION SCIENCE*
Grassman, W.K. / *COMPUTATIONAL PROBABILITY*
Pomerol, J-C. & Barba-Romero, S. / *MULTICRITERION DECISION IN MANAGEMENT*
Axsäter, S. / *INVENTORY CONTROL*
Wolkowicz, H., Saigal, R., & Vandenberghe, L. / *HANDBOOK OF SEMI-DEFINITE PROGRAMMING: Theory, Algorithms, and Applications*
Hobbs, B.F. & Meier, P. / *ENERGY DECISIONS AND THE ENVIRONMENT: A Guide to the Use of Multicriteria Methods*
Dar-El, E. / *HUMAN LEARNING: From Learning Curves to Learning Organizations*
Armstrong, J.S. / *PRINCIPLES OF FORECASTING: A Handbook for Researchers and Practitioners*
Balsamo, S., Personé, V., & Onvural, R./ *ANALYSIS OF QUEUEING NETWORKS WITH BLOCKING*
Bouyssou, D. et al. / *EVALUATION AND DECISION MODELS: A Critical Perspective*
Hanne, T. / *INTELLIGENT STRATEGIES FOR META MULTIPLE CRITERIA DECISION MAKING*
Saaty, T. & Vargas, L. / *MODELS, METHODS, CONCEPTS and APPLICATIONS OF THE ANALYTIC HIERARCHY PROCESS*
Chatterjee, K. & Samuelson, W. / *GAME THEORY AND BUSINESS APPLICATIONS*
Hobbs, B. et al. / *THE NEXT GENERATION OF ELECTRIC POWER UNIT COMMITMENT MODELS*
Vanderbei, R.J. / *LINEAR PROGRAMMING: Foundations and Extensions, 2nd Ed.*
Kimms, A. / *MATHEMATICAL PROGRAMMING AND FINANCIAL OBJECTIVES FOR SCHEDULING PROJECTS*
Baptiste, P., Le Pape, C. & Nuijten, W. / *CONSTRAINT-BASED SCHEDULING*
Feinberg, E. & Shwartz, A. / *HANDBOOK OF MARKOV DECISION PROCESSES: Methods and Applications*
Ramík, J. & Vlach, M. / *GENERALIZED CONCAVITY IN FUZZY OPTIMIZATION AND DECISION ANALYSIS*
Song, J. & Yao, D. / *SUPPLY CHAIN STRUCTURES: Coordination, Information and Optimization*
Kozan, E. & Ohuchi, A. / *OPERATIONS RESEARCH/ MANAGEMENT SCIENCE AT WORK*
Bouyssou et al. / *AIDING DECISIONS WITH MULTIPLE CRITERIA: Essays in Honor of Bernard Roy*

Early Titles in the
INTERNATIONAL SERIES IN
OPERATIONS RESEARCH & MANAGEMENT SCIENCE
(Continued)

Cox, Louis Anthony, Jr. / *RISK ANALYSIS: Foundations, Models and Methods*
Dror, M., L'Ecuyer, P. & Szidarovszky, F. / *MODELING UNCERTAINTY: An Examination of Stochastic Theory, Methods, and Applications*
Dokuchaev, N. / *DYNAMIC PORTFOLIO STRATEGIES: Quantitative Methods and Empirical Rules for Incomplete Information*
Sarker, R., Mohammadian, M. & Yao, X. / *EVOLUTIONARY OPTIMIZATION*
Demeulemeester, R. & Herroelen, W. / *PROJECT SCHEDULING: A Research Handbook*
Gazis, D.C. / *TRAFFIC THEORY*
Zhu/ *QUANTITATIVE MODELS FOR PERFORMANCE EVALUATION AND BENCHMARKING*
Ehrgott & Gandibleux/ *MULTIPLE CRITERIA OPTIMIZATION: State of the Art Annotated Bibliographical Surveys*
Bienstock/ *Potential Function Methods for Approx. Solving Linear Programming Problems*
Matsatsinis & Siskos/ *INTELLIGENT SUPPORT SYSTEMS FOR MARKETING DECISIONS*
Alpern & Gal/ *THE THEORY OF SEARCH GAMES AND RENDEZVOUS*
Hall/*HANDBOOK OF TRANSPORTATION SCIENCE - 2nd Ed.*
Glover & Kochenberger/ *HANDBOOK OF METAHEURISTICS*
Graves & Ringuest/ *MODELS AND METHODS FOR PROJECT SELECTION: Concepts from Management Science, Finance and Information Technology*
Hassin & Haviv/ *TO QUEUE OR NOT TO QUEUE: Equilibrium Behavior in Queueing Systems*
Gershwin et al/ *ANALYSIS & MODELING OF MANUFACTURING SYSTEMS*
Maros/ *COMPUTATIONAL TECHNIQUES OF THE SIMPLEX METHOD*
Harrison, Lee & Neale/ *THE PRACTICE OF SUPPLY CHAIN MANAGEMENT: Where Theory and Application Converge*
Shanthikumar, Yao & Zijm/ *STOCHASTIC MODELING AND OPTIMIZATION OF MANUFACTURING SYSTEMS AND SUPPLY CHAINS*
Nabrzyski, Schopf & Węglarz/ *GRID RESOURCE MANAGEMENT: State of the Art and Future Trends*
Thissen & Herder/ *CRITICAL INFRASTRUCTURES: State of the Art in Research and Application*
Carlsson, Fedrizzi, & Fullér/ *FUZZY LOGIC IN MANAGEMENT*
Soyer, Mazzuchi & Singpurwalla/ *MATHEMATICAL RELIABILITY: An Expository Perspective*
Chakravarty & Eliashberg/ *MANAGING BUSINESS INTERFACES: Marketing, Engineering, and Manufacturing Perspectives*
Talluri & van Ryzin/ *THE THEORY AND PRACTICE OF REVENUE MANAGEMENT*
Kavadias & Loch/*PROJECT SELECTION UNDER UNCERTAINTY: Dynamically Allocating Resources to Maximize Value*
Brandeau, Sainfort & Pierskalla/ *OPERATIONS RESEARCH AND HEALTH CARE: A Handbook of Methods and Applications*
Cooper, Seiford & Zhu/ *HANDBOOK OF DATA ENVELOPMENT ANALYSIS: Models and Methods*
Luenberger/ *LINEAR AND NONLINEAR PROGRAMMING, 2nd Ed.*
Sherbrooke/ *OPTIMAL INVENTORY MODELING OF SYSTEMS: Multi-Echelon Techniques, Second Edition*
Chu, Leung, Hui & Cheung/ *4th PARTY CYBER LOGISTICS FOR AIR CARGO*
Simchi-Levi, Wu & Shen/ *HANDBOOK OF QUANTITATIVE SUPPLY CHAIN ANALYSIS: Modeling in the E-Business Era*

**** A list of the more recent publications in the series is at the front of the book ****